Krause · Jäger · Resch (Eds.)
High Performance Computing in Science and Engineering '03

Springer-Verlag Berlin Heidelberg GmbH

Egon Krause · Willi Jäger · Michael Resch
Editors

High Performance Computing in Science and Engineering '03

Transactions of the High Performance Computing Center Stuttgart (HLRS) 2003

With 233 Figures, 121 in Color, and 41 Tables

Springer

Editors

Egon Krause
Aerodynamisches Institut
der RWTH Aachen
Wuellnerstraße zw. 5 u. 7
52062 Aachen, Germany
e-mail: ek@aia.rwth-aachen.de

Willi Jäger
Institut für Wissenschaftliches Rechnen
Universität Heidelberg
Im Neuenheimer Feld 368
69120 Heidelberg, Germany
e-mail: jaeger@iwr.uni-heidelberg.de

Michael Resch
High Performance Computing Center
Stuttgart - HLRS
Allmandring 30
70550 Stuttgart, Germany
e-mail: resch@hlrs.de

Cataloging-in-Publication Data applied for

A catalog record for this book is available from the Library of Congress.

Bibliographic information published by Die Deutsche Bibliothek
Die Deutsche Bibliothek lists this publication in the Deutsche Nationalbibliografie; detailed bibliographic data is available in the Internet at <http://dnb.ddb.de>.

Front cover figure: Simulated temperature increase for Central Europe caused by a doubling of global greenhouse gas concentrations (Lehrstuhl für Umweltmeteorologie, Brandenburgische Technische Universität, Cottbus).

Mathematics Subject Classification (2000): 65Cxx, 65C99, 68U20

ISBN 978-3-642-62486-5 ISBN 978-3-642-55876-4 (eBook)
DOI 10.1007/978-3-642-55876-4

springeronline.com

Originally published by Springer-Verlag Berlin Heidelberg New York in 2003
Softcover reprint of the hardcover 1st edition 2003

Typeset by the authors. Edited by Kurt Mattes, Heidelberg.
Cover design: *design & production* GmbH, Heidelberg

46/3142/LK - 5 4 3 2 1 0

Preface

Prof. Dr. Egon Krause

Aerodynamisches Institut
RWTH Aachen
Wüllnerstr. zw. 5 u. 7, D-52062 Aachen

Prof. Dr. Willi Jäger

Interdisziplinäres Zentrum für Wissenschaftliches Rechnen
Universität Heidelberg
Im Neuenheimer Feld 368, D-69120 Heidelberg

Prof. Dr. Michael Resch

Höchstleistungsrechenzentrum Stuttgart
Allmandring 30, D-70550 Stuttgart

The High-Performance Computing Center Stuttgart (HLRS) underwent dramatic changes during the last year. At the beginning of 2003 the Center was separated from the Computing Center of Stuttgart University and turned into an independent central unit of Stuttgart University. Last March the Land Baden-Württemberg shaped up her strategy of cooperation and formed of a new Center of Competence in High-Performance Computing in Baden-Württemberg by associating HLR Stuttgart with the SSC Karlsruhe.

At the present time the HLRS is preparing its decision for a new computer system to be planned to be operative in 2005. A safe decision is difficult to arrive at since the offers of the vendors of high-performance computers are continuously and rather rapidly changing. At the lower end the microprocessor based systems are attacked by clusters of PCs. The superior price-performance ratio makes such rather inexpensive systems attractive for a variety of applications, most of which are latency bound. At the upper end microprocessor based systems are pressured by vector-based systems. The latter have seen a revival nourishing on the success and political impact of the earth simulator project. With their superior memory bandwidth they attract a user community whose applications are memory bound. This is also reflected in work of the user community of the HLRS and the SSC Karlsruhe. As in the years before, the major results were reported at the Sixth Results and Review Workshop on High Performance Computing in Science and Engineering, which was held October 6–7, 2003 Stuttgart University.

This volume contains the written versions of the investigations presented. The papers were selected in an internal review from all projects processed at the HLRS and at the SSC Karlsruhe during the time period beginning October 2002. The various projects were initiated at a number of Universities in Germany, located in Aachen, Braunschweig, Bremen, Cottbus, Erlangen, Freiburg, Göttingen, Harburg, Heidelberg, Hohenheim, Jena, Karlsruhe, Köln, Konstanz, Mainz, Marburg, Münster, Nürnberg, Saarbrücken, Stuttgart, and Tübingen. In some of the projects several foreign Universities and Research Centers cooperated in the work: They are the Universities at Alma Ata, Budapest, Montpellier, Santa Cruz, Rome, Warsaw, and the Georgia Institute of Technology, the Research Center in Los Alamos, the Max-Planck Society in Stuttgart, the German Center for Aero- and Astronautics in Braunschweig and Köln, and one of the Fraunhofer Institutes in Freiburg.

Ten of the 34 papers contained in this volume deal with problems in physics, ten with problems in physics of fluids, and the rest of the papers is concerned with problems in other fields. Seven of these are concerned with atmospheric, astrophysical, and geophysical problems: For example, transport processes in the atmospheric boundary layer are studied with the method of large eddy simulation at the Universities in Bremen and in Hohenheim. Possible climate changes in central Europe are being simulated with high-resolution techniques at the University of Technology in Cottbus. In an investigation of the DLR Köln the global water distribution and its movement in the shallow subsurface of MARS is analyzed with a subsurface-atmosphere water cycle model. High-precision direct N-body integration is used to study questions of the thermodynamic behavior of dense stellar systems in a cooperative project of the Astronomical Computing Institute Heidelberg, the Astrophysical Institute in Alma-Ata, and the University of California at Santa Cruz. In the State Observatory Königstuhl near Heidelberg the formation and propagation of jets around compact astrophysical objects like the CH Cygni, R. Aquarii, and MWC 5000 are analyzed with a numerical solution of the conservation equations of ideal compressible flow, and in another study of the observatory the interaction of jets with a dense magnetized environment on a scale of more than 200 jet radii is studied. In the Institute of Geo-Sciences in Jena and at Los Alamos National Laboratory the influence of the viscosity stratification on the thermal evolution of the compressible earth mantle with time-dependent internal heating is investigated with a three-dimensional finite-element spherical-shell method for the solution of the differential equations describing the infinite Prandtl-number convection.

The contributions in physics cover also a wide range of topics. In a cooperative effort the deposition dynamics of clusters are simulated by scientists of the University of Freiburg, the Fraunhofer-Institute for Mechanics of Materials in Freiburg, and the School of Physics at Georgia Institute of Technology. Correlated bosonic and fermionic systems were investigated with quantum Monte-Carlo simulations at Stuttgart University and at the Max-Planck-Institute for Solid State Research in Stuttgart. The atomic and spectroscopic properties of P-rich InP(001)(2x1) surfaces grown in gas-phase epitaxy are explored with density-functional calculations based on finite-difference discretization and multigrid acceleration at the University in Jena. At the University in Münster, the structural and spectroscopic properties of porhyrin-derived polymers are studied within an ab-initio framework, and also in another investigation the transition from stationary to rotating bound states of dissipative solitons. The propagation of mode-I cracks in a three-dimensional model quasicrystal is studied by molecular dynamics simulations at Stuttgart University. Research on nanostructures is under way at the University of Konstanz, where new insights into pore condensates, phase transitions, and quantum effects in nano-systems in external potentials and reduced geometry are reported. At Mainz University the liquid-vapor transition of a Lennard-Jones fluid is studied in large-scale grandcanonical Monte Carlo simulations in conjunction with a multicanonical reweighting scheme, and in another project, carried out together with scientists from the University at Montpellier, the structure of surfaces and intersurfaces of silica (SiO2) is studied in large-scale molecular dynamics computer simulations. In another bilateral cooperation of Budapest University of Technology and of the University Erlangen-Nürnberg first attempts are described of systematic first principles quantum mechanical calculations of the characteristic vibration frequencies of all the Boron-interstitial clusters which have been suggested so far to be important in the de/reactivation process of boron.

In the problem area of physics of fluids the late stages of the laminar-turbulent transition process on a flat-plate boundary layer is investigated at Stuttgart University. Special emphasis is placed on describing the flow randomization process. At the University of Göttingen the Rayleigh-Bénard convection is simulated in a plane layer with periodic boundary conditions in the horizontal directions with a spectral method for Rayleigh numbers as high as 107, even for an aspect ratio of 10. Three-dimensional unsteady heat transfer on strongly deformed droplets is studied at high Reynolds numbers in a direct numerical simulation at Stuttgart University, where this method is also used to investigate turbulent premixed CO/H2/air-flames, and also semi-turbulent pipe flow. In another project, a numerical solution of the Reynolds-averaged Navier-Stokes equations for unsteady compressible flow is used to simulate three-dimensional chemically reacting supersonic flows, occurring in supersonic combustion ramjets. A finite-volume technique is used at the TU Hamburg-Harburg to numerically simulate the forced breakup of a liquid jet. In aerodynamics, the deformation of a high-speed transport aircraft-type wing is studied by direct aeroelastic simulation at the RWTH Aachen., while a numerical solution of the Reynolds-averaged Navier-Stokes equations is used at the DLR Braunschweig to predict the flow around transport aircraft configurations for high-lift conditions. At the University of Karlsruhe the influence of impinging wakes on the boundary layer of a thin-shaped turbine blade is studied in large-eddy simulations.

Two of the topics of the rest of the papers are devoted to the solution of problems in chemistry: At the University of Marburg the transition of metal complexes is studied in quantum chemical calculations, and the molecular transport through single molecules is investigated in a joint project of the Research Center and the University of Karlsruhe, and the RWTH Aachen. A problem of pharmaceutical and medical chemistry is analysed at the University of the Saarland in Saarbrücken, where the protonation states of methionine aminopeptidase are studied by QM/MM Car-Parrinello molecular dynamics simulations. In another project, which among others is in the center of attention of theoretical medical research, results of investigations are reported, that may lead towards a holistic understanding of the human genome. This investigation is carried out at the German Cancer Research Center Heidelberg.

The last three papers deal with problems of computer science. In the first, systems of the HLR Stuttgart were analyzed with the BAR-Bench benchmark system. In the following, two efficient and object-oriented libraries for particle simulations, developed at Tübingen University, are presented. Finally, in the last project reported, SKaMPI, the Special MPI-Benchmark, developed at the University of Karlsruhe, is described.

We gratefully acknowledge the continued support of the Land Baden-Württemberg in promoting and supporting high-performance computing. Grateful acknowledgment is also due to the Deutsche Forschungsgemeinschaft (DFG); many projects processed on the machines of the HLRS and the SSC could not have been carried out without the support of the DFG. Also, the activities of the WiR, strengthening scientific investigations in the State of Baden-Württemberg are gratefully acknowledged. Finally, we thank the Springer Verlag for publishing this volume and thus helping to position the local activities into an international frame. We hope that this series of publications is contributing to the global promotion of high performance scientific computing.

Stuttgart, August 2003

W. Jäger
M. Resch
E. Krause

Contents

Earth Sciences

High-Resolution Studies of Transport Processes in the Atmospheric Boundary Layer Using the Synergy of Large Eddy Simulation and Measurements of Advanced Lidar Systems

Tijana Janjić[1] and Volker Wulfmeyer[2]

[1] School of Engineering and Science
International University Bremen *t.janjic@iu-bremen.de*
[2] Institut für Physik und Meteorologie
Universität Hohenheim *wulfmeye@uni-hohenheim.de*

1 Introduction

On poorly resolved and sub-grid scales, turbulent processes may have significant impact on processes that are well represented on the computational grid. Even though the scales of motions of primary interest are generally much larger than those of turbulent motions, the turbulence profoundly affects all scales due to nonlinear interactions.

To date, turbulent processes cannot be explicitly resolved either in numerical weather prediction (NWP) or climate models. For this reason, extensive efforts have been made to "parameterize" the turbulent effects, i.e., to describe and quantify the summary effects of turbulent motions in terms of known, large scale parameters.

Traditionally, laboratory and field experiments have been used in order to study turbulence. Since recently, large eddy simulation (LES) (Moeng [9]) of turbulence has been introduced. The large eddy simulation models (LES) turned out to be an excellent tool for studying the atmospheric boundary layer. Such models have been applied for numerous studies of clear and cloudy boundary layers, and the results obtained in this way have been used for developing parameterization schemes for numerical models of the atmosphere (Holtslag and Moeng [6], Hong and Pan [7], Moeng [10]).

However, many open questions remain concerning the application of the LES approach in atmospheric sciences that cannot be resolved unless suitable instrumentation is developed that will allow comparisons of the observed and the LES data. In recent years, considerable progress has been made in the development, improvement and application of active remote sensing systems such as lidar and radar. In particular, advanced lidar systems can measure profiles of key variables in the atmospheric boundary layer (ABL) such as wind and water vapor with outstanding accuracy and resolution (Wulfmeyer [11]). Initial work demonstrated that these instruments provided excellent data that can be compared with the data obtained from high resolution atmospheric models (Mayor [8]).

In this work, lidar data collected during the Nauru99 field observation campaign will be compared with data obtained using the LES model. Lidar data has been collected with the High-Resolution Doppler Lidar (HRDL) of the Environmental Technology Laboratory (ETL) of the National Oceanic and Atmospheric Administration (NOAA) in Boulder, CO, USA (Wulfmeyer et al. [12], Grund et al. [5]). LES model used in this study is developed at the Max-Planck-Institute for Meteorology in Hamburg (Chlond [2], [3], [4]).

As the first step in our project, data collected during the Nauru99 campaign from other than lidar sources has been used for the initialization of the LES model. The outputs of these runs were compared with independent data including lidar in order to investigate the model performance.

2 Observations

The data set used in this study was collected during the Nauru99 campaign, which took place from June 15 to July 16, 1999 in the tropical Western Pacific. During campaign observations were collected with several lidar and radar systems as well as a suite of different in-situ sensors and radiosondes. These instruments were located on board of two research vessels, the Mirai and R/V Ronald H. Brown, and on Nauru Island.

The particular data which have been used here covered the period from June 23, 02:00 UTC to June 24, 08:00 UTC (June 23, 20 : 00 LT). During this period, the weather conditions were fair and no precipitation was observed. Fig. 1 shows sonde profiles of potential temperature and specific humidity from June 23, 02:00 UTC till June 24, 08:00 UTC. As can be seen from Fig. 1, well-mixed boundary layer varies in depth between 600 and 700 m. During the 24-h observation, the profiles (see left panel of Fig. 1) show a change in the strength of the inversion, gradually decreasing from 0.0166 to 0.0052 K/m. The dry layer in right panel of Fig. 1 is visible during the first three hours of the observation period and than disappears during the rest of the period.

Profiles of the horizontal wind are shown in Fig. 2. Surface horizontal wind speed was fairly low with values between 1.5 and 3 m/s. Relative humidity and surface pressure only varied slightly. Latent and sensible heat surface fluxes were quite small during this time period as can be seen from Fig. 3. The sensible heat flux was about 6 W/m^2 varying by about 2 W/m^2. The latent heat flux was about 70 W/m^2 and showed a slight indication for a diurnal cycle varying between 40 and 90 W/m^2. ECMWF analysis plots of the vertical velocity ω indicate that there was no significant effect of synoptic

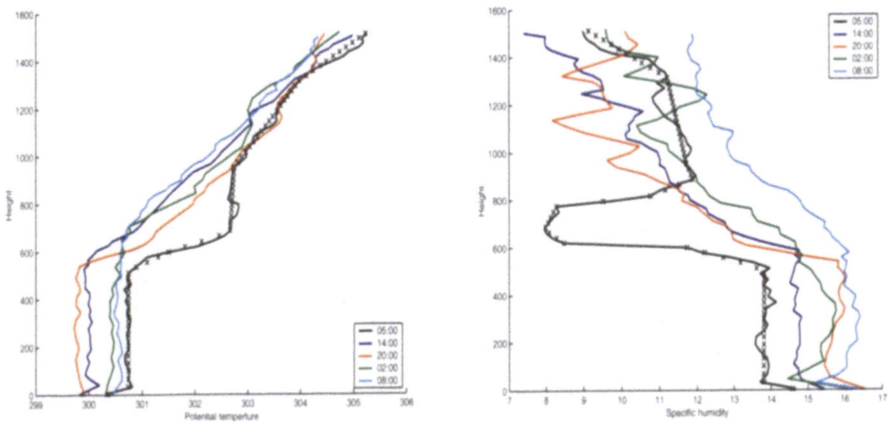

Fig. 1. Sonde profiles of potential temperature and specific humidity. The crosses indicate the initial profiles used in the simulations.

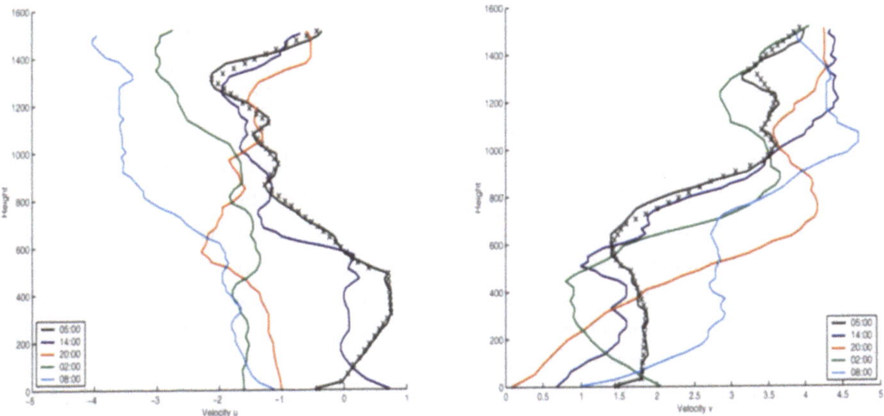

Fig. 2. Sonde profiles of horizontal wind.The crosses indicate the initial profiles used in the simulations.

scale motions on decrease and increase of the boundary layer height (figures omitted).

During a 24-h intensive observation period from June 23, 7:00 UTC to June 24, 7:00 UTC, data from the High-Resolution Doppler Lidar (HRDL) was also available. HRDL was located on the TAO buoy in the tropical Pacific at 2 S, 165 E.

HRDL was operated mainly in the vertical pointing mode. In this mode, high-resolution measurements of the vertical wind and backscatter intensity are available with a resolution of 1 s and 30 m. In addition, every 2 hours the HRDL configuration was switched to two VAD scans with 5 and 35 degree

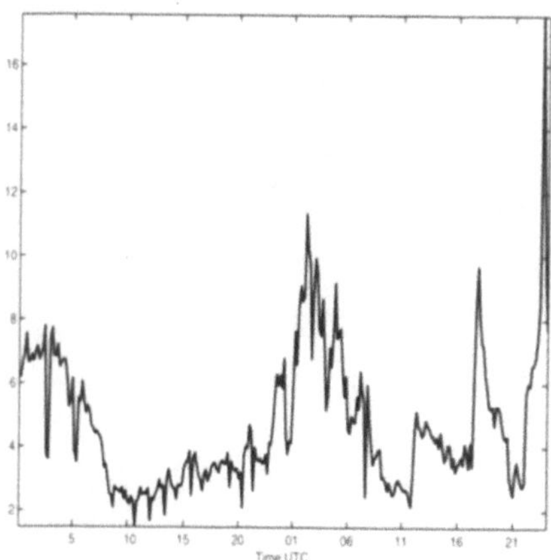

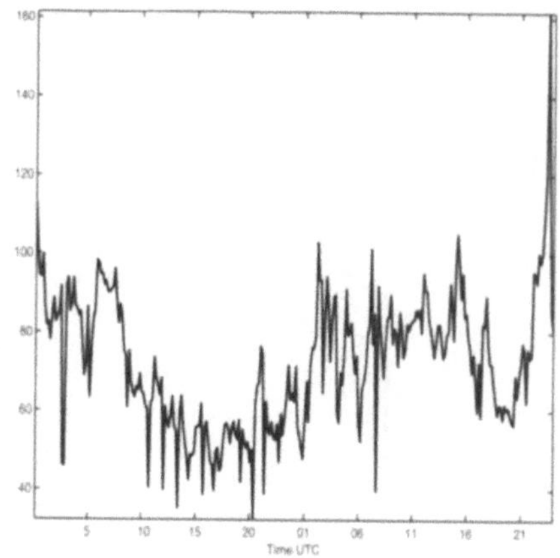

Fig. 3. Sensible and latent heat flux for June 23 and June 24, 1999 at 2 S, 165 E. The time period covered in this study is denoted with blue line.

elevation angles in order to obtain information about horizontal wind profiles from low levels to the top of the marine boundary layer (MBL).

An example of the backscatter intensity measured between 7:00 UTC and 14:32 UTC on June 23rd is presented in Fig. 4. Gradients in the backscatter signal can easily be detected and provide an excellent means for the determination of the height of the mixed layer. During the observation period, the MBL height could be detected by a positive or negative maximum of the backscatter intensity gradient. The MBL height started at about 800 m and decreased to about 600 m, and had increased again to 800 m by 5 : 30 UTC on June 24th. Within the whole 24-h observation period the MBL height varied from approximately 500 to 800 m.

An elevated aerosol layer with a vertical extension of about 200 m was observed during the first part of the measurement between 7:00 UTC and 11:00 UTC. It shows only low mixing with MBL air, as can be seen from the HRDL backscatter data (Fig. 4).

Vertical velocity measurements detected thermal structures with updrafts of 1 m/s. The HRDL velocity data will be used for the investigation of different turbulent parameters like vertical velocity variance, TKE and momentum fluxes. However, since the HRDL platform was not stabilized these data will not be used until corrections for the motion of the ship are done (work in progress).

From the available data it can be concluded, that on June 23rd and June 24th the weather situation was stable with presumably negligible advection. The boundary layer height and turbulence was mainly driven by surface forcing, wind shear at the top of the boundary layer (see Fig. 2) as well as gravity waves. Therefore this case is very interesting for investigating the performance of LES models.

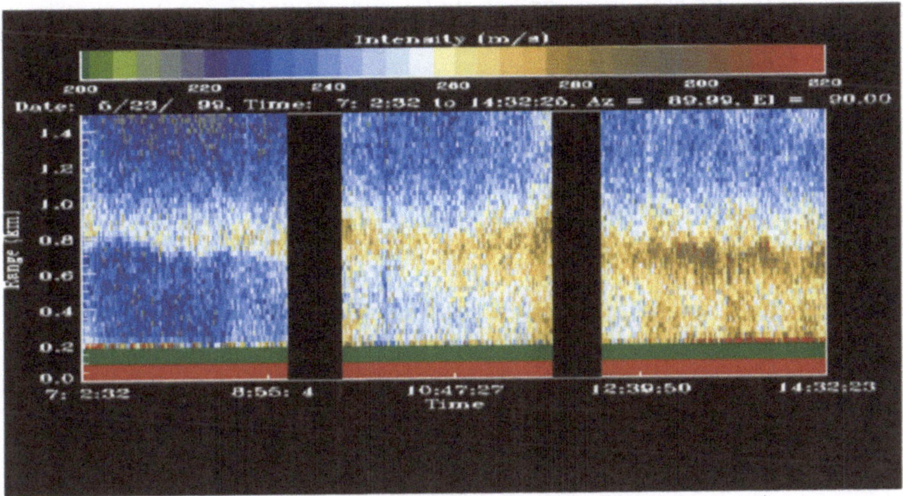

Fig. 4. Time-height cross section of backscatter intensity for June 23, 7:00 UTC to 14:32 UTC.

3 LES model

The model is based on the LES model developed at the Max-Planck-Institute for Meteorology in Hamburg (Chlond[2], [3], [4]). In this model, the equations governing the motion are the three-dimensional, spatially averaged, incompressible Boussinesq equations conserving mass and momentum. These spatially averaged equations can be solved once the subgrid scale (SGS) fluxes are determined in terms of the calculated fields. In the model, the subgrid fluxes are computed using a prognostic equation for the SGS kinetic energy.

For the purpose of our study, computation of the roughness length was modified using the Charnok ([1]) formula in the LES code. Initial potential temperature and specific humidity profiles were obtained by linear interpolation from sonde data profile at 5 : 00 UTC on June 23rd, as shown in Fig. 1. Initial profiles of horizontal velocity are shown in Fig. 2. Values for the geostrophic wind are set to $u_g = 0.1220$ m/s and $v_g = 1.3947$ m/s. Latent and sensible fluxes in the model change with time and are taken from 10 min resolution data shown in Fig. 3. Tests were performed with the resolution of $66 \times 66 \times 112$ points over the domain of $4800 \times 4800 \times 1562$ m,with the time step of 0.5 s. Turbulence was initiated by imposing random temperature perturbations, varying between ± 0.1 K at the surface. The magnitude of the perturbations decreased linearly with height to zero at 200 m. Initial subgrid TKE is set to $0.15 * (1.0 - z/150)$ for $z < 150$ m and to zero for $z \geq 150$. Large scale advection forcing was set to 0. As appropriate for the latitude of 2° S, Coriolis parameter was set to -0.51×10^{-5}.

The code was run on NEC SX-5 machine on four processors. CPU time required for one hour of calculations with resolution of $66 \times 66 \times 112$ points and time step of 0.5 s was 6 : 27 : 39. Memory used during one hour run was 192 MB.

4 The data intercomparison

4.1 Investigation of the temporal structure of the marine boundary layer

Boundary layer height from sonde data, together with buoyancy heat flux estimates were used in order to obtain a convective velocity scale w^* with the resolution of 30 min. Since the boundary layer height from the sonde data is available with the resolution of three hours the value of it was kept constant in the centered interval around the observation for our calculation. During the time period considered w^* varied between 0.45 and 0.72 as can be seen in Fig. 5 (black line). The values obtained using our LES model follow closely values obtained using sonde data up to 15 : 00 h. After this time, values obtained using LES model grow faster than sonde data.

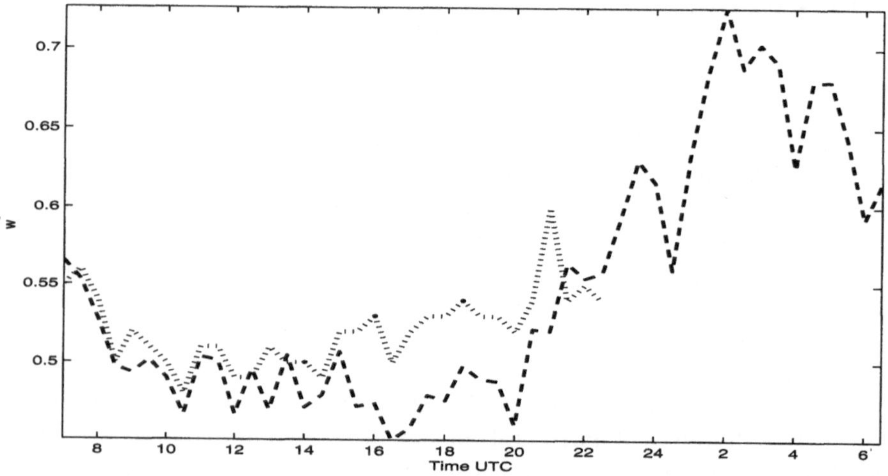

Fig. 5. Convective velocity scale for June 23 and June 24, 1999 at 2 S, 165 E, with z_i from sonde data (dashed black line) and with z_i from LES data (dotted black line).

In our simulation, initially well mixed boundary layer with the height of 600 m (Figs. 1 and 2) gradually increases to the boundary layer of height of 800 m. This can be seen from Figs. 6 to 9 which show observed versus simulated values at 14 : 00 and 20 : 00 UTC on June 23rd.

As can be seen from Figs. 6 and 7, the observed data and the data obtained in the LES simulation show a reasonable degree of qualitative agreement even

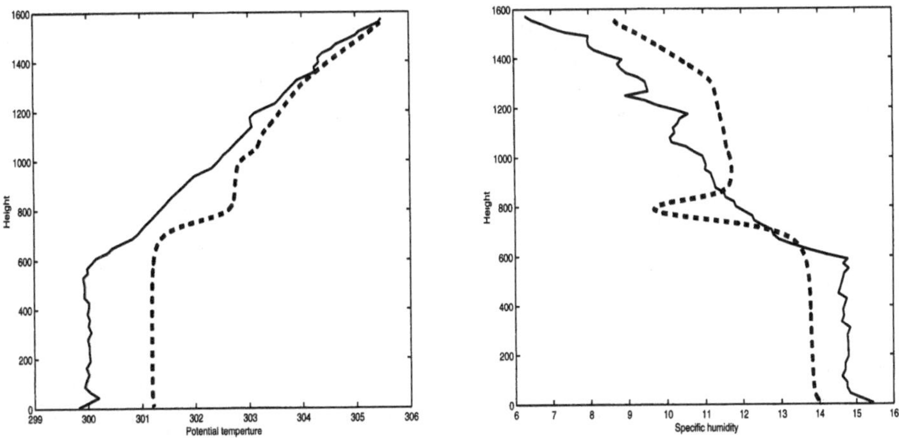

Fig. 6. Sonde (solid line) and LES (dashed line) profiles of potential temperature and specific humidity at 14 : 00 UTC.

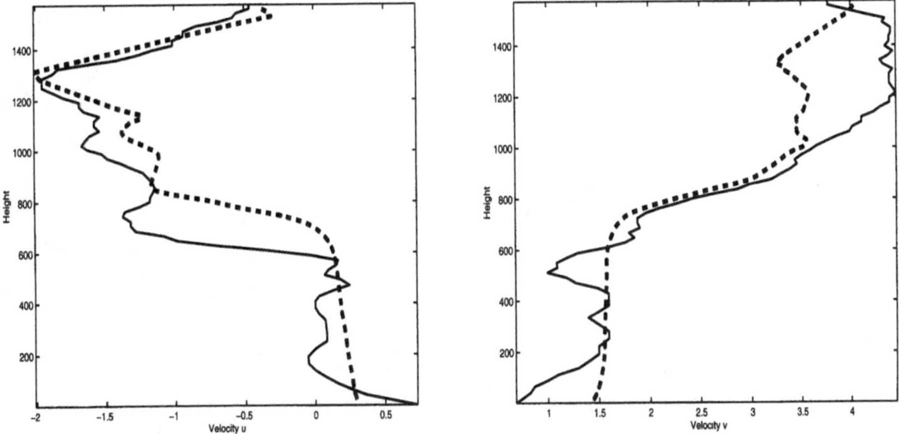

Fig. 7. Sonde (solid line) and LES (dashed line) profiles of horizontal wind at 14 : 00 UTC.

after the first seven hours of integration. Both the observations and the LES data depict a convective boundary layer with a well developed mixed layer. Perhaps the most disturbing deviation from the observations after the first seven hours is the difference between the observed and simulated profiles of potential temperature and specific humidity in the mixed layer. The left panel of Fig. 6 shows observed and simulated profiles of potential temperature after the first seven hours differing for more than 1 degree. Consistent with somewhat warmer mixed layer in the simulation, the simulated height of the PBL exceeds the observed one by about 100 m. On the other hand, as can be seen from the right panel of Fig. 6, the simulated profile of specific humidity is drier than the observed profile for about 1 g/kg. This indicates that the total

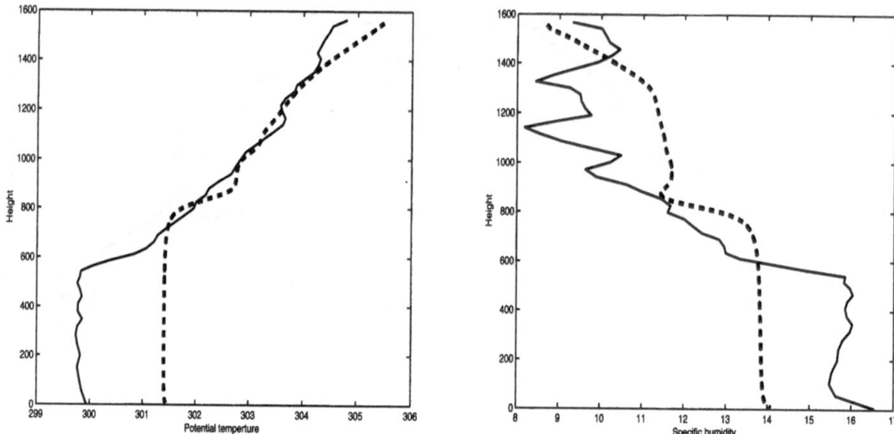

Fig. 8. Sonde (solid line) and LES (dashed line) profiles of potential temperature and specific humidity at 20 : 00 UTC.

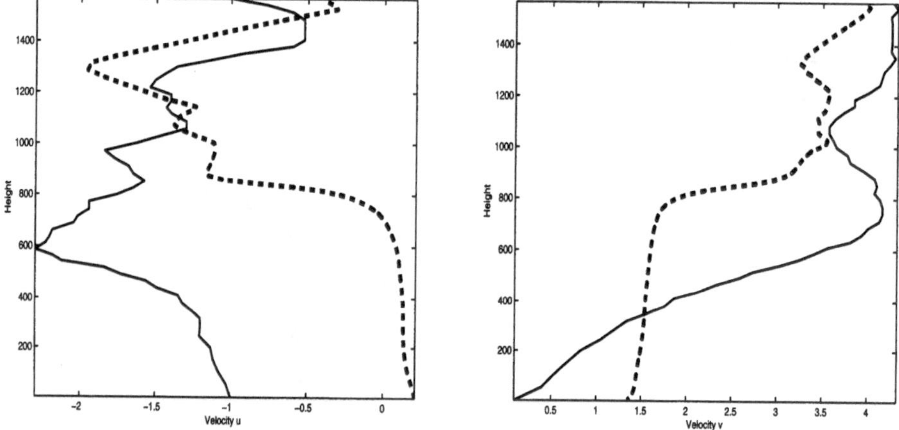

Fig. 9. Sonde (solid line) and LES (dashed line) profiles of horizontal wind at 20 : 00 UTC.

energy input by the sensible and latent heat fluxes is perhaps closer to reality than its sensible and latent heat components taken separately.

As can be seen from Figs. 8 and 9, the differences between the observations and the simulated data continue to grow with simulation time beyond the seven hour mark. After 15 hours, the difference in temperature is of the order of 1.5 degrees and in specific humidity of 2 g/kg. Also, the difference in height of the mixed layer is about 150 m, and the observed and simulated wind profiles show much less resemblance than before.

Possible explanation for the growing discrepancies between the observed and simulated data shown in Figs. 6 to 9, is that some relevant physical processes have not been taken into account in the model. For example, the radiation has been neglected, which may help explain the gradual overheating of the nocturnal marine PBL observed in the simulation. Also, in prognostic simulations beyond few hours, a change of the large-scale flow pattern and gravity waves can strongly influence the solution, even though the change itself may be subtle.

5 Discussion and conclusions

The idea of using synergy between the observations and the LES data is based on the assumption that there is a high level of agreement between the two sources of information about the PBL. If this condition is satisfied, the LES data can be used with a high degree of confidence in order to supplement observations of parameters that are not, or cannot be directly measured. In this sense, the run presented here has been only partially successful since a rather satisfactory agreement between the observed and simulated data persisted only during the first few hours. As can be inferred from Fig. 5, the simulation starts to deviate to a greater degree from the observations after about five hours.

In connection with this result, the question arises as to for how long it is realistic to expect that the observed and simulated data can agree in prognostic mode integrations similar to the one reported in this paper, and what is it that determines this time scale. Seeking the answers to these questions, it is natural to start from the differences between the experiment set-up and the real atmosphere. As discussed in the previous section, the LES model is not perfect. In addition to the numerical errors that can not be eliminated, physical processes such as radiation that have not been taken into account might have affected our results. Also, changes in large scale forcing after few hours of simulation might have influenced the solution. Although the present model set-up has not been fully successful in the extended prognostic mode simulation beyond the first five hours or so, the experience that has been accumulated strongly indicates that prognostic mode LES simulations may be possible up to the order of 12 hours. In order to use our LES results for the comparison with the turbulent statistics obtained from the lidar data we would like first to achieve a better agreement between the LES simulations presented here and the observations.

Acknowledgments: We would like to thank Dr. A. Chlond from Max-Planck-Institute for Meteorology in Hamburg for the LES model he provided to us. Dr. R. K. Newsom from Cooperative Institute for Research in the Atmosphere, Fort Collins, CO, USA and Dr. J. Hare from NOAA/ETL (Environmental Technology Laboratory) made the lidar and surface flux data used in this study available to us. Additional data were obtained from the Atmospheric Radiation Measurement (ARM) Program sponsored by the U.S.

Department of Energy, Office of Science, Office of Biological and Environmental Research, Environmental Sciences Division. We are also grateful to the staff of Stutgart Super Computing Center for many useful discussions and help concerning the computational aspects of our work.

References

1. Charnock, H.(1955): Wind stress on a water surface. Quart. J. Roy. Meteor. Soc., **81**, 639–640
2. Chlond, A. (1992): Three-dimensional simulation of cloud street development during a cold air outbreak. Bound.-Layer Meteor., **58**, 161–200
3. Chlond, A. (1994): Locally modified version of Bott's advection scheme. Mon. Wea. Rev., **122**, 111–125
4. Chlond, A. (1998): Large-eddy simulation of contrails. J. Atmos. Sci., **55**, 796–819
5. Grund, Christian J., Robert M. Banta, Joanne L. George, James N. Howell, Madison J. Post, Ronald A. Richter and Ann M. Weickmann (2001): High-Resolution Doppler Lidar for Boundary Layer and Cloud Research. Journal of Atmospheric and Oceanic Technology, **18**, 376–393
6. Holtslag, A. A. M., Chin-Hoh Moeng (1991): Eddy Diffusivity and Countergradient Transport in the Convective Atmospheric Boundary Layer. J. Atmos. Sci., **48**, No. 14, 1690–1700
7. Hong, Song-You, Hua-Lu Pan (1996): Nonlocal Boundary Layer Vertical Diffusion in a Medium-Range Forecast Model. Mon. Wea. Rev., **124**, No. 10, 2322–2339.
8. Mayor, S. D. (2001): Volume Imaging Lidar Observations and Large-Eddy Simulations of Convective Internal Boundary Layers. Ph.D. Thesis, University of Wisconsin-Madison, 177pp.
9. Moeng, Chin-Hoh (1984): A Large-Eddy Simulation Model for the Study of Planetary Boundary-Layer Turbulence. J. Atmos. Sci., **41**, No. 13, 2052–2062
10. Moeng, Chin-Hoh (2000): Entrainment Rate, Cloud Fraction, and Liquid Water Path of PBL Stratocumulus Clouds. J. Atmos. Sci., **57**, No. 21, 3627–3643
11. Wulfmeyer, V.(1999b): Investigation of humidity skewness and variance profiles in the convective boundary layer and comparison of the latter with large eddy simulation results. J. Atmos. Sci. **56**, 1077–1087
12. Wulfmeyer, V., M. Randall, A. Brewer, and R.M. Hardesty (2000): 2μm Doppler lidar transmitter with high frequency stability and low chirp. Opt. Lett., **25**, 1228–1230

High Resolution Climate Change Simulation for Central Europe

Klaus Keuler, Alexander Block, and Eberhard Schaller

Brandenburg University of Technology Cottbus, Chair for Environmental Meteorology, D-03013 Cottbus, Germany keuler@tu-cottbus.de

Summary. Two regional climate simulations are performed with a high-resolution limited area model for Central Europe, representing present-day climate conditions and a future climate change scenario according to a doubled global CO_2-concentration. Results of a global climate change simulation are used to initialize the regional model and to drive the simulations via time-dependent boundary values. The regional simulations show considerable changes in temperature and precipitation with noticeable geographical and seasonal modifications. For Germany, a significant warming of up to 4 K emerges but the simulated tendency of decreasing summer rainfall cannot be approved as significant.

1 Introduction

Numerous simulations with different global climate models executed during the last decade indicate a substantial change of the global climate conditions for the century ahead [H01]. In general, these models describe the physical processes of the coupled atmosphere–ocean system with a set of partial differential equations and integrate these equations in time on a global three-dimensional grid. The explicit numerical solution of the model equations in every grid-box requires a tremendous amount of computation time even on a present-day supercomputer. In order to perform climate simulations over a period of several hundred years within an acceptable timeframe, the horizontal resolution of these models must be limited to a few hundred kilometers. Therefore, these models are only able to provide information about the changes of global climate means and of large scale – hemispheric or continental – climate patterns. The assessment of possible impacts of global climate changes is, however, in most cases a regional problem and requires a more detailed resolution of the climate change patterns. In particular, we must face the possibility that in some regions climate changes may substantially differ from the mean global tendencies.

In order to meet the demand for high-resolution climate change signals of a particular region, an adequate technique is required which enables a so called 'downscaling' or regionalization of the results of a global climate simulation to the region of interest. In this project, the method of a dynamically nested regional climate simulation is used to produce a high-resolution regional climate scenario for Central Europe. The major scientific objective is to investigate in which way a probable global climate change induced by a further increase of greenhouse gas concentrations will affect the climate conditions in Europe. To achieve this, two regional climate simulations are performed representing present-day climate conditions on the one hand and a future climate scenario on the other hand. The differences between both simulations yield a first idea about possible impacts of a doubling of the global CO_2 concentration on the European climate.

In Sect. 2 the applied regionalization technique is explained. Then, the regional model and the simulation setup are described. In Sect. 4 some results of the climate change experiment are presented. A brief summary together with a first valuation of the scientific results follow in Sect. 5. The last section gives an overview of the performance of the model code on the HLRS computer system.

2 Regionalization of Global Climate Simuations

The regionalization of a global climate change is realized in this study in three subsequent steps. At first, a continuous global climate simulation with a coupled ocean–atmosphere model (OAGCM) is required. As indicated in Fig. 1, this simulation describes the response of the atmosphere and the ocean to a steady increase of greenhouse gas concentrations for the whole globe over a period of more than two centuries. We call this a transient climate change experiment. Due to the high computational expense of this simulation, the resolution is restricted to approximately 2.8°(about 300 km).

In a next step, two seperated periods (time-slices) are selected from this simulation, representing a present-day climate state (PDC) and and a future climate scenario (FCS) with a doubled CO_2 concentration. For both periods, separated climate simulations are performed with an atmospheric global climate model (AGCM) at a higher horizontal resolution than in the coupled OAGCM simulation. The ocean model is no longer used at this level. Information about the temporal development of the sea-surface temperature during the simulation periods as well as the initial atmospheric conditions for both time-slice experiments are interpolated from the results of the coarse-resolution transient simulation. The restriction to two limited simulation periods instead of a continuous run allows an increase of the horizontal resolution in the global model to 1.1°. As the time step for the numerical integration of the model equations has to be reduced by a factor of three, the total amount of computation time at this level is about 19 times higher than for the corresponding atmospheric part in the coarse-resolution coupled simulation.

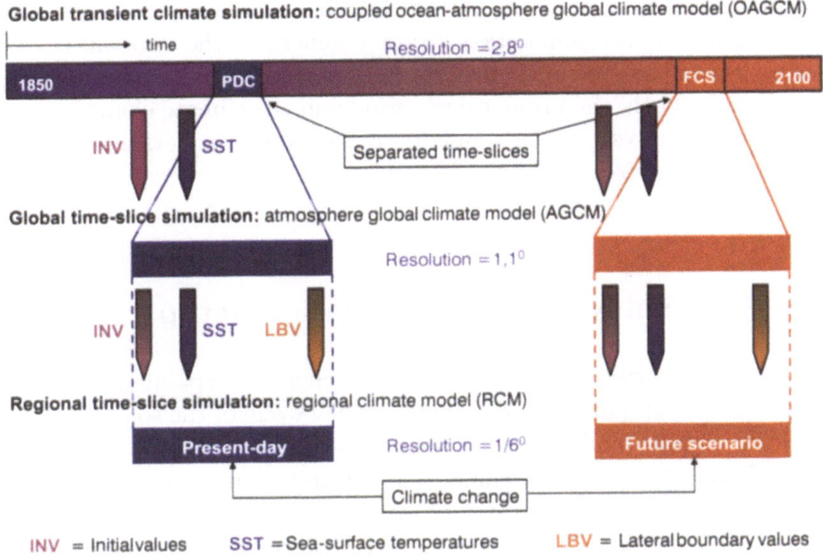

Fig. 1. Downscaling strategy for the regionalization of global climate change simulations

A further increase of the horizontal resolution in the global model is not practicable due to the substantial increase of the computation time. Therefore the model domain has to be restricted in the third step. With the usage of a so called regional climate model (RCM) the climate simulation is focused to a limited area of special interest within the global context. At this level, the model grid covers an area of typically five to ten million square kilometers with a horizontal resolution of about 18 km. Again only the atmospheric part of the climate system is simulated by the regional model but, as shown in Fig. 1, for the same periods as in the previous step. Besides the initial atmospheric conditions and the temporal variation of the sea-surface temperature, time dependant lateral boundary values for all prognostic model variables like temperature, moisture, and wind components are now required to run the model. All these data are interpolated in space and time from the results of the global atmospheric simulations executed in step two. This method of driving a high-resolution limited-area model by time-dependent boundary conditions, obtained from the results of a superior global model, is called a dynamically nested regional climate simulation.

The simulations at all three levels of resolution are carried out successively. The results of the three-dimensional atmospheric fields like air temperature, pressure, wind velocity, water vapor, and liquid water content as well as a set of time-dependent soil and surface parameters, e.g. the water content, the soil temperature, snow cover, and the land- and sea-surface temperatures, are stored in regular time intervals during the whole simulation. At the level of highest resolution, this simulation strategy generates the regional effects of global weather dynamics influenced by local and regional features of the considered area for both, the present-day period and the climate change scenario. The statistical analysis of the simulated development in both periods yields quantitative conclusions of the specific regional climate state and its projected change.

3 The Regional Model and Simulation Setup

Forced by the prescribed boundary values, which are interpolated from a superior global model simulation, and by the diurnal cycle of the solar irradiation, the regional climate model calculates the temporal development of wind, temperature, humidity, radiation, clouds and precipitation at every grid point of the model domain with a time step of two minutes. The model used for these simulations is the regional climate model REMO [JP97]. It has been developed at the Max-Planck Institute for Meteorology (MPIM) in Hamburg particularly for the purpose of continuous long-term simulations of atmospheric processes with a high horizontal resolution. The dynamical part of the model is based on the regional weather forecast model "Europa Modell" [M91] of the Deutscher Wetterdienst (DWD). The parameterizations of the physical processes like the radiative transfer, the formation of clouds and precipitation, and the vertical heat fluxes have been adopted from the corresponding formulations of the global climate model ECHAM4 [R96]. Some parameters, however, had to be adapted to the requirements of the higher horizontal resolution in the regional model.

The model has already been applied to various simulations of present-day climate over Europe and especially over the Baltic Sea area. Detailed validations of the model against available observations particularly show that REMO is able to reproduce the hydrological cycle [J01] and the turbulent heat fluxes at the surface [HJ02] rather well. A comprehensive model intercomparison study [JHA01] documents that the results achieved with REMO are in the range of other state–of–the–art regional climate models. However, for some parameters – e.g. cloud cover, radiative properties, precipitation, and runoff – significant differences still occur between the various models. This fact demonstrates that the results of regional climate simulations are still afflicted with a specific uncertainty and that some further model improvements seem to be necessary to strengthen the reliability of the predicted climate changes.

REMO was primarily developed and tested in a number of experiments on a Cray C90 vector computer and on a parallel processor system, an Origin 2000 by SGI, at the Deutsches Klimarechenzentrum (DKRZ) in Hamburg. Therefore, the transfer of the entire model code and of the automated job and I/O control to the NEC SX5 at HLRS in Stuttgart has required several modifications and adaptations [K02]. Meanwhile, the modified and adapted model system has been transferred back to DKRZ, where it is now running on the new NEC SX6. Section 6 gives a brief comparison between the performance of the code on both mainframes.

Two ten-year periods were simulated with the regional climate model REMO for Europe with a horizontal resolution of $1/6^{th}$ of a degree ($\approx 18km$) in a rotated geographical coordinate system. Figure 2 gives an impression of the selected regional model domain. As outlined in Fig. 1, the required time dependent boundary values for both simulations were provided by the results of two time-slice simulations with the global climate model ECHAM4 [WO00] at a horizontal resolution of T106 (about 1.1° in longitude and latitude). The time-slices span the periods 1971–1980 and 2041–2050, representing present-day climate and a scenario with a doubled CO_2 concentration. The greenhouse gas concentrations for these periods are specified according to the modified IPCC scenario IS92a [SAE95]. The corresponding global transient climate change experiment from 1850 to 2100, which on its part provides the initial values and the sea-surface temperatures for the T106 time-slice runs (Fig. 1), was simulated with the coupled model system ECHAM4/OPYC3 [RBF99] at the horizontal resolution T42 (about 2.8°).

The two global simulations with ECHAM4 were performed by colleagues at the ETH in Zuerich [WO00] and the DKRZ [RBF99] in Hamburg. The regional simulations were executed on the NEC SX5 at HLRS. They ran with a time step of two minutes to ensure numerical stability of the applied discretization schemes. The model requires a new set of input data every six hours to interpolate the time-dependent boundary values in-between. Every input data set has a volume of 12 MB. The results of the regional simulations are stored in the same interval claiming a disk space of 23 MB for each file. Some parameters like temperature and precipitation are already averaged or accumulated during the simulation so that continuous means or sums are available in the output files. In order to ensure a reasonable job and data management, as described in [K02], the ten-year simulation period is subdivided into a sequence of jobs, each of them covering a period of ten days .

Subsequently, daily, monthly, and yearly averages of interesting climate parameters are calculated from the six-hourly archived simulation results. The ten year means of these averaged parameters characterize the climate state of the two periods and are denoted in the following as the climatological monthly or yearly mean values. The differences between the calculated climatological means provide a first estimate of the projected climate change for Central Europe as a result of an increasing global greenhouse gas concentration.

4 Climate Change Signals

The two ten-year regional climate simulations produce a tremendous amount of data (see Sect. 6). An extensive and detailed analysis of different climatological parameters and their spatial distributions is necessary to assess the extent, the reliability, and the potential reasons of a possible regional climate change. Here, as a first step, the primary climate parameters precipitation and near surface air temperature are analyzed and discussed.

Figure 2 shows the difference between the climatological yearly mean temperatures of the scenario run and the control run (present-day climate). The increased greenhouse gas concentrations in the scenario run lead to a significant warming all over the model domain in a range between 2 K and 4 K. The warming is systematically smaller over the sea than over the land surface because of the damping effect of the ocean buffer. The temperature increase seems generally to be higher over the Baltic Sea than over the North Sea, which my be an effect of a stronger influence of the surrounding land mass. The maximum temperature increase between 3.5 K and 4 K occurs in the southern part of France, in northern Italy, and in the Pyrenees. For Central Europe the warming is generally higher than 2.5 K and is slightly decreasing towards eastern and south-eastern Europe. A second maximum exists in Scandinavia where the temperature rise partly exceeds 3 K. For comparison, the increase of the global mean temperature in the transient climate change experiment with the coupled model ECHAM4/OPYC3 of [RBF99] is about 2.8 K. For the European land area, the global simulation indicates a general warming around 3 K but with less regional structures than in the simulation here.

The simulated change of the annual rainfall is given in Fig. 3. The values denote the relative differences between the scenario and the control run in relation to the precipitation amount of the control run. The basically different development in the south-western and the north-eastern part of the model domain is remarkable. The annual precipitation decreases in the Mediterranean area by more than 30%. In the north-eastern part, however, the annual rainfall increases between 5 and slightly more than 10%. In some areas, in particular for most parts of Germany, the precipitation is modified by less than 5%.With respect to the accuracy of the regional model in simulating precipitation, we cannot expect a significant modification of the annual precipitation in these regions from the scenario simulation. The middle and eastern parts of Germany belong to this transition zone between the regions with decreasing and increasing annual precipitation as well as parts of the North Sea, southern Scandinavia and the Baltic Sea. The largest rise of precipitation with up to 30% occurs along the Norwegian coastline. However, this region is located close to the northern boundary of the model domain, so that we should not attach too much importance to this.

The next question is wether the effect of climate change, as derived from the annual mean values, is the same throughout the whole year, or does the

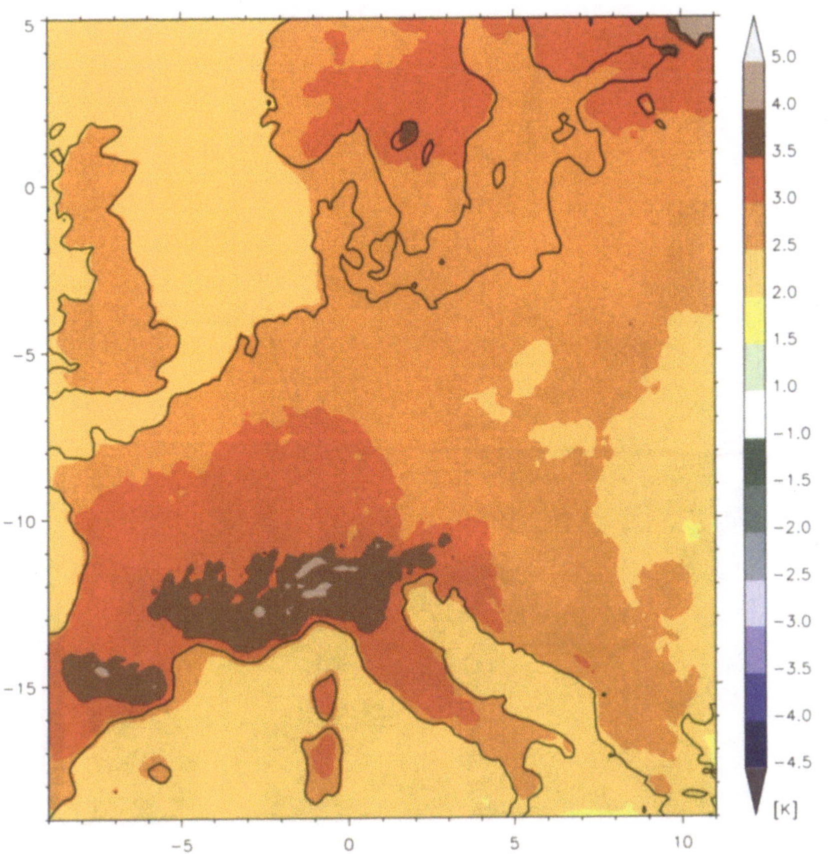

Fig. 2. Simulated climate change of the yearly mean of the near surface temperature; difference of scenario - present-day run in Kelvin

climate change signal show seasonal modifications? To answer this, we consider the annual cycle of the area mean of climatological monthly mean values. That means, we take the climatological mean values at every grid point for each month of the year and calculate the spatial average of these values over a selected region. The region of interest in this case is Germany. The results are given in Fig, 4.

For Germany the scenario run generally produces substantial higher temperatures in all months. This is indicated by the vertical bars in Fig. 4a, which give an estimate of the 95% confidence interval of the climatological mean values for both simulations. The differences to the control run vary, however, considerably over the year. The largest effects occur in late summer (Fig. 4c) with a maximum warming of 4.5 K in August. The climate change is smallest in March with a little bit more than 1.0 K. The warming during the winter season – December, January, February – ranges from 2 to 3 K. The

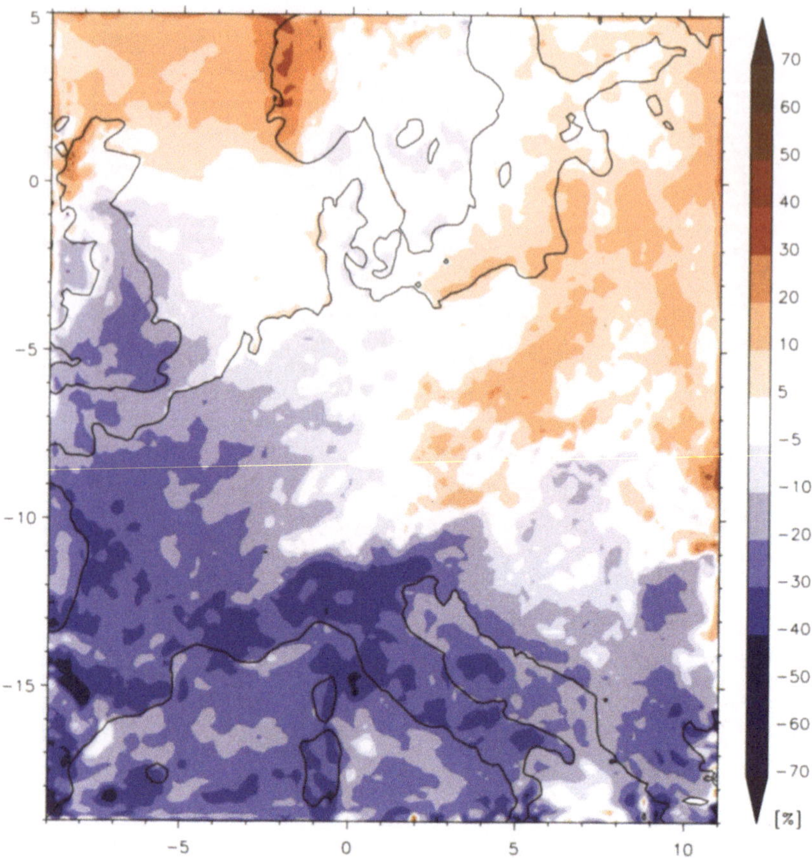

Fig. 3. Relative change of the simulated climatological annual precipitation; difference of scenario - present-day run in percent

annual mean temperature rises by 2.9 K from 10.0°C to 12.9°C. The increase during summer by more than 3.5 K leads to monthly averages of daily mean temperatures in July and August of more than 25°C. A t-test of the climatological mean temperature differences on the 95% significance level indicate that the temperature change is significant for all months other than March.

As demonstrated in Fig. 4b and 4d, the change of precipitation between the scenario and the control run is not uniformly distributed over the year. The 95% confidence intervals in Fig. 4b indicate, that the precipitation differences are substantially only for the period from July to September. Here the monthly values are reduced in the climate scenario by about 20 mm, which corresponds to a relative decrease of about 35%. The maximum increase appears in March and October with approximately 10 mm. From late spring to the beginning of autumn the scenario simulation tends to reduce the amount of precipitation. During wintertime only January shows a considerable de-

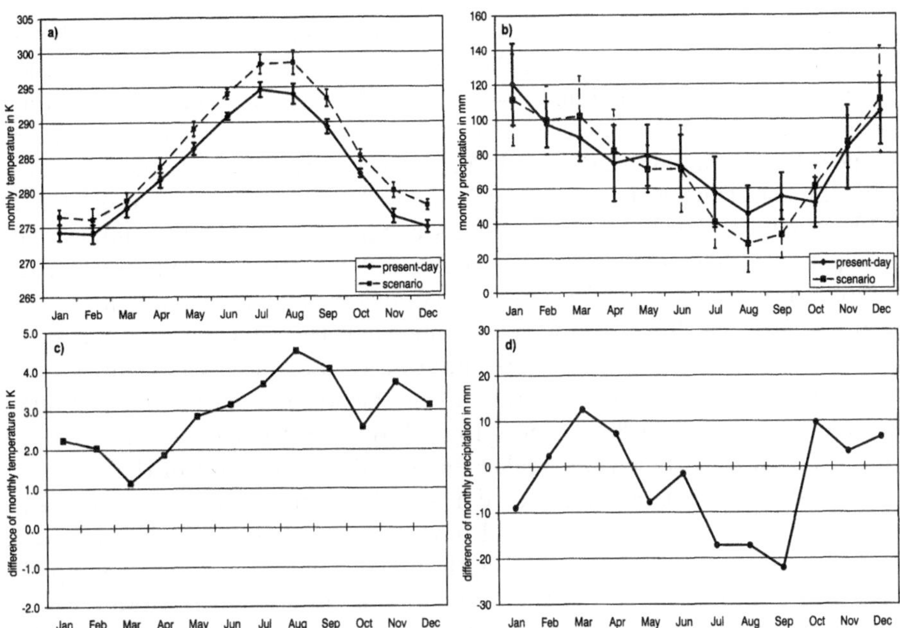

Fig. 4. Annual cycle of climatological monthly mean values: (**a**) air temperature in K and (**b**) precipitation in mm of the present-day (*solid lines*) and the scenario (*dashed lines*) simulation; (**c**) and (**d**) differences of the corresponding curves above in K and mm

crease of precipitation. The total annual precipitation amount for Germany is reduced by only 33 mm, which corresponds to an average monthly reduction of 2.7 mm. Despite the large relative decrease in late summer the statistical t-test yields a significant change with a remaining uncertainty of 5% for September only. The modification of the annual rainfall appears also not to be significant.

5 Summary and Assessment of Scientific Results

The dynamically nested regional climate model REMO is used for regional climate simulations over Europe. Two ten-year time-slice runs have been performed, representing present-day climate conditions and a future climate scenario, respectively. The underlying global change scenario is provided by global climate simulations with the ECHAM4/OYPC3 coupled model system. The climate change scenario corresponds to the IPCC scenario IS92a, which will lead to a doubling of the global CO_2 concentration around the middle of the 21st century. The climate change signal is determined as the difference of the monthly and annual climatological means of temperature and precipitation between the two regional simulations.

The horizontal resolution of the regional model is about 18 km. The results show detailed structures of the climate patterns as a consequence of the complex geographic and climatic situation in Europe. The annual mean temperature rise is in the range of 2 to 4 K with maximum values in southern France. For Germany, the scenario projects a warming of about 3 K in the annual mean. During the annual cycle the warming varies within the range of ± 1 K around this value and reaches its maximum during late summer. The precipitation change has a more irregular distribution over Europe than the temperature change. The strongest reductions of the annual rainfall are projected for the south-western part of France and the adjacent Mediterranean area. For Germany, the annual precipitation remains nearly unchanged with a weak tendency of increasing rainfall in the western parts. A substantial decrease of more than 30% occurs from July to September which is almost compensated by a somewhat weaker increase during the rest of the year.

The strong temperature increase in summer together with the simultaneous reduction of precipitation could have serious consequences for the water budget in some parts of Germany with further impacts, for instance, on the agricultural productivity and the potential risk of forest fires. However, only the temperature change turns out to be statistically significant so that further conclusions of the simulated changes should be drawn carefully.

The major problem in this context is the small ensemble size to calculate the climate means of the two simulated periods. Ten realizations for each month or year are still too small to narrow sufficiently the range of uncertainty of the climatological means so that a significant climate change signal for precipitation could be approved. Therefore, the sample size seems necessary to be expanded by additional scenario experiments and longer simulation periods. A task which will be addressed in upcoming studies.

6 Technical aspects

The regional model uses a three-dimensional grid with $121 \times 145 \times 24$ boxes and a time-step of two minutes. It requires a bit less than 500 MB of main memory. Both ten-year regional simulations together take a CPU-time of 1300 h on a single processor of the NEC-SX5 at HLRS. The simulations use 350 GB of input data and produce an output of more than 660 GB. Thus, the complete regional climate change experiment takes up a permanent storage capacity of 1.1 TB.

The model can also be compiled in parallel mode using auto-tasking directives on the shared memory architecture but does not perform exceptionally well in a multi-processor environment. The speed-up on 4 CPUs compared to a single CPU run of the SX5 is approximately 2.5. The performance on a single CPU with 1720 MFLOPS is satisfying. The averaged vector length of all calculations is 192 and the vector operation ratio reaches the remarkable value of 99%.

Due to the high utilization of the job-queues at HLRS, the real computation time was about five times higher than the effective CPU-time. As a consequence of the limited capacity of the SX5, the complete experiment took more than 270 days. Further model runs with a preferable larger model domain and an extended simulation period would require a substantial expansion of the computation capacity at HLRS. Some experiments with the same model code and experimental conditions could have been executed on the NEC-SX6 at DKRZ in Hamburg. A comparison of some performance characteristics are given in Tab. 1.

Table 1. Performance characteristics of the regional climate model on NEC super computers SX5 and SX6

Parameter	SX5	SX6
relative CPU-time	1.0	0.63
MIPS	60	99
MFLOPS	1700	2700
Vector length	190	192
Vector operation ratio	99%	99%

The conclusion of these experiments is, that a super computer with parallel vector processors is an adequate architecture for climate model simulations. Future scientific issues, like the execution of ensemble simulations or of continuous simulations over a century or more require, however, a considerable enhancement of computing power. In addition, the parallel performance of the model code on a shared memory system has to be improved too.

References

[H01] Houghton, J.T.: Climate Change 2001, the Scientific Basis. Contribution of Working Group I to the Third Assessment Report of the Intergovernmental Panel on Climate Change. Cambridge Univ. Press, Cambridge (2001)

[HJ02] Hennemuth, B., Jacob, D.: One year measurement and simulation of turbulent surface heat fluxes over the Baltic Sea. Meteorol. Z., **11**, 105–118 (2002)

[J01] Jacob, D.: A note to the simulation of the annual and inter-annual variability of the water budget over the Baltic Sea drainage basin. Meteorol. Atmos. Phys., **77**, 61–73 (2001)

[JHA01] Jacob, D., van den Hurk, B.J.J.M., Andræ, U. et al.: A comprehensive model inter-comparison study investigating the water budget during BALTEX–PIDCAP period. Meteorol. Atmos. Phys., **77**, 19–43 (2001)

[JP97] Jacob, D., Podzun, R.: Sensitivity studies with the regional climate model REMO. Meteorol. Atmos. Phys., **63**, 119–129 (1997)

[K02] Keuler, K: Regional climate simulation for central Europe. In: Krause, M., Jäger, W. (ed) High Performance Computing in Science and Engineering '02. Springer, Berlin Heidelberg New York (2002)

[M91] Majewski, D.: The Europa-Modell of the Deutscher Wetterdienst. ECMWF Seminar on numerical methods in atmospheric models, **Vol 2**, 147–191 (1991)

[R96] Roeckner, E., Arpe, K., Bengtsson, L., Christoph, M., Claussen, M., Duemenil, L., Esch, M., Giorgetta, M., Schlese, U., Schulzweida, U.: The atmospheric general circulation model ECHAM4: Model description and simulation of present-day climate. Report no. **218**, Max-Planck Institute for Meteorology, Hamburg (1996)

[RBF99] Roeckner, E., Bengtsson, L., Feichter, J., Lelieveld, J., Rodhe, H.: Transient climate change simulations with a coupled atmosphere-ocean GCM including the tropospheric sulfur cycle. J. Climate, **12**, 3004–3032 (1999)

[SAE95] Schimel, D., Alves, D., Enting, I., Heimann, M. et al.: Radiative Forcing of climate change. In: Houghton, J.T., Meira Filho, L.G., Callander, B.A., Harris, N., Kattenberg, A., Maskell, K. (ed) Climate Change 1995: The Science of Climate Change. Contribution of WG I to the Second Assessment Report of the Intergovernmental Panel on Climate Change. Cambridge Univ. Press, Cambridge (1955)

[WO00] Wild, M., Ohmura, A.: Change in mass balance of polar ice sheets and sea level from high-resolution GCM simulations of greenhouse warming. Ann. Glaciol., **30**, 197–203 (2000)

Water on Mars

Tetsuya Tokano

DLR-Institut für Raumsimulation, 51170 Köln, Germany, tetsuya.tokano@dlr.de

1 Introduction

A milestone in the understanding of the Martian hydrological evolution and water cycle is being marked by the Gamma-Ray Spectrometer (GRS) onboard Mars Odyssey ([1], [2], [4]). These gamma rays are supposed to be produced during inelastic collision and/or capture of secondary neutrons which were produced in the soil by cosmic rays. The signals are indicative of the presence of a large amount (at least 35 wt %) of hydrogen below a dry layer in this region. The Neutron Spectrometer [2] and the High Energy Neutron Detector (HEND) [4] measured leakage fluxes of fast, epithermal and thermal neutrons produced in the same way. As hydrogen has a strong ability to slow down the neutrons, depression of neutron fluxes, especially of faster ones, can be interpreted by a large abundance of hydrogen and hence H_2O. Both instruments detected a large region around the south pole southward of 60°S (with an area of 10^7 km^2) with strongly reduced epithermal fluxes indicative of large water abundance. In general the upper 15–20 cm of the soil at low and mid latitudes was inferred to be water-poor dust layer underlain by a slightly water-richer regolith [2]. The global map of fast neutrons (indicative of the upper layer) showed a somewhat different picture, particularly those measured by HEND. The area with a large deficit of fluxes was larger at northern high latitudes (10^7 km^2) than at southern latitudes (4×10^6 km^2). In all maps the global distribution is not symmetric about the equator and some distinct longitudinal variation is also discernible.

These preliminary results raise important tasks. First, the Mars Odyssey data themselves cannot tell us how the water is partitioned into phases, i.e. pore ice, adsorbed water, brine or, more importantly, chemically bound water in minerals. Secondly, it is useful to check whether the subsurface water distribution retrieved by Mars Odyssey is consistent with the present water cycle, or more specifically, the physical interpretation is meaningful from the viewpoint of hydrology. This has not been performed in previous studies with atmospheric water cycle models due to the lack of data.

In an effort to test and to understand the global water distribution in the shallow subsurface of Mars retrieved by the Mars Odyssey Gamma-Ray Spectrometer we investigate the present state and movement of water by a coupled global subsurface-atmosphere water cycle model. After some sets of simulation we examine whether the simulation can explain the observed feature and what additional factors may be important.

2 Model description

The numerical model developed for this study is a time-dependent, global model of water transport and phase change. The atmospheric and subsurface parts are coupled and they are treated with similar complexity to minimize a bias in the interpretation in favour of atmospheric or subsurface component.

The atmospheric part of the water cycle is treated two-dimensionally in the latitude-height plane. Rather than to predict the meteorological data by ourselves the model uses the output of a set of numerical simulations with sophisticated Mars GCMs compiled in the European Mars Climate Database [3]. The model domain extends vertically from the surface up to 60 km altitude and is represented by 20 grid points with a finer resolution near the surface. Laterally the model covers the entire globe and is represented by 36 equidistant grid points with an interval of 5° latitude. Water is treated as a passive tracer undergoing global transport and phase change. Water vapour in the atmosphere is subject to advective and diffusive transport. The meridional advection of the water vapour mixing ratio C is simulated with the instantaneous meridional wind v by means of the positive-definite advection algorithm of [5] with a constant timestep interval of 600 s. The vertical transport is approximated by eddy diffusion since the vertical wind speed is not provided in the climate database. We implement a simple condensation scheme to account for the water vapour abundance limitation by saturation. If the atmosphere gets supersaturated with water vapour, the supersaturated portion of water is immediately condensed out. Condensed water is not precipitated out, but is stored as cloud/fog until it sublimes as soon as the air gets drier. At the same time the cloud is advected by the wind in the same manner as water vapour.

When the residual H_2O cap at the pole is exposed to the atmosphere after the retreat of the seasonal CO_2 cap in early summer, water sublimation from the caps takes place. The northern H_2O cap is placed at gridpoints northward of 80°N, while the southern one will never be exposed to the atmosphere. The seasonal CO_2 caps, that overlie the H_2O caps, are provided by the climate database.

The energy equation in the subsurface is approximated by a one-dimensional heat conduction equation. We regard the latent heat exchange associated with phase change of water in the subsurface to be negligible compared to the diurnal temperature variation. The heat conduction equation is solved by the

Crank-Nicholson scheme with a time step interval of 600 s. The upper boundary condition is the surface temperature T_s provided by the meteorological input data. The lower boundary condition is given by a constant geothermal gradient resulting from the geothermal heat flux Q.

The subsurface water budget is predicted by a balance equation for the total water content σ (in kg m^{-3}) similar in the form to that of [6]. σ comprises all water phases not chemically bound in minerals, i.e. water vapour, pore ice and adsorbed water film ($\sigma = \phi\rho_{\text{vap}} + \sigma_{\text{ice}} + \alpha$). The vertical transport of water is assumed to take place solely as vapour diffusion through soil pores. We assume that the water vapour is always in equilibrium with the adsorbed water film and pore ice. In other words, the phase change takes place immediately. Adsorption of water molecules is calculated with the adsorption isotherms for palagonite and montmorillonite measured under Mars-like conditions by [7]. Phase changes are indirectly simulated by calculating the equilibrium partition into water vapour, adsorbed water and pore ice analogously to [6].

3 Computational performance

The model code is written in Fortran 77 and run on the IBM RS/6000 SP of the SSC Karlsruhe. The code is not parallelized, so it utilizes only 1 CPU simultaneously. The model contains 1400 grind points (2 dimensions). With an equidistant time step interval of 600 seconds constrained by the numerical stability condition approximately 10^4 time steps are required to simulate 1 Martian year. This takes approximately 1 CPU hour on this machine.

The most time-extensive components of the model are the advection scheme of the atmospheric vapour and clouds and the subsurface diffusion equations by the Crank-Nicholson scheme.

References

[1] Boynton, W. V., Feldman, W. C., Squyres, S. W., Prettyman, T., Brückner, J., Evans, L. G., Reedy, R. C., Starr, R., Arnold, J. R., Drake, D. M., Englert, P. A. J., Metzger, A. E., Mitrofanov, I., Trombka, J. I., d'Uston, C., Wänke, H., Gasnault, O., Hamara, D. K., Janes, D. M., Marcialis, R. L., Maurice, S., Mikheeva, I., Taylor, G. J., Tokar, R., Shinohara, C.: Distribution of hydrogen in the near-surface of Mars: evidence for subsurface ice deposits. Science, **297**, 81–85 (2002).

[2] Feldman, W. C., Boynton, W. V., Tokar, R. L., Prettyman, T. H., Gasnault, O., Squyres, S. W., Elphic, R. C., Lawrence, D. J., Lawson, S. L., Maurice, S., McKinney, G. W., Moore, K. R., Reedy, R. C.: Global distribution of neutrons from Mars: results from Mars Odyssey. Science, **297**, 75–78 (2002).

[3] Lewis, S. R., Collins, M., Read, P. L., Forget, F., Hourdin, F., Fournier, R., Hourdin, C., Talagrand, O., Huot, J.-P.: A climate database for Mars. J. Geophys. Res., **104** (E10), 24177–24194 (1999).

[4] Mitrofanov, I., Anfimov, D., Kozyrev, A., Litvak, M., Sanin, A., Tret'yakov, V., Krylov, A., Shvetsov, V., Boynton, W., Shinohara, C., Hamara, D., Saunders, R. S.: Maps of subsurface hydrogen from the high-energy neutron detector, Mars Odyssey. Science, **297**, 78–81 (2002).

[5] Smolarkiewicz, P. K.: A simple positive definite advection scheme with small implicit diffusion. Mon. Wea. Rev., **111**, 479–486 (1983).

[6] Zent, A. P., Haberle, R. M., Houben, H. C., Jakosky, B. M.: A coupled surface-boundary layer model of water on Mars. J. Geophys. Res., **98** (E2), 3319–3337 (1993).

[7] Zent, A. P., Quinn, R. C.: Measurement of H_2O adsorption under Mars-like conditions: effects of adsorbent heterogeneity. J. Geophys. Res. **102** (E4), 9085–9095 (1997).

Viscosity Stratification and a 3-D Compressible Spherical Shell Model of Mantle Evolution

Uwe Walzer[1], Roland Hendel[1], and John Baumgardner[2]

[1] Institut für Geowissenschaften, Friedrich-Schiller-Universität, Burgweg 11, 07749 Jena, Germany
[2] Los Alamos National Laboratory, MS B216 T-3, Los Alamos, NM 87545, USA

Summary. The viscosity stratification has a strong influence on the thermal *evolution* of a compressible Earth's mantle with time-dependent internal heating. The differential equations for infinite Prandtl-number convection are solved using a three-dimensional finite-element spherical-shell method on a computational mesh derived from a regular icosahedron with 1.351.746 or, alternatively, 10.649.730 nodes. We formulate a radial viscosity profile from solid-state physics considerations using the seismic model PREM. New features of this viscosity profile are a *high-viscosity transition layer* beneath the usual asthenosphere, a *second low-viscosity layer* below the 660-km endothermic phase boundary and a considerable *viscosity increase* in the lower 80 % of the lower mantle. To be independent of the special assumptions of this derivation, we vary the level and the form of this profile as well as the other physical parameters in order to study their consequences on the planforms and on the convection mechanism. The effects of the two mineral phase boundaries at 410 and 660 km depth proved to be smaller than effects of the strong variation of viscosity with radius. The latter had more influence on the convective style than all other parameters. Values of our material parameters are time independent and constant in the lateral directions, except for viscosity.

The focus of this paper is a variation of non-dimensional numbers as Rayleigh number, *Ra*, Nusselt number, *Nu*, the reciprocal value, *Ror*, of the Urey number, viscosity-level parameter, r_n, etc. We explored the parameter range for special solutions. For the wide parameter range $-0.5 \leq r_n \leq +0.3$, that includes our preferred viscosity profile, we obtain solutions characterized by reticular connected *thin cold sheet-like downwellings*. The downwellings are thinner than similar features in previous publications. They bear a resemblance to observed subducting slabs but are purely vertical. We find it remarkable that the downwellings penetrate the high-viscosity transition layer. They remain sheet-like to 1350 km depth. Below this depth they begin to lose definition but their locations are still visible at 1550 km depth. Such thin subducting sheets are notable since the viscosity is Newtonian. On the other hand, it is not surprising there are no transform-like features at the surface of the model. We compute laterally averaged heat flow at the Earth's surface, the ratio of heat output to radiogenic heat production, *Ror*, the Rayleigh number and the Nusselt number as a function of time. $Nu_{(2)}$ denotes the temporal average of the Nusselt number of a run for the last 2000 Ma of the evolution, $Ra_{(2)}$ is the temporal average of the Rayleigh number, respectively. For a wide parameter range, we obtain $Nu_{(2)} = 0.120\ Ra_{(2)}^{0.295}$ in this model.

1 Introduction

The focus of this paper is a parameter variation of the radial viscosity profile of the Earth's mantle and a variation of non-dimensional numbers that characterize thermal convection. The patterns of the planforms are relevant, too. We take an interest not only in convection of the geological present but also in the evolution of relevant parameters, e.g. Nusselt number and Urey number, as a function of time in the past.

For reasons explained in Section 3.4, we assume that the segregation of the metallic core and the primordial silicate mantle was finished $(4.49 \pm 0.03) \times 10^9$ a ago. The present model is essentially a viscous spherical shell heated mainly by radioactive decay from within and, only to a minor degree, from below. The spherical shell had a primordial initial heat and cooled down the course of the thermal evolution. For geochemical reasons, the start of the evolution of the system was fixed for an age of 4.49 Ga. In spite of this and other geochemical and geophysical specifications of the model, we are of course not able to model the Earth's mantle in a self-consistent manner, taking into account all the complexities necessary. E.g., it is not the focus of this paper to contribute to the problem of self-consistent generation of oceanic lithospheric plates (Bercovici, 1996; Christensen, 1996; Trompert and Hansen, 1998; Tackley, 2000a, 2000b, 2001, Bercovici et al., 2001a; etc.). Furthermore, the mantle of the model is chemically homogeneous. It contains no chemically different reservoirs. At the upper surface, there are no continents that would be able to modify the upper boundary conditions. Therefore the model is heated, spatially homogeneously and temporally decaying, essentially from within supposing the abundances of McCulloch and Bennett (1994) that do not essentially deviate from the primordial mantle abundances of Hofmann (1988). Although we fix further quantities and incorporate the two major mineral phase transitions into the model, we are able to characterize the essential features of the evolution of the system by a few non-dimensional numbers as the Rayleigh number, *Ra*, the Nusselt number, *Nu*, the reciprocal value, *Ror*, of the Urey number, the non-dimensional viscosity-level number, r_n, etc. We varied the mentioned numbers since many geophysical properties are not well constrained. The variation of the mantle's viscosity has the largest effect on the mechanism and on the non-dimensional numbers. Therefore, this variation is the center of this paper. Of course, the results directly refer only to the mechanism of the model and not immediately to the real mantle. Every geodynamical model can be criticized from two flanks. Firstly, it is feasible to refer to some non-included feature of the real mantle. In our model, e.g., volatiles are not explicitly included. In this case, a critical reader could demand a more complex model. On the other hand, another reader could require a simplification to allow that new understanding of fluid dynamics comes out of this paper. So, we would learn to understand each single mechanism. But unfortunately, a linear superposition of solutions is not possible since the equations are nonlinear. Therefore, it is legitimate to find a geophysical compromise between the two mentioned contradictory demands, provided that we can grasp the essence by non-dimensional numbers etc. The usual survey of related evolution and convection models occurs in Section 5.

2 Viscosity. General considerations

So far, the public discussion of the influence of viscosity on mantle convection was concentrated on temperature dependence. As is well known, Bénard convection in a constant viscosity fluid layer can be parameterized by the Nusselt number and the Rayleigh number only. If, additionally, the temperature dependence is taken into account then further parameters have to be introduced. If we have the simple expression

$$\eta = \eta_{\text{ref}} \cdot \exp[-E(T - T_{\text{ref}})] \tag{1}$$

then a further non-dimensional number, r_η, has to be introduced where

$$r_\eta = \exp(E \cdot \Delta T) \tag{2}$$

η is viscosity, η_{ref} is reference viscosity, E is a constant, T is temperature, T_{ref} is reference temperature, r_η is a viscosity ratio, and ΔT is the fixed total temperature difference across the fluid layer. In this simple case, convection is in the small viscosity contrast regime for $r_\eta \leq 10^2$, in the sluggish-lid regime for $10^2 < r_\eta < 10^4$, and in the stagnant-lid regime for $r_\eta \geq 10^4$ (Christensen, 1985, Solomatov, 1995, Reese et al., 1998, 1999, Schubert et al., 2001). The viscosity between the two thermal boundary layers is nearly constant. If it is intended to apply such simple temperature-dependent viscosity convection models to the mantle then we raise three principal objections to the effect that corresponding supplementations are necessary:

a) Viscosity is also pressure-dependent. In particular for minerals of the lower mantle, the rising pressure causes a strongly increasing viscosity. Even if the lower-mantle composition would not depend on depth, the product of activation volume and pressure generates a strong viscosity increase with depth, without jumps, so that the temperature effect is overcompensated except perhaps for D$''$ and plumes.

b) It is an obvious assumption that not only density, compressional velocity and shear velocity jump at the major phase boundaries of the mantle but also activation energy and activation volume of the relevant creeping mechanism (Karato et al., 1995, Karato, 1997a, Ranalli, 1998). Since the activation enthalpy is in the exponent of the viscosity equation, major viscosity jumps are to be expected. We will show that the influences of a) and b) on the planforms are essential.

c) We take an interest not only in mantle convection of a short time interval but also in the thermal evolution all the 4490 Ma round. Based on geological observations, it is a realistic approach to assume a constant upper surface temperature. However, there are good reasons and observations to assume that the volumetrically averaged temperature, T_a, of the mantle diminishes as a function of time. Therefore, the temperature at the core-mantle boundary (CMB) is probably *not* a constant *with respect to time.* For *evolution* models, the difference ΔT cannot be a constant. This conclusion applies equally to viscous layer models and spherical shell models. Further arguments are given in Section 3.4.

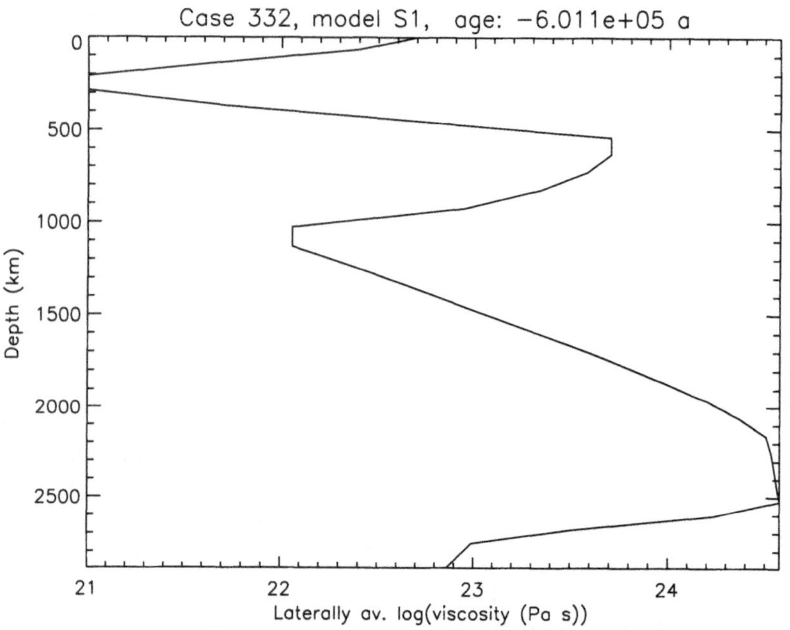

Fig. 1. The basic radial viscosity profile, $\eta_1(r)$ for $r_n = 0$.

Because of a) and b) we offer now a survey of possible radial viscosity profiles for the Earth's mantle. We then present the reasons for the viscosity profile, we favor most, as shown in Fig. 1 and specified by Eq. (5). It is this profile we than apply in our three-dimensional thermal evolution model of the mantle. Starting from this preferred viscosity distribution, we vary our parameters to learn the range of validity of the corresponding numerical solutions. Increasing pressure reduces the number of defects, but increasing temperature generates an increased number of defects and dislocations. The change in viscosity with pressure is due not only to the number of defects or dislocations, but also to the change in their mobility, i.e., the energy required to jump from one site to another increases with pressure. Therefore, the solid-state viscosity increases with increasing pressure and decreases with increasing temperature. Because the P, T-dependence of the other physical parameters is much smaller, we assume only a radial dependence in these other quantities.

Unfortunately, the viscosity of the mantle is relatively poorly constrained. There are two classes of methods for estimating mantle viscosity as a function of depth. The *first* class of methods relies upon geophysical observables such as post-glacial rebound data, the geoid, global seismic tomography models, and observed surface heat flow. The *second* class relies upon physical investigations of the creep mechanisms in monocrystalline and polycrystalline silicates and oxides at high pressure and high temperature. The *first* class is exemplified by investigations that derive depth-dependent mantle viscosity from deformation of spherical-earth models by surface loads, e.g., by a changing ice-water distribution (Lambeck and Johnston, 1998). In addition there are anelastically compressible internal loading theories. Tomography derived differences in the seismic compressional wave velocity, v_p, and the seismic shear wave velocity, v_s, can be converted into density differences using an empirical relation. The observed aspherical geoid anomalies as well as the observed free-air gravity field ought to be identical to the corresponding theoretical quantities arising from such density differences. Such studies typically assume mantle convection is driven by density differences derived from the tomography. In a convecting mantle, the actual geoid depends not only on the density heterogeneity but also on the mantle's viscosity structure. Typically, however, calculations of this sort that seek to recover the mantle's viscosity structure have yielded wildly different viscosity distributions as a function of depth. Various alternatives have been explored to reduce this ambiguity. Pari and Peltier (1998), for example, sought to mitigate the extreme degree of nonuniqueness by constraining the radial component of the internal-load driven flow velocity to be linearly related to the surface heat flow.

In the seismic velocity-to-density scaling it is usually assumed that the seismic velocity anomalies have a thermal origin. Since this premise is clearly wrong for the continental lithospheric mantle and partly wrong for subducted oceanic slabs, Kido and Čadek (1997) restricted this procedure to only the *oceanic* geoid at intermediate degrees *(l= 12–15)* and to areas at a sufficiently large distance from slabs. As a starting point, they explored three different seismic tomography models and five rather different radial profiles for the velocity-to-density scaling factors $(\partial \ln \rho / \partial \ln v_p)$ and $(\partial \ln \rho / \partial \ln v_s)$. Their inversion technique sought to maximize the correlation between observed and predicted oceanic geoid at degrees 12–25. Kido and Čadek (1997) used a genetic algorithm to find the maximum correlation coefficient. They found *two* low-viscosity layers from their inversions for *all* 15 combinations of the three seismic tomography models and the five velocity-to-density scaling profiles. One of these layers was the usual asthenosphere immediately beneath the lithosphere, while the other one was situated within the upper quarter of the lower mantle.

We suggest that the most likely profiles have a second low viscosity layer that begins *immediately* beneath the 660-km phase boundary since the activation energy and the activation volume of the relevant minerals must jump at the phase boundary independent of the prevailing creeping mechanism. Moreover, the grain-size reduction of material flowing downward through the 660-km boundary is a good argument in favor of a second low-viscosity layer just below the 660 (Kubo et al., 2000). Čadek and van den Berg (1998) discussed a parameterization of the radial temperature profile of the mantle where the geoid and the seismic velocities have been taken as observables. The well-known non-unique nature of the geoid inversions leads to two different sets of viscosity profiles. The first set, called A,B, has a stiff lithosphere at the top that is underlaid by a thick asthenosphere between 100 and 400 km depth and a viscosity maximum around 500 km depth. The next layers consist of a narrow low-viscosity zone *above* the 660-km discontinuity and a uniformly high viscosity through most of lower mantle. Only near the CMB does the viscosity decrease. A second set of viscosity profiles, called C,D, also display a stiff lithospheric lid. Beneath this lid is a flat viscosity minimum down to 660 km. At 660 km depth, the viscosity increases abruptly by 2-3 orders of magnitude followed by a second low-viscosity layer between 800 and 1000 km depth. A remarkable feature of the C,D viscosity profile is a broad viscosity maximum between 2000 and 2500 km depth. Finally, as in the A,B case, the viscosity decreases near the CMB.

A *second* class of methods for constraining the mantle's viscosity profile comes from mineral physics. Based on broad consensus, Ranalli (1998) concluded that the rheology of the asthenosphere between 100 and 410 km depth is determined primarily by the creep properties of olivine, that of the transition layer primarily by the creep properties of spinel and garnet phases, and that of the lower mantle primarily by the properties of perovskite. If there is no drastic reduction in grain size within the transition layer, the creep properties of spinel and garnet appear to exclude low viscosity in this zone. Therefore we conclude the A,B set of viscosity profiles by Čadek and van den Berg (1998) must be excluded because of their narrow low-viscosity layer *above* the 660-km discontinuity. The main features of the C,D profiles seem more probable than those of the A,B profiles since solid-state physics considerations argue for a broad viscosity maximum in the lowermost 1000 km of the mantle apart from the D$''$ layer. For each of the different creep mechanisms, Ranalli (1998) specified values for the activation energy, E_0, and the activation volume, V_0. Even if some details are uncertain, it is evident that the viscosity must show *considerable jumps* at the phase boundaries in 410, 520, and 660 km of depth since E_0 and V_0 are in the exponent of the function that describes the P,T-dependence of the viscosity.

Karato (1997a) emphasizes that garnet has a unique crystal structure that does not allow easy dislocation glide. An increase in density and seismic wave velocities does not mean that the viscosity continuously rises with depth since the resistance to plastic deformation in garnets is considerably higher than that of perovskite, olivine, spinel and most of the remaining mantle minerals. Karato et al. (1995) studied the deformation properties of both oxide and silicate garnets from room temperature up to $T = 0.95\ T_m$, where T_m denotes the melting temperature. They used an analogue materials approach to search for systematics since majorite is stable only at very high pressures. In their indirect approach, they used uvarovite, spessartine, grossular and some other garnets that belong to the same isomechanical group. All samples had high strength both at high temperatures (creep) and at low temperature (hardness). Karato et al. (1995) concluded that "the difference in creep strength between garnet and olivine compared at the same T/T_m is about a factor of 10, which would translate into a difference in diffusion coefficient of a factor of $\sim 10^3$, if creep were diffusion controlled." A similar statement applies for the comparison between majorite and perovskite.

This is one of the reasons why we introduced the radius-dependent part of the viscosity in the upper half of the mantle as shown in Fig. 1. For the lower part of the mantle, for depths between 1234.4 and 2891 km, we assume that perovskite controls the shear creeping and that the activation enthalpy can be estimated using the pressure dependence of the melting temperature of perovskite according to Zerr and Boehler (1993, 1994). Of course, we are conscious of the fact that the viscosity should jump abruptly at 410, 520 and 660 km depth. However, to avoid numerical difficulties we smooth the viscosity at these discontinuities.

There are two reasons to conclude the existence of a high-viscosity transition layer. The *first* reason was discussed by Karato et al. (1995). If the upper mantle consists of pyrolite with 20-40 % garnet, then the garnet does *not* control the rheology. For a model with more than 50 % garnet, the garnet would cause a maximum of the effective viscosity in the transition layer. Already more than a decade ago, Ringwood (1990) concluded more than 90 % garnet should exist at the bottom of the transition layer. Weidner (2001) showed that the stress capacity of the majorite, pyrope, wadsleyite and ringwoodite of the transition layer is much higher than that of the olivine above and the MgO below of it. The possible influence of water on the viscosity of the transition layer is discussed by Karato(2003). A *second* reason for a high-viscosity transition layer is derived in Section 3.1.

3 Model

In the following, the parts of the present model are consecutively presented.

3.1 The viscosity of the model

In this Section, we treat only that aspects of viscosity which are necessary for computation. In Section 2 we have put forward an argument for viscosity jumps at the major phase transitions in the mantle. A reason for the assumption of a high-viscosity transition layer was introduced in a paper by Walzer and Hendel (2003). In that paper, the authors derived the viscosity as a function of depth based on a method independent of prior geophysical approaches and also independent of approaches based on the distribution of minerals in the mantle.

The authors started from a self-consistent theory using the Helmholtz free energy, an equation of state (EoS), the free-volume Grüneisen parameter and Lindemann's law. Viscosity is determined as a function of melting temperature obtained from Lindemann's law. Walzer and Hendel (2003) applied the Ullmann-Pan'kov EoS, whereas in the present paper we use the Birch-Murnaghan EoS. We note that Birch-Murnaghan is a special case of the Ullmann-Pan'kov for $K_0' = 4$, where K_0' is the pressure derivative of the zero pressure bulk modulus. To obtain the relative variation in radial viscosity distribution, Walzer and Hendel (2003) relied upon the pressure, P, the bulk modulus, K , and $\partial K/\partial P$ from PREM (Dziewonski and Anderson, 1981) that is derived from seismic observations. For the absolute scale of the viscosity profile, Walzer and Hendel (2003) used the standard postglacial-uplift viscosity of the asthenosphere beneath the continental lithosphere. Two important conclusions of Walzer and Hendel (2003) are that not only the asthenosphere, but also the upper part of the lower mantle, has low viscosity and that a high-viscosity transition layer tends to divide the mantle into two principal reservoirs relative to concentrations of incompatible elements and volatiles. The transition layer acts as a barrier, but a permeable one, to flow across the mantle.

In the present approach, we do not apply the values of P, K and $\partial K/\partial P$ directly from PREM but rather values that were first smoothed by means of the Birch-Murnaghan EoS (Cf. Fig.2). Of course, we do not smooth across the phase boundaries. Apart from this difference, in the present paper, we derive our radial viscosity profile for 0 to 1234.4 km depth using the method of Walzer and Hendel (2003). This procedure avoids complex assumptions regarding the depth distribution of mineral phases but nevertheless allows derivation of the viscosity from observables. For depths between 1234.4 km and 2891 km, we derive the radial viscosity profile from the melting curve of perovskite (Zerr and Boehler, 1993, 1994; Boehler 1997). The resulting profile is displayed in Fig. 1 where viscosity jumps at the phase boundaries have been replaced by steep viscosity gradients to avoid numerical difficulties in our convection code. The general form of the P,T-dependence of viscosity, η, is given by the following expression where the pressure dependence is hidden in the melting temperature, T_m.

$$\eta = A \cdot \exp\ (c \cdot T_m/T) \tag{3}$$

This relation is equivalent to

$$\eta = A \cdot \exp\left(c\frac{T_m}{T_{av}}\right) \cdot \exp\left[cT_m\left(\frac{1}{T} - \frac{1}{T_{av}}\right)\right] \tag{4}$$

where T_{av} is the laterally averaged temperature. The formula has been slightly simplified by interpreting the first two factors as the radial factor, $\eta_1(r)$, of the viscosity.

$$\eta(r,\theta,\phi,t) = 10^{r_n} \cdot \eta_1(r) \cdot \exp\left[c_t \cdot T_m(r) \cdot \left(\frac{1}{T(r,\theta,\phi,t)} - \frac{1}{T_{av}(r,t)}\right)\right] \quad (5)$$

where r is the radius, θ the colatitude, ϕ the longitude and t the time. The viscosity therefore depends on these four independent variables.

The introduction of the factor 10^{r_n} is a generalization of the equation. For our first estimate of the viscosity, $r_n = 0$ has been chosen. The variation of the non-dimensional parameter r_n generates a systematic shift of the viscosity level. In Section 4.4, the form of the initial radial viscosity profile has been additionally varied to be independent of the specific derivation of $\eta_1(r)$.

3.2 Conservation of mass, momentum and energy

The model is based on the numerical solution of the equations expressing conservation of mass, momentum, and energy. The equation describing the conservation of mass,

$$\frac{\partial \rho}{\partial t} + \nabla \cdot (\rho \vec{v}) = 0 \quad (6)$$

under the anelastic-liquid approximation (i.e., neglecting the $\frac{\partial \rho}{\partial t}$ term) simplifies to

$$\nabla \cdot \vec{v} = -\frac{1}{\rho}\vec{v} \cdot \nabla \rho \quad (7)$$

where ρ is density, t time, and $\vec{v}$ is velocity.

The conservation of momentum can be expressed by

$$\rho\left(\frac{\partial \vec{v}}{\partial t} + \vec{v} \cdot \nabla \vec{v}\right) = -\nabla P + \rho \vec{g} + \frac{\partial}{\partial x_k}\tau_{ik} \quad (8)$$

where P is the pressure, $\vec{g}$ is the gravity acceleration, and τ_{ik} is the deviatoric stress tensor. For spherical symmetry, we have $\vec{g} = -g\vec{e}_r$ and the hydrostatic pressure gradient may be written

$$-\frac{\partial P}{\partial r} = \rho g \quad (9)$$

By definition $K_S = -V\left(\frac{\partial P}{\partial V}\right)_S$ and $\frac{V}{V_0} = \frac{\rho_0}{\rho}$, where K_S is the adiabatic bulk modulus, Vvolume, S entropy, r the radial distance from the Earth's center. Hence

$$K_S = \rho\left(\frac{\partial P}{\partial \rho}\right)_S = \rho\left(\frac{\partial P}{\partial r}\right)_S\left(\frac{\partial r}{\partial \rho}\right)_S \quad (10)$$

Substituting Eq. (9) into Eq. (10) we obtain

$$\left(\frac{\partial\rho}{\partial r}\right)_S = \frac{-\rho^2 g}{K_S} \tag{11}$$

Upon neglecting horizontal spatial variations in ρ Eqs. (7) and (11) yield

$$\nabla\cdot\vec{v} = -\frac{1}{\rho}\vec{v}\cdot\nabla\rho \cong -\frac{1}{\rho}v_r\frac{\partial\rho}{\partial r} = \frac{\rho g v_r}{K_S} \tag{12}$$

It is well-known that

$$K_S = \frac{c_p}{c_v}K_T = (1+\alpha\gamma_{th}T)K_T \tag{13}$$

where K_T is the isothermal bulk modulus, c_p the specific heat at constant pressure, c_v the specific heat at constant volume, α the coefficient of thermal expansion, γ_{th} the thermodynamic Grüneisen parameter and T the absolute temperature.

Eq. (8) can be rewritten as

$$\rho\frac{dv_i}{dt} = \rho g_i + \frac{\partial\sigma_{ki}}{\partial x_k} \tag{14}$$

Using this equation, the conservation of energy can be expressed as follows

$$\rho\frac{du}{dt} + \frac{\partial q_i}{\partial x_i} = \mathcal{Q} + \sigma_{ik}\dot{\varepsilon}_{ik} \tag{15}$$

where u is the specific internal energy, $\mathcal{Q}$ is the heat generation rate per unit volume; v_i, g_i, q_i, x_i, σ_{ik}, $\dot{\varepsilon}_{ik}$ are the components of velocity, gravity acceleration, heat flow density, location vector, stress tensor and strain-rate tensor, respectively.

Another formulation of Eq. (15) is

$$\rho\left[\frac{\partial}{\partial t} + \vec{v}\cdot\nabla\right]u = \nabla\cdot(k\nabla T) + \mathcal{Q} - P\nabla\cdot\vec{v} + 2W_D \tag{16}$$

where

$$2W_D = \sigma_{ik}\dot{\varepsilon}_{ik} + P\nabla\cdot\vec{v} \tag{17}$$

and

$$q_k = -k\frac{\partial T}{\partial x_k} \tag{18}$$

k is the thermal conductivity. Using

$$du = T\,ds - P\,dv \tag{19}$$

and

$$du = T\left(\frac{\partial s}{\partial T}\right)_P dT + T\left(\frac{\partial s}{\partial P}\right)_T dP - Pdv \tag{20}$$

we eliminate the specific internal energy in Eq. (16) and obtain the equation

$$\rho c_p \frac{dT}{dt} = \nabla \cdot (k \nabla T) + \mathcal{Q} + \alpha T \frac{dP}{dt} + 2W_D \tag{21}$$

since

$$c_p = T \left(\frac{\partial s}{\partial T}\right)_P \quad \text{and} \quad \left(\frac{\partial s}{\partial P}\right)_T = -\left(\frac{\partial v}{\partial T}\right)_P = -v\alpha \tag{22}$$

Here is s the specific entropy, v the specific volume, c_p the specific heat at constant pressure and α the coefficient of thermal expansion.

Next let us derive a less well known version of the conservation of energy equation: Eq. (16) is equivalent to

$$\rho \left(\frac{du}{dt} + P\frac{dv}{dt}\right) = \tau_{ik}\frac{\partial v_i}{\partial x_k} + \nabla \cdot (k \nabla T) + \mathcal{Q} \tag{23}$$

because of Eq. (7) and $\frac{1}{\rho} = v$.

Inserting Eq. (19) into Eq. (23), we obtain

$$\rho T \frac{ds}{dt} = \tau_{ik}\frac{\partial v_i}{\partial x_k} + \frac{\partial}{\partial x_j}\left(k\frac{\partial}{\partial x_j}T\right) + \mathcal{Q} \tag{24}$$

On the other hand,

$$ds = \left(\frac{\partial s}{\partial T}\right)_v dT + \left(\frac{\partial s}{\partial v}\right)_T dv \tag{25}$$

and

$$\left(\frac{\partial s}{\partial T}\right)_v = \frac{c_v}{T} \quad , \quad \left(\frac{\partial s}{\partial v}\right)_T = \alpha K_T \tag{26}$$

This implies

$$Tds = c_v dT + \alpha K_T T d\left(\frac{1}{\rho}\right) \tag{27}$$

or

$$Tds = c_v dT - \frac{c_v \gamma T}{\rho} d\rho \tag{28}$$

where

$$\gamma_{th} = \frac{\alpha K_T}{c_v \rho} \tag{29}$$

is the thermodynamic Grüneisen parameter.

Inserting Eq. (28) into Eq. (24) we obtain

$$\rho c_v \frac{dT}{dt} - c_v \gamma T \frac{d\rho}{dt} = \tau_{ik}\frac{\partial v_i}{\partial x_k} + \frac{\partial}{\partial x_j}\left(k\frac{\partial}{\partial x_j}T\right) + \mathcal{Q} \tag{30}$$

From Equations (6) and (30)

$$\rho c_v \frac{dT}{dt} = -\rho c_v \gamma T \frac{\partial v_j}{\partial x_j} + \tau_{ik} \frac{\partial v_i}{\partial x_k} + \frac{\partial}{\partial x_j}\left(k \frac{\partial}{\partial x_j} T\right) + \mathcal{Q} \tag{31}$$

or

$$\frac{\partial T}{dt} = -v_j \frac{\partial}{\partial x_j} T - \gamma T \frac{\partial v_j}{\partial x_j} + \frac{1}{\rho c_v}\left[\tau_{ik} \frac{\partial v_i}{\partial x_k} + \frac{\partial}{\partial x_j}\left(k \frac{\partial}{\partial x_j} T\right) + \mathcal{Q}\right] \tag{32}$$

or

$$\frac{\partial T}{\partial t} = -\frac{\partial (T v_j)}{\partial x_j} - (\gamma - 1) T \frac{\partial v_j}{\partial x_j} + \frac{1}{\rho c_v}\left[\tau_{ik} \frac{\partial v_i}{\partial x_k} + \frac{\partial}{\partial x_j}\left(k \frac{\partial}{\partial x_j} T\right) + \mathcal{Q}\right] \tag{33}$$

Thus we have an alternative expression for the *energy conservation.* Although c_v appears in Eq. (33), the latter is equivalent to Eq. (21) where c_p is used.

The deviatoric stress tensor can be expressed by

$$\tau_{ik} = \eta \left(\frac{\partial v_i}{\partial x_k} + \frac{\partial v_k}{\partial x_i} - \frac{2}{3}\frac{\partial v_j}{\partial x_j}\delta_{ik}\right) \tag{34}$$

in the Eqs. (8) and (33), where η is the viscosity. For the *equation of state* we choose

$$\rho = \rho_r \left[1 - \alpha(T - T_r) + K_T^{-1}(P - P_r) + \sum_{k=1}^{2} \Gamma_k \Delta\rho_k/\rho_r\right] \tag{35}$$

where the index r refers to the adiabatic reference state, $\Delta\rho_k/\rho_r$ or f_{ak} (see Table 1) denotes the non-dimensional density jump for the kth phase transition. Γ_k is a measure of the relative fraction of the heavier phase where $\Gamma_k = \frac{1}{2}\left(1 + \tanh \frac{\pi_k}{d_k}\right)$ with $\pi_k = P - P_{0k} - \gamma_k T$ describing the excess pressure π_k. The quantity P_{0k} is the transition pressure for vanishing temperature T. A non-dimensional transition width is denoted by d_k (see Table 1). γ_k (see Table 1) represents the Clapeyron slope for the kth phase transition. Γ_k and π_k have been introduced by Richter (1973) and Christensen and Yuen (1985).

Because of the very high Prandtl number, the left-hand side of Eq. (8) vanishes. So, we have the following version of the equation of *conservation of momentum*:

$$0 = -\frac{\partial}{\partial x_i}(P - P_r) + (\rho - \rho_r) g_i(r) + \frac{\partial}{\partial x_k}\tau_{ik} \tag{36}$$

The final version of the equation of *conservation of mass* is

$$0 = \frac{\partial}{\partial x_j}\rho v_j \tag{37}$$

which stems from Eq. (7). The Equations (33), (35), (36) and (37) are a system of six scalar equations we use to determine six scalar unknown functions, namely T, ρ, P and the three components of v_i.

3.3 The calculation of the functions P, ρ, K, γ, α, c_p and c_v as a function of radius

In this paper we assume *compressible* convection in a spherical shell. For our starting model with $r_n = 0$, we choose our parameters to approach earth-like conditions so far as limits on spatial resolutions allow. The viscosity, η, is the most important material parameter. It depends not only on the pressure, P, and therefore on the radius, r, but also on the temperature, T. Therefore, the viscosity has lateral and temporal dependence (see Section 3.1). As Zhang and Yuen (1996), our model includes some other quantities that weakly depend on radius. The first panel of Fig. 2 displays pressure as a

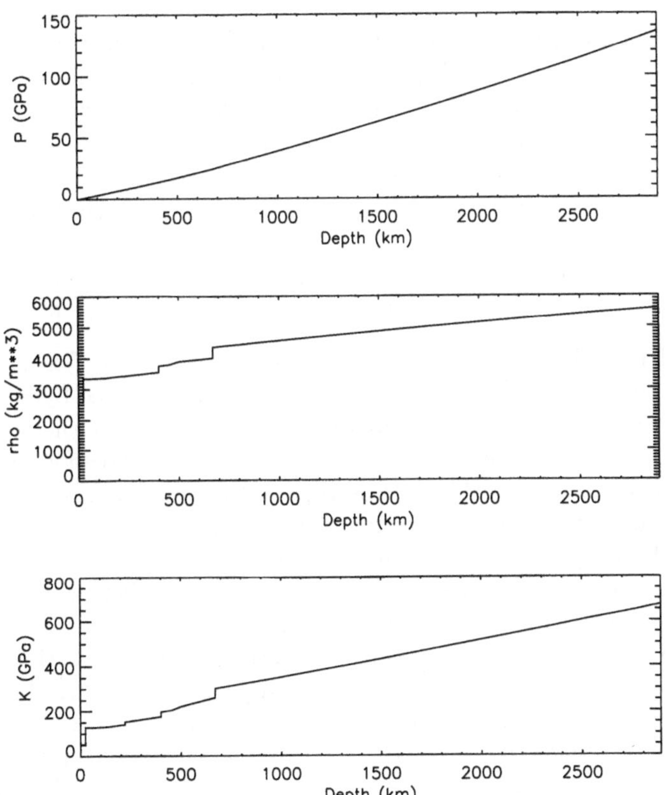

Fig. 2. The pressure, P, the density, ρ, and the bulk modulus, K, as a function of depth.

function of depth. The corresponding numerical values are taken from Table II of PREM (Dziewonski and Anderson, 1981). Our profiles for density, ρ, and bulk modulus, K, we obtain by a Birch-Murnaghan EoS, to smooth the discrete PREM fields, separately for each depth interval. Because there are only a few numerical values for ρ and K in the upper shells of the mantle, we used a second-order Birch-Murnaghan EoS for the four shells between 0, 24.5, 220, 400, and 500 km depth. However, for the rest of the mantle we apply a third-order Birch-Murnaghan EoS. The resulting density profile is shown in the second panel of Fig. 2, while the bulk modulus profile is depicted in the third panel. The relative density jumps, f_{a1} and f_{a2}, for the major phase transitions at the top and bottom of the transition zone are given in Table 1. We use these in the calculation to make the model consistent. Irvine and Stacey (1975) calculated the free-volume Grüneisen parameter from coupled three-dimensional vibrations in a fcc crystal. We used their result

$$\gamma = \frac{\frac{1}{2} \cdot \frac{dK}{dP} - \frac{5}{6} + \frac{2}{9} \cdot \frac{P}{K}}{1 - \frac{4}{3} \cdot \frac{P}{K}} \tag{38}$$

to evaluate γ from the seismically determined but smoothed P and K. In this connection, we assume Eq. (38) may be applied to other closely packed lattices. The first panel of Fig.3 shows the Grüneisen parameter computed in this way as a function of depth. This procedure has the advantage that it involves no assumptions concerning the mantle's chemical composition. Of course, it is possible to use the more exact formula by Barton and Stacey (1985)

$$\gamma = \left[\frac{1}{2} \cdot \frac{dK}{dP} - \frac{1}{6} - \frac{1}{3} f \left(1 - \frac{1}{3} \cdot \frac{P}{K}\right)\right] / \left[1 - \frac{2}{3} f \frac{P}{K}\right] \tag{39}$$

in place of Eq. (38). In this case, for perovskite the value f is 2.27 (Stacey, 1996), assuming 6 % of the Mg atoms in $MgSiO_3$ are replaced by Fe. Using Eq. (39), however, would force us to add assumptions concerning the mineralogical composition of the mantle and to estimate $f(r)$ for the corresponding mixtures of minerals. We can show this procedure alters our model only in minor ways. We prefer to keep our model independent of the chemical composition and therefore choose $f = 2.0$ corresponding to Eq. (38).

There is no generally accepted model for the coefficient of thermal expansion, α, for the Earth's mantle. It is clear, however, that for all relevant minerals α decreases with increasing pressure. Chopelas and Boehler (1992) showed that the expansion coefficient for MgO is $5 \times 10^{-5} K^{-1}$ at adiabatically decompressed mantle conditions *(1 atm, 1700 K)*. But at the CMB pressure, α for MgO decreases to $1 \times 10^{-5} K^{-1}$. Tackley (1997) estimated $\alpha = 3.3 \times 10^{-5} K^{-1}$ at the surface and $\alpha = 1.1 \times 10^{-5} K^{-1}$ at the CMB. Stacey (1998) investigated the thermoelasticity of mineral composites for the mantle and found $\alpha = 3.35 \times 10^{-5} K^{-1}$ for the adiabatically decompressed state and $\alpha = 1.07 \times 10^{-5} K^{-1}$ at a depth of 2741 km. We choose a profile for α that varies from $\alpha = 3.46 \times 10^{-5} K^{-1}$ at the surface to $\alpha = 1.15 \times 10^{-5} K^{-1}$ at the CMB. The curvature of the profile was obtained from the formula of van den Berg and Yuen (1998, p. 223). The second panel of Fig. 3 displays the coefficient of thermal expansion of our model as a function of depth. Obviously, this is a simplification. In reality jumps in α should be expected at the upper and lower boundaries of the transition layer that are comparable to

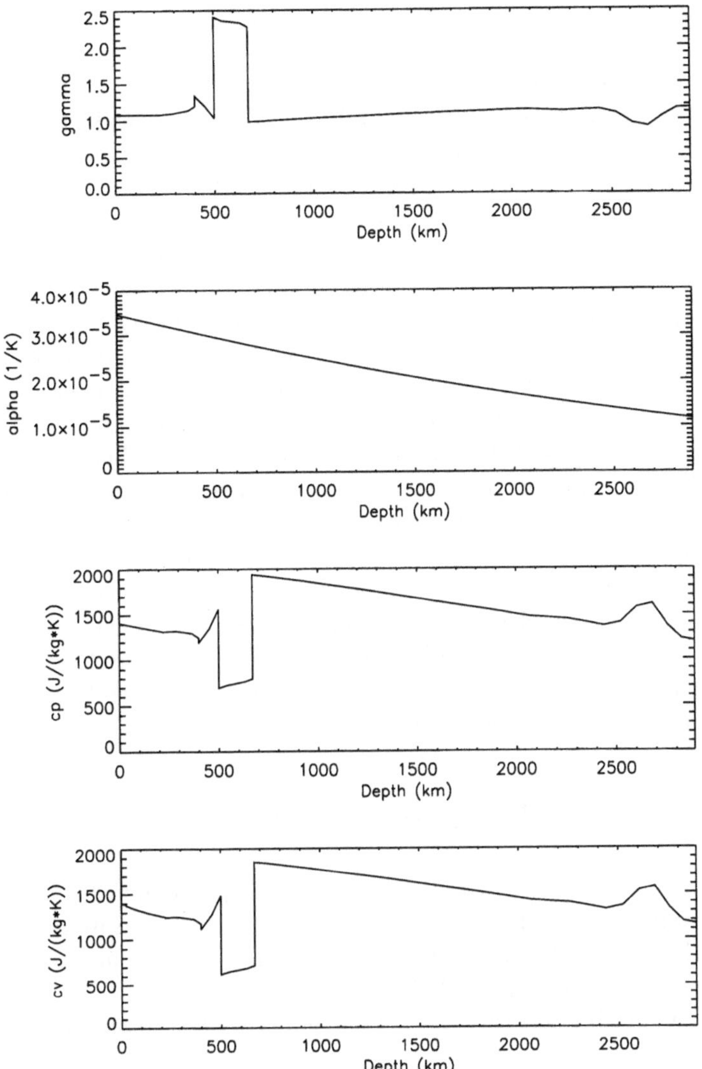

Fig. 3. The Grüneisen parameter, γ, the coefficient of thermal expansion, α, the specific heat at constant pressure, c_p, and the specific heat at constant volume, c_v, as a function of depth.

the corresponding jumps in the Grüneisen parameter. From expressions (13) and (29), we obtain the specific heat at constant pressure, c_p, and the specific heat at constant volume, c_v, as a function of depth. These are plotted in the third panel and fourth panel of Fig. 3, respectively.

3.4 Heating, initial and boundary conditions

Here, we describe some additional assumptions of the present model, S1. The age of the most meteorites is between 4.56 and 4.54 Ga (Dalrymple, 1991). It is assumed that the radioactive decay of the most important heat-producing elements began 4.56 Ga ago. If 0.07 Ga are needed for the segregation into metallic core and silicate mantle then the primordial silicate mantle was finished before $(4.49 \pm 0.03) \times 10^9$ a (McCulloch and Bennett, 1998). There is no perceivable change in the siderophile element concentrations of the rocks from the Archean to the present time (O'Neill and Palme, 1998). So, there is a high probability that the Earth's mantle and the core had no essential compositional exchange since $(4.49 \pm 0.03) \times 10^9$ a. Therefore, this age represents the beginning of the thermal mantle evolution for the present model. Walzer and Hendel (1999) computed thermal mantle convection and chemical segregation together in a 2-D model where the generation of the depleted mantle and the growth of the continental crust were obtained as a function of time. In the present paper, however, we confine ourselves to the *thermal* evolution problem alone: the silicate mantle is homogeneously heated from within and also from the CMB. It is assumed that the concentrations, $a_{\mu\nu}$, of the radioactive elements of the primordial mantle according to McCulloch and Bennett (1994) determine the internal heating (Cf. Table1). The time-dependent specific heat production, H, is computed from

$$H = \sum_{\nu=1}^{4} a_{\mu\nu} a_{if\nu} H_{0\nu} \exp(-t/\tau_\nu) \tag{40}$$

where τ_ν denotes the decay time or the $1/e$ life, $H_{0\nu}$ the initial heat generation rate per unit volume of the νth radionuclide, $a_{if\nu}$ the isotopic abundance factor, ν the indices of the four major heat-producing elements. The numerical values we use are listed in Table 2.

The heat production rate per unit volume, $\mathcal{Q}$, introduced in Section 3.2, is given by

$$\mathcal{Q} = H \cdot \rho \tag{41}$$

We assume free-slip and impermeable boundary conditions for both the Earth's surface and the CMB. What about the thermal boundary conditions? The solar luminocity has increased by some 25 % since 4.56 Ga. On the other hand, there are fluviatile and organic sediments in the Archean. Evidence of that kind shows that the average surface temperature was constant with respect to time. Obviously, some kind of thermostatic mechanism is working at the Earth's surface, possibly the outer carbon cycle. In the contrary case, the surface temperature ought to rise, since the average distance between sun and Earth is constant because of the conservation of moment of momentum. Investigations of Archean komatiites (Sleep, 1979) and theoretical estimates of the Earth's secular cooling rate (see Schubert et al., 2001) demonstrate a decrease of about 100 K Ga^{-1} for the volumetrically averaged temperature, T_a,

Table 1. Model parameters

Parameter	Description	Value	
r_{min}	Inner radius of spherical shell	3.480×10^6	m
r_{max}	Outer radius of spherical shell	6.371×10^6	m
	Temperature at the outer shell boundary	288	K
γ_1	Clapeyron slope for the olivine-spinel transition	$+3 \times 10^6$	$Pa{\cdot}K^{-1}$
γ_2	Clapeyron slope for the spinel-perovskite transition	-4×10^6	$Pa{\cdot}K^{-1}$
f_{a1}	Non-dimensional density jump for the olivine-spinel transition	0.0547	
f_{a2}	Non-dimensional density jump for the spinel-perovskite transition	0.0848	
	Begin of the thermal evolution of the Earth's silicate mantle	4.490×10^9	a
d_1	Non-dimensional transition width for the olivine-spinel transition	0.05	
d_2	Non-dimensional transition width for the spinel-perovskite transition	0.05	
	Begin of the radioactive decay	4.565×10^9	a
c_t	Factor of the lateral viscosity variation	2	
k	Thermal conductivity	12	$W{\cdot}m^{-1}{\cdot}K{-}1$
q_{CMB}	Heat flow at the core-mantle boundary	28.9	$mW{\cdot}m^{-2}$
$nr+1$	Number of radial levels	33	
	Number of grid points	1.351746×10^6	
$a_{\mu\nu}(U)$	Concentration of uranium	0.0203	ppm
$a_{\mu\nu}(Th)$	Concentration of thorium	0.0853	ppm
$a_{\mu\nu}(K)$	Concentration of potassium	250	ppm

Table 2. Data on major heat-producing isotopes

Isotope	^{40}K	^{232}K	^{235}U	^{238}U
ν	1	2	3	4
τ_ν[Ma]	2015.3	20212.2	1015.4	6446.2
$H_{0\nu}[Wkg^{-1}]$	0.272×10^{-3}	0.0330×10^{-3}	47.89×10^{-3}	0.1905×10^{-3}
$a_{if\nu}$	0.000119	1	0.0071	0.9928

of the mantle. Since there is evidently no thermostatic mechanism at CMB, we ought to expect that the CMB temperature is decreasing *as a function of time.* Some authors write that it is well known that the core imposes a uniform temperature at the base of the mantle. This is right for the dependence of the location vector. For short-term convection investigations, it is a good approach to fix the CMB temperature with respect to time, too. For the *evolution* as a whole, it is very improbable that the CMB temperature is a constant *with respect to time.* Instead we assume that the CMB heat flow has been constant. Our justifications are given by Stacey's (1992) Section 6.7.5

entitled Constancy of the Core-to-Mantle Heat Flux and by Appendix A of Walzer and Hendel (1999). Anderson (1998) estimated a conductive power of $(4.4 \pm 1) \times 10^{12}$W flowing out of the core. Therefore, we assume a heat flow of 28.9 mW·m^{-2} at the CMB. Consequently, the mantle in our model, S1, is mainly heated from within but also somewhat from below. For the overall mechanism of our model, this assumption is not very important. If we use Tackley's (2000b) assumption of zero heat flow at CMB the planforms in the upper half of the mantle stay very similar. Only the plumes are reduced. We could be tempted to replace the assumption of a constant core-to-mantle heat flow, q_{CMB}, by a prescribed decaying q_{CMB}. However, Stevenson et al. (1983) showed that the onset of inner core freezeout leads to slightly rising or nearly constant q_{CMB}-values of about 23 mW·m^{-2} for the second half of the Earth's history (see also Schubert et al., 2001, pp.607-609). The temporally and laterally averaged surface temperature of the Earth was 288 K for the last 40 years. We assume this value as the constant surface temperature of the model.

3.5 Numerical method and implementation

The solutions of the system of differential equations of convection in a compressional spherical shell, Eqs. (33) to (37) with the additional Eqs. (5), (40) and (41), are obtained using a three-dimensional finite-element method, a fast multigrid solver and the second-order Runge-Kutta procedure. The mesh is generated by projection of a regular icosahedron onto a sphere to devide the spherical surface into twenty spherical triangles or ten spherical diamonds. A dyadic mesh refinement procedure connects the midpoints of each side of a triangle with a great circle such that each triangle is subdivided into four smaller triangles. Successive grid refinements generate an almost uniform triangular discretization of the spherical surface of the desired resolution. Corresponding mesh points of spherical surfaces at different depths are connected by radial lines. The radial distribution of the different spherical-surface triangular networks is so that the volumes of the cells are nearly equal. More details are given by Baumgardner (1983, 1985), Bunge et al. (1997) and Yang (1997). For the multitude of runs we needed for our parameter study, we employed a mesh with 1351746 nodes. Some runs were made with 10649730 nodes to check the convergence of the lower resolution runs. The result is that the laterally averaged heat flow, the ratio of heat outflow to radiogenic heat production, the Rayleigh number, and the Nusselt number as functions of the time show hardly any discernable differences (< 0.5 %). Calculations were performed on 128 processors of a Cray T3E. A scalability test showed a scaling degree of nearly 90 %. The code uses all processor–processor communications with the same load. This is optimal for the T3E architecture. A hierarchically organized processor connection could not be effectively used by the code without a basic time wasting reconstruction. The code was benchmarked for constant viscosity convection by Bunge et al. (1997) with numerical results of Glatz-

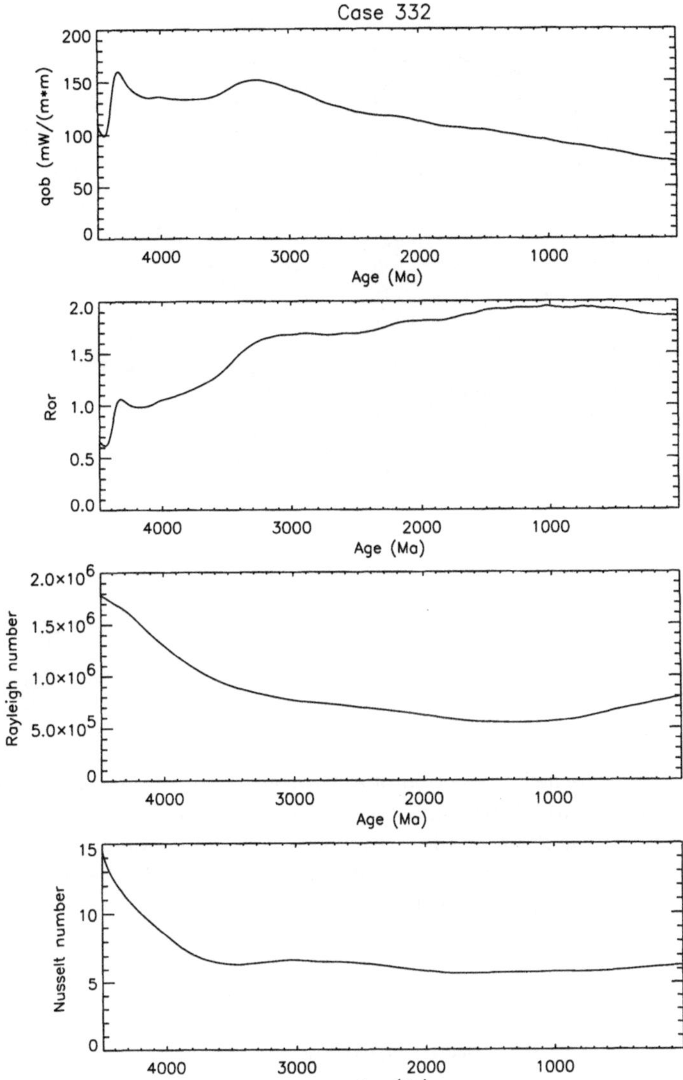

Fig. 4. The evolution of the laterally averaged surface heat flow, *qob*, of the ratio of the surface heat outflow per unit time to the mantle's radiogenic heat production per unit time, *Ror*, of the Rayleigh number, *Ra*, and of the Nusselt number, *Nu*.

maier (1988) for Nusselt numbers, peak temperatures, and peak velocities. A good agreement (≤ 1.5 %) was found.

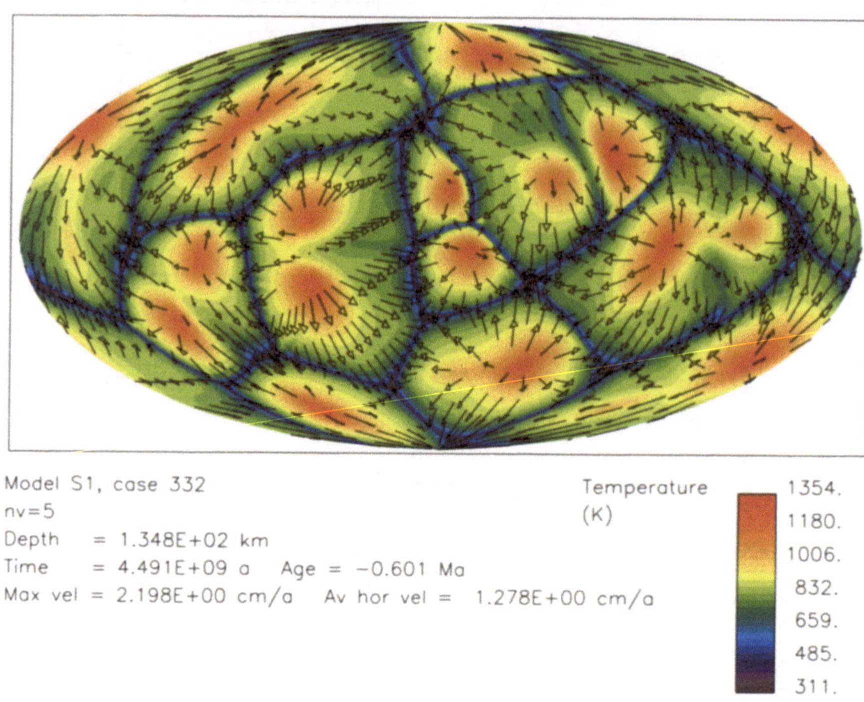

Fig. 5. (a) Spherical-shell convection of a Newtonian fluid heated from within and slightly from below with depth- and temperature-dependent viscosity. The radial factor of the viscosity is given by Fig.1. An equal-area projection of the planforms at various depths: (a) 134.8 km, (b) 632.9 km, (c) 1130 km, (d) 1551 km. Temperature is denoted by colors, creeping velocity by arrows.

4 Results

4.1 Thin cold sheet-like downwellings

Before we examine results from our parameter study, we will describe the numerical results from our reference model with $r_n = 0$. We anticipate that the main features of the solutions, especially of the planforms that we describe here, apply also for a wider range of Rayleigh numbers. The first panel of Fig. 4 shows the averaged heat flow at the Earth's surface as a function of time in the past. The present-day heat flow from the starting model is 74.0 $mW \cdot m^{-2}$. By comparison Pollack et al. (1993) found a mean global heat flow of 87 $mW \cdot m^{-2}$ for the present Earth based on 24774 heat flow measurements. Apart from some initial transient behavior, the heat flow of the model declines in a smooth fashion. The *r*atio of the heat *o*utflow per unit of time to the *r*adiogenic heat produced in the mantle per unit of time, *Ror*, is displayed in the second panel of Fig. 4. The present-day value of *Ror* is 1.851. Note that since about

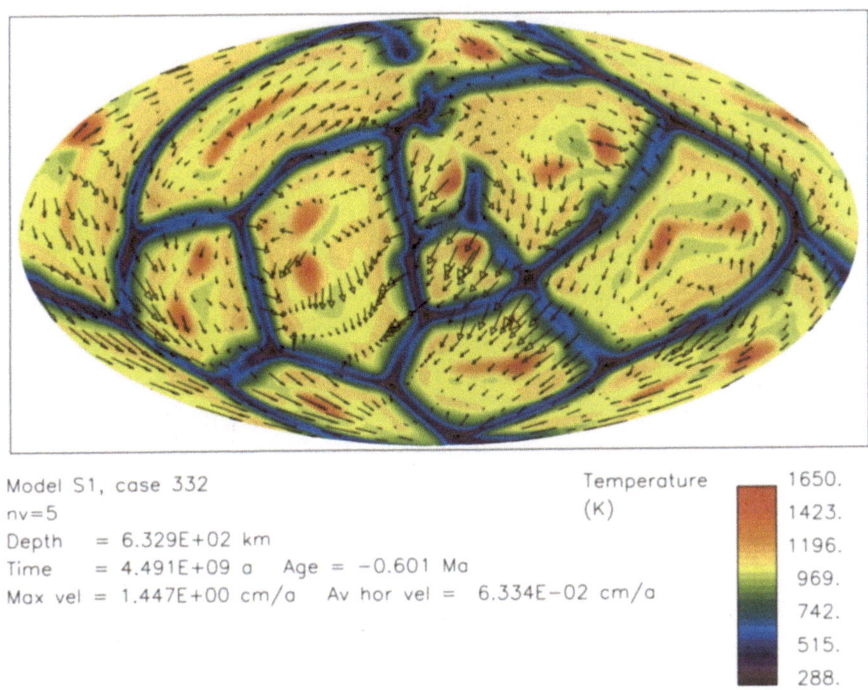

Fig. 5. (b) Text see Fig. 5. (a). Here: 632.9 km depth

2000 Ma the value of *Ror* changes only slightly. By way of comparison, Stacey and Stacey (1999) estimate 14.4×10^{30} J for the total heat loss of the Earth and 7.8×10^{30} J for the radiogenic heat of the Earth. The quotient of these two numbers is 1.846. *Ror* is the reciprocal value of the Urey number. The third panel of Fig. 4 shows the time history of the Rayleigh number, while the fourth panel of Fig. 4 depicts the time history of Nusselt number.

Because of Section 3.4, the Rayleigh number

$$Ra = \left\langle \frac{\rho\alpha g h^3}{\kappa\eta_{al}} \cdot \frac{(\mathcal{Q}h + q_{CMB})h}{k} \right\rangle \tag{42}$$

is better adapted to the problem than

$$Ra_B = \left\langle \frac{\rho\alpha g h^3}{\kappa\eta_{al}} \right\rangle \cdot \Delta T \tag{43}$$

since the mantle is mainly heated from within. The bracket $\langle\ \rangle$ denotes a volumetric average but not a temporal average. The temperature difference, ΔT, across the mantle is not constant for the *evolution* problem (see Section 3.4.). h is depth of CMB, κ is thermal diffusivity. $\mathcal{Q}$ is defined by Eqs. (40)

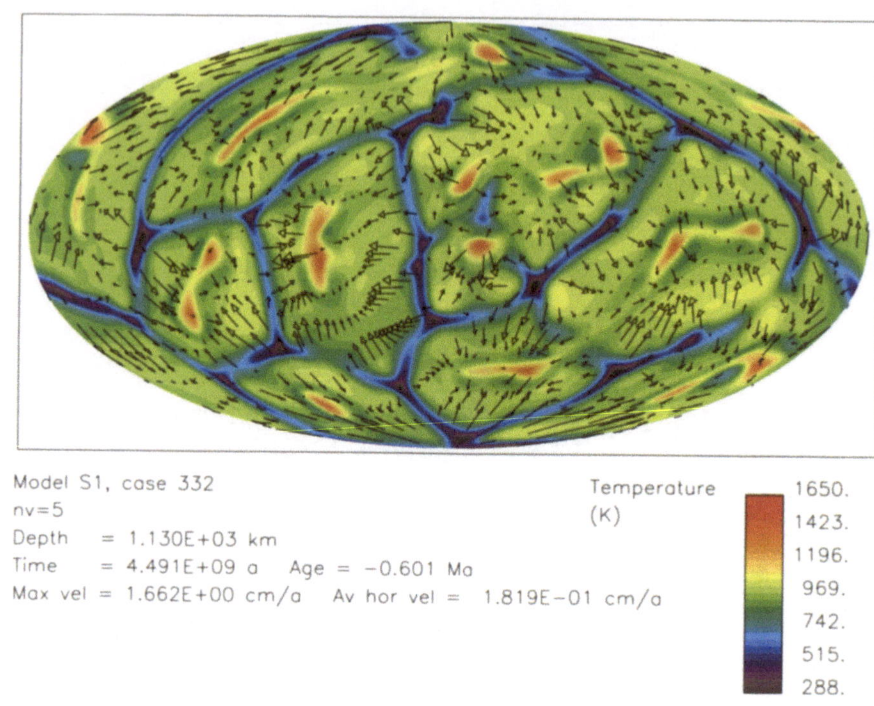

Fig. 5. (c) Text see Fig. 5. (a). Here: 1130 km depth

and (41). Therefore, the heat generation rate per volume, Q , is monotonously declining as a function of time. The quantity η_{al} is given by

$$\log \eta_{al} = \langle \log \eta \rangle \tag{44}$$

The quantity η_{al} is also a function of time because of Eqs. (5) and (44). Therefore Ra is a function of time. The Nusselt number, Nu, is the ratio of the actual heat flow to the purely conductive heat flow down the superadiabatic temperature gradient. Nu is a function of time, too.

Fig. 5 presents the temperature distribution (multicolored) and the solid-state creep velocities (arrows) on equal-area projections for different depths. These pictures are computed for the geological present using the radial viscosity profile of Fig. 1. The *narrow blue stripes* of Fig. 5(a) at 135 km depth are remarkable and represent a network of interconnected downwelling sheets of cold material. Fig. 5(b) shows that these sheets maintain their identity at a depth of 633 km, and are sinking vertically with neglible horizontal displacement. This reflects the simplicity of our treatment of the cold upper boundary layer, a treatment that lacks buoyant continents, one-sided subduction, and other plate-like features. Fig. 5(c) reveals that the thin downwelling sheets persist to 1130 km depth. By 1550 km depth, however, the sheet-like charac-

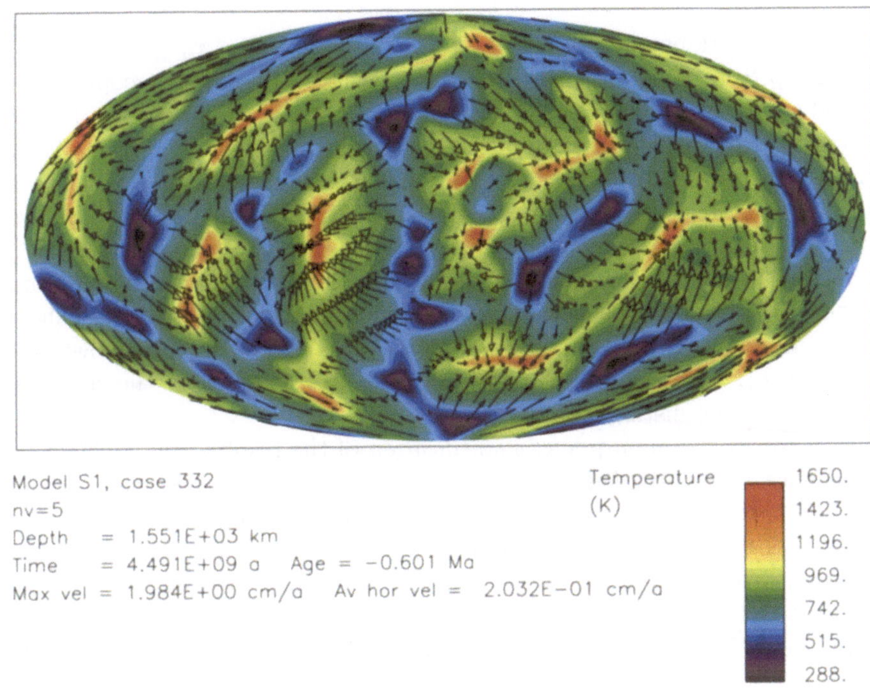

Fig. 5. (d) Text see Fig. 5. (a). Here: 1551 km depth

ter of the downwellings has largely vanished as shown in Fig. 5(d). This fact is probably a consequence of the viscosity hill of the lower mantle (Cf. Fig. 1). Figs. 4 and 5 apply for a run with $r_n = 0$. For higher values of r_n, however, the thin sheets dip deeper into the mantle. Two features of our solutions are noteworthy:

a) Other published convection models with Newtonian rheology display much wider zones of downwelling near the surface and

b) the relatively thin downwelling sheets in the present model freely penetrate the high-viscosity transition layer.

For lack of space, we do not include a temporal sequence of plots of temperature and velocity from the early Archean (4000 Ma) to the present. However, the pattern of a network of thin downwelling sheets appears very early in the calculation and changes only slowly with time. In the first 2000 Ma, some new sheets form to subdivide originally oblong convective cells. In the last 2000 Ma, no new sheets appear, but some cells slowly grow, while other cells slowly shrink. In general the positions of the cell centers are nearly stable.

4.2 Variation of non-dimensional numbers

We vary the parameter r_n of Eq. (5) in the interval $-0.5 \leq r_n \leq +1.5$. This is a first step to become independent of the special results of the geophysically derived viscosity model. First of all we don't alter η_1, i.e., only the level of the radial reference viscosity profile is shifted as a whole without the altered temperature influence on the viscosity that is included in the further computation, too. The thickness of the subducting downwelling sheets decreases with decreasing negative values of r_n down to -0.5, i.e. for higher Rayleigh numbers. In this way, the blue lines on the equal-area projection plots become thinner and thinner with decreasing r_n. Up to $r_n = +0.3$, the cold sheet-like downwellings are rather thin yet, but the thickness slightly grows with increasing r_n. The more the quantity r_n increases, the deeper the sheet-like structures can be distinguished.

Now comes the main point: *quantification* of the results. Fig. 6 shows the dependence of the Rayleigh number, *Ra*, on the non-dimensional viscosity level parameter, r_n. The definition of *Ra* is given by Eq. (42). It is an essential feature of the evolution of the Earth that its direction is defined by irreversible processes, especially that the heat generation rate per volume, $\mathcal{Q}$, decreases with time and that the Earth cools down, i.e., that the volumetrically averaged temperature, T_a, diminishes. The latter process would happen, of course, also

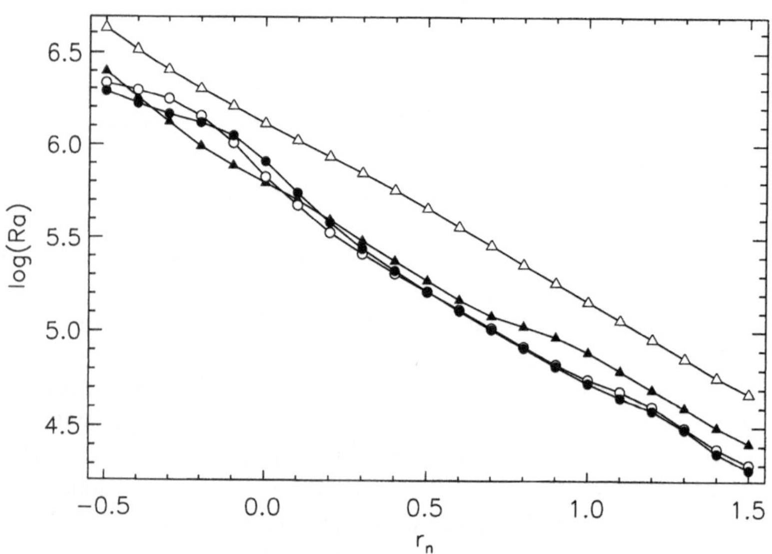

Fig. 6. The Rayleigh number, $Ra(t)$, as a function of the non-dimensional viscosity-level parameter, r_n. Open triangles represent an age of 4000 Ma, filled triangles stand for 2000 Ma, open circles for 500 Ma, filled circles for 0 Ma.

if the internal heating would take place only at the beginning by accretion. Therefore η_{al} rises as a function of time (Cf. Eqs. (5) and (44)). On the other hand, $\mathcal{Q}$, grows less because of the exponentially decreasing radioactive decay (Cf. Eg. (40)). So for all r_n, the Rayleigh number decreases as a function of time (Cf. Fig. 6 and capture). Initially, Ra diminishes quickly, then only slowly, nearly independent of r_n. On that account we computed the temporal average of Ra only for the last 2000 Ma and called it $Ra_{(2)}$. In a similar way, Fig. 7 represents the Nusselt number, Nu, for different ages as a function of r_n. For the interval $-0.3 \leq r_n \leq +0.1$, Nu, decreases somewhat in the beginning and fluctuates only slightly around a rather stable value for the main part of the evolution. For the same r_n-interval, the sheet-like downwellings are not only

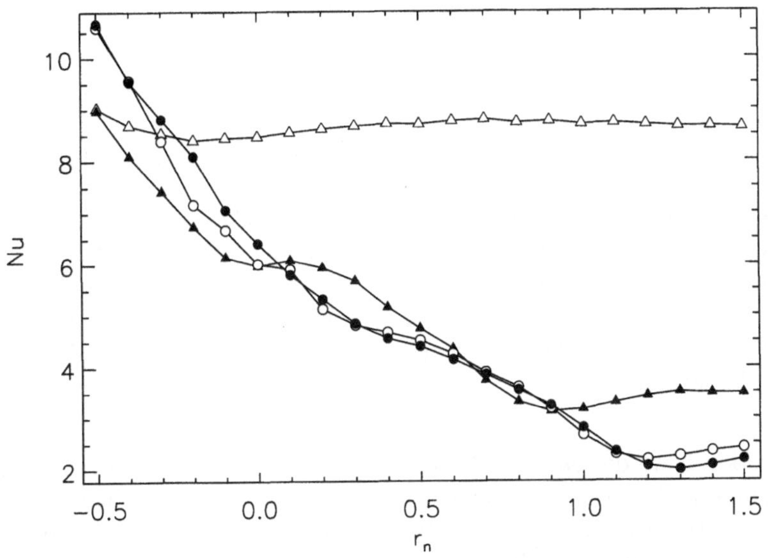

Fig. 7. The Nusselt number $Nu(t)$, vs. r_n. For explanation of symbols see Fig. 6.

marked but they stretch deeper into the mantle than for runs with smaller r_n and higher Ra. Fig.8 displays Ror for different ages as a function of r_n. For the last 2000 Ma and $-0.1 \leq r_n \leq +0.1$, the reciprocal value, Ror, of the Urey number is rather near the 1.85 value that has been derived by Stacey and Stacey (1999). The interval $+1.2 \leq r_n \leq +1.4$ has to be excluded since the Nusselt numbers are too low in this range. For runs with $-0.5 \leq r_n \leq +0.7$, Ror does not vary about very much for the last 2000 Ma. Although formulae like

$$Nu = cRa^{\beta} \tag{45}$$

apply only for convection with constant viscosity and since certain parameterized steady-state convection models employ Eq. (45), we carry out the

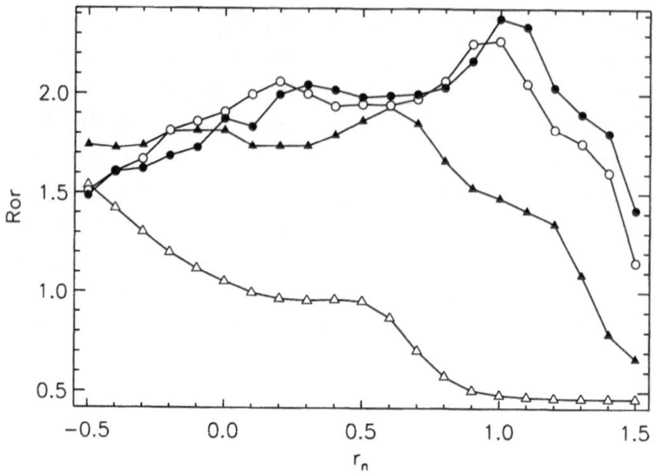

Fig. 8. The reciprocal value, $Ror(t)$, of the Urey number, versus r_n. For explanation of symbols see Fig. 6.

following simple evaluation. Of course, Nu and Ra are functions of the time in our more complex non-steady evolution model. We obtain $Nu_{(2)}$ and $Ra_{(2)}$ calculating the temporal average of $Nu(t)$ and $Ra(t)$, respectively, for the last 2000 Ma. $Nu_{(2)}$ is plotted versus $Ra_{(2)}$ in Fig. 9. Each black square represents one run. The solid curve is a least square fit to a power law similar to Eq. (45). We obtained

$$Nu_{(2)} = 0.120 Ra_{(2)}^{0.295} \tag{46}$$

In Section 5 we discuss why Eq.(46) describes the $Nu_{(2)} - Ra_{(2)}$ relation of the numerical experiments in a rough approximation and why Eq. (46) is similar to the $Nu - Ra$ relation of simple steady-state convection models with constant viscosity. A further set of runs showed that the radial profiles of γ, α, c_ν and c_p (Cf. Fig. 3) have only minor influence: Planforms as well as the non-dimensional numbers Ra, Nu and Ror alter only slightly. Therefore and for lack of space the corresponding plots are not presented here. Furthermore, we introduced a volumetrically averaged Grüneisen parameter, γ_b, instead of the radial gamma profile (Cf. Fig.3, first panel). Subsequently, this γ_b-value has been varied. The results of the runs, e.g. the non-dimensional numbers and temperature-plus-velocity plots, proved to be rather robust. As an example, Fig. 10 shows Nu as a function of γ_b and reveals that the temporal dependence of Nu is much stronger than the γ_b-dependence. Fig. 11 displays $Ror_{(2)}$ versus r_n where $Ror_{(2)}$ is the temporal average of Ror for the last 2000 Ma. The value $Ror = 1.85$ according to Stacey and Stacey (1999) can be found at about $r_n = -0.1$ and $r_n = +1.1$. Figs. 6 and 7 show that $r_n = +1.1$ must drop out since otherwise Ra and Nu would be extremely low.

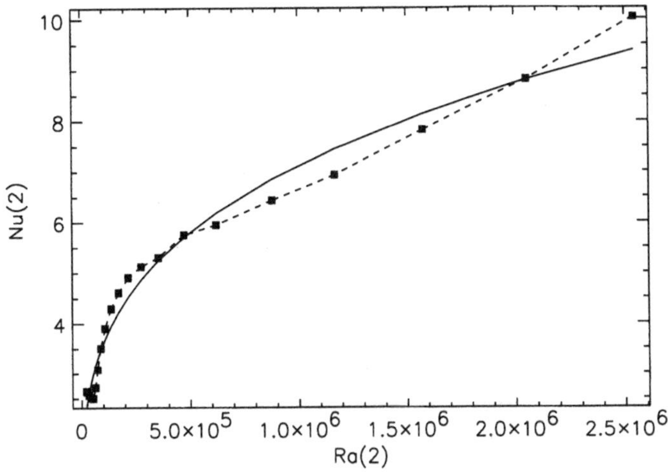

Fig. 9. An average Nusselt number, $Nu_{(2)}$, versus an average Rayleigh number, $Ra_{(2)}$. The supplement (2) stands for the time average of the proper quantity over the last 2000 Ma. Filled squares represent the results of different runs. The solid curve is a non-linear least square fit to a power law of the form $Nu_{(2)} = c\ Ra_{(2)}^{\beta}$ where $c = 0.120$ and $\beta = 0.295$.

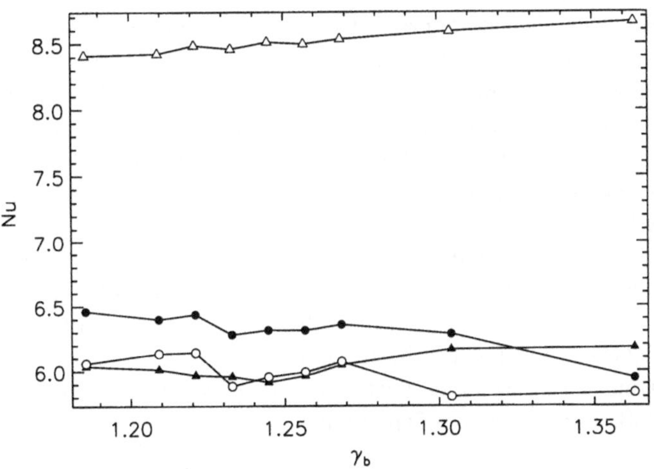

Fig. 10. The Nusselt number, $Nu(t)$, as a function of the volumetrically averaged Grüneisen parameter, γ_b. For explanation of symbols see Fig. 6.

4.3 Effects of r_n-variations on planforms. Further variation of model parameters

If, because of the last remarks, $r_n = -0.1$ is regarded as the optimum value of r_n then we should have a look at the other results of the relevant run.

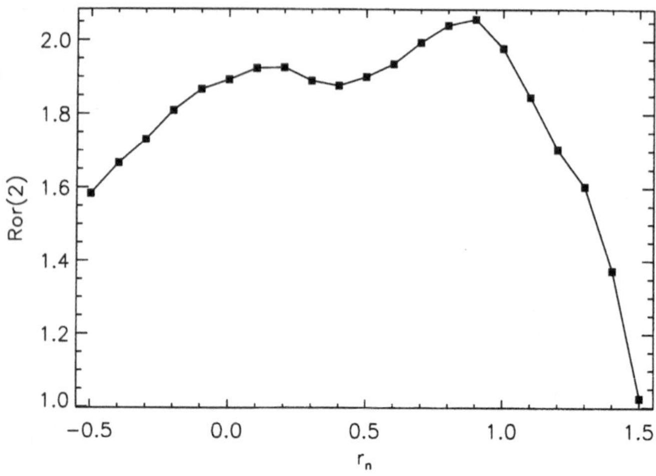

Fig. 11. The reciprocal value, $Ror_{(2)}$, of the average Urey number versus r_n. The supplement (2) denotes the time average of *Ror* over the last 2000 Ma.

The first panel of Fig. 12 shows a laterally averaged surface heat flow that decreases during nearly the whole Earth's history and has a present value of 69.1 mWm^{-2}. The variations of *Ror* during the last 2000 Ma are small. The present value is 1.73. The curve of the Rayleigh number arrives at 1.12×10^6 for today, and the present-day Nusselt number is 7.07. The last panel of Fig. 12 reveals that Nu is nearly a constant since about 4000 Ma. Fig. 13 represents the present-day distribution of temperature and creeping velocity in 134.8 km depth for $r_n = -0.1$. As expected, there is no plate-like velocity distribution near the surface since it is a Newtonian-fluid model. However, the thickness of the blue lines is smaller yet than in the case with $r_n = 0.0$ (Cf. Fig. 5(a)). The network of the sheet-like downwellings of both pictures is related but in Fig. 13 some sheets are missing. The network of plate-like subducting zones shows distinct outlines in 1150 km depth, yet. In greater depths, the tendency to oblong isothermal lines gradually vanishes.

A few runs were performed with equivalent parameters but with an incompressible EoS. The structure of the planforms was similar but the downwellings were broader.

We also examined the effect of switching off the lateral temperature dependence of the viscosity. The results confirm the prevailing wisdom that the radial viscosity variation (Cf. Fig.1) is the decisive variation. We did not vary the heating since the various chemical models for the Earth show only minor differences regarding the concentration of the heat-producing elements in the primordial mantle.

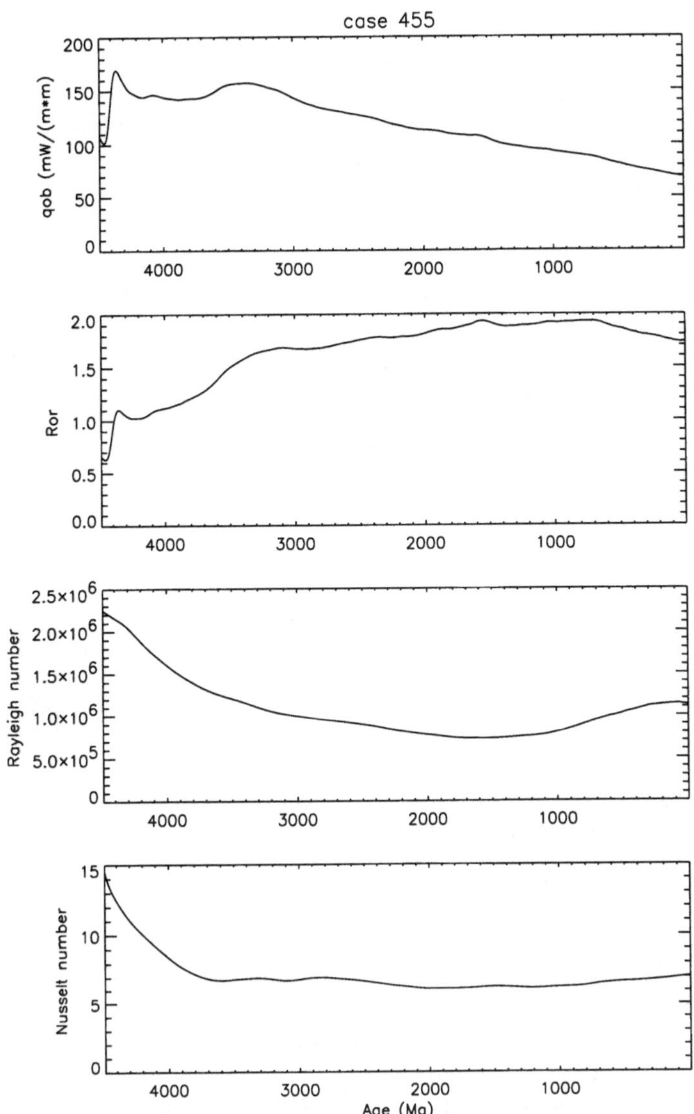

Fig. 12. For $r_n = -0.1$, the time dependence curves of the following quantities are shown: laterally averaged surface heat flow, *qob*, the reciprocal value, *Ror*, of the Urey number, the Rayleigh number, *Ra*, and the Nusselt number, *Nu*.

4.4 Runs with radial viscosity profiles without a second low-viscosity layer

After confirming that variation in the viscosity level had the largest influence on the results we sought to explore the effects of altering the profile. Although the radial profile of Fig.1 seemed realistic on the basis of solid-state physics and seismology, we decreased the viscosity minimum at the top of the lower mantle stepwise until it vanished, to check its influence on the fluid mechanics.

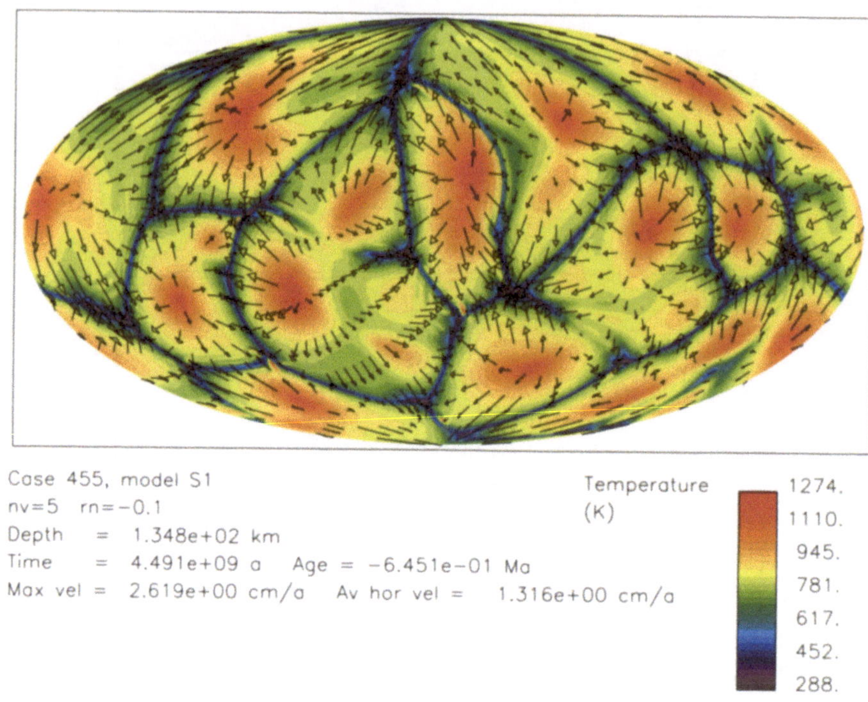

Fig. 13. For $r_n = -0.1$, temperature (color) and the creeping velocity (arrows) are depicted on an equal-area projection of a spherical surface in 134.8 km depth for the present time.

Fig. 14 is the extreme case of such a profile. We ran a case with this viscosity profile but everything else identical to our reference model. Fig. 15 reveals the evolution of the laterally averaged surface heat flow, *qob*, the reciprocal value, *Ror*, of the Urey number, the Rayleigh number, *Ra*, and the Nusselt number, *Nu*. Fig. 16 is the counterpart of the Fig. 5(a). Fig. 16 shows broader downwelling sheets than Fig.5(a). The number of cells is reduced. For the depths 632.9 and 1130 km, the downwelling zones are considerably less slab-like. The dissolution of these structures at 1550 km depth was very pronounced in Fig. 5(d), but is less so in the counterpart plot without second low-viscosity layer beneath the transition zone. The explanation seems to be that long wavelength downwelling structures can penetrate deeper into the viscosity hill of the lower mantle than short wavelength ones.

5 Discussion and conclusions

We begin with some remarks on our starting radial viscosity profile. In spite of the runs we performed in connection with Sections 4.2., 4.3. and 4.4., it

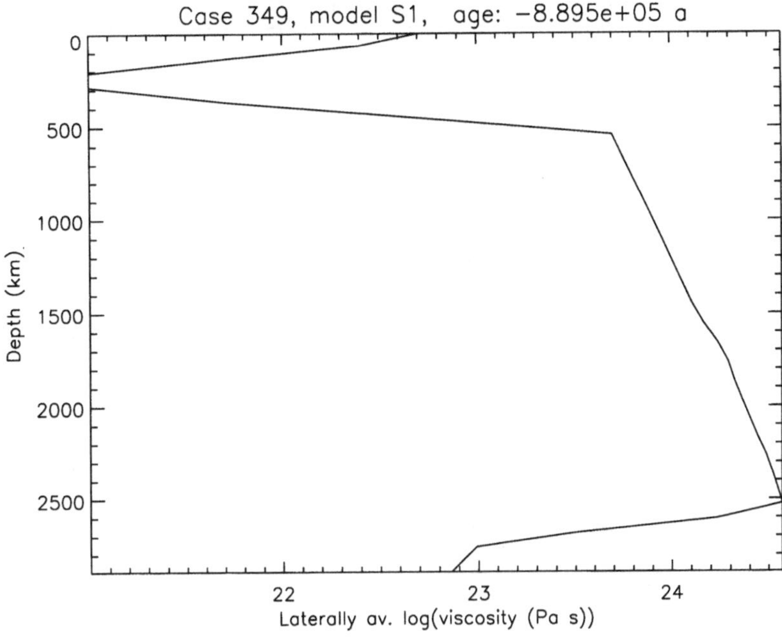

Fig. 14. An alternative basic radial viscosity profile, $\eta_2(r)$, for $r_n = 0$.

is our opinion that a viscosity profile with two minima in the upper half of the mantle is realistic based on solid-state physics and PREM (Cf. Fig.1). We shall review briefly some relevant papers that provide hints in this direction but have been obtained using other methods. Spada et al. (1992) investigated effects on post-glacial rebound from a high-viscosity transition layer. They found that a hard layer between 410 and 660 km depth diminishes ones ability to fit polar wander data with a small viscosity increase across the lower mantle. Therefore, a prominent maximum in viscosity in the lower mantle (Cf. Fig.1) appears necessary not only because the melting temperature increases with increasing pressure (and thereby represents an estimate of the activation enthalpy) but also for purely geophysical reasons. The seismic activity along the passive margins of Fennoscandia requires a remarkable viscosity increase below the asthenosphere, most likely in the transition layer. Čížková et al. (1996) and Kido and Čadek (1997) also argued for the existence of a high viscosity transition layer and of two low viscosity layers, one above and one below of it. They started from a distribution of the seismic velocities by Li and Romanowicz (1996), transformed the seismic velocities into densities, and used a genetic algorithmic inversion that excluded those parts of the mantle where the density differences have predominantly compositional causes. Using intermediate-wavelength geoid inversions, Kido and Yuen (2000) studied the viscosity profile of the mid-mantle in detail. They found it to be of no major consequence whether they introduced an impermeable boundary at 660, 700, 750, or even 1000 km depth or whether there was no impermeable boundary. In

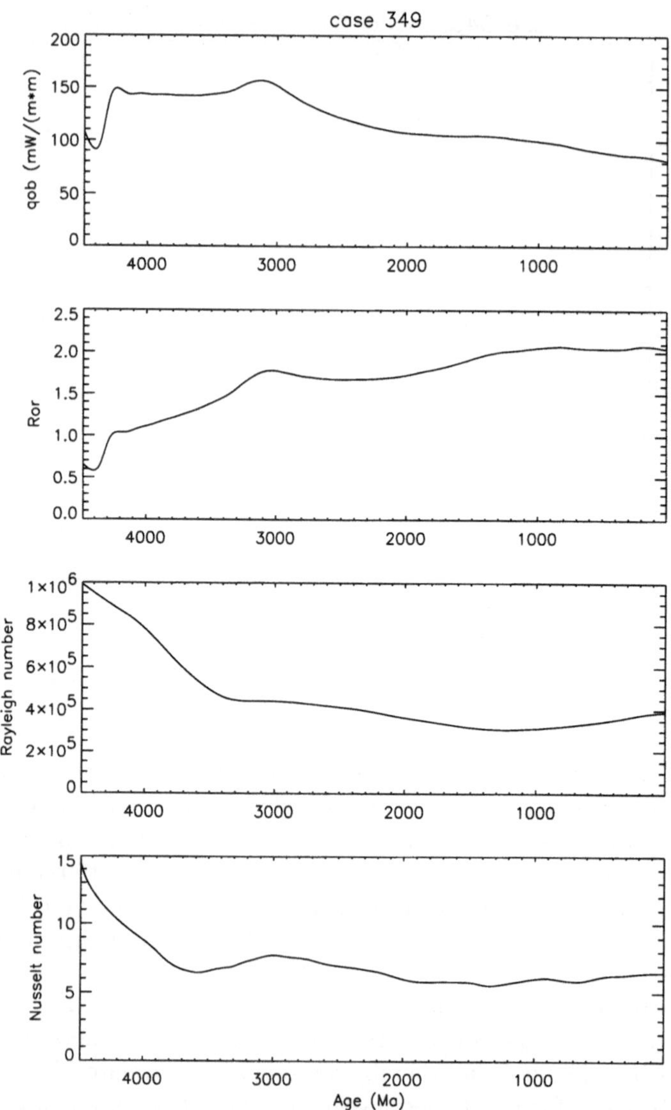

Fig. 15. For $\eta_2(r)$, the time dependence of the following quantities is shown: laterally averaged surface heat flow, *qob*, the reciprocal value, *Ror*, of the Urey number, the Rayleigh number, *Ra*, and the Nusselt number, *Nu*.

all cases they obtained similar viscosity profiles, namely, a high viscosity lithosphere, followed by a low viscosity asthenosphere, followed by a high-viscosity transition layer, followed by a second low viscosity layer in the uppermost part of the lower mantle. Their curves are similar to the upper half of Fig.1. They did not show the profile for the lower half of the mantle because of low resolution. Forte (2000) derived a set of viscosity profiles through a formal inversion procedure. His profiles produce a best fit to the observed free-air anomalies

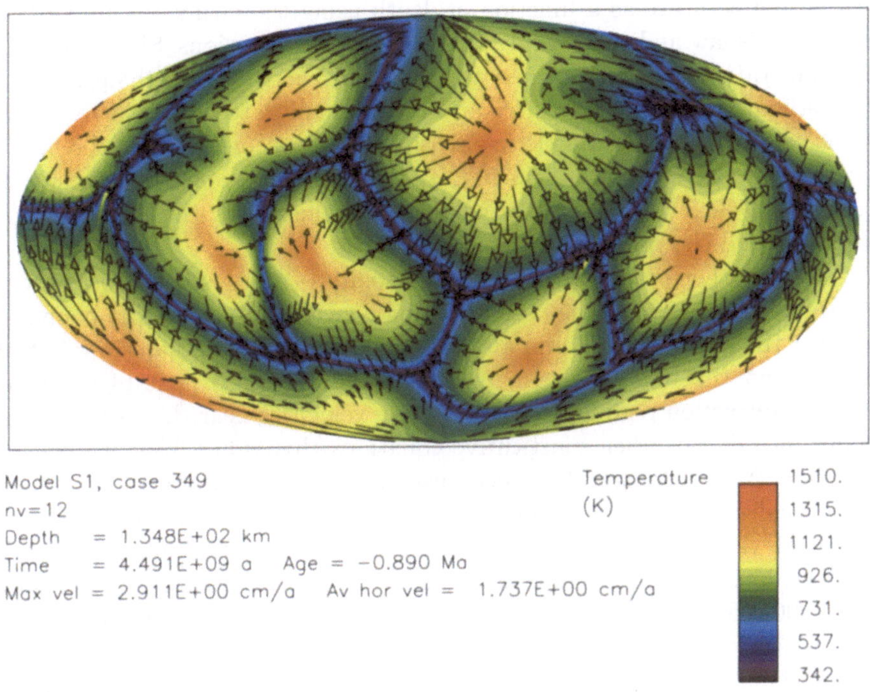

Fig. 16. For $\eta_2(r)$, temperature (color) and the creeping velocity (arrows) is displayed on an equal-area projection of a spherical surface in 134.8 km depth for the present time.

and to the observed motions of the lithospheric plates where his starting point was a pair of seismic tomography models. When an impermeable boundary is assumed at 670 km depth, the resulting viscosity-depth curve is similar to our Fig.1. Whithout this impermeable boundary the viscosity peak, corresponding to the transition layer, is shifted down by 240 km. As a final remark on the starting profile, the observed peak in deep-focus seismicity at transition zone depths (Kirby et al., 1996), and the non-occurence of earthquakes deeper than 700 km, despite tomographic evidence for slab penetration into the lower mantle, supports our inference of a high-viscosity transition layer with low-viscosity layer immediately below it. However, in the course of the work we gained a certain independence of the starting viscosity profile by variation of the parameters.

The present paper describes a set of numerical experiments with an infinite Prandtl number fluid in a compressible spherical shell heated mainly from within. We used the anelastic liquid approximation with Earth-like material parameters. The Birch-Murnaghan EoS was employed to derive the Grüneisen parameter, isothermal bulk modulus and its pressure derivative as a function of depth from observational values provided by PREM. We computed the melting temperature as a function of depth using the Grüneisen parameter, Lindemann's law and some solid-state physics considerations. Since the melting temperature can be regarded as a proxy for activation enthalpy, we use it to estimate the radial viscosity variation. We calibrated this relative profile by the value of the asthenospheric viscosity that matches post-glacial rebound observations. In addition, we include the lateral dependence of viscosity on temperature such that the viscosity depends on the radius, the colatitude, the longitude, and the time.

A lot of other models reduce the variability of the viscosity to the temperature dependence. In the corresponding convection models, the boundary layers at the upper surface and at the CMB control the mechanism (Cf. Section 2.) and the bulk of the mantle is isoviscous. Since none of the three possible convection regimes shows plate-like velocities near the surface, some groups introduced other constitutive equations that differ from the Newtonian fluid. New rheologies have been used in order to reproduce lithospheric strain localization and weakening. E.g., Tackley (2000a) investigated the influence of viscoplastic yield stress where a part of it depended on depth. Tackley (2000b) studied the effect of increasing strain weakening. Richards et al. (2001) found that the generation of plates is facilitated by plastic yield stress and a low-viscosity asthenosphere beneath the lithosphere. Bercovici et al. (2001b,c,2002) and Ricard et al. (2001) examined a two-phase mixture to achieve a continuum description of weakening and shear localization and to explain the plate boundary formation. Continuing the work of Trompert and Hansen (1998), the Hansen group is working on the problem of the formation of oceanic lithosphere and subduction slabs, too. It is fully legitimate that this main problem attracts so much attention and efforts.

However, the focus of the present paper is the influence of the physical material properties, especially of the viscosity, *inside* the mantle on the thermal *evolution*, hence preferably on the development of convection during a long time interval. Karato (1997a) concluded that in the typical mantle, where temperature is moderately high and stress is relatively low, the dominant creeping mechanism is either diffusion creep or power law. We suppose that a Newtonian fluid is an acceptable approximation for the main part of the mantle since the prevailing gradients of the creeping velocity are probably sufficiently small in the lower mantle and in the transition zone. First we report briefly on the less important point: We found by variation of the parameters that the influence of γ, α, c_ν and c_p is small compared with the viscosity effect. This statement applies for constant values of γ, α, c_ν and c_p as well as for the radial profiles of these quantities that have been derived from physics and geophysics.

It is however most important that, for the reasons a) and b) of Section 2., the viscosity profile is stronger dependent on the radius than often assumed. Our preferred viscosity profile displays not only a high-viscosity lithosphere and a viscosity hill in the lower portion of the lower mantle but, also a prominent *high-viscosity transition layer* inferred to arise from a high garnet content. Moreover, there is not only the usual asthenosphere but also a *second low-viscosity zone just below the 660-km* phase boundary. We note that our preferred viscosity profile bears a similarity to the successful viscosity profile LVZ of Cserepes et al. (2000, p. 139) and to the profile VL of Forte (2000) that is based on mantle flow calculations in which the density contrasts were estimated from long-wavelength seismic heterogeneity assuming density differences are caused exclusively by temperature differences. The computed flows were constrained to match observed free-air gravity anomalies and observed tectonic plate motions.

Systematic variation of the viscosity-level parameter, r_n, shows the following result within wide boundaries ($-0.5 \le r_n \le +0.3$): *The existence of two low-viscosity layers inside the mantle causes networks of very thin tabular-shaped downwellings.* These sheets are considerably thinner than other downwellings of known Newtonian-fluid models. The quantity r_n can be transformed into the Rayleigh number using Fig.6 or, more exactly, Eq. (5). *The plate-like downwelling sheets are thin down to about 1350 km depth.* Hence, they are able to penetrate the high-viscosity transition layer. The sheets begin to assume the form of large drop shaped features at a depth of about 1550 km. Of course, our solutions have no plates at the surface. A natural way to get plates would probably be to take into account the different oceanic-lithospheric layers with different constitutive equations, from top to bottom brittle, semi-brittle and viscoplastic (Kohlstedt et al., 1995). But the incorporation of this stratification in a flow model is a problem of node distance. Furthermore, the incorporation of water in the oceanic crust seems to be an important cause for the possibility of subduction: Terrestrial planets without oceans don't have plate tectonics. Another important effect is the dehydration of the harzburgite that occurs during partial melting. This is likely to increase the slab viscosity by 2-3 orders of magnitude (Braun et al., 2000). All these effects are not included in our model.

We found extremely well developed thin sheet-like downwellings are in the surrounding of $r_n = -0.1$. This corresponds to the *Ror* value by Stacey and Stacey (1999). For the interval $-0.5 \le r_n \le +0.7$, *Ror* varies only slightly as a function of time during the last 2000 Ma. An interesting result is the *nearly time-independent Nusselt number for* $-0.3 \le r_n \le +0.1$ *and the last 4000 Ma of evolution. All the mentioned major trends in our results are robust at least for* $-0.3 \le r_n \le +0.1$.

The transformation of the sheet-like downwellings into broader drop-like structures can be demonstrated by a comparison between Figs. 5 (c) and 5 (d). The depth of this transformation increases with growing r_n. The mentioned transformation can be observed also in the real mantle: Seismic tomographic models change their statistical character between 1400 and 1900 km depth. P-wave and S-wave velocity deviations are decorrelated in the deep mantle (van der Hilst and Kárason, 1999; Davaille, 1999). Most subducting slabs do not extend deeper than 1700 km (van der Hilst et al., 1997). *This is well compatible with a high-viscosity lower part of the lower mantle.*

The more the Rayleigh number has been augmented and the nearer the approach is therefore to Earth-like Rayleigh numbers the more sheet-like are the downwellings of our model. For reasons of solid-state physics, we think that steeper viscosity gradients near 410, 520 and 660 km depth are unavoidable for the real Earth. For numerical reasons, however, we have been forced to diminish these viscosity gradients. Our numerical experiments show: The steeper these gradients are the less vertical mass flow can be observed at the corresponding depth. Therefore we anticipate that the real transition layer is a kind of barriere for mantle convection but a permeable one. However, for the *present* viscosity profile we obtained whole-mantle convection that is sluggish in the lower part of the lower mantle. The large radial variations of the viscosity proved to be very important for the horizontal length scales of the flow patterns. They were more relevant than the temperature dependence of the viscosity. The latter result corresponds with that of Ratcliff et al. (1997).

Howard (1966), Parsons and McKenzie (1978) and Kenyon and Turcotte (1983) performed a simple boundary stability analysis for an incompressible viscous fluid in a horizontal layer with constant material parameters, with prescribed boundary temperatures, without internal heating and free-slip boundary conditions. The heat loss at the surface is conductive. Non-zero thermal gradients exist only in the boundary layers. The boundary layers will gradually thicken and break away if enough buoyancy is collected. For this problem

$$Nu = 0.112 Ra^{0.333} \tag{47}$$

has been deduced. It is surprising that our non-steady compressible-shell evolution model with depth- and temperature-dependent viscosity and mainly internal, temporally decreasing heating has led to the result that the numerical experiments can be summarized by Eq. (46) and Fig.9. Stagnant-lid convection regimes established other Nu– Ra parameterizations which could be fit for Mars and Moon. For Earth, however, Eq. (46) or related equations seem to be appropriate. This can be taken as a confirmation of the Nu – Ra parameterization for the Earth's evolution of Schubert et al. (1979, 1980, 2001). The other results on non-dimensional numbers are to be found in Section 4.3.

Acknowledgements

We want to thank Woo-Sun Yang for his kind help and interesting discussions. Two of us (U.W. and R.H.) gratefully acknowledge the hospitality of Charles Keller, LANL, Los Alamos, NM. This research was supported by the Volkswagenstiftung through the grant I75474, by the Rechenzentrum der Universität Stuttgart (HLRS), and by the John von Neumann Institute of Computing, Forschungszentrum Jülich, through the supply of computing time.

References

Anderson, O.L., 1998. The Grüneisen parameter for iron at outer core conditions and the resulting conductive heat and power in the core. Phys. Earth Planet. Int. 109, 179-197.

Barton, M.A. and Stacey, F.D., 1985. The Grüneisen parameter at high pressure: a molecular dynamical study. Phys. Earth Planet. Int. 39, 167-177.

Baumgardner, J.R., 1983. A three-dimensional finite element model for mantle convection. Thesis, Univ. of California, Los Angeles.

Baumgardner, J.R., 1985. Three-dimensional treatment of convective flow in the Earth's mantle. J. Stat. Phys. 39 (5-6), 501-511.

Bercovici, D., 1996. Plate generation in a simple model of lithosphere-mantle flow with dynamic self-lubrication. Earth Planet. Sci. Lett. 144, 41-51

Bercovici, D., Ricard, Y., Richards, M.A., 2001a. Relation between mantle dynamics and plate tectonics. In: Richards, M.A., Gordon, R., van der Hilst, R. (Eds.), History and Dynamics of Global Plate Motions. American Geophysical Union, Washington, DC, pp. 5-46

Bercovici, D., Ricard, Y., Schubert, G., 2001b. A two-phase model for compaction and damage. 1. General theory. J. Geophys. Res. 106, 8887-8906.

Bercovici, D., Ricard, Y., Schubert, G., 2001c. A two-phase model for compaction and damage. 3. Application to shear localization and plate boundary formation. J. Geophys. Res. 106, 8925-8939.

Bercovici, D., Ricard, Y. 2002. Energetics of a two-phase model of lithospheric damage, shear localization and plate-boundary formation, Geophys. J. Int., submitted

Boehler, R., 1997. The temperature in the Earth's core. In: Crossley, D. J. (Ed.), Earth's Deep Interior. Gordon and Breach Sci. Publ., Amsterdam, pp. 51-63.

Braun, M.G., Hirth, G., Parmentier, E.M., 2000. The effects of deep damp melting on mantle flow and melt generation beneath mid-ocean ridges. Earth Planet. Sci. Lett, 176 (3-4), 339-356.

Bunge, H.-P., Richards, M.A., Baumgardner, J.R., 1997. A sensitivity study of three-dimensional spherical mantle convection at 10^8 Rayleigh number: effects of depth-dependent viscosity, heating mode, and an endothermic phase change. J. Geophys. Res. 102, 11991-12007.

Čadek, O., van den Berg, A.P., 1998. Radial profiles of temperature and viscosity in the Earth's mantle inferred from the geoid and the lateral seismic structure. Earth Planet. Sci. Lett. 164, 607-615.

Christensen, U.R., 1984. Heat transport by variable viscosity convection and implications for the Earth's thermal evolution. Phys. Earth Planet. Int. 35, 264-282.

Christensen, U.R., 1985. Thermal evolution models for the Earth. J. Geophys. Res. 90, 2995-3007.

Christensen, U.R., 1996. The influence of trench migration on slab penetration into the lower mantle. Earth Planet. Sci. Lett. 140, 27-39.

Christensen, U.R., Yuen, D.A., 1985. Layered convection induced by phase transitions. J.Geophys. Res. 90, 10291-10300.

Chopelas, A., Boehler, R., 1992. Thermal expansivity in the lower mantle. Geophys. Res. Lett. 19, 1983-1986.

Čížková, H., Čadek, O., Yuen, D.A., Zhou, H., 1996. Geoid slope spectrum and constraints on mantle viscosity stratification. Geophys. Res. Lett. 23, 3063-3066.

Cserepes, L., Yuen, D.A., Schroeder, B.A., 2000. Effect of the mid-mantle viscosity and phase-transition structure on 3D mantle convection. Phys. Earth Planet. Int. 118, 135-148.

Dalrymple, G.B., 1991. The Age of the Earth, Stanford University Press, Stanford.

Davaille, A., 1999. Simultaneous generation of hotspots and superswells by convection in a heterogeneous planetary mantle. Nature 402, 756-760.

Dziewonski, A.M. & Anderson, D.L., 1981. Preliminary reference Earth model. Phys. Earth Planet. Int. 25, 297-356.

Forte, A.M., 2000. Seismic-geodynamic constraints of mantle flow: implications for layered convection, mantle viscosity, and seismic anisotropy in the deep mantle. In: Karato, S.-I., Forte, A.M., Liebermann, R.C., Masters, G., Stixrude, L. (Eds.), Earth's Deep Interior. American Geophys. Union, Washington, DC, pp.3-36.

Glatzmaier, G.A., 1988. Numerical simulations of mantle convection: Time-dependent, three-dimensional, compressible, spherical shell. Geophys. Astrophys. Fluid Dyn. 43, 223-264.

Hofmann, A.W., 1988. Chemical differentiation of the Earth: the relationship between mantle, continental crust, and oceanic crust. Earth Planet. Sci. Lett. 90, 297-314.

Howard, L.N., 1966. Convection at high Rayleigh number. In: Grtler, H. (Ed.), Proc. 11th Intl. Congr. Appl. Mech. Springer-Verlag, Berlin, pp.1109-1115.

Irvine, R.D. & Stacey, F.D., 1975. Pressure dependence of the thermal Grüneisen parameter, with application to the lower mantle and outer core. Phys. Earth Planet. Int. 11, 157-165.

Karato, S.-I., Wang, Z., Liu, B., Fujino, K., 1995. Plastic deformation of garnets: Systematics and implications for the rheology of the mantle transition zone. Earth Planet. Sci. Lett. 130, 13-30.

Karato, S.-I., 1997a. Phase transformations and rheological properties of mantle minerals. In: Crossley, D.J. (Ed.), Earth's deep interior: The Doornbos memorial volume. Gordon and Breach Sci. Publ., Amsterdam, pp. 223-272.

Karato, S.-I., 1997b. On the separation of crustal component from subducted oceanic lithosphere near the 660km discontinuity. Phys. Earth Planet. Int. 99, 103-111.

Karato, S.-I., 2003 Mapping water content in the upper mantle. In preparation.

Kenyon, P.M., Turcotte, D.L., 1983. Convection in a two-layer mantle with a strongly temperature-dependent viscosity. J.Geophys. Res. 88, 6403-6414.

Kido, M., Čadek, O., 1997. Inferences of viscosity from the oceanic geoid: Indication of a low viscosity zone below the 660-km discontinuity. Earth Planet. Sci. Lett., 151, 125-137.

Kido, M., Yuen, D.A., 2000. The role played by a low viscosity zone under a 660km discontinuity in regional mantle layering. Earth Planet. Sci. Lett. 181, 573-583.

Kirby, S.H., Stein, S., Okal, E.A., Rubie, D.C., 1996. Metastable mantle phase transformations and deep earthquakes in subducting oceanic lithosphere. Rev. Geophys. 34, 261-306.

Kohlstedt, D.L., Evans, B., Mackwell, S.J., 1995. Strength of the lithosphere: Constraints imposed by laboratory experiments. J. Geophys. Res. 100, 17587-17602.

Kubo, T., Ohtani, E., Kato, T., Urakawa, S., Suzuki, A., Kanbe, Y., Funakoshi, K., Utsumi, W., Fujino, K., 2000. Formation of metastable assemblages and mechanisms of the grain-size reduction in the post spinel transformation of Mg_2SiO_4. Geophys. Res. Lett. 27, 807-810.

Lambeck, K., Johnston, P., 1998. The viscosity of the mantle: Evidence from analyses of glacial-rebound phenomena. In: Jackson, I. (Ed.), The Earth's mantle: Composition, structure and evolution. Cambridge Univ. Press, Cambridge, UK, pp. 461-502.

Li, X.D., Romanowicz, B., 1996. Global mantle shear velocity model developed using nonlinear asymptotic coupling theory. J. Geophys. Res. 101, 22245-22272.

McCulloch, M.T., Bennett, V.C., 1994. Progressive growth of the Earth's continental crust and depleted mantle: geochemical constraints. Geochim. Cosmochim. Acta 58, 4717-4738.

McCulloch, M.T. & Bennett, V.C., 1998. Early differentation of the Earth: an isotopic perspective. In: Jackson, I. (Ed.), The Earth's mantle. Cambridge Univ. Press, Cambridge, pp.127-158.

O'Neill, H.S.C., Palme, H., 1998. Composition of the silicate Earth: implications for accretion and core formation. In: Jackson, I. (Ed.), The Earth's Mantle. Cambridge Univ. Press, Cambridge, pp. 3-126.

Pari, G., Peltier, W.R., 1998. Global surface heat flux anomalies from seismic tomography-based models of mantle flow: Implications for mantle convection. J.Geophys. Res. 103, 23743-23780.

Parsons, B., McKenzie, D., 1978. Mantle convection and the thermal structure of the plates. J. Geophys. Res. 83, 4485-4496.

Pollack, H.N., Hurter, S.J., Johnson, J.R., 1993. Heat flow from the Earth's interior: analysis of the global data set. Rev. Geophys. 31, 267-280.

Ranalli, G., 1998. Inferences on mantle rheology from creep laws. GeoResearch Forum 3-4, 323-340.

Rattcliff, J.T., Tackley, P.J., Schubert, G., Zebib, A., 1997. Transitions in thermal convection with strongly variable viscosity. Phys. Earth Planet. Int. 102, 201-212.

Reese, C.C., Solomatov, V.S., Moresi, L.-N., 1998. Heat transport efficiency for stagnant lid convection with dislocation viscosity: Application to Mars and Venus. J. Geophys. Res. 103, 13643-13657.

Reese, C.C., Solomatov, V.S., Moresi, L.-N., 1999. Non-Newtonian stagnant lid convection and magmatic resurfacing on Venus. Icarus 139, 67-80.

Ricard, Y., Bercovici, D., Schubert, G., 2001. A two-phase model for compaction and damage. 2. Applications to compaction, deformation and the role of interfacial surface tension. J. Geophys. Res. 106, 8907-8924.

Richards, M.A., Yang, W.-S., Baumgardner, J.R., Bunge, H.-P., 2001. Role of a low-viscosity zone in stabilizing plate tectonics: Implications for comparative terrestrial planetology. Geochemistry, Geophysics, Geosystems vol. 2, paper no.2000GC000115

Richter, F.M., 1973. Finite amplitude convection through a phase boundary. Geophys. J. R. Astron. Soc. 35, 265-276.

Ringwood, A.E., 1990. Slab-mantle interactions: petrogenesis of intraplate magmas and structure of the upper mantle. Chem. Geol. 82, 187-207.

Schubert, G., Cassen, P., Young, R.E., 1979. Subsolidus convective cooling histories of terrestrial planets. Icarus 38, 192-211.

Schubert, G., Stevenson, D., Cassen, P., 1980. Whole planet cooling and the radiogenic heat source contents of the Earth and Moon. J. Geophys. Res. 85, 2511-2518.

Schubert, G., Turcotte, D.L., Olson, P., 2001. Mantle Convection in the Earth and Planets. Cambridge Univ. Press, Cambridge, 940 pp.

Sleep, N.H., 1979. Thermal history and degassing of the Earth: Some simple calculations. J. Geology 87, 671-686.

Solomatov, V.S., 1995. Scaling of temperature- and stress-dependent viscosity convection. Phys. Fluids 7, 266-274.

Spada, G., Sabadini, R., Yuen, D.A., Ricard, Y., 1992. Effects on post-glacial rebound from the hard rheology in the transition zone. Geophys. J. Int. 109, 683-700.

Stacey, F.D., 1992. Physics of the Earth, 3rd edn., Brookfield Press, Brisbane, 513 pp.

Stacey, F.D., 1996. Thermoelasticity of (Mg,Fe)SiO_3 perovskite and a comparison with the lower mantle. Phys. Earth Planet. Int. 98, 65-77.

Stacey, F.D., 1998. Thermoelasticity of a mineral composite and a reconsideration of lower mantle properties. Phys. Earth Planet. Int. 106, 219-236.

Stacey, F.D., Stacey, C.H.B., 1999. Gravitational energy of core evolution: implications for thermal history and geodynamo power. Phys. Earth Planet. Int. 110, 83-93.

Stevenson, D.J., Spohn, T., Schubert, G., 1983. Magnetism and thermal evolution of the terrestrial planets. Icarus 54, 466-489.

Tackley, P.J., 1997. Effects of phase transitions on three-dimensional mantle convection. In: Crossley, D.J. (Ed.), Earth's deep interior: the Doornbos memorial volume. Gordon and Breach Sci. Publ., Amsterdam, pp. 273-335.

Tackley, P.J., 2000a. Self-consistent generation of tectonic plates in time-dependent, three-dimensional mantle convection simulations. 1. Pseudoplastic yielding. Geochem. Geophys. Geosyst., 1, Paper no. 2000GC000036

Tackley, P.J., 2000b. Self-consistent generation of tectonic plates in time-dependent, three-dimensional mantle convection simulations. 2. Strain weakening and asthenosphere. Geochem. Geophys. Geosyst., 1, Paper no. 2000GC000043

Tackley, P.J., 2001. The quest for self-consistent generation of plate tectonics in mantle convection models. In: Richards, M.A., Gordon, R., van der Hilst, R. (Eds.), History and Dynamics of Global Plate Motions. American Geophysical Union, Washington, DC, pp. 47-72

Trompert, R., Hansen, U., 1998. Mantle convection simulations with rheologies that generate plate-like behavior. Nature 395, 686-689.

van den Berg, A.P. & Yuen, D.A., 1998. Modelling planetary dynamics by using the temperature at the core-mantle boundary as a control variable: effect of rheological layering on mantle heat transport. Phys. Earth Planet. Int. 108, 219-234.

van der Hilst, R.D.; Widiyantoro, S., Engdahl E.R., 1997. Evidence for deep mantle circulation from global tomography. Nature 386, 578-584

van der Hilst, R.D., Kárason, H., 1999. Compositional Heterogeneity in the Bottom 1000 Kilometers of Earth's Mantle: Toward a Hybrid Convection Model. Science 283, 1885-1888.

Walzer. U., Hendel, R., 1999. A new convection-fractionation model for the evolution of the principal geochemical reservoirs of the Earth's mantle. Phys. Earth Planet. Int. 112, 211-256.

Walzer, U., Hendel, R., 2003. Chemical differentiation, viscosity and the thermal evolution of the mantle. In preparation.

Weidner, D.J., Chen, J., Xu,Y., Wu,Y., Vaughan,M.T., Li,L., 2001. Subduction zone rheology. Phys. Earth Planet. Int. 127,67-81.

Yang, W.-S., 1997. Variable viscosity thermal convection at infinite Prandtl number in a thick spherical shell. Thesis, Univ. of Illinois, Urbana-Champaign.

Zerr, A. & Boehler, R., 1993. Melting of (Mg,Fe)SiO_3-perovskite to 625 kilobars: indication of a high melting temperature in the lower mantle. Science 262, 553-555.

Zerr, A. & Boehler, R., 1994. Constraints on the melting temperature of the lower mantle from high-pressure experiments on MgO and magnesiowstite. Nature 371, 506-508.

Zhang, S., Yuen, D.A., 1996. Intense local toroidal motion generated by variable viscosity compressible convection in 3-D spherical-shell. Geophys. Res. Lett. 23, 3135-3138.

Physics

PD Dr. Hans-Peter Nollert and Prof. Dr. Hanns Ruder

Theoretische Astrophysik, Universität Tübingen, Auf der Morgenstelle 10, 72076 Tübingen

Life in the world of computers is fast: What used to be last year's fastest and biggest machine may be superseded by the next model today. However, many of the problems tackled with these computers require much more time to solve than the life time of the machines that are used for solving them. Defining the questions, developing algorithms, obtaining results and interpreting them takes not only ingenuity and knowledge, but a lot of hard work, and therefore time.

A typical example represented in this book is the collisional dynamics of black holes, star clusters and galactic nuclei. Major efforts to simulate these processes in the context of general relativity have been going on worldwide for more than ten years. Problems encountered along the way lie mostly in fundamental mathematics and numerical techniques. While bigger and faster computers will be helpful, they will not suffice to solve these problems. We will likely see contributions to this subject for many more years to come, with steady progress towards an ability to treat these systems and learn more about their behaviour.

The situation is very similar for the project on large scale simulations of jets in astrophysical systems. This has been a long-standing, extremely complex problem in astrophysics. It involves the numerical treatment of the magnetohydrodynamic equations in a fully three-dimensional way.

Not only systems of cosmological dimensions require a considerable amount of time (fortunately, still many orders of magnitude less than cosmological time scales!) until solutions can be obtained. At the other end of the length scale, in the microscopic world, equally challenging problems are waiting for us. As simple as cracking a crystal may appear to be, simulating the process on a computer presents considerable demands. Here, the increase in computer power has made it possible to treat a molecular object with a size approaching the micrometer scale.

To the delight of the researchers involved in these projects, the next generation of computers is just around the corner, marking the start of the Teraflop age in Stuttgart. We may all look forward to new spurs in scientific creativity inspired be the new technological potential of these new machines.

Collisional Dynamics of Black Holes, Star Clusters and Galactic Nuclei

Emil Khalisi[1], Chingis Omarov[1,2], R. Spurzem[1], M. Giersz[3], D.N.C. Lin[4]

[1] Astronomisches Rechen-Institut, Mönchhofstrasse 12-14, D-69120 Heidelberg
[2] Fessenkov Astrophysical Institute, 480068 Alma-Ata, Kazakhstan
[3] Nicolaus Copernicus Astronomical Centre, Polish Academy of Science, ul. Bartycka 18, 00-716 Warsaw, Poland
[4] University of California at Santa Cruz, CA 95064, USA

Summary. We use high-precision direct N-body integration to study questions of the thermodynamic behaviour of dense stellar systems. The processes examined include mass segregation and equipartition processes, the study of planetary orbits in dense star clusters, and stellar orbits in galactic nuclei with thick accretion disks.

1 General Introduction and Supercomputing

In this paper we describe results of three ongoing intertwined subprojects. They are all based on the numerical simulation of dense stellar systems (systems of gravitating point masses) by direct accurate orbit integrations. We use the direct parallel N-body code NBODY6++ [36]. The first subproject (studies of mass segregation and equipartition) had already been started earlier and was mentioned in the previous year's report. Now the Phd thesis of Dr. Emil Khalisi [17] has been completed and can be obtained (in English) by
`http://www.ub.uni-heidelberg.de/archiv/3096`

The second subproject (orbit studies of stellar orbits in galactic centres with thick accretion disks) is mainly performed by Dr. Chingis Omarov, who is working in our institute as a DAAD fellow; this work is related to SFB439 "Galaxies in the Young Universe" at the Univ. of Heidelberg. The third project (studies of planetary orbits in dense star clusters) is a new application of our code to study the changes of orbital parameters of planets (very small mass bodies bound to a host star) under perturbations from field stars in a dense star cluster, where most stars and planetary systems ought to form.

For all these and other related projects the highly accurate determination of gravitational and other forces acting on particles on their orbits is very important. There are three main problems which require such accuracy:

- the correct modelling of relaxation processes (e.g. mass and energy transport due to small angle star star gravitative encounters) determines the rates with which stars in a dense stellar cluster are transported into a massive accretion disk and also further down to the massive black hole. Gravity is not shielded at something like a Debye sphere so all interactions (even those with large impact parameters) must be followed by high precision (i.e. relative energy error of 10^{-5} over hundred crossing times, to make a specific example).

- the larger the star clusters are the longer the relaxation time becomes as compared to the orbital or crossing time. Therefore simulations consist of tens of billions of individual steps and this is another reason why any secular force errors have to be kept at the lowest technically and numerically possible level.
- there is an intrinsically large dynamic range even at any given moment, for example planetary orbits have time scales of a few days in the worst case, while the relaxation in a star cluster may be up to a few Gigayears. That's around ten orders of magnitude, and another reason why none of the approximative methods used elsewhere in N-body dynamics (TREE, series expansions, mesh based Fourier transformation) can be used.

The only massively parallel code able to fulfil all requirements from the three items above is NBODY6++ [36], which uses an Ahmad-Cohen neighbour scheme, direct force computations (asymptotically the algorithms scales with N^2, but individually blocked time steps and regularisations of close encounters and bound subsystems (such as the planetary systems). The reader interested in more details about this family (very often called Aarseth-codes) is referred to the above cited paper and two more overview papers by Aarseth [1, 2].

For any particle number larger than a few 10^4 significant simulations require the use of massively parallel computers with extremely fast communication (of the order of Gbit/s simultaneously between many pairs of nodes). In Fig. 1 we can see how the use of the CRAY T3E turned the N^2 scaling of our algorithm down into a linear dependence. We can also see that a recent PC cluster obtained at our institute for SFB439 (20 P4 CPU's, 2.2 Ghz, Myrinet 2000) is for less than 50k particles meanwhile faster than the T3E. However, we need particle numbers much larger than 50k in the future, up to one million, in particular for all problems related to galactic nuclei. Here we will still be requiring the large scale parallel computing facilities of HLRS, in particular we are looking forward what will be the performance of the new generation of supercomputers (IBM, NEC, Hitachi) with our codes. Similarly, the Japanese GRAPE-6 special purpose computers seem to beat all other machines by a large factor. However, it is only able to outplay this performance for pure gravitational problems. Any additional complexities such as interactions with gas or planetary or binary system will make it impossible to use it with that performance. Rather, the general purpose supercomputers can be used much more efficiently here (that includes also the PC clusters.

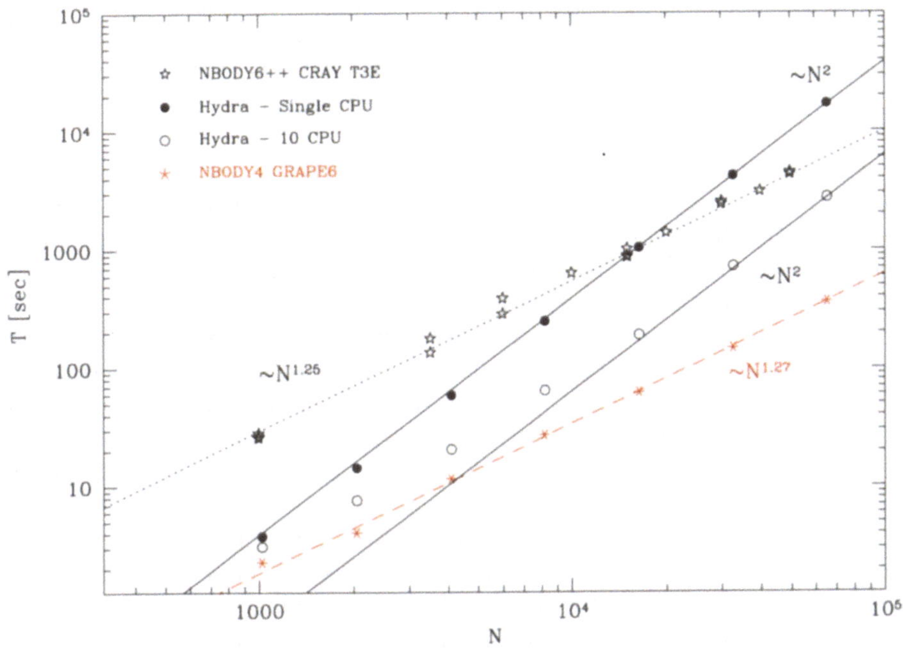

Fig. 1. Wallclock time for one third of a crossing time used on different hardware for our direct NBODY codes.

2 Limits of mass segregation in two–component models

E. Khalisi, R. Spurzem, D.N.C. Lin

2.1 Introduction

During the past two decades the theoretical and numerical study of the relaxation driven evolution of star clusters has seen an enormeous progress. One of the important reasons was a dramatic increase in the capabilities of fast general and special purpose computers such as GRAPE [41, 28], but also direct N-body modelling software for them [36, 1, 2]. In a large number of studies the evolution of stellar clusters has been studied and the validity of simplified theoretical models assessed, including many astrophysical effects such as stellar evolution, realistic tidal fields, primordial binaries, mergers and collisions (compare the reviews [27, 37]).

While it is still impossible to directly model globular star clusters or galactic nuclei with realistic particle numbers, there is another class of clusters which have just a particle number feasible for a direct N-body study: young open star clusters with relaxation effects going on. They have been out of focus for the pure dynamicists in many cases, although new deep infrared observations of the Trapezium Cluster in Orion [26, 13, 14] allow a detailed measurement of the kinematics and distribution of their stars. The results show a considerable mass segregation of heavy stars to the centre; according to [14] the average mass of stars within the core radius of the cluster (0.205 pc) is three times larger than the average mass within 2 pc. The cluster is less than 1 Myr old.

Collisional star cluster evolution naturally leads to a segregation of heavy masses in the core due to two-body relaxation, an effect already known for long time [34]. If the amount of heavy stars is relatively minor, an accelerated core collapse takes place, including all stars, but leading to an increase of the average stellar mass in the centre. Systems with a large amount of heavy masses tend to separate the heavy masses out completely, the latter undergoing a dynamically decoupled core collapse on their own. Such distinction has already been made in Spitzer's [34] mass segregation instability. If in a system with only a few thousand or ten thousand stars, it could be more appropriate to describe the process as dynamical friction, which brings the few most heavy stars to the centre.

Surprisingly, there has been only little quantitative study of mass segregation in the pre-collapse phase of star clusters, probably because more interest was prevailing in the post-collapse evolution of globulars. Bonnell & Davies [4] delivered a study using ensemble averaged results of clusters of $N = 150$ to $N = 1500$. Their conclusion was that the mass segregation time scale is not sufficiently small to account for the observed segregation in young clusters. However, they were using an N-body code with softening, thus cutting off all close encounters and binary formation, which affect the cluster evolution strongly in late collapse. Also, the maximum particle number of 1500 does not allow statistically secure conclusions for clusters in which the most massive stars are 20 or 30 times more massive than the average star, because there is only a very small number of massive stars.

One variant of star formation theory in clusters by Murray & Lin [29] predicts the formation of the most massive stars already near the cluster centre, since their proto–stellar clouds require many dissipative mergers with low–mass cloudlets for the growth. They are to happen, of course, most likely in the dense regions. Such newly formed clusters will remain gravitationally bound with the massive stars remaining preferentially in the inner parts of the cluster. On the other side, Podsiadlowski & Price (1992) draw a scheme in which massive stars might also form in some cold gaseous clumps in the outskirts of a star–forming region. These individual massive stars sink to the centre (possibly while still forming), and are likely to affect the formation of other stars, e.g. by heating or disrupting their gas cloud. It is only the later phases of star formation that will be dominated by a sequential formation of massive stars surrounded by previously formed low–mass stars. However, the different results are conflicting.

Regarding the present unclear situation our approach is rather to reduce the complexity of the system, but to study the physical mechanism of mass segregation in young star clusters more systematically. The results will be related to the above papers and observations. As a completely new study, we investigate the effect of initial segregation and "anti"–segregation on the evolution of the system. We explore the range of the segregation process and compare its saturation according to the different initial configurations. Comparisons with other simplified models are given.

2.2 The models

All simulations start as Plummer spheres, which appear to be very similar to a King $W_0 = 6$ model in its core collapse parameters (Quinlan 1996). They start from a global virial equilibrium, and the particles are treated as point masses, i.e. without softening. An external tidal field, primordial binaries, and stellar evolution are neglected here. Our model parameters were chosen to closely match the observational results of Hillenbrand & Hartmann [14] for the Orion cluster: 5000 stars, in two mass components (subscripts 1 for the light stars, and 2 for the heavies), with $q := M_2/M_{\rm tot} = 0.26$, and $\mu = m_2/m_1 = 20$, where M_1, M_2 are the total masses of the two components, and m_1, m_2 are the individual stellar masses, respectively. The total number of particles is $N = N_1 + N_2 = 4914 + 86 = 5000$.

The spatial distribution of each mass m_1 or m_2 is chosen randomly to initialize a different setup of positions and velocities at the start of the simulation. These physically equivalent runs were repeated 10 times and averaged into a mean model — we shall call this the "RND"–model. The goal of such "ensemble averaging" is to increase the statistical significancy of the global output data [8].

Additionally, we tested two exoctic setup configurations by placing the heavy masses m_2 artificially at the two extremė parts: First, all of them in the inner regions such that the two mass components would already be maximum segregated and no further process of stratification expected — those models will be called "INS"; second, in the outside regions foothills of the Plummer density profile, in order to construct a most possible "anti–segregation" — these models are named "OUT". Though unrealistic, such extreme cases are of special interest for investigating the full range of the segregation process. The limits give us for example the shortest and longest time scale. As before, both kind of models are ensemble averaged over 10 individual runs.

Besides these models, we also varied the parameter of the mass ratios in the following steps: $\mu = 1.25, 1.5, 2.0, 3.0, 5.0, 10.0, 20.0, 40.0, 75.0, 100.0$. They serve as a check how the course of the core collapse times changes with respect to Spitzer's accelerated evolution.

2.3 Selected Results

To discuss the segregation of masses, we have divided the system in spherical mass shells containing 1, 2, 5, 10, 20, 30, 40, 50, 75, 90, 100% of the total mass, respectively. The changes of the mean mass inside such a shell is plotted in Figure 2. The innermost shells quickly gather the heavy masses and undergo a core collapse. In the post–collapse phase, the mean mass remains in the centre is rather constant indicating that the global segregation has come to an end. The intermediate shells still [anhäufen] the heavy masses, though somewhat the progress has slowed down in the core bounce. The degree of segregation in the shells can be understood as the mean mass achieved with regard to the maximum possible mass (i.e. the case if a shell would consist of the heavy particles only), $\langle m \rangle / m_2$. For this model the degree in the innermost shell reaches about $11 M_\odot / 20 M_\odot \approx 0.55$, the lower next shell about 0.4, etc.

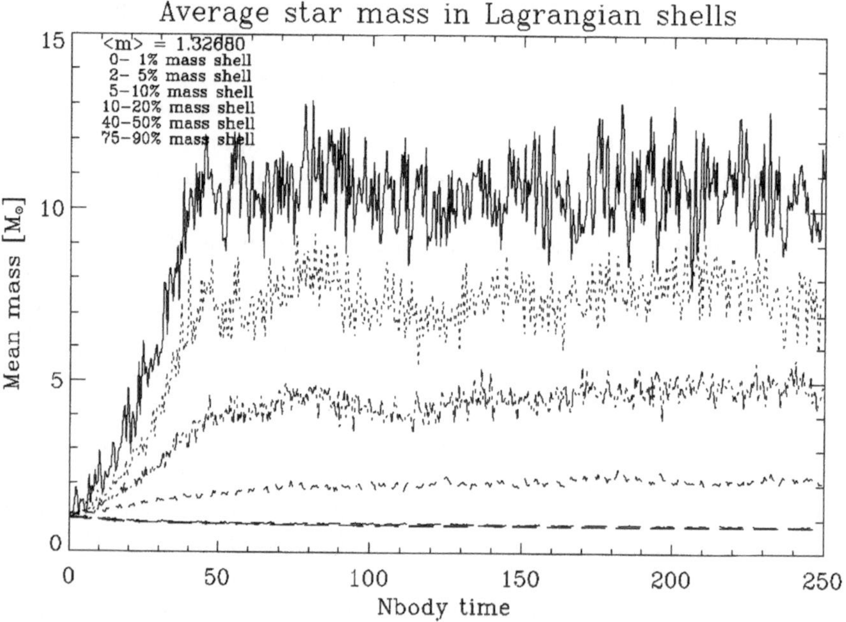

Fig. 2. Evolution of the mean mass within Lagrangian shells of the two–mass model RND-G. The raw data of 10 independent runs was averaged into one single ensemble–model.

The segregation of the heavy stars is to be compared with the range of "possible" segregation processes. In order to exploit a minimum and maximum time scale, we performed two more cases of the same model in which the heavy stars have been artificially placed either in the inside regions such that a pre–segregated state is given (we call it an INS–model), or in the outer halo for constructing an "anti–segregated" state (OUT–model). Though unrealistic, such extreme cases disclose the width of the segregation time scale. Figure 3 illustrates the evolution of the mean mass in both models: In the left panel, the cluster starts from the INS–configuration: The 1%–Lagrangian shell has got a mean mass that is μ–times higher than the outer layers. The light stars penetrate immediately after the start into the innermost shell and reduce $\langle m \rangle$, as seen at the very left margin. The sphere of heavy bodies expands a little and raises the mean mass in the outer spheres (lower curves). After the short settling period of some few crossing times, the model turns into a configuration as in the RND–model.

In the right panel of Figure 3, it is visible how slowly the heavy masses sink from the outskirts to the centre; the decreasing dashed line is the mean mass of the 75–95% shell. Its slow decline indicates that fair number of heavies remains in the halo for a long time before they start falling inwards. Their orbital velocity is also small, and their motion starts from a rather inert state. The collapse is performed a by those few heavy bodies that by chance moved quicker to the centre. As seen from the lower $\langle m \rangle$ in the innermost shell, the number of the heavies being present at the time of the collapse is smaller than in the INS– and RND–models.

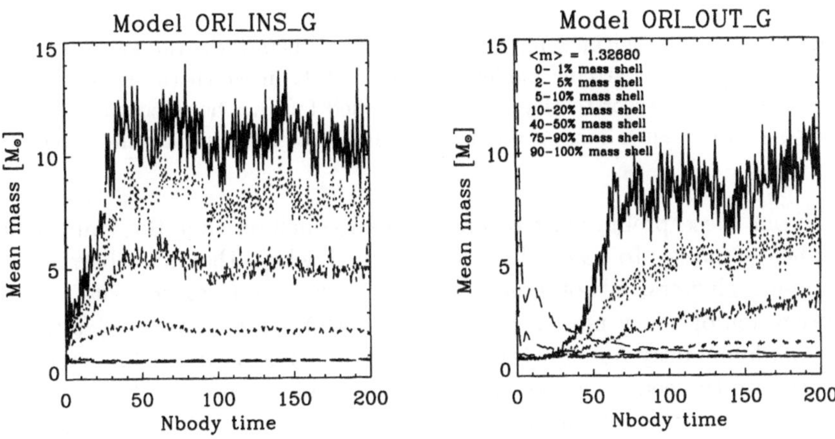

Fig. 3. Comparison of the mean mass between Lagrangian radii for Models G–INS (*left*) and G–OUT (*right*).

In addition to this model that matches the parameters of the Orion Cluster, we expoited the core collapse times for other relative masses m_2/m_1, i.e. for a wide range of μ's from 1 to 75. The maximum value might correspond to an imaginary cluster with two different masses like $0.08M_{\odot}$ and $6.0M_{\odot}$, but a constant fraction of 26% heavies. The core collapse times and their ranges are summarized in Figure 4. The triangles indicate the values from the random setup and the dashed lines are the INS– and OUT–models.

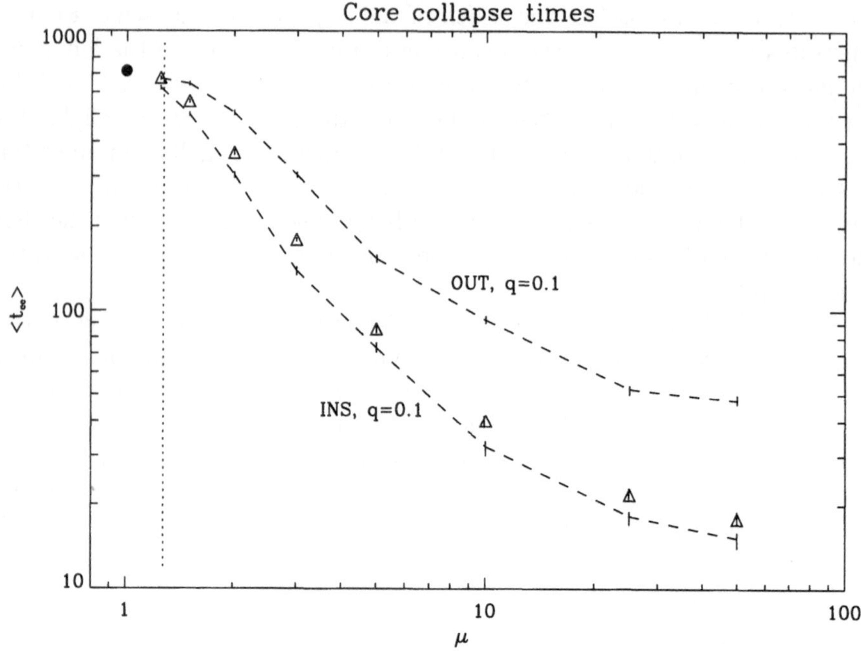

Fig. 4. Core collapse times for models with $N = 5000$ of the random configuration (triangles) and the extreme models INS and OUT, in which the heavy stars were placed in the center or to the outskirts (dashed lines). The filled circle represents the value of the equal–mass system. The small error bars are also indicated.

From these preliminary results we can conclude that the time scale for the heavy masses to reach the cluster centre, where they are observed today in young clusters, is not much changed, even if one starts with an initial segregation of heavy masses inwards (model INS). If, however, heavy masses are predominantly outside (model OUT) the time is much longer for them to get into the centre. So we can constrain possible models of star formation by looking at the dynamical evolution of the cluster. More details will be published soon[18].

3 Stellar orbits in galactic centres with accretion disks

Ch. Omarov, R. Spurzem

The main physical subsystems of active galactic nuclei (hereafter AGNs), containing most of the mass of AGN are the compact stellar cluster (CSC) and the massive black hole (MBH) at the center of the cluster. The last one is usually surrounded by an accretion disk (AD), providing most of the luminosity of bright AGNs. Apparently, the dissipative interaction of the CSC with the AD determines the stellar dynamics in the central part of the cluster and, consequently, the AGN evolution. Nevertheless, the task of the evolution taking into account both stellar two-body gravitation interaction and the star-disk interaction still is not solved, though many works were devoted to the stellar dynamics in central parts of AGNs. The interaction of the compact stellar cluster with a massive central object in the forms of super- star and MBH was considered by Vilkoviski, Hara, and Hills [43, 15, 11]. The evolution of the dense non-rotating stellar cluster was studied by Spitzer, Saslaw, Bisnovatyi-Kogan among others [35, 5], and the evolution of the gas sphere was considered by Langbein and collaborators [23]. Stellar interactions with accretion disks were as well considered by Vilkoviskij and Syer [44, 45, 42]. More detailed investigations of the stellar orbits, crossing accretion disks were presented in the works by Artymovicz, Karas, Vokrouhlicky [3, 47, 40]. Finally Rauch [31, 32] considered and numerically calculated the cases when non-elastic star-disk interactions or pure star-star collisions dominate, and Vilkoviskij and Cherny [46] calculated an analytical model of the joint action of the star-disk and star-star interactions.

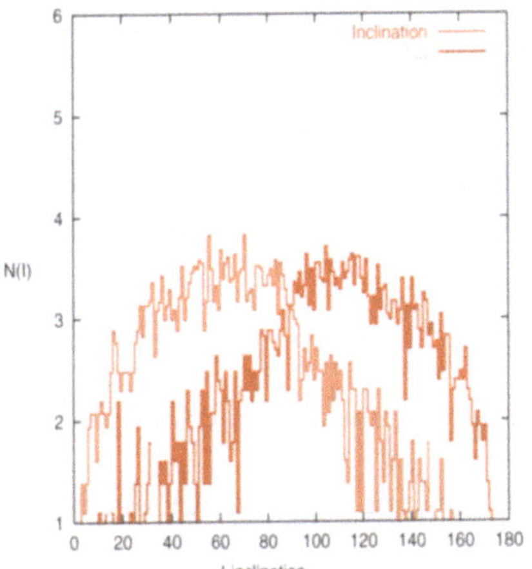

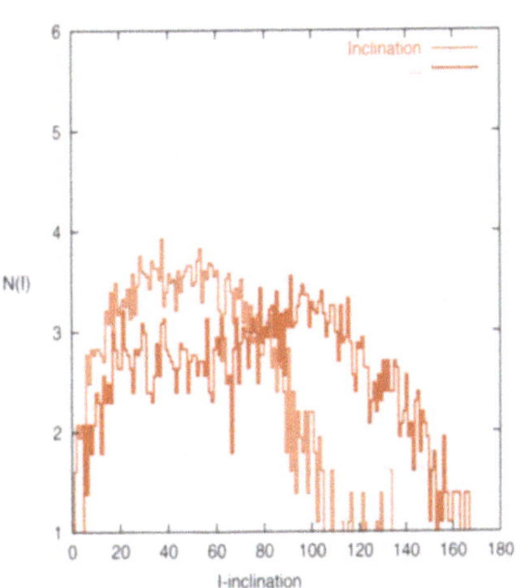

Fig. 5. Distribution of inclinations of stellar orbits in a galactic nucleus with a massive accretion disk. Thin curves co-rotating stars, thick curves counter-rotating stars. Left panel initial model. The original distribution should be like a sine function (if both components are addded). Right panel after a few crossing times, where the interaction with the disk has depleted the counter-rotating orbit family.

This ongoing projects is examining by direct N-body simulation the problem of the interaction of compact stellar cluster with the accretion disk in active galactic nuclei. We have accomplished the inclusion of a standard ram pressure force into the Hermite scheme of the N-body code NBODY6++. This is the first time that such high precision direct N-body simulation considers non-gravitational forces. However, our disk is yet stationary, with a Keplerian rotation velocity. We include as free parameters of order unity the interaction

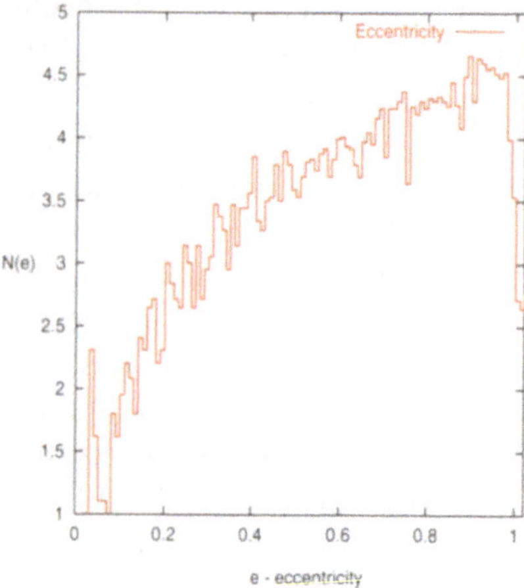

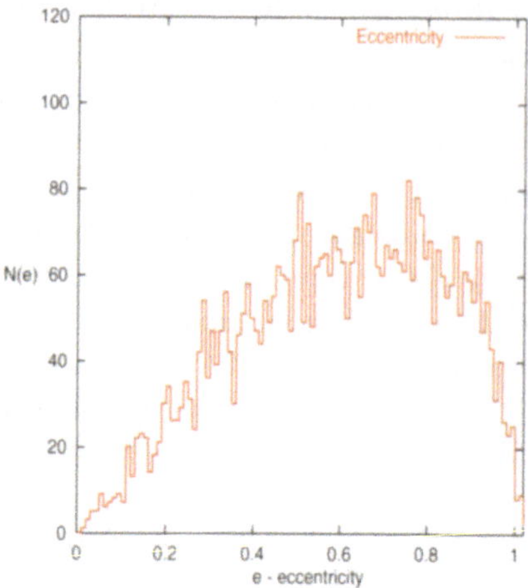

Fig. 6. As in Fig.5, but here showing the eccentricities of all the orbits, not separated for their sense of motion. We can see that high eccentricities are depleted.

coefficient of the ram pressure force and the Keplerian velocity at the inner edge of the disk.

Vilkoviskij & Cherny [46] have compared the star-star two-body interactions with the star-disk interactions and concluded that the last is more strong in the inner parts of the ADs. The rate of the change of a star energy E due to the two-body star-star interaction is

$$dE^{(ss)}/dt =.4\pi G^2 M_s^3 V_s^{-1} \ln \Lambda, \tag{1}$$

where G is the gravitation constant, M_s and V_s- mass and velocity of the star, $\ln \Lambda$ - the Coulomb logarithm.

It can be compared with the rate of the change of a star energy due to the non-elastic star-disk interaction:

$$dE^{(sd)}/dt = \pi q(i) R_s^2 \Sigma_d V_s^2 / T, \tag{2}$$

where $q(i) < 1$ is dissipation parameter, depending on the inclination angle i, R_s – the radius of the star, Σ_d – the surface mass density of the accretion disk, and $T-$ the orbital period of the star.

One can see that in the close-to-Kepler potential of the BH in the inner part of the AD, $dE^{(sd)}/dt > dE^{(ss)}/dt$ dew to the different velocity dependance of the both values. On the other hand, the star-disk interaction leads to inclination of the stellar orbits to the plane of the disk, which diminish $dE^{(sd)}/dt$. Without the star-star scattering, the inclination angle and the energy dissipation could go to zero, but if the stellar density is large enough, the scattering will increase the inclination angle. As a result, both $dE^{(sd)}/dt$ and $dE^{(ss)}/dt$ are keeping in some equilibrium, regulating the inclination angle and the inflow of the stars to the centrum of the AD.

A simple model for the stellar distribution and the inflow was calculated in the work by Vilkoviskij and Cherny (2002), and it was shown that stars can influence the AGN variability and the emission broad line regions properties.

As first examples of our results we show the initial and final (after a few orbital times) inclinations and eccentricities of 5000 stars in sample calculation. One can see how the interaction with the disk creates a lack of high inclinations and also influences co- and counterrotating stellar orbits in a different way. This will affect the global shape and angular momentum distribution of the system. With more simulation data we will be able to predict data of the central accretion disk (not resolved by direct observations) from the stellar kinematics much further outside where it can be observed.

These data will be completed using different particle numbers in the present computing period, and a refereed publication will be completed describing results from a larger parameter study using different particle numbers, if the visit of Dr. Omarov can be prolonged until Sep. 30, 2003.

4 Planetary Orbits in Dense Star Clusters

R. Spurzem, M. Giersz, D.N.C. Lin

4.1 Introduction

Recent detection of Jupiter-mass extrasolar planets indicate that they are attached to at least $\sim$ 8% of nearby solar-type stars and that they have diverse dynamical properties (`http://www.exoplanets.org`). An important challenge to the theory of planetary formation and evolution is to account for the dynamical diversity of extrasolar planets. The standard theoretical models of runaway and oligarchic growth of planets from kilometer size protoplanets [19, 20] 2000) predict that most planets have small eccentricities. Observations, however, reveal that there are many even massive planets with highly eccentric orbits. One of a number of explanations would be the influence of encounters between host stars of planets and field stars [6, 7, 24]. This hypothesis arises naturally because most young stars are formed in clusters [21]. Strong dynamical perturbations resulting from close encounters between stars may also have led to the loss of their planetary companions. Using the equation of restricted three body motion to approximate encounters between planetary systems and single stars, Smith and Bonnell [33] inferred that only a small fraction of planets would become detached from their host stars during the characteristic life spans of open and young stellar clusters. Dynamical instability resulting from the interaction between planets may lead to both eccentricity excitation and orbital migration [25].

4.2 Method and Model Parameters

We use for small star clusters (up to 20000 particles) the direct integration code NBODY6++ [36], which is based on full computations of individual pairwise particle forces, and is an offspring of the family of NBODY codes developed by Aarseth (1999). In addition to that for very large star clusters (few hundred thousand stars) we use a new stochastic Monte Carlo code [9, 10], which models the stellar system using a gaseous model and follows all planetary systems by a Monte Carlo method, and all encounters are directly integrated. For all models, we adopt an isotropic Plummer model for the phase space distribution function. In most runs all stars had equal masses, whereas in one case for comparison we adopt an initial mass function following the prescription of Kroupa, Tout and Gilmore [22]. All models are in dynamical equilibrium initially, all stars are assumed to be point masses neglecting stellar evolution effects. As model units we take $G = 1$, $M = 1$, $E = -0.25$, for the gravitational constant, initial total mass and energy, respectively (standard Nbody units).

Table 1. Model Parameters

Model	N_*	$M_*(M_\odot)$	N_p	a_p(AU)	e_p
1	$5 \cdot 10^3$	1	10^3	3-50	0.01
2	$5 \cdot 10^3$	1	10^3	3-50	0.1
3	$5 \cdot 10^3$	1	10^3	3-50	0.3
4	$5 \cdot 10^3$	1	10^3	3-50	0.6
5	$5 \cdot 10^3$	1	10^3	3-50	0.9
6	$5 \cdot 10^3$	1	10^3	3-50	0.99
7	$5 \cdot 10^3$	1	10^3	3-50	thermal
8	$5 \cdot 10^3$	(ktg)	10^3	3-50	thermal
9	$5 \cdot 10^3$ 100 stellar binaries	1	10^3	3-50	thermal
10	$5 \cdot 10^3$ 100 stellar binaries	(ktg)	10^3	3-50	thermal
MC1	$3 \cdot 10^5$	1	$3 \cdot 10^4$	3-50	thermal
MC2	$3 \cdot 10^5$	1	$3 \cdot 10^4$	0.01-10	thermal

Models labelled 1-9 are direct N-body models, models labelled MC are stochastic Monte Carlo models. Columns are number of stars N_*, number of planets N_p, and initial semi-major axes and eccentricities a_p, e_p of planets. "thermal" refers to a thermal eccentricity distribution function $f(e) = 2e$, and "ktg" refers to a mass spectrum using the Kroupa, Tout and Gilmore [22] mass function.

4.3 Numerical Results, Cross Sections and Discussion

Here we plot the direct changes of e_p and a_p in individual encounters (as in the direct N-body models left) but also statistically determined differential cross sections for them, comparing with standard cross sections for stellar encounters. While for eccentricity changes there is a fairly good model provided

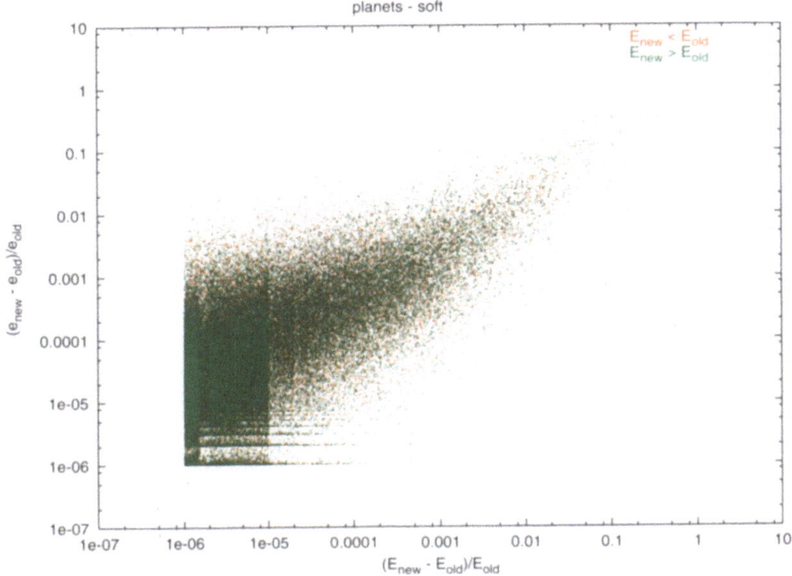

Fig. 7. Direct record of individual scattering events in our Monte Carlo model, plotting changes of eccentricities versus changes of the energies of the orbits. The same kind of display can be found in the following figures obtained from direct N-body models.

by the semi-analytic results of Heggie & Rasio [12], predicting a scaling of the differential cross section as $\propto \delta e^{-5/3}$ in fair agreement with our measurements, we find for the changes of the semi-major axis a scaling with $\propto \delta a^{-3/2}$, which is theoretically not yet understood. It is interesting to note that for a wide parameter range the cross sections are symmetric for positive and negative changes, so at the same time some planetary orbits get wider and others get tighter. Such diffusion process can lead to the formation of "hot" Jupiters. On the other hand side a large number of planets is liberated. A "tail" of large orbital changes exists, which belongs to closer encounters and exhibits an asymmetry in the cross-section (preference of positive changes in e and a. These results are preliminary. For more details see Spurzem & Lin [38] and Spurzem, Giersz & Lin [39].

Acknowledgements

CO acknowledges financial support by a DAAD scholarship A/01/08529 and DAAD NATO grant A/03/19445. RS is grateful for the kind and generous hospitality of Doug Lin and the UC Santa Cruz during his visits there. MG and RS acknowledge financial support by DFG international cooperation grant 436 POL 113/103/0-1 and followups. This work has been partly supported

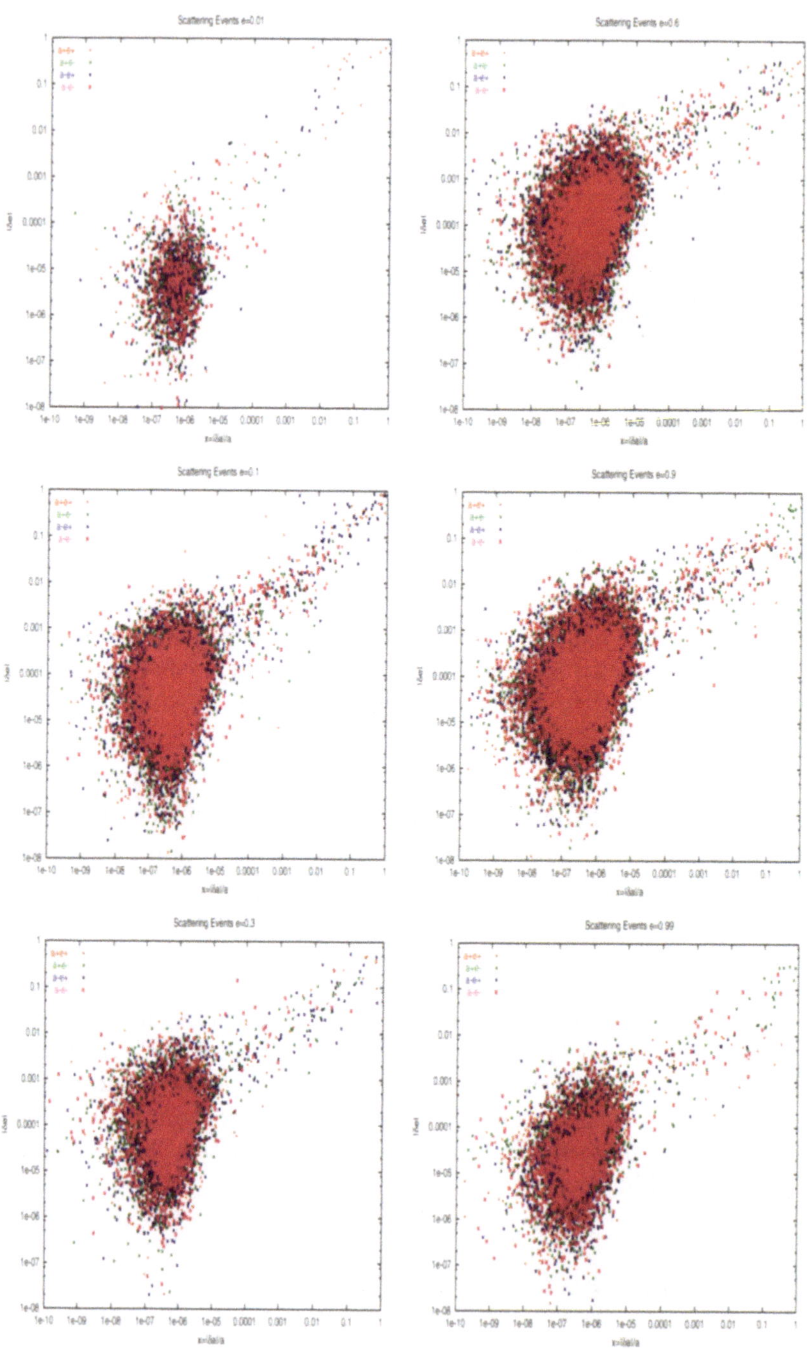

Fig. 8. Scattering events as a function of semi-major axis change $x = \delta a/a$), key indicates whether it is a negative or positive change. Left to right, up to down, for $e_0 = 0.01$, 0.1, 0.3, 0.6, 0.9, 0.99, respectively, as indicated in the figure, which corresponds to models 1-6 in Table 1.

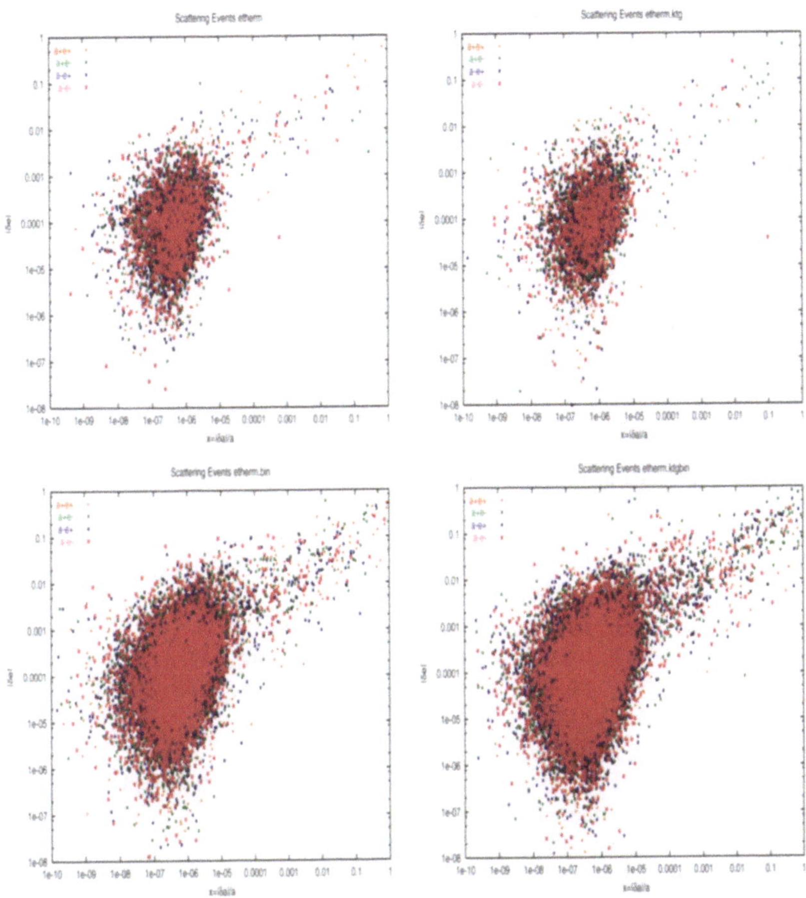

Fig. 9. Scattering events as a function of semi-major axis change ($x = \delta a/a$), key indicates whether it belongs to a negative or positive change. Left to right, up to down, for model 7 ("etherm"), 8 ("etherm.ktg"), 9 ("etherm.bin"), 10 ("etherm.ktgbin"), models explained in Table 1.

by SFB439 of the German Science Foundation at the Univ. of Heidelberg. We kindly acknowledge the use of supercomputing facilities at HLRS Stuttgart and NIC Jülich, and the use of the Heidelberg Beowulf PC cluster at Astronomisches Rechen-Institut, financed by the State of Baden-Württemberg for SFB439.

References

1. Aarseth S.J., 1999a, Publ. Astr. Soc. Pac. 111, 1333
2. Aarseth S.J., 1999b, Cel. Mech. Dyn. Astron. 73, 127

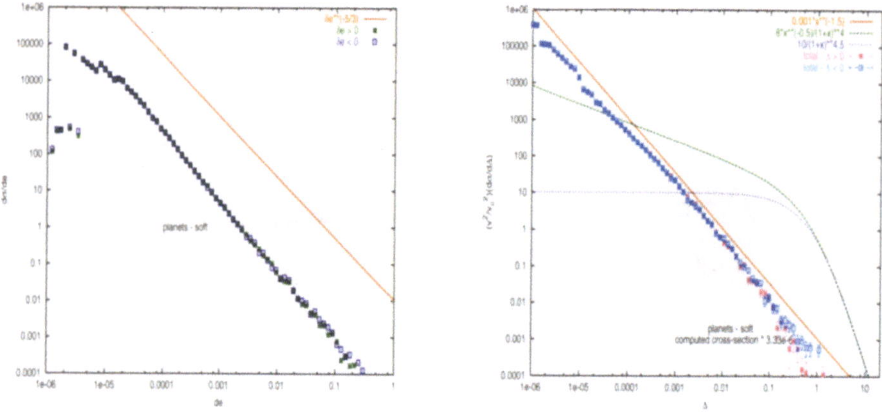

Fig. 10. Cross Sections obtained from measuring orbital changes of planets in a star cluster, see explanation in main text.

3. Artymowicz P., Lin, D.C.N., Wampler E.J., 1993. ApJ 409,592
4. I. A., Davies M. B., 1998, MNRAS, 295, 691
5. Bisnovatyi-Kogan G.S., 1978. Astronomy Lett. 4,130
6. de La Fuente Marcos C., & de La Fuente Marcos R. 1997, A&A, 326, L21
7. de La Fuente Marcos C., & de La Fuente Marcos R. 1998, New Astron., 4, 21
8. Giersz M., Heggie D. C., 1994, MNRAS, 268, 257
9. Giersz, M., Spurzem, R., 2000, MNRAS, 317, 581
10. Giersz, M., Spurzem, R., 2003, MNRAS, in press, astro-ph/0301643
11. Hara T., 1978. Prog. Theor. Phys. 60, 711
12. Heggie D.C., Rasio F.A., 1996, MNRAS, 282, 1064
13. Hillenbrand L. A., 1997, AJ, 113, 1733
14. Hillenbrand L. A., Hartmann L. W., 1998, ApJ, 492, 540
15. Hills J. G., 1975, AJ, 80, 1075
16. Karas V., Vokrouhlicky D., 1983. MNRAS 265, 365
17. Khalisi E., Mass Segregation and Equipartition in Two-Component Star Clusters, Diss. Univ. Heidelberg, 2003.
18. Khalisi E., Spurzem R., Lin D.N.C., 2003, MNRAS, in preparation.
19. Kokubo E., Ida S., 1998, Icarus, 131, 171
20. Kokubo E., Ida S., 2000, Icarus, 143, 15
21. Kroupa P., 1995, MNRAS, 277, 1491
22. Kroupa P., Tout C.A., Gilmore G., 1993, MNRAS, 262, 545
23. Langbein T., Spurzem R., Fricke K.J., Yorke, H.W., 1990, A&A 227, 333
24. Laughlin, G., & Adams, F.C. 1998, ApJ, 508, L171
25. Lin, D. N. C. & Ida, S., 1997, ApJ, 477, 781
26. McCaughrean M. J., Stauffer J. R., 1994, AJ, 108, 1382
27. Meylan G., Heggie D.C., 1997, A&A Rev., 8, 1
28. Makino J., Taiji M., Ebisuzaki T., Sugimoto D., 1997, ApJ, 480, 432
29. Murray S. D., Lin D. N. C., 1996, ApJ, 467, 728
30. Podsiadlowski P., Price N. M., 1992, Natur, 359, 305
31. Rauch K.P., 1995. MNRAS 275, 628
32. Rauch K.P., 1999. ApJ 514, 725

33. Smith, K. W., Bonnell, I. A. 2001, MNRAS, 322, L1
34. Spitzer L., ApJ 158, L139.
35. Spitzer L. J., Saslaw W. C., 1966, ApJ, 143, 400
36. Spurzem R., 1999, J. Comp. Appl. Math. 109, 407
37. Spurzem R., Deiters S., Fiestas J., 2003, in "New Horizons in Globular Cluster Astronomy", G. Piotto, G. Meylan, G. Djorgovski, M. Riello (eds.), ASP Conf. Ser., in press
38. Spurzem R., Lin D.N.C., 2003, in Scientific Frontiers in Research on Extrasolar Planets, D. Deming & S. Seger (eds.), ASP Conf. Ser., in press.
39. Spurzem R., Giersz M., Lin D.N.C., 2003, MNRAS, in prep.
40. Subr L., Karas V., 1999, A&A 352, 452
41. Sugimoto D., Chikada Y., Makino J., Ito T., Ebisuzaki T., Umemura M., 1990, Nature, 345, 33
42. Syer D., Clarke C.J, and Rees M.J., 1991, MNRAS 250, 505
43. Vilkoviskij E.Y., 1975, Astroph. Letters, 1, 8
44. Vilkoviskij E.Y., and Bekbosarov N., in: KazGU Proc., Alma-Ata, 1982, 100
45. Vilkoviskij E.Y., 1983, Astroph. Letters, 9, 405
46. Vilkoviskij E.Y., Cherny B., 2002, A&A 387, 804
47. Vokrouhlicky D., Karas V., 1998, MNRAS 298, 53.

Formation and Propagation of Jets Around Compact Objects

Matthias Stute and Max Camenzind

Landessternwarte Königstuhl, D-69117 Heidelberg
M.Stute@lsw.uni-heidelberg.de
M.Camenzind@lsw.uni-heidelberg.de

1 Introduction

The phenomenon of jets is very common among astrophysical objects. Jets appear around supermassive black holes in radio galaxies and quasars, stellar black holes in microquasar or black hole X-ray binaries (BHXBs), around young stellar objects, but also around stellar compact sources as white dwarfs in symbiotic stars and neutron stars in low mass X-ray binaries (LMXBs). In interacting binaries containing compact objects, the secondary – a main-sequence star or red giant – loses matter through a stellar wind or Roche lobe mass overflow. This matter is forced by the gravitational field of the compact primary to form an accretion disc. On the surface of the compact object, explosive nuclear burning causes nova-like outbursts. Through interactions between the accretion disc and the magnetic field of the compact object, matter is accelerated towards the polar regions and ejected as jets. For all objects, the exact mechanism, how jets are launched and the material is accelerated, is not understood very well. A few analytic approaches are developed, but to solve this problem, time-dependant numerical magnetohydrodynamics (MHD) simulations with very high spatial resolution have to be used. Blandford & Payne [BlP82] examined the magnetically forced extraction of energy and angular momentum from the disc and found centrifugally driven outflows, if the angle between the poloidal magnetic field and the disc surface is less than 60°. With increasing distance from the star-disc-system, the toroidal field becomes important, collimating the outflow to jets. Camenzind [Cam90] discovered the first self-consistent model considering all parts of a protostellar star-disc-system. He assumed that this system is dominated by the rotating magnetosphere of the star. Due to interactions of the magnetic field of the compact object with the accretion disc, a gap between the disc and the surface of the star is created. The plasma from the inner edge of the disc is accelerated following the field lines. Two topologies of the resulting magnetic field of the whole system can occur which are distinguished by the conductivity of the disc.

While the propagation of jets around stellar black holes and protostars is examined by many groups, simulations of jets with parameters in the range, suitable for those around compact objects, are only rarely performed. In contrast to extragalactic jets, galactic jets are less extended – only to a few 100–1000 AU – and their velocity is only several 100 km/s (YSO) to several 1000 km/s (compact objects). Additionally, the jets of compact objects are probably much denser than those of young stars.

The most prominent symbiotic systems which produced jet-like outflows are CH Cygni, R Aquarii and MWC 560. In the first two objects the observer is viewing the system from the side, while in MWC 560 the jet axis is parallel to the line of sight [SKC01]. Structural investigations using VLA radio and CHANDRA X-ray observations with high spatial resolution and kinematic examinations using spectroscopic monitoring were made for these objects.

Although a huge amount of observational data is available for jets of symbiotic systems, simulations in the parameter range, suitable for them, have been neglected until now.

2 The Codes *NIRVANA_CP* and *NIRVANA2.0*

2.1 *NIRVANA_CP*

The set of the hyperbolic differential equations of ideal hydrodynamics

$$\frac{\partial \rho}{\partial t} + \nabla(\rho \mathbf{v}) = 0 \tag{1}$$

$$\frac{\partial (\rho \mathbf{v})}{\partial t} + \nabla(\rho \mathbf{v} \otimes \mathbf{v}) = -\nabla p \tag{2}$$

$$\frac{\partial e}{\partial t} + \nabla(e \mathbf{v}) = -p \nabla \mathbf{v} + \Lambda(T; \rho_i) \tag{3}$$

$$p = (\gamma - 1)\, e \tag{4}$$

with gas density ρ, energy density e, velocity $\mathbf{v}$, temperature T and the ratio of the specific heats at constant pressure and volume γ is solved with the code *NIRVANA* [ZiY97] which uses second order accurate finite-difference and finite-volume methods and explicit time-stepping. This code was modified by M. Thiele [Thi00] to calculate radiative losses due to non-equilibrium cooling by line emission. The microphysics is introduced via a cooling term Λ in the energy equation and interaction equations for the species.

When cooling is important, the above equations are supplemented by an atomic network

$$\frac{\partial \rho_i}{\partial t} + \nabla(\rho_i \mathbf{v}) = \sum_{i=1}^{N_s} \sum_{j=1}^{N_s} k_{ij}(T)\, \rho_i\, \rho_j + \sum_{i=1}^{N_s} I_i(\nu)\, \rho_i \tag{5}$$

with ρ_i the species densities satisfying $\rho = \sum_{i=1}^{N_s} \rho_i$ for the total density. k_{ij} are the rate coefficients for two-body reactions which are functions of the fluid temperature T and I_i are frequency-integrated photoionization and dissociation rates [DoS03]. The summations go over the N_s atomic and molecular species included in a chemistry model. *NIRVANA_C* can handle up to 36 species [Thi00]. In the simplest case, we have to include HI, HII and e^-. In the optically thin case, the cooling rate is a function of the species densities and temperature, $\Lambda = \Lambda(T;\rho_i)$ [SuD93]. When cooling is very efficient, the chemistry network is solved in a time-implicit way [Thi00]. This code has been extensively tested in Thiele [Thi00] and Krause [Kra01].

When chemistry is solved dynamically with the full set of non-equilibrium equations, the various ionization states and concentration densities ρ_i of each element are calculated from the chemistry equations. They are used explicitly in the cooling functions as

$$\Lambda(T;\rho_i) = \sum_{i=1}^{N_s} \sum_{j=1}^{N_s} e_{ij}(T)\, \rho_i\, \rho_j + \sum_{i=1}^{N_s} J_i(\nu)\, \rho_i + \Lambda_{BS}(T) \tag{6}$$

with e_{ij} the cooling rates from two-body reactions between species i and j, and J_i the frequency-integrated photoionization and dissociation heating. In this case, the equation of state has to be given in the form

$$T = \frac{\gamma - 1}{k_B} \frac{e}{\sum_{i=1}^{N_s} n_i}. \tag{7}$$

The code in its present form can handle collisional excitation, collisional ionization, recombination, metal-line cooling, Bremsstrahlung and photoionization heating.

NIRVANA_CP is the parallelized version using the automatic and directive-based parallelization of the NEC compilers.

2.2 *NIRVANA2.0*

NIRVANA2.0 [Zie02] solves the equations of visco-resistive, compressible MHD

$$\frac{\partial \rho}{\partial t} + \nabla(\rho \mathbf{v}) = 0 \tag{8}$$

$$\frac{\partial(\rho \mathbf{v})}{\partial t} + \nabla(\rho \mathbf{v} \otimes \mathbf{v}) = -\nabla p + \frac{1}{\mu}(\nabla \times \mathbf{B}) \times \mathbf{B} + \nabla \cdot \sigma - \rho \nabla \Phi \tag{9}$$

$$\frac{\partial e}{\partial t} + \nabla(e \mathbf{v}) = -p \nabla \mathbf{v} + \sigma : \nabla \mathbf{v} + \frac{\eta}{\mu^2} |\nabla \times \mathbf{B}|^2 + \nabla \cdot (\kappa \nabla T) \tag{10}$$

$$p = (\gamma - 1)\, e \tag{11}$$

$$\frac{\partial \mathbf{B}}{\partial t} = \nabla \times (\mathbf{v} \times \mathbf{B} - \frac{\eta}{\mu} \nabla \times \mathbf{B}) \tag{12}$$

$$\sigma_{ij} = \nu \, (\nabla \mathbf{v}_{ij} + \nabla \mathbf{v}_{ji} - \frac{2}{3} \nabla \mathbf{v} \, \delta_{ij}) \tag{13}$$

with the magnetic field $\mathbf{B}$, the gravitational potential Φ, dynamic viscosity ν, electrical resistivity η, magnetic permeability μ and thermal conductivity κ. This additional property is very important to study the acceleration, initiation and launching mechanisms of jets. All parts of the system – the accretion disc, the star and the stellar magnetic field – play equally important roles in these activities. To work out a global model, the accretion disc has to be modeled self-consistently which requires a code which is able to solve not only the equations of ideal hydrodynamics, but also those of visco-resistive, compressible hydrodynamics. The interplay of the disc and the stellar magnetic field demands additionally the extension of hydrodynamics towards magneto-hydrodynamics.

To be able to use the advantages of both codes, *NIRVANA2.0* has to be extended with the same tools to consider the microphysics. Due to personal preferences, the possibility to use debugger and profiler tools as *assureview* and *guideview* and the portability to local multiprocessor machines, the code was parallelized using the OpenMP programming model and vectorized. Also the extensions dealing with microphysics were already transformed to OpenMP. Therefore the inclusion of them in *NIRVANA2.0* will be done within the next weeks.

3 Future Work

After including the microphysics into *NIRVANA2.0*, several test calculations will be performed. Then the code *NIRVANA_CP* will be used to continue the examination of the propagation of the jets of symbiotic stars – especially the systems MWC 560 and R Aquarii will demand the project members' attention. Previous work has shown the importance of cooling effects to model the jet of MWC 560 correctly. Therefore *NIRVANA_CP* is the only tool which is predestinated to continue the research on the propagation of jets in symbiotic binaries. To achieve a reasonable spatial resolution of the propagation simulations, the computational domain should consist of at least 1000 grid cells in the direction of the jet axis. To resolve the internal structure of the jet, the other two spatial coordinates should be represented by 500 grid cells each. The absolute resolution will be 20 grid cells per jet radius which is the basic length scale of the system. The total sum of grid cells is then some 10^8. As the effects of cooling were already studied in 2D, it is not a great loss to approximate the cooling functions which seems to be necessary for 3D models. For each grid cell, twelve variables have to be stored leading to a typical memory requirement of about 50 GB. The jet will be injected on one side of the computational domain and will reach the other side after approximately 500 days depending on the basic parameters of the system. Using the CFL condition and the cooling timestep, this calculation will end after a few 10^6 timesteps. Estimating 5×10^{10} floating point operations per timestep, this leads to required CPU times per run of 20000 CPU hours in total with 8 processors and 1 GFlops per processor which seems to be achievable with the code.

Simultaneously the formation of collimated jets around white dwarfs and related compact objects will be studied with *NIRVANA2.0*. A global consistent model will be developed which should be in accordance with observational data. The investigation of the possible parameter region needs several simulations with high spatial resolution including all relevant physics. The requirements of the jet formation simulations can only be estimated, as this project is still at its beginning. The numerical resolution should be of the order of 256^2 grid cells – a 2D model should be sufficient preliminarily. The memory requirement is then about 5 GB with cooling. One calculation should end after also $10^6 - 10^7$ timesteps which lead to required CPU times per run of about 8000 h.

Each simulation run will produce some 100 files with a few GB in the 3D case. The total amount of data is then of the order of 500 GB which has to be stored and transfered to local machines for visualization and interpretation.

References

[DoS03] Dopita, M. A., Sutherland, R. S.: Astrophysics of the diffuse universe. Springer, Berlin, New York (2003)

[Kra01] Krause, M.: A 3D Hydrodynamic Simulation for the Cygnus A Jet as a Prototype for High Redshift Radio Galaxies. In: Krause, E., Jäger, W. (ed) High Performance Computing in Science and Engineering 2001. Springer, Berlin, Heidelberg, New York (2001)

[BlP82] Blandford, R. D., Payne, D. G.: Hydromagnetic flows from accretion discs and the production of radio jets. Mon. Not. Roy. Astr. Soc., **199**, 883–903 (1982)

[Cam90] Camenzind, M.: Magnetized Disk-Winds and the Origin of Bipolar Outflows. Rev. Modern Astron., **3**, 234–265 (1990)

[SKC01] Schmid, H. M., Kaufer, A., Camenzind, M., et al.: Spectroscopic monitoring of the jet in the symbiotic star MWC 560. I. Spectroscopic properties, general outflow structure and system parameters. A & A, **377**, 206–240 (2001)

[SuD93] Sutherland, R. S., Dopita, M. A.: Cooling functions for low-density astrophysical plasmas. Ap. J. S., **88**, 253–327 (1993)

[ZiY97] Ziegler, U., Yorke, H.: A nested grid refinement technique for magnetohydrodynamical flows. Comp. Phys. Comm., **101**, 54–74 (1997)

[Zie02] Ziegler, U.: Box simulations of rotating magnetoconvection. Spatiotemporal evolution. A & A, **386**, 331–346 (2002)

[Thi00] Thiele, M.: Numerical simulations of protostellar jets. PhD Thesis, University of Heidelberg (2000)

Large Scale Simulations of Jets in Dense and Magnetised Environments

Martin Krause and Max Camenzind

Landessternwarte Königstuhl 69117 Heidelberg, Germany
M.Krause@lsw.uni-heidelberg.de

We have used the vectorised and parallelised MHD code NIRVANA on the NEC SX-5 in parallel mode to simulate the interaction of jets with a dense environment on a scale of more than 200 jet radii. A maximum performance of 0.75 GFLOP per processor could be reached.

One simulation is axisymmetric and purely hydrodynamic, but with a resolution of 20 points per beam-radius (ppb). The bipolar jet is injected in the center of a spherically symmetric King profile, initially underdense to its environment by a factor of 10,000. As expected from our previous work, the jet starts with producing a spherical bubble around it, bounded by the bow shock. The bubble slowly elongates, first with roughly elliptical shape, and then forms narrower extensions in beam direction. The final aspect ratio of the bow shock is 1.8. We have transformed the results on a 3D-rectangular grid and integrated the emission properties to compare the results with observed central cluster radio galaxies. In the particular case of Cygnus A, we come to convincing consistency, morphologically, regarding the size of the influenced region by the jet, size, and cylindrical shape of the radio cocoon, and source age. This strongly supports our earlier hypothesis on the nature of the jet in Cygnus A, and the derived constraints on other jet parameters like a power of 8×10^{46} erg/s and an age of 27 Myr. But, the simulation also clearly shows the shortcoming of the model: The jet's beam is very unstable, reaching the tip of the bow shock only very seldom. Also, the contact discontinuity between shocked beam plasma and shocked ambient gas is quite disrupted by the action of the Kelvin-Helmholtz-instability. This is not seen in observations, and necessitates the presence of dynamically important magnetic fields or an at least moderately relativistic flow, or both.

The other simulation was designed to explore the impact of a jet on a randomly magnetised environment. A bipolar jet is injected into a King profile on a 3D Cartesian grid, with a resolution of 3 ppb. This simulation was not successful, because the timestep became too low after 0.7 Myr. The preliminary results show no increase of the magnetic fields reversal scale yet, which was expected from an unpublished 2D slab jet simulation, but not that early.

1 Introduction

Several billion years ago, at redshifts in excess of two, the centers of galaxy clusters, typically hosting already a relaxed elliptical galaxy with an old population of stars, were usually equipped with a powerful radio jet. There are many different lines of evidence for this (Carilli et al. 2001). The host galaxies of radio jets are the brightest ones at their redshift. Also, there have been found dozens of line emitting objects around five high redshift radio galaxies, so far (Venemans et al. 2003). This is a significant overdensity. Furthermore, the space density of galaxy clusters at low redshift agrees with that of high redshift radio galaxies. Hence it is clear that radio galaxies pinpoint the most massive structures in the high redshift universe.

Contrary to the situation at high redshift, the brightest cluster galaxies (BCGs) in the local universe are generally associated only with weak radio jets. The only exception being Cygnus A (Fig. 2a). This classical double radio galaxy has an outstanding power, only reached by sources with redshifts in excess of roughly unity. The affected gas surrounding the radio jet can be observed in great detail in the X-ray regime by the Chandra satellite. Therefore, Cygnus A can serve as a model for the interaction of the jet with the intergalactic medium (IGM), for powerful, classical double radio jets. Much work has been done on the propagation of extragalactic jets (compare e.g. Krause & Camenzind 2001; Krause 2003, and references therein). Of particular interest for the present study are simulations that include all the external gas that is affected by action of the jet (Clarke et al. 1997; Reynolds et al. 2001, 2002; Saxton et al. 2002a,b; Krause & Camenzind 2002b,c; Krause 2002a, 2003; Zanni et al. 2003). These simulations did not yet reach the actual size of the jet in Cygnus A, but could be extrapolated to derive a hydrodynamical model of the jet (Krause 2002b; Krause & Camenzind 2002c). In this model, the jet in Cygnus A resides in a stratified galaxy cluster atmosphere, with a central density contrast (jet/IGM) of roughly 10^{-4}. We have tested this model with a large scale simulation that we report in sect. 2. The radio emission of jets has been applied to study the large scale coherence of magnetic fields in galaxy clusters (e.g. Dolag et al. 2002). We propose the idea that the magnetic field is randomly oriented, but with roughly the same plasma β everywhere, before the jet passes. We have already shown, that a slab jet with similar parameters as used here, produces a coherent field structure in the shocked ambient gas (unpublished), with reversal scales of 10 to 50 R_j, when simulated up to a diameter of 400 R_j. In order to check the validity of this result in 3D, we tried to accomplish such a simulation. The results are described in sect. 3.

1.1 Numerics

For the computations in this contribution, the magneto-hydrodynamic (MHD) code *Nirvana_CP* was employed. The main part of this code (*NIRVANA*) was written by Udo Ziegler Ziegler & Yorke (1997). In that version, it solves the

MHD equations in three dimensions (3D) for density ρ, velocity $\mathbf{v}$, internal energy e, and magnetic field $\mathbf{B}$:

$$\frac{\partial \rho}{\partial t} + \nabla \cdot (\rho \mathbf{v}) = 0 \tag{1}$$

$$\frac{\partial \rho \mathbf{v}}{\partial t} + \nabla \cdot (\rho \mathbf{v}\mathbf{v}) = -\nabla p - \rho \nabla \Phi + \frac{1}{4\pi}(\mathbf{B} \cdot \nabla)\mathbf{B} - \frac{1}{8\pi}\nabla \mathbf{B}^2 \tag{2}$$

$$\frac{\partial e}{\partial t} + \nabla \cdot (e\mathbf{v}) = -p\, \nabla \cdot \mathbf{v} \tag{3}$$

$$\frac{\partial \mathbf{B}}{\partial t} = \nabla \times (\mathbf{v} \times \mathbf{B}) \ , \tag{4}$$

where Φ denotes an external gravitational potential.

NIRVANA can be characterised by the following properties:

1. explicit Eulerian time–stepping,
2. operator–splitting formalism for the advection part of the solver,
3. method of characteristics–constraint–transport algorithm to solve the induction equation and to compute the Lorentz forces;
4. artificial viscosity has been included to dissipate high–frequency noise and to allow for shock smearing in case the flow becomes supersonic.

The code was vectorised and parallelised by OpenMP like methods, and successfully tested on the SX-5 (Krause & Camenzind 2002a). All the significant loops could be vectorised. The number crunching part scales without significant performance loss. This is also true for the MHD part of the solver. We show a typical profile output below (Tables 1 and 2), for a run without data output, which indicates an optimum in vectorisation and parallelisation efficiency. The average performance in the 2D simulation was only 434 cumulative MFLOPS with eight processors, probably because in this run, about 500 GB of data had to be dumped to the hard disk, which is a serial process.

2 Simulation of a Very Light Jet to Large Scale

We have performed a bipolar axisymmetric simulation – in the following called run A – of a very light jet in a King type galaxy cluster atmosphere with a

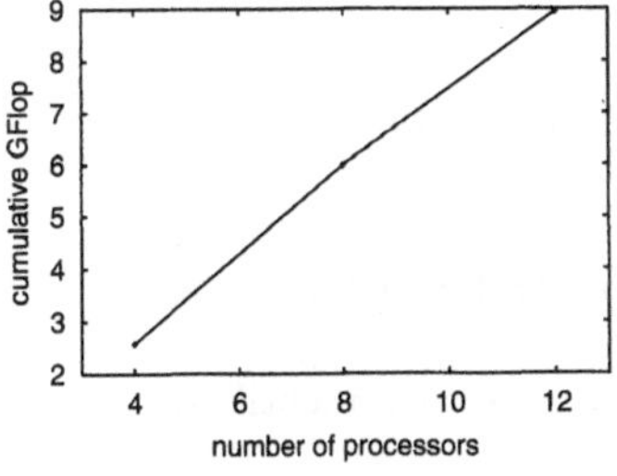

Fig. 1. Performance for the 3D-MHD problem. The vectorised loops contained 512 cells, and the parallelised ones 224 cells. With that parameters, the code scales very good.

Table 1. Typical profile output: Program Information

Real Time (sec)	: 196.546238	User Time (sec)	: 1121.746862
Sys Time (sec)	: 5.748649	Vector Time (sec)	: 941.991913
Inst. Count	: 29681221986	V. Inst. Count	: 12290935628
V. Element Count	:3133399958530	FLOP Count	:851661050792
MOPS	: 2808.824657	MFLOPS	: 759.227487
MOPS (concurrent)	: 33086.532028	MFLOPS (concurrent)	: 8943.315309
A.V. Length	: 254.935837	V. Op. Ratio (%)	: 99.448066
Memory Size (MB)	: 3216.000000	Max Concurrent Proc.	: 12
Conc. Time(>=1) (sec)	: 95.228785	Conc. Time(>=2) (sec)	: 94.162890
Conc. Time(>=3) (sec)	: 94.106573	Conc. Time(>=4) (sec)	: 94.077965
Conc. Time(>=5) (sec)	: 94.062470	Conc. Time(>=6) (sec)	: 94.044669
Conc. Time(>=7) (sec)	: 94.026083	Conc. Time(>=8) (sec)	: 94.009968
Conc. Time(>=9) (sec)	: 93.985855	Conc. Time(>=10)(sec)	: 93.883754
Conc. Time(>=11)(sec)	: 93.346376	Conc. Time(>=12)(sec)	: 86.836165
Event Busy Count	: 0	Event Wait (sec)	: 0.000000
Lock Busy Count	: 1223	Lock Wait (sec)	: 0.945273
Barrier Busy Count	: 0	Barrier Wait (sec)	: 0.000000
MIPS	: 26.459822	MIPS (concurrent)	: 311.683300
I-Cache (sec)	: 1.106138	O-Cache (sec)	: 33.629114
Bank (sec)	: 23.186143		:

Table 2. Typical profile output: Multitasking Information

	Seconds				Seconds			Thread/Macro[tid]
%Res.	Res.	T/M	Micro	%CPU	CPU	CPUcum.	Wait	-micro[n]
0.0	93.66	0.04		0.0	0.04	0.04	0.01	Root1
100.0	-	-	93.62	8.4	93.62	93.66	0.77	-micro1
99.5	-	-	93.15	8.3	93.15	186.81	0.28	-micro2
99.9	-	-	93.61	8.4	93.61	280.42	0.00	-micro3
99.5	-	-	93.16	8.3	93.16	373.57	0.32	-micro4
99.4	-	-	93.11	8.3	93.11	466.68	0.22	-micro5
99.4	-	-	93.14	8.3	93.14	559.82	0.21	-micro6
99.3	-	-	93.02	8.3	93.02	652.84	0.30	-micro7
99.4	-	-	93.13	8.3	93.13	745.97	0.31	-micro8
99.6	-	-	93.26	8.3	93.26	839.22	0.27	-micro9
99.5	-	-	93.16	8.3	93.16	932.38	0.32	-micro10
99.2	-	-	92.93	8.3	92.93	1025.31	0.26	-micro11
99.5	-	-	93.18	8.3	93.18	1118.49	0.30	-micro12

%Res.	:The residence ratio.
Res.	:The residence time.
T/M	:The CPU time of thread/macrotask.
Micro	:The CPU time of microtask.
%CPU	:The CPU ratio of thread/macrotask or microtask.
CPU	:The CPU time of thread/macrotask or microtask.
CPUcum.	:The cumulative CPU time.
Thread/Macro[tid]	:The thread/macrotask and microtask identifier.

final jet size of > 200 jet radii. The simulation was run for 6200 CPU hours on eight processors of the SX-5 at the HLRS.

2.1 Simulation Setup

The jet was injected in both directions (bipolar) in the center of a cylindrical grid with 8300 and 2000 points in axial (Z) and radial (R) direction, respectively. The basic length scale of the problem is the jet radius which is represented by 20 points. With that resolution global parameters like the bow shock velocity on the Z-axis or energy and momentum conservation are accurate to $\approx 10\%$ (Krause & Camenzind 2001). As our unit of length we choose the observed jet radius in Cygnus A, i.e. the jet radius is set to $R_\mathrm{j} = 0.5$ kpc. The total grid size is therefore $[207.5 \times 50]$ kpc. The environmental density (ρ_e) is given by an isothermal King profile:

$$\rho_\mathrm{e}(R,Z) = \rho_{\mathrm{e},0}\left(1+\frac{r^2}{a^2}\right)^{-3\beta/2} , \tag{5}$$

where $r^2 = R^2 + Z^2$. This means that the density is constant up to a core radius a that was set to $a = 10$ kpc. Then it starts to decrease, asymptotically reaching $\rho_\mathrm{e} \propto r^{-3\beta}$ ($\beta = 3/4$). The central density was set to $\rho_\mathrm{e} = m_\mathrm{H}\mathrm{cm}^{-3}$, with the hydrogen mass $m_\mathrm{H} = 1.67 \times 10^{-24}$ g. The temperature was set to $T = 30$ Mio. K. This atmosphere is kept in hydrostatic equilibrium by the gravity of a dark matter halo:

$$\Phi = \frac{3\beta kT}{2\mu m_\mathrm{H}} \log\left(1+\frac{r^2}{a^2}\right) . \tag{6}$$

μ is the number of particles per proton mass. Here, $\mu = 0.5$ for an ionised medium. In order to brake the symmetry, density perturbations were included, i.e. with 10% probability, the density in a cell was increased by a random factor between 0 and 40%. The jet density was set to $\rho_\mathrm{j} = \eta_0\rho_\mathrm{j}$, with the density contrast $\eta_0 = 10^{-4}$. The sound speed in the jet was set to $20\%c$, c being the the speed of light, and the jet's Mach number to $M = 3$.

The simulation was run for 20 Myr, in total. During that time, the jet reached an extention of 110 kpc, i.e. 220 jet radii, on the axis.

2.2 Results

We present logarithmic density plots of the simulation results for four different simulation times (5, 10, 15, 20 Myr) in Fig. 3. The morphology that appears in these figures is a continuation of previous simulations that could not reach the size shown here. Fig. 3a shows the state that was reached by Krause (2003), and extensively discussed therein. In this early phase, the bow shock is spherical, its radius following an expansion law given by the force balance

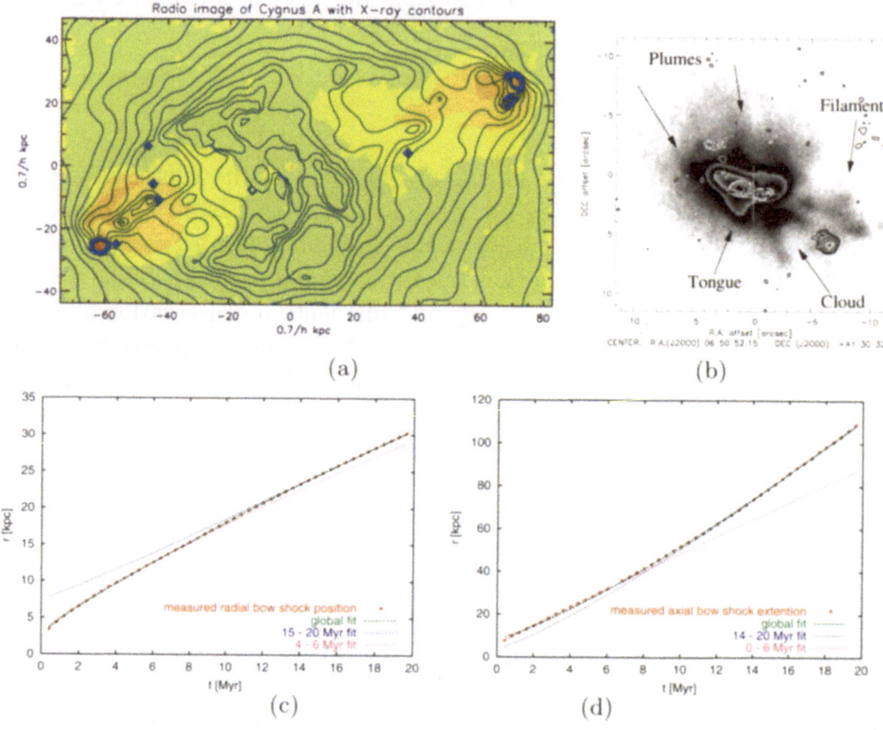

Fig. 2. **(a)**The nearby radio galaxy Cygnus A. Colours show the logarithmic intensity of the 5 GHz radio image (VLA, credits: NRAO / AUI / NSF). Contours show the adaptively smoothed X-ray emission (Chandra satellite, credits: NASA / UMD / A.Wilson et al., courtesy: P. Strub). The axis shows the length scale at the luminosity distance of Cygnus A (246 $(h/0.7)$ Mpc), where h denotes the Hubble constant in units of 100 km/s/Mpc. The jet beam is created at the origin of the coordinate system where the active center of the host galaxy is located. Two barely visible, narrow beams emerge from there in opposite directions, powering the hotspots at (-60,-25) and (70,30). The beam plasma then assembles in a cylindrical cocoon, at lower radio surface brightness. Lower frequency images show that the cocoon continues through the empty region in the center. It removes and compresses the IGM, shaping it elliptically. **(b)**The radio galaxy 4C 41.17 at redshift 3.8, corresponding to a lookback time of 12 billion years. The gray scale shows the Lyman α emission line nebula, the contours represent the 5 GHz radio emission. The white cross indicates the radio core, i.e. the region where usually the active center of the galaxy is located. Adopted from astro-ph: 0303637. Courtesy: Wil van Breugel. **(c)**Bow shock position at Z=0 versus time for the simulation in sect. 2. The three fits are: $2.57017+2.13717t^{0.859365}$ (global fit),$7.34264+0.960744t^{1.06464}$ (15-20 Myr), and $2.59878+2.27161t^{0.821546}$ (4-6 Myr). **(d)**Bow shock extention on the axis versus time for the simulation in sect. 2. The three fits are: $8.39571+2.51264t^{1.23664}$ (global fit),$3.73811+2.92394t^{1.20278}$ (14-20 Myr), and $5.39429+4.71728t^{0.956775}$ (4-6 Myr).

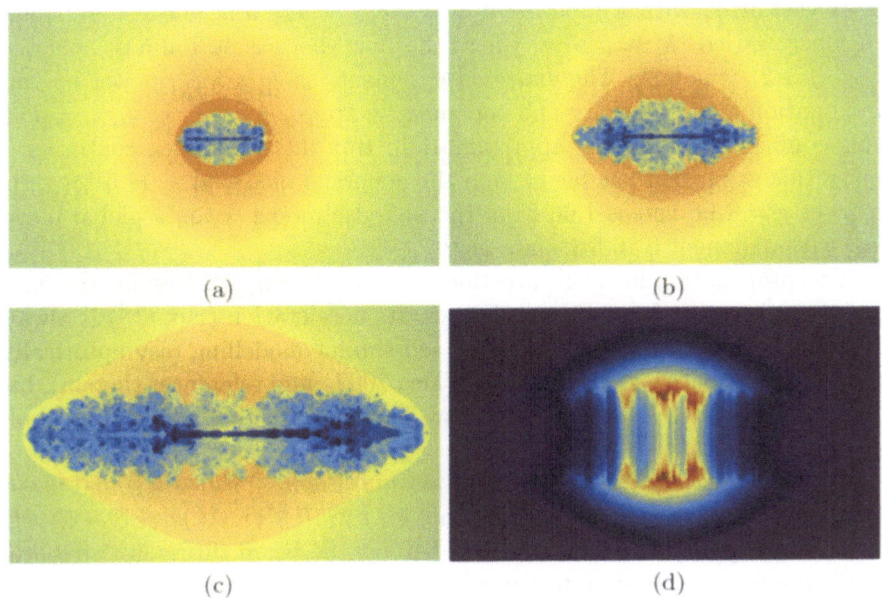

Fig. 3. a-c: Three snapshots of run A. The logarithm of the density is shown, dark blue indicates the lowest values, red the highest ones. The times for the snapshots are 5,10,20 Mio. years, for figure a,b, and c, respectively. The same part of the grid is shown in each case. The jet forms first a spherical bow shock, associated with a spherical cocoon. The cocoon then transforms via a conical state to a cylindrical one. The bow shocks aspect ratio (length/width) also grows, up to 1.8 in c.
d: Ray-traced synthetic bremsstrahlung image of (c).

equation which can be integrated to yield, for arbitrary mass distribution $\mathcal{M}(r)$ and energy injection $E(t)$:

$$\int_0^r \mathcal{M}(r_1) r_1 dr_1 = 2 \int_0^t dt_1 \int_0^{t_1} E(t_2) dt_2. \tag{7}$$

For the given matter profile (5), (7) can be integrated numerically. We only discuss the asymptotic power law parts here. For a power law density distribution ($\rho = \rho_0 (r/r_0)^\kappa$) and constant energy injection ($E = Lt$), the solution is:

$$r = \sqrt[\kappa+5]{\frac{(\kappa+3)(\kappa+5) r_0^\kappa L t^3}{12\pi\rho_0}} \tag{8}$$

The density profile used here has the asymptotic power law approximations: $\lim_{r \mapsto 0}(\rho) = \rho_0$, and $\lim_{r \mapsto \infty}(\rho) \propto r^{-9/4}$. Therefore, the bow shock should expand with $r \propto t^{0.6}$ in the beginning, steepening towards $r \propto t^{1.09}$, at least as long as it remains spherical. The radial bow shock position was determined every 0.4 Myr (compare Fig 2c).

It was fitted with a function $r = a + bt^c$, where a,b, and c were simultaneously varied. At $\approx$ 5 Myr, where the bow shock is still almost spherical (aspect 1.2), c is 0.82. The expected exponent, using a local power law approximation for the density and the analytic approximation, is: $c_{\rm theo} = 0.79$. This confirms the analytic approximation. But the bow shock continues to follow this expansion law far beyond the spherical phase: at $\approx$ 17.6 Myr, the fit gives $c = 1.06$ versus 1.00 from the analytic model. Even a global power law, with exponent 0.86 fits quite good.

The propagation in axial direction is shown in Fig 2d. Besides the early times, the bow shock always accelerates. Its acceleration ($v \propto t^{0.2}$) is always significantly higher than predicted by self-similar modelling (asymptotically: $v \propto t^{0.09}$) (compare e.g. Carvalho & O'Dea 2002, and references therein), but does not reach the prediction for narrow beams of constant radius that would be super-exponential for $\kappa < -2$.

This reflects the state of the beam plasma. Figure 3 shows a sound beam, reaching to the tip of the bow shock, only for $t < 10$ Myr. At later phases , the beam is disrupted towards the head. Thereby, the beam thrust is distributed over a greater area, leading to lower bow shock velocity compared to the case of a narrow beam with constant radius.

The cocoon transforms gradually from conical to cylindrical at later times. Its width does not grow much for the second half of the simulation. Towards the center, there is a turbulent region, where the ambient gas is entrained into the cocoon by the action of gravity, and mixed with the shocked beam plasma.

2.3 Comparison to Observations

The prime target to compare the simulation data to is the X-ray data of Cygnus A (compare Fig. 2a). The simulation was run to almost the size of the real source (110 kpc and 140 kpc, for simulation and observation, respectively). It is evident from Fig. 3 that this is necessary in order to get the correct cocoon morphology. Many other details can be understood by this new simulation: Fig. 2a shows that the radio cocoon is bordered by the brightest X-ray emission. We show the line-of-sight integrated X-ray emission for the simulation together in Fig. 3d. These can be compared directly to the observations in Fig. 2. The simulation clearly shows the peaks in emission next to the radio cocoon. This was already predicted analytically, in self-similar models (Alexander 2002). Outwards of this peak, the emissivity is increased with respect to the undisturbed cluster emission, by a constant factor in the plane of symmetry. The emission falls suddenly at the bow shock. This detail cannot be seen in the observational data, because there are not enough photons (but compare Fig. 4 in Smith et al. (2002)). Fig. 3d shows that the emission of the shocked ambient gas has been shaped by the bow shock in an elliptical way. Smith et al. (2002) have shown that the elliptical isophotal fit is better than the spherical fit inside a radius of $66/(h/0.7)$ kpc, which should indicate the bow shock's radial position in Cygnus A.

The simulation shows significant emission in the mixing region in the central parts of the cocoon. However, this is less than observed. The reason is numerical mixing with cocoon gas in that region. Also, the observation shows dominantly non-axisymmetric modes, which cannot be represented in our axisymmetric simulation. We note that this region has been found to be non-axisymmetric in 3D simulation (Krause & Camenzind 2002a).

It has been suggested that repeated jet episodes could heat the surrounding gas to the observed X-ray temperatures during cosmological timescales. A lower limit to the heat injection in the cluster gas was derived by Krause (2003), equ. (21). For the simulation parameters here, this lower limit amounts to 3.19×10^{58} erg. The cluster gas was followed by a passive tracer that was set to unity. Counting only cells where the tracer variable is above 0.1, we measure 8.17×10^{59} erg internal energy injected by the jet into the cluster gas by the end of the simulation. The cluster gas has also gained 2.37×10^{60} erg of potential energy. In total, 69% of the energy injected by the jet has been put into the ambient gas. This is enough to power the X-ray emission of the cluster in Cygnus A for ≈500 Mio. years. This makes it entirely plausible that the cluster gas is heated by that mechanism.

We have proposed to measure jet parameters based on the radial bow shock propagation. The radial bow shock velocity in Cygnus A can be constrained from the shock temperature measured by Chandra (Krause & Camenzind 2002c). This procedure gives a jet power of $L = 8 \times 10^{46}$ erg/s and an age of $t = 27$ Myr for Cygnus A. In order to demonstrate the validity of the procedure, we determine the jet parameters for the simulation in the same way. The simplest approximation for the external density is taking it to be constant. Then, from (8) it follows with the bow shock velocity of 1220 km/s, the radial bow shock position 30.17 kpc and the external density of 10^{-25} g/cm^3: $L = 9.96 \times 10^{45}$ erg/s and $t = 19.7$ Myr. This is to be compared to the true jet power, $L_{\mathrm{true}} = \pi R_{\mathrm{j}}^2 \rho_{\mathrm{j}} v_{\mathrm{j}}^3 = 7.72 \times 10^{45}$ erg/s and the true jet age of 20 Myr. This very good agreement demonstrates that the exact shape of the cluster atmosphere is not critical in determining jet parameters. The agreement can even be improved by taking better approximations to the density profile.

The simulation result shows an important difference to the observation: In the simulation, the beam is very unstable, reaching the tip of the bow shock not even once, after 10 Myr. The reason for this is the low Mach number in the beam that is a consequence of the strong interaction with the cocoon and the entrained shocked ambient gas therein. Higher Mach numbers can only be reached by a relativistic jet. However, the beam could also be stabilised by an appropriate, significant magnetic field. The magnetic field is also demanded in order to preserve the contact discontinuity near the tip of the bow shock from the action of the Kelvin-Helmholtz-instability, because the strong disruption found in the simulation can not be found in the radio data. This result is in good agreement with with magnetic field determinations in Cygnus A's hot spots via the self-synchrotron-Compton assumption (Wilson et al. 2000) and our earlier suggestion, based on a jet power argument, that the jet's mean Lorentz factor is $\Gamma \approx 20$.

2.4 Comparison to High Redshift Radio Galaxies

So far, high redshift radio galaxies have not been observed long enough in order to study the cluster gas emission in great detail. But the region where the shocked ambient gas could be expected is typically bright in emission lines. The emission line regions often have the same cone shaped structure as the X-ray emission in Cygnus A (compare Fig 2a and 2b). The line emission is brightest in the region, corresponding to the mixing region in the simulation. This could be interpreted in the way that in these objects the line emission is caused by material that was entrained into the radio cocoon.[1] This gas is subject to the combined thermal and Rayleigh-Taylor instability, which may cool some gas to the appropriate temperatures (Basson 2002; Krause 2002b).

3 A Jet in a Randomly Magnetised Environment

We have accomplished a 3D simulation of a jet in a randomly magnetised environment. The simulation run for 500 CPU hours on eight processors of the SX-5. Unfortunately, the timestep became too low, before significant evolution of the jet could be seen.

3.1 Numerical Setup

We injected the jet in the center of a Cartesian grid ($[X \times Y \times Z] = [170 \times 74 \times 74]$kpc $= [512 \times 224 \times 224]$) in the X direction. The jet radius was set to$R_\mathrm{j} = 1$ kpc and resolved with 3 points. The magnetic field at the jet inlet was set to zero for the Y and the Z direction, and $\partial B_\mathrm{X}/\partial x = 0$ on the jet nozzle. In the surrounding King atmosphere ($\rho_{\mathrm{e},0} = m_\mathrm{p}/10$, $a = 10$ kpc , and $\beta = 2/3$, $T = 3 \times 10^7$ K), a random magnetic field was established. The vector potential was randomly determined, folded with the King distribution in order to get an average plasma β of $\beta = 8\pi p/B^2 = 12$, i.e. sub-equipartition fields. The initial density contrast was set to $\eta = 10^{-3}$ and the Mach number to $M = 5$.

[1] The alternative interpretation is that cooling is significant in the whole shocked ambient gas region. This would lead to very large cocoon width and turbulent mixing of radio plasma and emission line gas (Krause 2002a). Most recent X-ray data for 4C 41.17 (Scharf et al. 2003), showing probably inverse Compton cocoon emission, point in this second direction for that particular object.

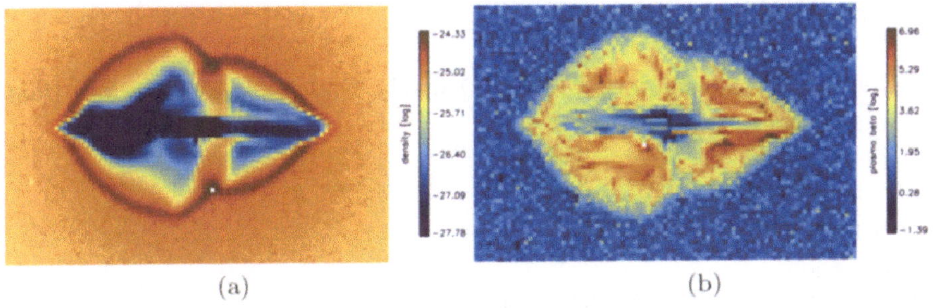

(a) (b)

Fig. 4. Results of the 3D simulation with randomly magnetised environment at $t = 0.7$ Myr. a: density slice at Z=0, b: plasma beta at Z=0.

3.2 Results and Discussion

A slice of the density distribution and the plasma β is shown in Fig. 4. In that phase, the bow shock is already weak. Weak MHD shocks reduce the field strength and refract the field towards the shock normal. The simulation shows a decrease of the magnetic field of typically an order of magnitude, behind the bow shock. yet. Unfortunately the simulation could not be run for longer time, since the timestep became too low. In that phase, no amplification of the reversal scale could be found, as in a slab simulation (compare sect. 1). This probably happens only when the shocked ambient gas expands to several times the cocoon diameter, like e.g. in Fig. 3d. If this effect could be confirmed, it would be a good candidate for the explanation of the high reversal scale in galaxy clusters, i.e. the coherence of the field is created by the weak bow shock.

4 Conclusions

Employing the NEC SX-5 supercomputer at the HLRS, we could simulate an axisymmetric jet to the large extention of more than 200 jet radii. On that scale, the results can be directly compared to observational data of Cygnus A. Many details, like elliptically shaped emission, morphology of cocoon and shocked ambient gas, or low aspect, are well reproduced. This confirms the hypothesis that the jet is underdense with respect to its environment by a factor of $\approx 10,000$, and the consequences on the jet parameters discussed above. We also find differences to the observational data, concerning stability of the beam and the contact discontinuity, which we ascribe to the missing magnetic fields and the disregard of relativistic physics. We speculate that emission line halos of high redshift radio galaxies might be identified with the centrally concentrated mixing region in the cocoon.

We also tried to simulate the influence of the jet on a randomly magnetised environment. Unfortunately, the timestep became too low, and the simulation had to be stopped, before significant progress has been made.

Acknowledgements

This work was also supported by the Deutsche Forschungsgemeinschaft (Sonderforschungsbereich 439).

References

Alexander, P. 2002, MNRAS, 335, 610

Basson, J. F. 2002, Ph.D. Thesis, University of Cambridge

Carilli, C. L., Miley, G., Röttgering, H. J. A., Kurk, J., & et al. 2001, in Gas and Galaxy Evolution, ASP Conference Proceedings, Vol. 240. eds.: John E. Hibbard, Michael Rupen, and Jacqueline H. van Gorkom

Carvalho, J. C. & O'Dea, C. P. 2002, ApJ Supplement, 141, 337

Clarke, D. A., Harris, D. E., & Carilli, C. L. 1997, MNRAS, 284, 981

Dolag, K., Bartelmann, M., & Lesch, H. 2002, A&A, 387, 383

Krause, M. 2002a, A&A, 386, L1

—. 2002b, Ph.D. Thesis, Universität Heidelberg

—. 2003, A&A, 398, 113

Krause, M. & Camenzind, M. 2001, A&A, 380, 789

Krause, M. & Camenzind, M. 2002a, in High Performance Computing in Science and Engeneering '01, eds.: Krause, E. and Jäger, W., Springer, 329+

Krause, M. & Camenzind, M. 2002b, in Active Galactic Nuclei: from Central Engine to Host Galaxy, meeting held in Meudon, France, July 23-27, 2002, Eds.: S. Collin, F. Combes and I. Shlosman. To be published in ASP Conference Series, p. 42.

Krause, M. & Camenzind, M. 2002c, in The Physics of Relativistic Jets in the CHANDRA and XMM Era , meeting held in Bologna, Italy, September 23-27, 2002, Eds.: G. Brunetti, D.E. Harris, R.M. Sambruna and G. Setti. To be published in New Astronomy Review

Reynolds, C. S., Heinz, S., & Begelman, M. C. 2001, ApJ Letter, 549, L179

—. 2002, MNRAS, 332, 271+

Saxton, C. J., Bicknell, G. V., & Sutherland, R. S. 2002a, ApJ, 579, 176

Saxton, C. J., Sutherland, R. S., Bicknell, G. V., Blanchet, G. F., & Wagner, S. J. 2002b, A&A, 393, 765

Scharf, C. A., Smail, I., Ivison, R., Bower, R. G., van Breugel, W., & Reuland, M. 2003, ArXiv Astrophysics e-prints astro-ph/0306314

Smith, D. A., Wilson, A. S., Arnaud, K. A., Terashima, Y., & Young, A. J. 2002, ApJ, 565

Venemans, B., Miley, G., Kurk, J., Rottgering, H., & Pentericci, L. 2003, The Messenger, 111, 36

Wilson, A. S., Young, A. J., & Shopbell, P. L. 2000, ApJ Letter, 544, L27

Zanni, C., Bodo, G., Rossi, S., Massaglia, S., Durbala, A., & Ferrari, A. 2003, A&A, in press

Ziegler, U. & Yorke, H. W. 1997, Computer Physics Communications, 101, 54

Crack Propagation in Icosahedral Model Quasicrystals

Christoph Rudhart, Frohmut Rösch, Franz Gähler, Johannes Roth, and Hans-Rainer Trebin

Institut für Theoretische und Angewandte Physik, Universität Stuttgart, D-70550 Stuttgart, Germany

Summary. Propagation of mode-I cracks in a three-dimensional model quasicrystal is studied by molecular dynamics simulations. The samples are endowed with an atomically sharp crack and subsequently loaded by linear scaling of the displacement field. The response of the system is then monitored during the simulation. In particular, the crack surface morphology is investigated in dependence of the orientation of the fracture plane. For this purpose, fracture surfaces perpendicular to two- and fivefold axes are compared. For both directions, brittle fracture with rough fracture surfaces is observed.

1 Introduction

Quasicrystals are intermetallic alloys whose diffraction patterns display sharp Bragg peaks with non-crystallographic point symmetries. Therefore their mass density is quasiperiodic rather than periodic. Most concepts used to predict the response of a material to an applied load are (at least on an atomic scale) based on the periodicity of the underlying structures, and thus do not apply to quasicrystals.

Although it is possible to grow single quasicrystals of centimeter size, experiments on crack propagation in single quasicrystals are rare. Most of the available experiments are indentation tests where the fracture toughness is estimated from the geometry of the indentations, the applied force and the length of the microcracks that are emitted from the corners [1]. The values for the fracture toughnesses are about 1 MPa $m^{1/2}$ [1, 2], which is close to that obtained for brittle ceramics or silicon.

Cracks in the vicinity of microhardness indentations are observed to propagate predominantly along well defined crystallographic planes [1], suggesting that crack propagation in quasicrystals is, as in periodic crystals, influenced by the plane structure. On the other hand, investigations of cleavage surfaces by scanning tunneling microscopy show that the morphology of fracture surfaces is strongly influenced by the cluster-based structure of quasicrystals [3].

In previous studies [4, 5, 6, 7] we have performed numerical simulations of crack propagation in decagonal model quasicrystals. Decagonal quasicrystals show quasiperiodic order in two dimensions. They consist of quasiperiodically ordered planes which are arranged periodically in the third dimension. For decagonal systems it is thus possible to use simple two-dimensional models to investigate the characteristic features and elementary processes which dominate the fracture of real decagonal quasicrystals.

Icosahedral quasicrystals, however, show quasiperiodic order in three dimensions, which cannot be reduced to simple two-dimensional model systems. In this article, we report on large scale molecular dynamics simulations of crack propagation in three-dimensional icosahedral model quasicrystals.

2 Simulation Method

2.1 Model

The simulations are carried out for a three-dimensional binary model quasicrystal proposed by Henley and Elser [8] as a structure model for icosahedral $(Al,Zn)_{63}Mg_{37}$. As we do not distinguish between Al and Zn, the model consists of two types of atoms, larger ones that represent Mg and smaller ones that represent Al or Zn atoms. The atomic interactions are modelled by Lennard-Jones (LJ) potentials [9], originally derived for the van der Waals type interaction of inert gases. The depths of the LJ potentials are ϵ_0 and $2\epsilon_0$ for atoms of the same and different types, respectively. As unit of length we use the nearest neighbour distance r_0 of two small atoms in the structure. All masses are set to unity, and the time is measured in units of $t_0 = r_0\sqrt{m/\epsilon_0}$. This is thus a very simplistic model quasicrystal, but it nevertheless should produce the correct qualitative behaviour of crack propagation in close-packed quasicrystals like icosahedral $(Al,Zn)_{63}Mg_{37}$.

2.2 Method

Since we are interested in the morphology of fracture surfaces we use a geometry that allows us to follow the dynamics of the running crack for a long time. For this purpose, a strip geometry is used to model crack propagation with constant energy release rate [10]. The samples consist of about 4 million atoms, with dimensions of approximately $450r_0 \times 150r_0 \times 60r_0$. Periodic boundary conditions are applied in the direction parallel to the crack front. For the remaining directions, all atoms in the outermost boundary layer of width $2.5r_0$ remain fixed during the simulation. The strip is homogeneously strained perpendicular to the fracture plane, and an atomically sharp crack is inserted from one short side, to about one quarter of the strip length. Subsequently the sample is relaxed to obtain the displacement field of the stable crack at zero temperature. The strip is chosen long enough to ensure that the crack does not feel the boundary conditions at the two ends of the strip, so that the dynamics is independent of the crack tip position. The crack thus feels a constant driving force and propagates at constant energy release rate.

The system is initially strained to the Griffith load where the energy release rate G is equal to the surface energy of the two crack surfaces, 2γ. In this work, we will concentrate on exploring brittle fracture without thermal fluctuations. Thus we set the initial temperature to 10^{-4} of the melting temperature T_m, which is close to zero temperature. Afterwards the crack is loaded by adding a fraction of the displacement field to the stable crack. The answer of the system is followed by molecular dynamics simulations. The overloads are given by ΔK^* in the following, which is the relative fraction of the stress intensity factor due to the displacement field that is added to the stable crack.

According to the Griffith criterion, planes with low surface energy are potential fracture planes. To identify those planes we relax a specimen and split it into two regions. Subsequently, the two parts are shifted rigidly by a distance of $10r_0$ perpendicular to the cutting plane. The surface energy is then calculated from the difference of the artificially cleaved and the undisturbed specimen.

For simple crystal structures like the face centered cubic structure of noble metals the surface energy only depends on the crystallographic orientation of the surface. In quasicrystals, however, it even varies for crystallographically equivalent but structurally distinct surfaces. Fig. 1 shows the surface energy for three different orientations as a function of the position of the cutting

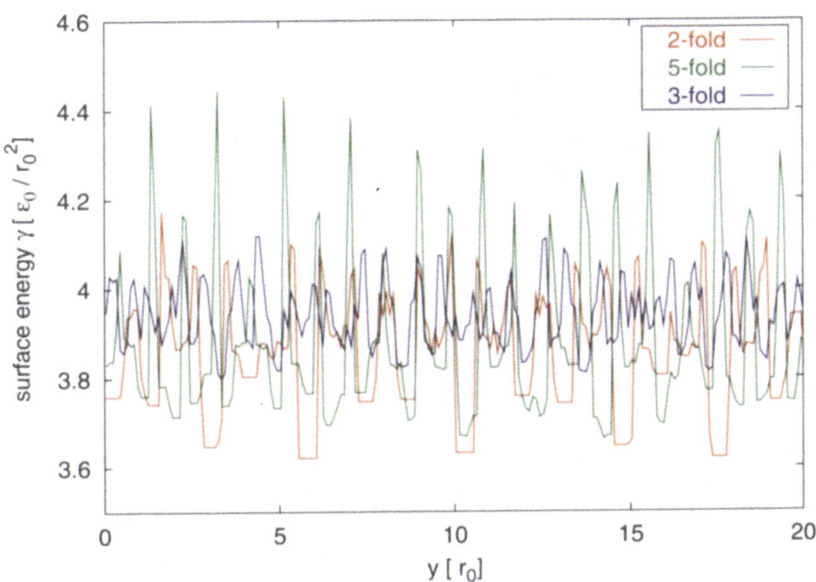

Fig. 1. Surface energy in dependence of the position of the cutting plane, for plane orientations perpendicular to two-, three- and fivefold directions.

plane. We find a pronounced plane structure of low and high surface energies along twofold directions. The planes of lowest surface energy occur with two separations, forming a Fibonacci chain. Along the fivefold direction the plane structure is less pronounced, but we still find planes of low surface energy, while for the threefold direction there is no such distinct plane structure. For our simulations we select as initial fracture planes surfaces of lowest energy perpendicular to two- and fivefold directions.

All molecular dynamics simulations were done with our own code IMD [11, 12], which performs well on a large variety of hardware, including single and dual processor workstations and massively parallel supercomputers.

2.3 Visualization

There are two essentially different types of data that can be used for the visualization of a molecular dynamics simulation. The first possibility is to compute the distribution of certain scalar quantities like the kinetic energy density, which is evaluated on a regular grid and then displayed with a volume renderer. Fig. 2 shows such a volume data set of the kinetic energy. Regions of low intensity are rendered with high transparency. Sound waves emitted by the propagating crack are clearly visible. In most cases, however, volume data are often too homogenous and show little contrast, so that not much can be seen.

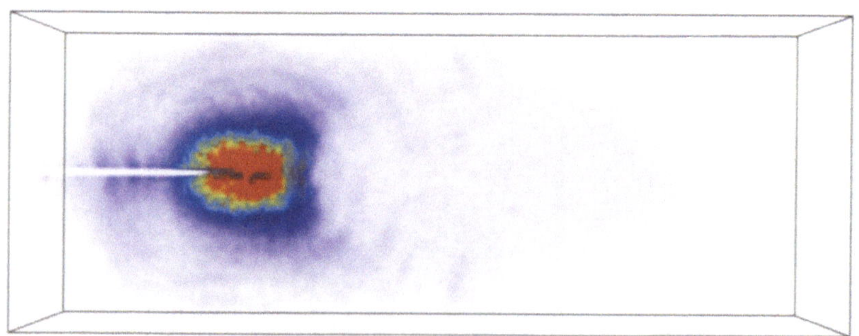

Fig. 2. Kinetic energy density, displayed with the Virvo volume renderer [13]. Sound waves emitted by the propagating crack are clearly visible.

Volume data sets represent, by their very nature, a continuous distribution of some locally averaged quantity. For this reason, they are not suited for the elucidation of microscopic processes. To study crack propagation on an atomistic level, it is necessary to render selected atoms only. Due to the large number of atoms required for the study of crack propagation in three dimensional systems, the selection and reduction of data is of crucial importance. It is not feasible to always write out the positions of all atoms, and even less to display them all. One simply would not recognize any useful information. In periodic crystals, defects can be visualized by plotting only those atoms whose potential energy exceeds a certain threshold [14]. In quasicrystals, however, atoms may have largely varying local environments. Their potential energy thus varies significantly from atom to atom, even for atoms of the same type in a defect-free sample. Defects can therefore not be visualized by applying a simple energy cut-off as in periodic systems.

A more promising method is to display only those atoms whose coordination number is smaller than a certain threshold. The coordination number is evaluated by counting the number of atoms within a certain distance. This cut-off distance is configurable and may depend on the type of the atoms involved. Like the potential energy, the coordination number varies from atom to atom, but to a much smaller degree. In a perfect sample, it is 12 or 13 for the small atoms, and ranges from 14 to 16 for the large atoms. Fortunately, atoms near a defect have a significantly lower coordination number, so that it is possible to visualize fracture surfaces and dislocation cores by displaying only atoms below a suitable threshold in the coordination number. Fig. 3 shows a snapshot of a simulation with 4 million atoms, where atoms are displayed if their coordination number is less than 12 for small atoms, and less

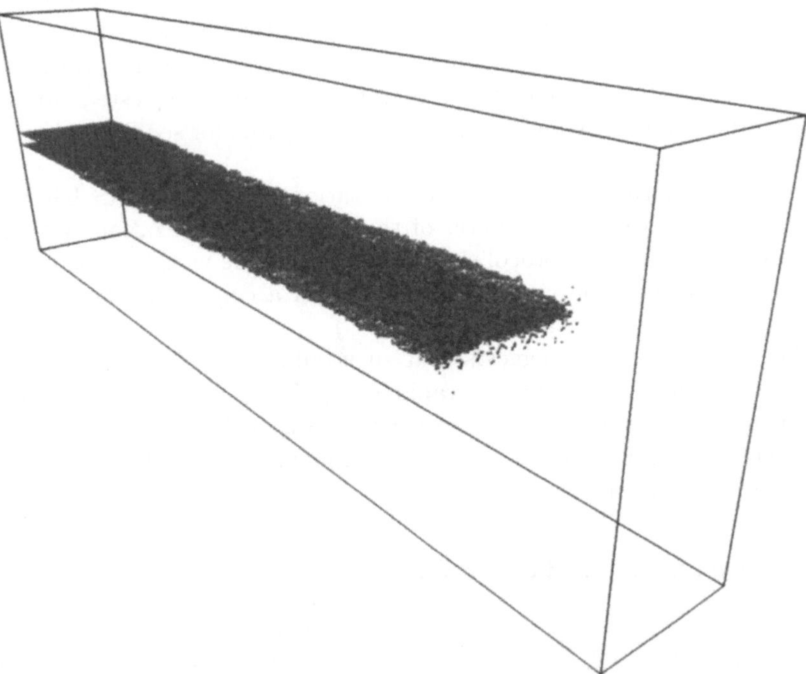

Fig. 3. Snapshot of a simulation with 4 million atoms. Only atoms with low coordination number are displayed.

than 14 for large atoms. With this method, the number of atoms to write to the output files could be reduced by three orders of magnitude, which allows to take more frequent snapshots instead.

2.4 Online Visualization

In addition to the offline visualization of data written to files during the simulation, an online visualization interface has been developed. A visualization application can request data from the simulation via a socket connection. By the same mechanism, it is also possible to interactively steer the simulation by sending requests to change certain simulation parameters. For online visualization, it is even more important to carefully select the data before it is sent to the visualization. Otherwise, the communication overhead and rendering time could become overwhelming. Currently supported are requests for changing certain simulation parameters, and request for sending selected atoms, along with some of their properties. The selection criteria for those atoms, and the atom properties that shall be sent with them, can be chosen interactively. Atoms can be selected by requesting that their position is inside some rectangular box, and that their type, kinetic energy, potential energy, and number of neighbours are within some chosen interval. All these selection criteria can be switched on and off. The data sent with each selected atom can include the type, the position, the velocities, the kinetic and the potential energy, and the number of neighbours. Each of these quantities can be selected or deselected. This allows to keep the amount of requested data small. The interface can easily be extended, e.g. by requests for sending volume data instead of atom data.

Jürgen Schulze-Döbold of the visualization group at HLRS has written a plugin for the COVER renderer of the COVISE visualization system [15], which implements the protocol sketched above on the visualization side. This plugin can be used for visualizations both on a computer screen and in a virtual reality environment like a CAVE.

The first experience suggests that such online visualization facilities can be a very valuable tool for the rapid determination of suitable simulation parameters. The visualization of large systems in real time requires, however, considerable computational power on the simulation side, which is not always available interactively.

2.5 Performance and Load Balancing

IMD is known to perform very well on PC processors [16]. On a 2-processor Athlon MP1900+ PC, a typical crack simulation run with 4 million atoms takes about 48 hours per 10'000 time steps. Often, some 50'000 to 60'000 time steps are necessary, so that such a simulation takes 10 to 12 days. Since we have several such machines available, a reasonable throughput can be obtained for parameter studies, but a turnaround time of almost two weeks is not optimal.

As IMD scales very well up to large CPU numbers [16], the solution seems to be massively parallel processing. The only available machine with very many CPUs was the Cray T3E. Its processors are about 8 times slower than the AMD processors on the PCs, however. With 96 T3E CPUs, a simulation run takes some 8 hours per 10'000 time steps. The problem here is, that the maximal time limit of the queues is 12 hours, so that a simulation of 60'000 time steps has to be split into 4 consecutive runs, where the later runs require the output of the previous ones as input. In principle, it is possible to start the later runs automatically from the previous ones, but such a scheme is complicated and error prone. Starting the later runs by user intervention means a lot of work and reduces the turnaround time considerably. The entire simulation would fit into a single run only if about three quarters of the machine could be used, but so many processors are usually not available. With only 96 or 128 CPUs, the restarting of the jobs is sufficiently inconvenient, that most of the simulations have actually been performed on our 2-CPU PCs. The T3E has mainly been used for the shorter relaxation runs which fit into the 12 hour queue limit. It would be highly desirable to reduce the turnaround time considerably, but this seems possible only with a more powerful machine and/or larger queue time limits. In a sense, the problems we want to study have become too large for the T3E.

If a large number of CPUs is used for a crack simulation, the load balancing problem deserves particular attention. IMD uses a geometric domain decomposition scheme for the work sharing [11]. More than 95% of the time is spent in the force computation, which depends on the number of neighbours of an atom. In a crack simulation, fixed boundary conditions are used in two directions, and periodic boundary conditions in the third direction. Atoms near fixed boundaries have less neighbours, so that the corresponding CPUs have less work to do. A similar effect occurs in the middle of the sample along the crack. On the other hand, by the widening of the crack some atoms are moved to the boundary CPUs, which means more work for them. In practice, it is very hard to arrange things such that these competing effects exactly cancel each other. If some CPUs have less work to do than others, they have to wait (at least once every time step) until the other CPUs catch up. This usually occurs in a collective communication routine, or in the collecting of the forces from the neighbouring CPUs. It turns out that for a crack simulation the problem of an unbalanced work load can mostly be avoided by a clever choice of the dimension of the CPU grid. In a normal bulk simulation, it is usually most efficient to assign to each CPU a block of material that is approximately cubic. This reduces the surface of the block, and thus the communication overhead. In a crack simulation, it is more efficient to use only two CPUs in the direction perpendicular to the crack surface. With such a scheme, all blocks contain a similar amount of work, so that the communication overhead (including waiting time) is reduced to as little as 3-4%. If three or four CPUs are used in this direction, which would result (at a fixed number of 96 CPUs) in a better aspect ratio of the blocks, the communication overhead is increased up to 10-12% (which seems still acceptable).

3 Results

For our simulations we have set up initial fracture planes perpendicular to five- and twofold axes. For both orientations a plane of lowest surface energy is chosen, that corresponds to one of the deepest minima in Fig. 1. The propagation direction is along a twofold symmetry axis. In both cases we have performed a series of simulations with overloads in a range from $\Delta K^* = 0.1$ to $\Delta K^* = 0.8$.

We observe brittle fracture without any crack tip plasticity irrespective of the orientation of the fracture plane. This is in good agreement with simulations of dislocation motion in the same model [17]. These simulations show clearly that the plasticity is very limited in this model, in particular at low temperatures.

For small overloads up to $\Delta K^* = 0.2$, the crack propagates only a few atomic distances r_0, and then stops for all orientations of the fracture plane. The minimal velocity for brittle crack propagation is about 10% of the shear wave velocity v_s. For loads $\Delta K^* > 0.2$ the velocity increases monotonically with the applied load. The crack velocities are in a range of 10-45% of v_s.

To analyse the morphology of the fracture planes, the height of the fracture surfaces, $h(\mathbf{r})$, is calculated as a function of the two lateral coordinates $\mathbf{r} = (x, y)$. Fig. 4 shows examples of such surface profiles. The crack propagation direction is from the left to the right. The initial fracture surface is flat, as can be seen from the homogeneous regions on the left. The surfaces resulting from the propagation of the crack, however, show a pronounced pattern of

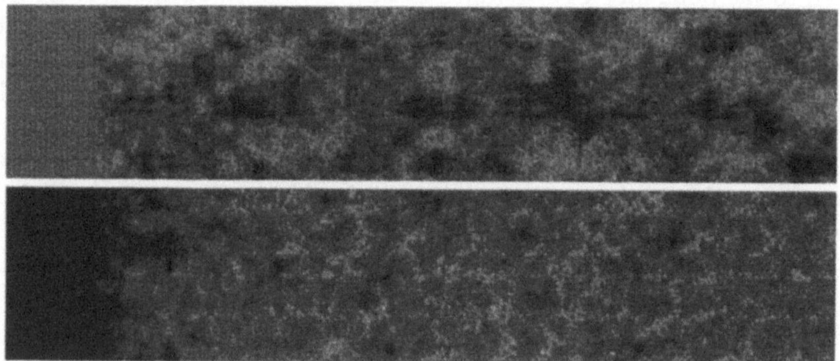

Fig. 4. Height profile of fracture surfaces perpendicular to twofold (top) and fivefold (bottom) axes, for $\Delta K^* = 0.6$. The height increases from blue ($-2r_0$) via cyan ($-1r_0$), green (average height), and yellow ($+1r_0$) to red ($+2r_0$).

regions with different heights. The average vertical roughness is of the order of $4r_0$.

4 Conclusions

In this article we have reported on molecular dynamics simulations of crack propagation in icosahedral model quasicrystals. For this purpose the quasicrystal stucture was endowed with an atomically sharp crack on fracture planes perpendicular to five- and twofold directions. Subsequently the crack was loaded by linear scaling of the displacement field of the stable crack, and the response of the system was followed by molecular dynamics simulations. For both directions we find brittle fracture with rough fracture surfaces.

Acknowledgements

We would like to thank Jürgen Schulze-Döbold for providing us with software modules interfacing between IMD and the visualization software COVISE. This work was supported in part by Deutsche Forschungsgemeinschaft (DFG) through the Priority Programme 1310 "Quasicrystals: Structure and Physical Properties" and the Collaborative Research Centre 382 "Methods and Algorithms for Simulating Physical Processes on Supercomputers".

References

1. C. Deus, B. Wolf, and P. Paufler, Phil. Mag. A **75**, 1171–1183 (1997).
2. U. Koester, W. Liu, H. Liebertz, and M. Michel, J. Non-Cryst. Solids **153 & 154**, 446–452 (1993).
3. Ph. Ebert, M. Feuerbacher, N. Tamura, M. Wollgarten, and K. Urban, Phys. Rev. Lett. **77**, 3827–3830 (1996).
4. F. Krul, *Molekulardynamik–Simulationen von Rissen in ebenen Quasikristallen*. Dilpoma thesis, Universität Stuttgart (1996).
5. R. Mikulla, F. Krul, P. Gumbsch, and H.-R. Trebin, in *New horizons in quasicrystals: Research and applications*, eds. A. I. Goldman, D. J. Sordelet, P. A. Thiel, and J. M. Dubois, Singapore, World Scientific 1997, pp. 200–207.
6. R. Mikulla, J. Stadler, P. Gumbsch, and H.-R. Trebin, in *Proceedings of the 6th International Conference on Quasicrystals*, eds. S. Takeuchi and T. Fujiwara, Singapore, World Scientific 1997, pp. 485–492.
7. R. Mikulla, J. Stadler, F. Krul, H.-R. Trebin, and P. Gumbsch, Phys. Rev. Lett. **81**, 3163–3166 (1998).
8. C. L. Henley and V. Elser, Phil. Mag. B **53**, L59–L66 (1986).
9. J. E. Lennard-Jones, Proc. R. Soc. London, Ser. A **106**, 441–462 (1924).
10. P. Gumbsch, B. L. Holian, and S. J. Zhou, Phys. Rev. B **55**, 3445–3455 (1997).
11. J. Stadler, R. Mikulla, and H.-R. Trebin, Int. J. Mod. Phys. C **8**, 1131–1140 (1997).

12. IMD, the ITAP Molecular Dynamics Program. `http://www.itap.physik.uni-stuttgart.de/~imd`
13. Virvo, the virtual reality volume renderer. `http://www.hlrs.de/organization/vis/people/schulze/virvo/`
14. S. J. Zhou, D. L. Preston, P. S. Lomdahl, and D. M. Beazley, Science **279**, 1525–1527 (1998).
15. COVISE, the Collaborative Visualization and Simulation Environment. `http://www.hlrs.de/organization/vis/covise/`
16. F. Gähler, C. Kohler, J. Roth, and H.-R. Trebin, in *High Performance Computing in Science and Engineering '2002*, eds. E. Krause and W. Jäger, Springer, Heidelberg 2003, pp. 3–14.
17. G. Schaaf, *Numerical simulation of dislocation motion in icosahedral quasicrystals*. Ph.D. thesis, Institut für Theoretische und Angewandte Physik, Universität Stuttgart (2002).

Structure and Spectrum of Poly-Porphyrin

Michael Rohlfing

Institut für Festkörpertheorie, Universität Münster, Wilhelm-Klemm-Str. 10, 48149 Münster, Germany

Summary. We discuss the structural and spectroscopic properties of porphyrin-derived polymers within an ab-initio framework. The polymer is characterized by small effective masses of the relevant electronic bands, accompanied by significant electron-hole interaction. This results in a small fundamental band gap and strong optical absorption in the infrared.

1 Introduction

Porphyrins are large cyclic organic molecules of about 1 nm diameter. They are characterized by a conjugated π-electron system that extends over the entire molecule, leading to strong optical absorption and emission in the visible spectrum.

Recently, the formation of polymers from porphyrins has attracted a lot of interest. In particular, Tsuda and Osuka [1] have succeeded in linking porphyrins to large chains, in which each porphyrin is bound to its neighbor by three chemical C-C bonds. The resulting structure can be considered as a "stripe" or "tape" of 1 nm width, extending over a total length of as much as 12 nm. Its electronic structure is characterized by a corresponding "stripe" of π states extending over the entire tape, only limited by the boundaries of the polymer. This enlargement of the π-conjugated electronic structure is accompanied by a strong shift of optical activity from the visible into the infrared energy range, thus opening the possibility to create polymers for infrared optoelectronic applications.

In this paper we investigate the porphyrin monomer and polymer systems within an ab-initio framework, aiming at their electronic and optical properties. Based on a density-functional theory (DFT) description of the electronic ground state, we apply many-body perturbation theory (MBPT) to investigate the excited electronic states, including electron ionization processes and coupled electron-hole excitations, i.e. excitons. The geometric structure of the gas-phase molecule and of the polymer are calculated and compared with one another; they are, in fact, quite similar. The electronic single-particle spectrum (i.e., the band structure in the case of the polymer) results from MBPT. Due to strong coupling between electronic states of linked monomers, the band structure of the polymer is distinctly different from the spectrum of the gas-phase molecule. In fact, strong band dispersion occurs, leading to a much smaller HOMO-LUMO gap than in the molecule. The optical spectrum of the molecule, as well as that of the polymer, is characterized by sharp peaks in the absorption spectrum. In the case of the molecule, the onset of absorption occurs in the visible spectrum. In the case of the polymer, the onset is strongly red-shifted due to the much larger conjugation length as compared to the molecule.

2 Computational Approach

The calculations presented here have been carried out by combining several computational approaches to condensed-matter physics. In particular, we employ ab-initio many-body perturbation theory (MBPT) specifically designed for the investigation of excited electronic states [2, 3, 4, 5, 6].

The structure of porphyrin and of its polymer is addressed within density-functional theory (DFT), which yields the total energy of the system depending on the geometric arrangement, as well as the forces on the atoms. By relaxing the forces to mechanical equilibrium we obtain the ground-state structure. In the case of the porohyrin molecule, this is done for a single molecule in the gas phase, containing 37 atoms. In the case of the polymer, we assume a periodic polymer of infinite length in vacuum, which enables us to work with conventional supercell techniques. Each monomer consists of 31 atoms in this case.

The wave functions of the DFT are represented by Gaussian-orbital basis sets, with 332 basis functions in total. A Fourier transformation of $81 \times 81 \times 57$ is employed for the representation of the charge density and the local potential. In the case of the periodic polymer, 4 $\mathbf{k}$ vectors are employed to sample the one-dimensional Brillouin zone.

Based on the DFT, we employ many-body perturbation theory (MBPT) to describe the excited electronic states. The determination of the single-particle spectrum (i.e., the band structure) of electron and hole states is carried out by solving the equation of motion of the single-particle Green function. As the crucial quantity, the electron self energy must be evaluated, describing the exchange and correlation effects among the electrons. This is done within the so-called *GW* approximation [7]. The key aspect of this approximation is the inclusion of dielectric screening effects, that dominate the Coulomb interaction between charged particles in condensed matter.

Optical spectra, on the other hand, can only be described if electron-hole correlation is included in the excitation process. This leads to the problem of solving the equation of motion of a two-particle Green function (given

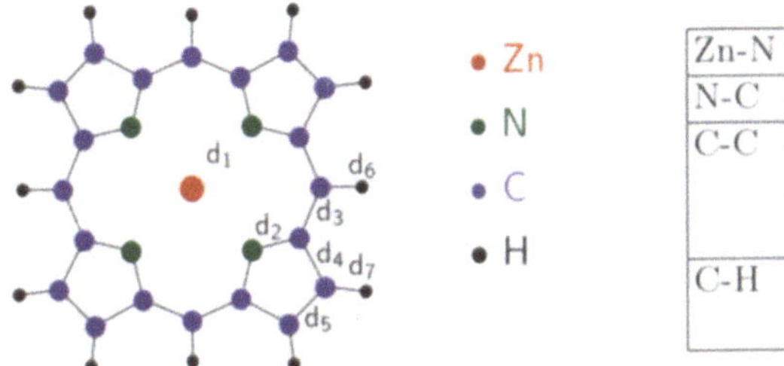

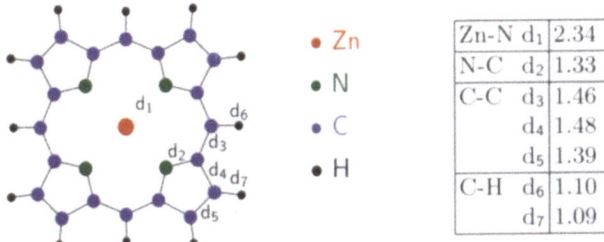

Zn-N	d_1	2.34 Å
N-C	d_2	1.33 Å
C-C	d_3	1.46 Å
	d_4	1.48 Å
	d_5	1.39 Å
C-H	d_6	1.10 Å
	d_7	1.09 Å

Fig. 1. Structure of the porphyrin molecule in the gas phase, as resulting from DFT-LDA geometry optimization.

by the Bethe-Salpeter equation, BSE); this is a consequent extension of the many-body perturbation theory. The "perturbation" is again dominated by the electron self-energy operator. This method allows to investigate the entire linear optical spectrum, both in the frequency range of bound excitons and in the range of resonant states above the fundamental energy gap. Of particular importance is the analysis of individual excitonic states. We thus gain detailed insight in the correlation between electrons and holes on a microscopic scale.

The MBPT calculations are carried out by again employing Gaussian-orbital basis sets, with as much as 764 basis functions. The calculation of the dielectric screening and of the self-energy operator includes 57 valence bands (54 in the case of the polymer) and 400 conduction bands. Finally, the Bethe-Salpeter equation is solved in a subspace of 7 occupied and 7 empty states. In the case of the polymer, 48 k-points in the one-dimensional Brillouin zone are used to represent the spatial properties of the excitons, thus resulting in a dimension of 2352×2352 for the BSE Hamiltonian.

The most time-demanding part of the work is the MBPT of the periodic polymer, containing 31 atoms. Systems of this size form the present limit of the method. Fortunately, the MBPT codes can be efficiently parallelized, running on up to 64 processors. The total computation time amounted to several thousand CPU hours.

3 Results and Discussion

We first discuss the gas-phase porphyrin molecule before addressing the formation of the polymer.

3.1 The porphyrin molecule

Figure 1 displays the structure of the porphyrin molecule. For simplicity, no functional side groups are considered. Instead, all C atoms at the ring

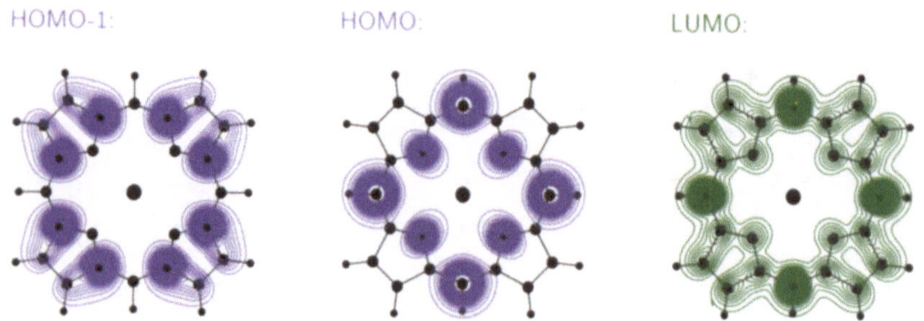

Fig. 2. Charge density of three characteristic conjugated electronic states (HOMO–1, HOMO, and LUMO) of the porphyrin molecule.

Table 1. Single-particle excitation energies of the HOMO and HOMO–1 state (ionization energy) and of the LUMO state (electron affinity) of the porphyrin molecule, as obtained within DFT-LDA, Hartree-Fock, and GWA.

[eV]	LDA	HF	GWA
HOMO-1	–5.05	–6.25	–5.86
HOMO	–4.51	–5.85	–5.36
LUMO	–2.77	0.53	–0.36
HOMO-LUMO gap	1.74	6.38	5.00

boundary have been saturated by hydrogen. The molecule consists of four five-membered C_4N elements that are linked by additional carbon atoms. In the center, a Zn atom stabilizes the nitrogens by providing its two valence electrons. As expected for a conjugated electron system, the bonds within the five-membered ring are significantly shortened compared to simple single bonds, thus indicating significant double-bond character. One exception is given by d_4 which is of single-bond character.

The states near the fundamental gap are of particular importance for the electronic and optical properties of the molecule. Figure 2 shows the HOMO–1, HOMO, and LUMO states of the molecule. These states extend over the entire molecule, thus indicating the delocalized character of conjugated π states. Table 1 compiles the single-particle energies of the three states, as obtained within three different approaches. DFT-LDA yields a quite small HOMO-LUMO gap of 1.74 eV, while Hartree-Fock (i.e., GWA without dielectric screening effects) yields a much larger gap of 6.38 eV. The GWA, which includes the dielectric screening effects in the self energy, results in a HOMO-LUMO gap of 5.00 eV.

The onset of charge-neutral electron-hole excitations of the molecule, on the other hand, occurs at much lower energy than the HOMO-LUMO gap. Due to the attractive electron-hole interaction, the lowest singlet-to-singlet

Table 2. Electron-hole excitation energies (singlet-to-singlet) of the porphyrin molecule, as obtained within the GW-BSE approach. The experimental data are from Ref. [1].

Energy	Composition	Dipole	Energy [exp.]
2.16	(HOMO/HOMO-1) → LUMO	in-plane	2.29
2.92	(HOMO-2) → LUMO	0	
3.17	(HOMO-3/HOMO-4) → LUMO	~0	
3.34	(HOMO-3/HOMO-4) → LUMO	perp.	3.04
3.48	(HOMO-3/HOMO-4) → LUMO	~0	
....			

excitation of Porphyrin is found at 2.16 eV within our GW-BSE approach, i.e. in the visible spectrum. Experiment observes the first strong optical absorption peak at 2.29 eV. Table 2 compiles the lowest singlet excitations, together with a discussion of their orbital composition and optical dipole moments. Among the dipole moments, "in-plane" refers to a dipole within the plane of the molecule, while "perp" means perpendicular to this. Experiments on gas-phase molecules cannot distinguish between the two orientations. Two optically active excitations occur in the visible spectrum, i.e. at 2.16 eV and at 3.34 eV. Both transitions are attributed to the conjugated π electrons. The experimental spectrum [1] shows two corresponding peaks at 2.29 eV and 3.04 eV that are, however, significantly broadened due to coupling to molecular vibrations. Vibrational broadening is not considered in our present theory. This makes it slightly difficult to quantitatively compare our theoretical data with the experimental ones and may account for the slight differences between our calculated data and the measured peak positions.

3.2 Poly-porphyrin tapes

The three carbon atoms at the left-hand and right-hand side of the porphyrin, that are saturated by hydrogen in the gas-phase molecule (see Fig. 1), can be used to form three covelent bonds to another porphyrin. Such three-fold bonding has been achieved experimentally [1], leading to a polymer structure as showh in Fig. 3. In here, we have assumed a periodic chain of infinite length, with a unit cell as indicated by the brackets. Table 3 compiles the optimized interatomic distances of the polymer, in comparison with those of the free molecule. Apparently, the two structures are quite similar to one another. Note that the three carbon atoms on, e.g., the right-hand side of the free molecule have slightly different x coordinates. When forming the polymer, the three covalent bonds between two monomers would thus have slightly different bond length: the middle bond would have to be 0.23 Å longer than the two outer bonds. In order to form three covalent single bonds of equal length, slight distortions have to be invoked within the monomer, i.e. the five-

Table 3. Calculated bond lengths (in Å) of the porphyrin polymer "tape", as resulting from DFT-LDA geometry optimization (see Fig. 3).

	polymer	free mol.
Zn-N d_1	2.35 Å	2.34 Å
N-C d_2	1.32 Å	1.33 Å
d_2'	1.36 Å	1.33 Å
C-C d_3	1.51 Å	1.46 Å
d_3'	1.46 Å	1.46 Å
d_4	1.49 Å	1.48 Å
d_4'	1.48 Å	1.48 Å
d_5	1.41 Å	1.39 Å
C-H d_6	1.10 Å	1.10 Å
d_7	1.09 Å	1.09 Å
C-C d_8	1.45 Å	
link d_9	1.46 Å	
a_{latt}	8.77 Å	

membered C_4N rings are slightly deformed and are rotated with respect to one another.

The three-fold bonding between the monomers results in strong coupling between the conjugated π electrons of neighboring monomers. Consequently, significant band dispersion occurs, and a semiconducting band structure is formed. This band structure is shown in Fig. 4. As expected for a conjugated polymer, the effective masses (i.e., the inverse curvature at the valence-band maximum and conduction-band minimum) are quite small. In DFT-LDA, they amount to 0.120 m_e for the valence band and 0.109 m_e for the conduction band. In GWA, the band dispersion and curvature is again increased, resulting in even smaller effective masses of 0.097 m_e and 0.090 m_e, respectively. Therefore, the formation of spatially extended Wannier excitons can be expected.

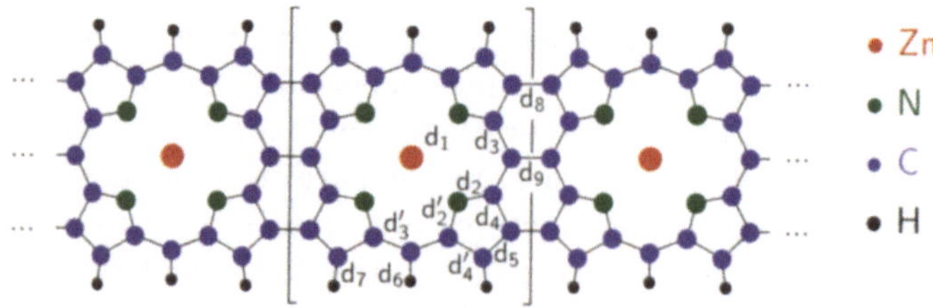

Fig. 3. Structure of the porphyrin polymer "tape", as resulting from DFT-LDA geometry optimization. The bond lengths are listed in Tab. 3.

As a last step, we have solved the Bethe-Salpeter equation of the band structure of Fig. 4, including the electron-hole interaction. The resulting optical spectrum is shown in Fig. 5, together with experimental data from Ref. [1]. The spectra are characterized by a strong absorption peak below 1 eV, i.e. in the infrared regime. In the upper panel, the theoretical spectrum with and without the electron-hole interaction is shown. Apparently, significant differences occur due to the interaction. The correlated spectrum (i.e., with the interaction included) is characterized by three peaks at 0.6 eV, 1.3 eV, and 2.2 eV. In the measured data, absorption is observed at 0.5 eV, 1.4 eV, and 3 eV. Our results for the low-energy part of the spectrum (below 2 eV) is

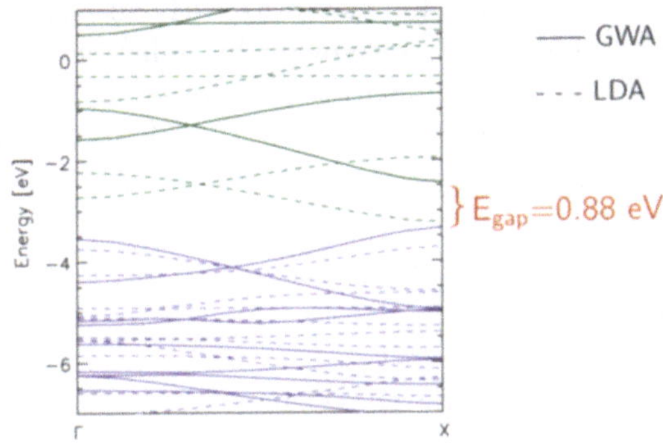

Fig. 4. Band structure of the porphyrin polymer "tape", as resulting from DFT-LDA (dashed lines) and GWA (solid lines), respectively.

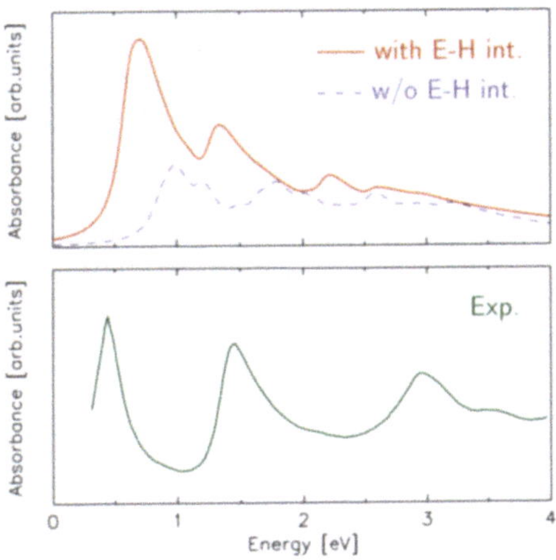

Fig. 5. Optical absorption spectrum of the porphyrin polymer "tape". Upper panel: Calculated data, obtained with the electron-hole interaction (solid line) and without it (dashed line). Lower panel: experimental data (of Por_{12}) from Ref. [1].

in good agreement with the measured data. The differences at higher energy may be related to more complex transitions, like double excitations, that are not described by our present aproach.

The most important excitation is the one at 0.6 eV, which is due to an exciton with a binding energy of 0.28 eV. This state has a strong optical dipole due to coherent superposition of the optical transition matrix elements of the contributing interband transitions.

The occurrence of absorption below 1 eV is a direct consequence of the chemical bonds between the polymers, that enlarge the effective conjugation length. In the gas-phase molecule, the conjugation length was limited by the size of the molecule of about 1 nm. In fact, such quantum confinement results in peak shifts of several eV for electrons with an effective mass of 0.1 m_e. If the quantum confinement is reduced by enlargement of the conjugation length, the onset of optical absorption is lowered. In the present case, the lowering amounts to about 1.5 eV (from 2.16 eV to 0.6 eV) in our theory, in good agreement with the experimental value of 1.8 eV.

4 Acknowledgments

This work was supported by the Deutsche Forschungsgemeinschaft under Grant No. Ro-1318/4-1 and Ro-1318/5-1 Computational resources have been provided by the Bundes-Höchstleistungsrechenzentrum Stuttgart (HLRS).

References

1. A. Tsuda and A. Osuka, Science **293**, 79 (2001).
2. M.S. Hybertsen and S.G. Louie, Phys. Rev. Lett. **55**, 1418 (1985).
3. M. Rohlfing, P. Krüger, and J. Pollmann, Phys. Rev. Lett. **75**, 3489 (1995).
4. S. Albrecht, L. Reining, R. Del Sole, and G. Onida, Phys. Rev. Lett. **80**, 4510 (1998).
5. L.X. Benedict, E.L. Shirley, and R.B. Bohn, Phys. Rev. Lett. **80**, 4514 (1998).
6. M. Rohlfing and S.G. Louie, Phys. Rev. Lett. **81**, 2312 (1998); Phys. Rev. B **62**, 4927 (2000).
7. L. Hedin, Phys. Rev. **139**, A796 (1965).

How Do Droplets Depend on the System Size? Droplet Condensation and Nucleation in Small Simulation Cells

P. Virnau, L. González MacDowell, M. Müller and K. Binder

Institut für Physik, WA331
Johannes Gutenberg Universität
D55099 Mainz, Germany
Marcus.Mueller@uni-mainz.de

Summary. Using large scale grandcanonical Monte Carlo simulations in junction with a multicanonical reweighting scheme we investigate the liquid-vapor transition of a Lennard–Jones fluid. Particular attention is focused on the free energy of droplets and the transition between different system configurations as the system tunnels between the vapor and the liquid state as a function of system size. The results highlight the finite size dependence of droplet properties in the canonical ensemble and free energy barriers along the path from the vapor to the liquid in the grandcanonical ensemble.

1 Introduction

Transitions from one equilibrium state to another in response to a change of an intensive thermodynamics variable (like pressure or temperature) are of pivotal importance for many technical procedures [1, 2, 3, 4, 5]. In principle, the kinetics of phase transition is a non-equilibrium process. If it is sufficiently slow however, one can obtain a useful description via the excess free energy of spatially inhomogeneous systems. Roughly, two pathways can be distinguished. In spinodal decomposition, phase transformation occurs spontaneously and simultaneously everywhere in the system via spatially extended small variations of the order parameter. In nucleation, small droplets of the new stable phase form on the background of the mother phase. The kinetics of the early stages is determined by the droplet's free energy cost as a function of its size.

Much insight into the properties of droplets can be gained from computer simulations [6, 7, 8, 9, 10, 11]. In contrast to most experiments which can only observe nucleation by indirect means, computer simulations enable us to study and visualize clusters directly. The properties of small droplets might differ significantly from those of macroscopic drops. The latter can be conceived as spherical domains of the new stable phase which are separated from the mother phase by a bulk-like interface. The interior of microscopic droplets and their interface might depend on the droplet size and the deviation of the intensive thermodynamical variable from its coexistence value (e.g., the supersaturation).

Most simulation studies do not observe the kinetics of droplet formation but determine the excess free energy of droplets from which the concomitant nucleation barrier can be inferred. The latter quantity is a key ingredient in predicting the rate of phase transformation. In our Monte Carlo study on the CRAY T3E at the HLR Stuttgart we investigate the nucleation of droplets from a supersaturated vapor within a Lennard-Jones model, paying due attention to finite size effects. Our report is arranged as follows: In the next section we will provide some background about the droplet condensation/evaporation transitions and the free energy in the canonical ensemble. Then we shall describe our computational model and technique and present our results. The paper closes with a discussion and an outlook.

2 Background

In the following we consider the condensation of a supersaturated vapor. At a given supersaturation, which is quantified by the pressure difference $\Delta p = p - p_{\text{coex}} > 0$ or the deviation from the coexistence chemical potential $\Delta\mu = \mu - \mu_{\text{coex}} > 0$, the vapor is only metastable and will condense into the thermodynamically stable liquid. If the supersaturation is sufficiently small the phase transformation will proceed via the nucleation of a drop. In the framework of the classical nucleation theory, the excess free energy can be decomposed into a surface and a volume contribution:

$$\Delta F(R) = 4\pi\gamma R^2 - \frac{4\pi}{3} R^3 \Delta p. \tag{1}$$

γ denotes the interface free energy (per unit area) and the second term describes the free energy reduction by the formation of the thermodynamically stable phase. As it is well known, the droplet free energy $\Delta F(R)$ exhibits a maximum as a function of the droplet size. The properties of this transition state, the so-called the critical droplet, are given by:

$$R^\star = \frac{2\gamma}{\Delta p} \qquad \text{and} \qquad \Delta F^\star = \frac{16\pi\gamma^3}{3\Delta p^2} \tag{2}$$

If the nucleation barrier $\Delta F^\star$ is sufficiently large, a single large critical drop can be conceived as the transition state and the rate is proportional to the Boltzmann factor $\exp(-\Delta F^\star/k_B T)$. Here T denotes the temperature and k_B Boltzmann's constant. In what follows we measure free energies in units of $k_B T \equiv 1$. In view of the strong dependence of the transition rate on the value of the interface tension it is clearly warranted to assess the validity of this phenomenological concept. Therefore, we measure the free energy of droplets in our Monte Carlo simulations.

In the grandcanonical ensemble all droplets but the critical one are unstable. For $R < R^\star$ the size of the droplet shrinks (subcritical droplets), while it grows for supercritical sizes $R > R^\star$. The critical droplet is a saddle point in the grandcanonical ensemble. It can be observed in mean field theories but not in computer simulations for small size fluctuations lead to a growing or shrinking of a droplet.

Therefore, droplets can only be investigated in the canonical ensemble. Rather than working at a fixed value of the supersaturation (i.e. the thermodynamically intensive variable), one constrains the number of particles in the finite volume simulation cell. For a range of particle numbers, a single droplet is the thermodynamically stable configuration.

Let ρ_v and ρ_l denote the coexisting densities of the vapor and the liquid respectively, and V the volume of the simulation cell. If the excess number of particles $\Delta N = (\rho - \rho_v)V$ is small, they will distribute homogeneously throughout the simulation cell (supersaturated vapor). In this case the free energy is given by:

$$\Delta F_{\text{vapor}}(\Delta N) = \frac{V}{2\kappa}\Delta\rho^2. \tag{3}$$

κ denotes the compressibility of the vapor and $\Delta\rho = \frac{\rho-\rho_v}{\rho_l-\rho_v}$. Increasing the excess number of particles (or $\Delta\rho$) we increase the free energy quadratically.

If ΔN grows, it becomes favorable to condense the excess number of particles into a droplet [5, 6, 7, 12, 13, 14, 15, 16, 17]. The transition from the homogenous supersaturated vapor to a droplet configuration is termed droplet condensation. To the simplest approximation we assume that (i) all excess particles condense into a single droplet and (ii) the density of the droplet's interior corresponds to the liquid ρ_l. Its interface tension is given by the macroscopic value γ. It should be noted that these assumptions are only valid for larger clusters. Then the free energy takes the form:

$$\Delta F_{\text{drop}}(\Delta N) = 4\pi\gamma R^2 \equiv g(V\Delta\rho)^{2/3} \qquad \text{with} \qquad \frac{4\pi}{3}R^3 = V\Delta\rho. \tag{4}$$

and $g = 3^{-2/3}(4\pi)^{1/3}\gamma$. Expanding the free energy difference $\delta F = \Delta F_{\text{drop}} - \Delta F_{\text{vapor}}$

$$\delta F \approx -\frac{3}{4}\left([2\kappa]^{-1/3}gV\right)^{3/4}\left(\Delta\rho - (2\kappa g)^{3/4}V^{-1/4}\right), \tag{5}$$

we readily read off that droplet condensation occurs at $\Delta\rho_{\text{dc}} = (2\kappa g)^{3/4}V^{-1/4}$. Of course, for any finite simulation cell going from a supersaturated vapor to a droplet is not a sharp transition but rounded over a range $\delta\rho = \Delta\rho - \Delta\rho_{\text{dc}}$ where $\delta F \sim \mathcal{O}(1)$. This estimate yields $\delta\rho \sim \left(\kappa^{-1/3}gV\right)^{-3/4}$. As we increase the system size, the excess density at which the supersaturated homogeneous vapor is stable decreases like $V^{-1/4}$ and the smallest droplets that are observable at that density are of radius $R_{\text{min}} \sim V^{1/4}$.

If ΔN grows further, the droplet grows until its size becomes comparable to the linear dimension $V^{1/3}$ of the simulation cell. At that point it might become favorable to form a liquid slab which is separated from the vapor by two interfaces of area $V^{2/3}$. In this case the free energy

$$\Delta F_{\text{slab}} = 2\gamma V^{2/3} \tag{6}$$

is independent from the excess density $\Delta\rho$. As both, the droplet and the slab free energy scale like $\gamma V^{2/3}$, the transition from a droplet to a slab occurs at a fixed $\Delta\rho$ which depends on the aspect ratio of the simulation cell, but which is independent from its volume or the interface tension. Hence, the largest droplet that are observable have a radius $R_{\text{max}} \sim V^{1/3}$.

These considerations have two important consequences for simulational studies of nucleation phenomena: (i) In a simulation cell of volume V and fixed excess number of particles one can only observe equilibrium drops with radii $V^{1/4} \sim R_{\text{min}} < R < R_{\text{max}} \sim V^{1/3}$. Consequentially, to study the dependence of the free energy on the drop's radius one has also to vary the system size [6, 7, 12, 13]. (ii) The transitions from the supersaturated vapor to the droplet and from the droplet to the slab configurations represent barriers in the configuration space which are not removed by the multicanonical weighting scheme (cf. below). Hence, they limit the applicability of reweighting methods in the study of phase equilibria [14].

3 Simulations of droplets in a Lennard–Jones fluid

Grandcanonical Monte Carlo simulations were carried out in $d = 3$ dimensions using a truncated and shifted LJ potential:

$$U_{LJ} = 4\epsilon\left[\left(\frac{\sigma}{r}\right)^{12} - \left(\frac{\sigma}{r^6}\right) + \frac{127}{16384}\right], \text{ if } \mathrm{r} \leq \mathrm{r_c} = 2(2^{1/6}\sigma) \text{ and 0 else.} \tag{7}$$

Of course, configurations at a fixed particle number have the statistical weight of the canonical ensemble. The grandcanonical scheme simply allows us to compute the free energy difference as a function of the number of excess particles. In order to sample the pertinent interval of excess particles we employ a multicanonical weighting scheme [18], i.e. we modify the original Hamiltonian by adding a term $k_B T \ln w(\rho)$. The bias of the weighting function $w(\rho)$ is removed in the subsequent analysis. w is generated by adopting Wang-Landau sampling [19] to our off-lattice (μVT) ensemble [20]. The interval of particles can be divided into subintervals which, in turn, are distributed onto different processing elements.

The free energy difference of system configurations is obtained from the logarithm of the probability distribution $P_{\mu VT}$ at the coexistence value of the chemical potential:

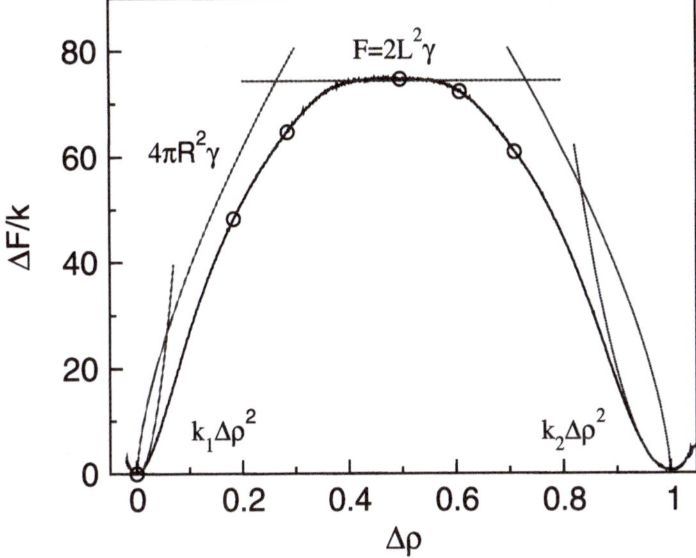

Fig. 1. Logarithm of the probability distribution in grandcanonical simulations for $L = 11.3\ \sigma$ and $T = 0.78\ \epsilon/k_B$. The phenomenological expressions (3-6) for the free energy are also indicated using $\gamma = 0.291\ \frac{\epsilon}{\sigma^2}$. $\circ$ denote densities at which typical configurations were visualized in Fig. 2.

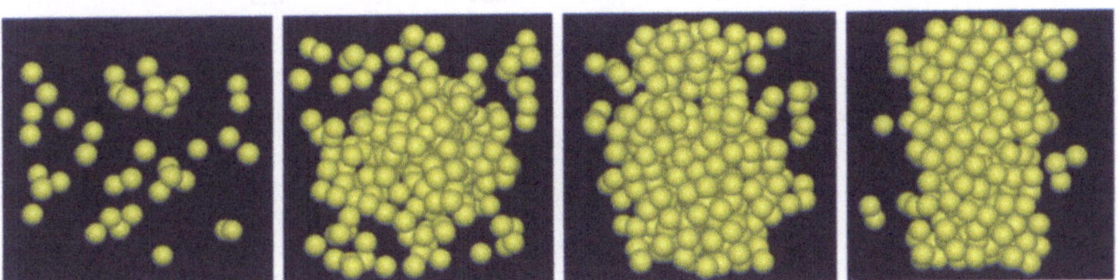

Fig. 2. Typical system configurations for the same parameters as in Fig. 1. The density is $\Delta\rho = 0, 0.184, 0.286, 0.498$. Only a thin slice of the simulation box is shown in the right most image. Each monomer is represented by a sphere of diameter $1.12\ \sigma$, which corresponds to the minimum of the Lennard-Jones potential. The pictures are in qualitative agreement with our phenomenological discussion.

$$\Delta F = -k_B T \ln \left(\frac{P_{\mu VT}(\Delta\rho)}{P_{\mu VT}(\Delta\rho = 0)} \right) \tag{8}$$

The result for $L = 11.3\ \sigma$ and $T = 0.78\ \epsilon/k_B$ are presented in Fig. 1. The dependence of the free energy according to the phenomenological considerations in Sec. 2 are also indicated. Typical snapshot of the system configurations are shown in Fig. 2. These corroborate the correct identification of the dominant system configurations.

In order to locate the droplet condensation more accurately, we regard the derivative of the free energy $\Delta\mu \equiv \partial\Delta F/\partial N$. As indicated by our analytical results, the first–order transition becomes 'sharp', for low temperatures and large systems. The results for such a system are shown in Fig. 3. The equilibrium droplets observed in the canonical ensemble correspond to critical droplets in the nucleation theory at a supersaturation $\Delta\mu$. Inside the miscibility gap and for finite V, $\Delta\mu$ exhibits an s-shaped variation with density $\Delta\rho$ which indicates the droplet condensation. The turning point of the curve yields an estimate for the location of the transition [21].

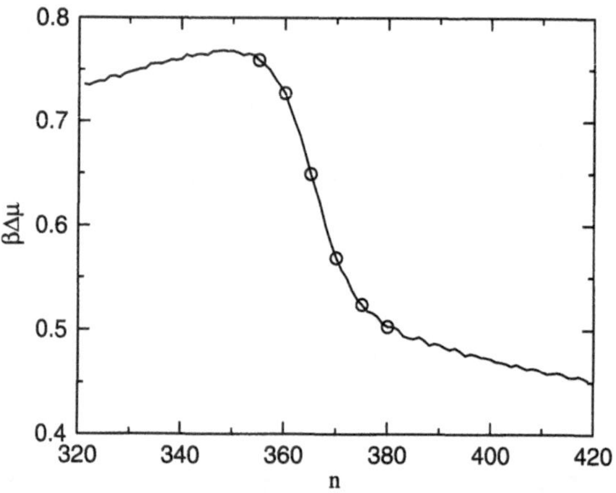

Fig. 3. Plot of $\beta\Delta\mu$ vs. N, for a LJ fluid in a box of $L = 22.5\ \sigma$ at $T = 0.68\ \epsilon/k_B$ ($\Delta\mu = \mu_{NVT} - \mu_{\text{coex}}$). $\circ$ denotes states at which configurations were stored for analysis (Fig. 4 and 5).

At densities marked by $\circ$ we store configurations for further analysis. After simulation, the cluster size distribution $p(N_c)$ was determined. Any ensemble of atoms whose distance is smaller than 1.5σ is assumed to belong to the same cluster (Stillinger criterion)[22]. The chemical potential was measured by the Widom particle insertion method [23]. For $N \leq 350$, i.e. on the ascending branch of the $\Delta\mu$ vs. N curve in Fig. 3, $p(N_c)$ is monotonically decreasing with N_c. For $N \approx 355$ a peak near $N_c \approx 120$ appears. It becomes more pronounced and moves to larger N_c as N increases. This peak represents a single large liquid droplet, present in the descending part of the $\Delta\mu$ vs. N curve. However, the liquid droplet cannot be found in all sampled configurations: when we sample the distribution of the largest cluster $N_c^{\max}$ in the system [18], we find a bimodal distribution with one peak near $N_c^{\max} \approx 20$ in the whole range of N studied in Fig. 4 (left). This peak corresponds to configurations which consist of supersaturated gas with small clusters but no single large droplet. The second peak is identical to the peak of the single droplet as

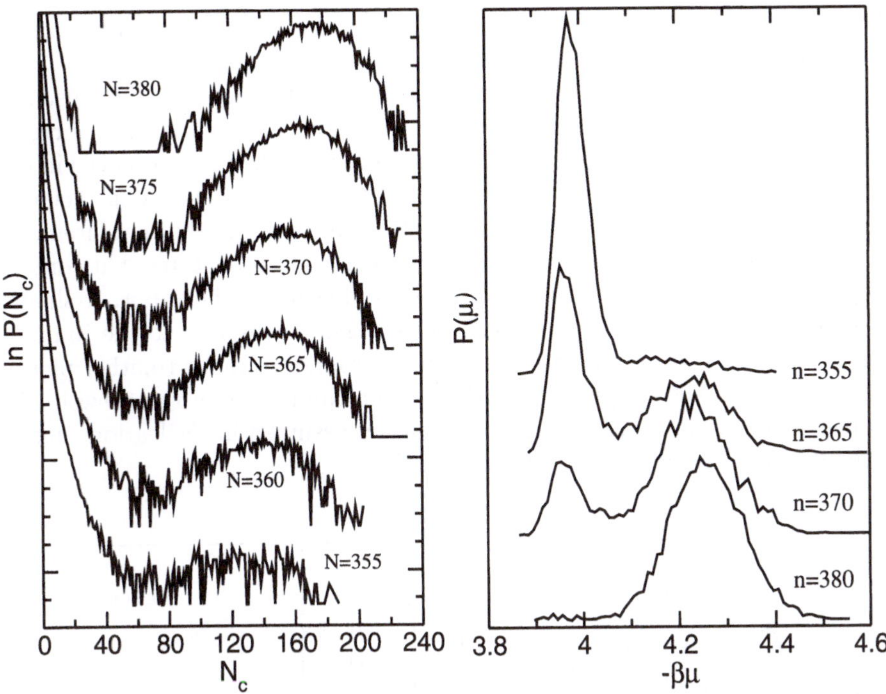

Fig. 4. Distribution $p(N_c)$ of the cluster size N_c for several choices of N (left part) and the corresponding distribution of the chemical potential of the supersaturated gas $p(\mu)$ (right part). System parameters are the same as in Fig. 3.

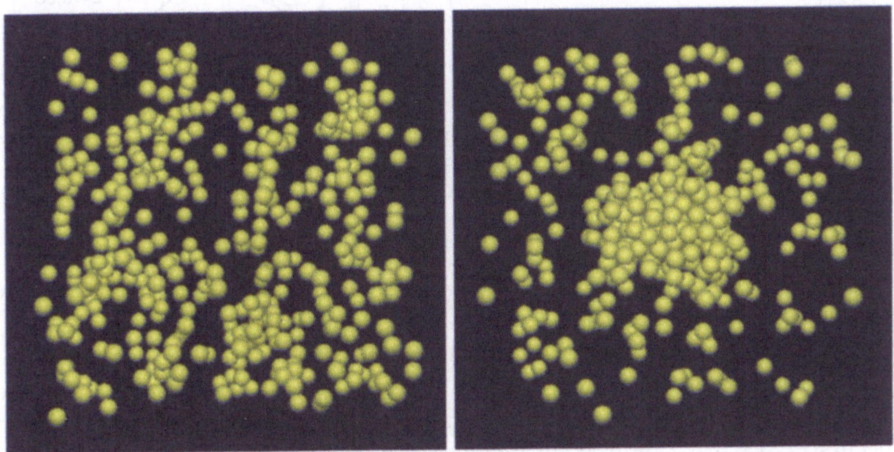

Fig. 5. Two snapshots of configurations at the transition point ($N = 365$ particles). System parameters are the same as in Fig. 3.

shown in Fig. 4(left). Snapshots of the supersaturated vapor and the droplet at the transition point are shown in Fig.5.

This clear cut evidence for the "evaporation" of the liquid droplet also appears in the double peak - distributions of the chemical potential (Fig. 4 right) in the vapor phase. To this end we have measured the chemical potential in the vapor via Widom's insertion test method. The droplet volume was excluded from the measurement to avoid the interface to influence the measurement. The peak with the lower chemical potential corresponds to states that contain droplets, and a concomitant reduction of the density of the vapor. It should be noted that the difference between both peaks correspond to the jump in Fig. 3. At fixed volume, the density of the surrounding vapor (mother phase) does depend on the droplet's size. Hence, one cannot study the growth of a droplet at fixed supersaturation, the situation assumed in nucleation theory, without a systematic study of finite size effects.

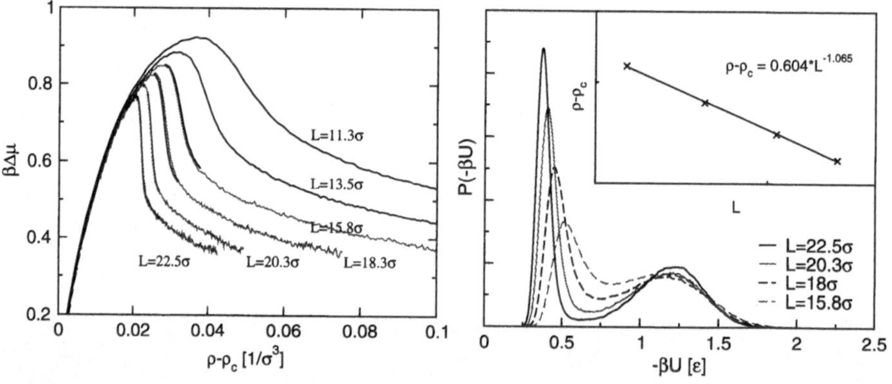

Fig. 6. Distribution of the energy per particle for different system sizes at the droplet condensation. Inset: $\rho - \rho_c$ as a function $L = V^{1/3}$. ($\rho - \rho_{\text{coex}}(L = 15.8\ \sigma) = 0.0320\ \frac{1}{\sigma^3}, \rho - \rho_{\text{coex}}(L = 18\ \sigma) = 0.0277\ \frac{1}{\sigma^3}, \rho - \rho_{\text{coex}}(L = 20.3\ \sigma) = 0.0244\ \frac{1}{\sigma^3}, \rho - \rho_{\text{coex}}(L = 22.5\ \sigma) = 0.0220\ \frac{1}{\sigma^3}$).

The dependence on the system size is explored in Fig. 6. In left panel we plot the chemical potential vs density for system sizes $L = 11.3\ \sigma - 22.5\ \sigma$. As predicted, the turning point of the curves shifts closer to the density of the vapor binodal as we increase the system size. Also the maximum slope increases with increasing L, indicating that for $L \to \infty$ a sharp transition occurs[12, 13, 14]. The second panel presents the probability distribution of the energy at the transition for different system sizes. In qualitative agreement with the expectations the transition becomes sharper and both states (the supersaturated vapor and droplet) become more separated as we increase the system size.

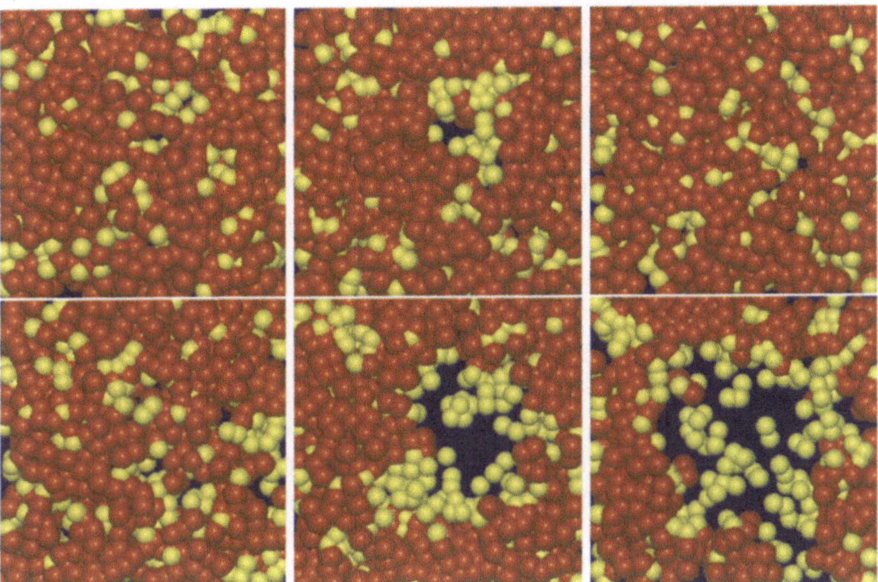

Fig. 7. Time evolution of bubble formation after a pressure jump in a polymer-solvent mixture. Dark spheres correspond to pentamer segments while grey spheres represent a monomeric solvent. Only a thin slice of 2σ is shown. Note the enrichment of the monomeric solvent at the interface between the bubble and the mother phase.

4 Discussion

The inset of Fig. 6 (right) also shows the dependence of $\Delta\rho_{\mathrm{dc}}$ on the system size. The data are compatible with an effective exponent $\Delta\rho_{\mathrm{dc}} \sim V^{-0.35}$, while phenomenological considerations suggest an exponent $-1/4$. Nevertheless, similar studies on the two-dimensional Ising model [14] suggest that the $L^{-\frac{3}{4}}$ domain is only reached for large system sizes. Similarly, the gross qualitative features of $\Delta F(\Delta\rho)$ in Fig. 1 are captured by equations (3,4,6), but there are quantitative differences. The phenomenological considerations give only an upper bound of the free energy. If one insists on an identification of the droplet radius according to Eq. (4), these deviations correspond to a reduction of the interface tension of small droplets of the order 20%. Given that the nucleation barrier $\Delta F^{\star} \sim \gamma^{3}$ and the nucleation rate, in turn, depends exponentially on $\Delta F^{\star}$ we infer a significant increase of the phase transformation rate. The detailed structure of the droplet (i.e. the density profiles of the droplet's liquid/vapor interface and the density in its interior) shall be investigated in the future. We expect the deviations from the phenomenological considerations to decrease for larger drops (and larger system sizes). These studies are also potentially very relevant for applications: The excess free energy of small droplets is pertinent to nucleation in highly supersaturated vapors, a situation long-sought for producing nano–porous foams. Moreover, in future work we explore also the behaviour of polymer-solvent mixtures which are important for practical applications. In mixtures, one component may enrich at the liquid vapor interface and thereby decrease the interface tension. Such a behavior is illustrated in Fig. 7, where we present the time sequence of bubble formation in a polymer-solvent mixture. All investigations require large scale simulations with varying temperatures and system sizes and are ideally suited for a supercomputer like a CRAY T3E.

Acknowledgments

Financial support from the BASF AG (P.V.) and from the Deutsche Forschungsgemeinschaft (DFG) via a Heisenberg fellowship (M.M.) and by grant No Bi 314/17-3 (L.G.M) is gratefully acknowledged. LGM would like to thank Ministerio de Ciencia y Tecnologia (MCYT) and Universidad Complutense (UCM) for the award of a Ramon y Cajal fellowship and for financial support under contract BFM-2001-1420-C02-01. It is a pleasure to thank the HLR Stuttgart for providing ample CPU time on the CRAY T3E supercomputer.

References

1. A.C. Zettelmoyer (ed.): *Nucleation* (Marcel Dekker, New York, 1969)
2. J.D. Gunton, M. San Miguel, and P.S. Salni, in: *Phase Transitions and Critical Phenomena, Vol. 8.* Eds. C. Domb and J.L. Lebowitz (Academic Press, London, 1983) p. 267
3. K. Binder: Rep. Progr. Phys. **50**, 783 (1987)
4. F.F. Abraham: *Homogeneous Nucleation Theory* (Academic Press, New York, 1974)
5. K. Binder and D. Stauffer: Adv. Phys. **25**, 343 (1976)
6. K. Binder and M.H. Kalos: J. Stat. Phys. **22**, 363 (1980)
7. H. Furukawa and K. Binder: Phys. Rev. A**26**, 556 (1982)
8. R. P. Ten Walde and D. Frenkel: J. Chem. Phys. **109**, 9901 (1998)
9. P. Virnau, M. Müller, L.G. MacDowell, and K. Binder, New J. Phys., preprint cond-mat/0303642
10. K. Binder, in: *Computational Methods in Field Theory.* Eds. H. Gausterer and C.B. Lang, (Springer, Berlin, 1992) p. 59.
11. D.P. Landau and K. Binder: *A Guide to Monte Carlo Simulation in Statistical Physics* (Cambridge Univ. Press, Cambridge, 2000)
12. K. Binder: Physica A**319**, 99 (2003)
13. M. Biskup, L. Chayes, and R. Kotecky: Europhys. Lett. **60**, 21 (2002)
14. T. Neuhaus and J. Hager: J. Stat. Phys. (in press)
15. T. Müller and W. Selke: Eur. Phys. J. B**10**, 549 (1999)
16. M. Pleimling and W. Selke: J. Phys. A **33**, L199 (2000)
17. M. Pleimling and A. Hüller: J. Stat. Phys. **104**, 971 (2001)
18. B. Berg and T. Neuhaus: Phys. Rev. Lett. **68**, 9 (1992)
19. F. Wang and D. Landau: Phys. Rev. Lett. **86**, 2050 (2001)
20. P. Virnau and M. Müller, in preparation

21. L.G. MacDowell, P. Virnau, M. Müller, and K. Binder; in preparation
22. F. H. Stillinger: J. Chem. Phys. **38**, 1486 (1963)
23. B. Widom: J. Chem. Phys. **39**, 2808 (1963)

Solid State Physics

Prof. Dr. Werner Hanke

Institut für Theoretische Physik und Astrophysik
Universität Würzburg
Am Hubland, D-97074 Würzburg

The contributions from solid-state physics to the high-performance simulation results in Stuttgart can be divided into surface physics contributions, contributions dealing with strongly correlated bosonic and fermionic systems and Car-Parrinello simulations of clusters and surfaces. This latter project was also partly carried through at the SSC in Karlsruhe.

Nano-structures in reduced geometry have been dealt with by Prof. P. Nielaba from the Physics Department in Konstanz. Nano-structures in such a reduced geometry have become an extremely interesting research domain in particular for supercomputing in the last few years, despite the fact that many structural and other properties of these systems in the size of a few nano-meters have been obtained and clarified. The theoretical investigations are still at an initial stage. The basic reason is that these systems are not accessible to analytical methods, which are suitable for systems with either infinitely many particles or very few particles (2-5). Therefore, in this field, computer simulations have become more and more important since the nano-systems in reduced geometry contain typically of the order between 10 to 10^{10} particles. The Nielaba group basically followed two schemes to obtain new insights into phase transitions and quantum effects in nano-systems. In the first one, a path integral formulation of the partition function of a quantum particle is represented by a classical chain of so-called Totter particles. In this path integral simulation, performed on the T3E, a very efficient algorithm could be utilized, which allowed to approach the quantum limit properly by using 64 processes in parallel. These studies were complemented at the SSC in Karlsruhe by Carr-Parinello simulations of clusters and self-assembled monolayers. Very interesting results were obtained from these computations e. g. for the thiol molecules on Au surfaces. The calculations demonstrated that only by utilizing the SP2/SP3 combination such computations are feasible and also that further computer time is clearly required for the computation of this interesting research project.

In a project of the F. Bechstedt group in Jena, density functional calculations were used to explore the atomic and spectroscopic properties of

InP surfaces grown in gas-phase eptitaxy. These calculations seem to have resolved a long standing puzzle about the microscopic structure of the InP growth plane relevant to standard gas-phase epitaxy conditions where it was argued before in previous work that the electron counting principle on the surface could possibly be violated. This was attributed to strong electronic correlations. The Jena group does not agree with that conclusion. A question here, which still has to be answered in the future, is clearly that the density functional calculations used are basically appropriate for weakly correlated electrons and not for the strong electron correlation physics, which was previously made accountable for the violation of the electron counting rule. So also here, further studies in the future are required.

An interesting project was carried out at the supercomputing center by the K. Binder group from Mainz in a collaboration with a French group from Montpellier on amorphous silica at surfaces and interfaces. Here the first goal was to compare the results of more standard classical molecular dynamics simulations with the so-called Carr-Parrinello molecular dynamics technique. The idea was then to check the accuracy of the model potential that underlies the classical simulation. This is important because the predicted power of a molecular dynamics simulation depends strongly on the quality on the potential with which one models the interactions between the particles. This check came out favourably in the calculations of the Mainz and Montpellier groups. The second step was then the use of the classical molecular dynamics for the investigation of non-bulk behavior of amorphous SiO_2 at surfaces and interfaces. New results have been obtained for the structure of a silica melt between walls studied in equilibrium and under shear.

The last example deals with a subject presented and studied by the Stuttgart group around A. Muramatsu and F. F. Assaad (now Würzburg). This project is concerned with Quantum-Monte-Carlo (QMC) calculations of correlated bosonic and ferminonic systems. This subject has produced essentially three groups of new results. The first one concerns the verification of a third type of elementary excitations in the so-called one-dimensional t-J model, which is the existence of anti-holons in addition to the known spinons and holons. The second topic deals with the Bose-Einstein condensation of alkali atoms experimentally studied in an optical trap. The numerical simulations in the Stuttgart group in one dimension have helped to further understand the corresponding new states of matter, which allow for an transition from the Mott insulating state into the superfluid state by changing the confining potential and in the experiment the interfering laser beams. Last not least in the third topic by considering multi-flavored models it could be shown that the transition of these bosonic and fermionic systems from one to two and three dimensions may become very feasible and first results have been presented for the fermionic phase.

Numerical Studies of Collective Effects in Nano-Systems

M. Dreher, D. Fischer, K. Franzrahe, G. Günther, P. Henseler, J. Hoffmann, W. Strepp, and P. Nielaba

Physics Department (Theory), University of Konstanz, 78457 Konstanz, Germany
peter.nielaba@uni-konstanz.de

Summary. We have studied quantum effects, structures and phase transitions in Nano-systems. In the following sections an overview is given on the results of our computations on atomic wires, clusters, pore condensates, Bose fluids, elastic properties of model colloids and model colloids in external fields.

1 Introduction and general remarks

Nanostructures in reduced geometry have become an interesting research domain in the last years. Despite the fact that by experimental techniques many structural-, elastic-, electronic- and phase- properties of systems in the size of a few nanometers have been obtained, the theoretical investigations and analyses are still in an initial stage. This is partly due to the fact that systems which are far away from the thermodynamic limit (with infinitely many particles) due to their finite size are difficult to handle by analytical methods which are suitable for systems with either few particles (2-5) or in the limit of infinitely many particles. In this field computer simulations have become more and more important since nano-systems in reduced geometry contain about 10–10.000 particles, which is nearly ideal for the application of computer simulation methods. Many important results have been obtained by the support of HPC centers (HLRS, SSC, NIC) [1, 2, 3, 4, 5].

Our research on nanostructures is embedded in the Sonderforschungsbereich 513 with two projects (A11 and B10), in the Transregio-SFB TR6 with project C4, and with a project in a European Graduate College on soft matter. Besides this our activities are linked to recent research goals in the European-Science-Foundation programme "Challenges in Molecular Simulations: Bridging the Length- and Time- Scale Gap (SIMU)" (http://simu.ubl.ac.be).

The present work contains several new insights into pore condensates and phase transitions and quantum effects in nano-systems in external potentials and reduced geometry. In the path integral formulation of the partition function a quantum particle is represented by a classical chain of P "Trotter particles" ($P \rightarrow \infty$), where each Trotter particle interacts harmonically with its neighbors on the chain and interactions between the quantum particles are always at the same Trotter particle index. In the path integral simulations performed on the T3E a very efficient parallel algorithm along the chain-coordinate could be utilized, putting the system at a given Trotter particle index on one processor, which allowed us to approach the quantum limit properly by using 64 processors in parallel ($P=128$), the algorithm only scaling with P. This good scaling property with P allowed us to compute full phase diagrams of the systems, which otherwise had been a hopeless task. In addition the Monte Carlo procedure employed requires the computation of statistical averages which can be done very efficiently if averages of system replicas with different initial conditions are computed in parallel on several processors. The SSC in Karlsruhe also granted time for the Car-Parrinello simulations of clusters and self assembled monolayers. In these computations a very efficient algorithm, optimized for the SP2 could be used [6, 7, 8].

2 Car Parrinello Simulations of Clusters and Surface Structures (SSC Karlsruhe)

In the HLRS/SSC-project three systems in the field of surface- and cluster physics have been studied. An important goal in nano physics research is the determination of building blocks which can be used for structuring of surfaces. Another important goal is the development of methods for the technological applicability of such building blocks. Besides this it is of great interest to improve the understanding of adsorption phenomena on small particles. In particular nano particles consisting of gold parts have promising properties which could be used in catalysis, for example. In this context the systems studied in the SSC project have been hydrogenated gold clusters Au_nH [6], Si_4 clusters deposited on a solid surface [7] and self-assembled monolayers with alkanethiol chains on a gold surface [8].

The method for the study of such systems is the density functional theory (DFT). This theory permits the quantum mechanical investigation of atomic systems by computer simulation methods. The essential advantage of this method is that no empirical parameters are required. Besides this by DFT systems in the order of about 1000 atoms can be studied. For the study of Au_nH clusters and deposited Si_4 clusters the DFT standard implementation has been used, in case of the self assembled monolayers additional classical interactions had to be taken care of, which required the development of a "QM/MM" approach, in which parts of the system interactions have been described by quantum mechanics (QM) and classically (MM), respectively.

Since the discovery of the "supermagic" cluster C_{60} [9] the possibility of the synthesis of new materials consisting of highly stable clusters fascinates many researchers. In case of C_{60} and similar fullerenes like C_{70} and $La@C_{82}$ such materials exist and, e.g., fullerite – the bulk material formed by weakly interacting C_{60} "soccer balls" – represents a new form of carbon beside diamond and graphite [10].

The question arises, whether "magic" clusters of other elements like Si or Al might be suitable as building blocks of new cluster materials. However, many of these clusters are much more reactive compared to the rather inert fullerenes and the chemical methods [10] which were used for the generation and separation of fullerene materials cannot be used for clusters of most other elements. In experimental studies of free clusters in the gas phase many other very stable clusters have been found like C_{32} [11] or Si_4 [12, 13]. In addition, there are also theoretical predictions of possible building blocks of new materials like e.g. $Al_{13}H$ [14, 15] and Si_{45} [16]. For the theoretical and experimental studies published so far the relative stability is the only criterion making a "magic" cluster to a candidate as building block for new cluster material. However, even more important is the interaction between neighboring clusters. In cluster materials neighboring clusters "touch" each other and there must be a barrier against fusion.

In combination with experimental studies in the group Ganteför in Konstanz we analyzed by Car-Parrinello methods the electronic and structural properties of Si_4- clusters [7]. Experimentally, the clusters are mass-selected and soft landed on an inert van-der-Waals surface [17]. They are probably highly mobile on this surface at room temperature and will immediately form large islands of bulk Si if there wouldn't be a barrier against fusion [18, 19, 20]. The samples are studied using XPS and, in contrast to an earlier study of Si_{10} on amorphous carbon [21], the spectra contradict the formation of large islands supporting the existence of a barrier. The interaction potential between two approaching Si_4 clusters is calculated for two geometries and for both channels a barrier against fusion is found which is large compared to room temperature. Accordingly, a new bulk of pure Si consisting of Si_4 clusters should be existing.

The Car-Parrinello type computational approach is based on density functional theory [22, 23, 24]. Our density functional (DFT) calculations [7] for the Si_4 clusters have been performed with the approximative gradient-corrected exchange-correlation (xc) functionals of Perdew, Burke and Ernzerhof (PBE) [23]. This choice for the xc- functional should give reliable results whenever both localized and extended electron states appear. The computational details are similar as in the studies of Au [25, 26, 6, 27, 24], modified to the case of Si [28]. In a first step the ground state structure of an isolated Si_4 cluster was determined. The isolated Si_4 clusters form planar rhomboedric structures with two sharp and two flat corners. Fixing the distance R between two Si- atomic centers on the x-axis in two different Si_4- clusters the potential energy surface was calculated. Two different "reaction channels" have been considered (see Fig.1) starting with structures with symmetry about the x-axis. In Fig.1(a) the distance between the two Si-atoms at the flat angles is fixed, in Fig.1(b) the distance of the atoms at the sharp angles. The total energy is displayed in Fig. 1 as a function of R. These calculations correspond

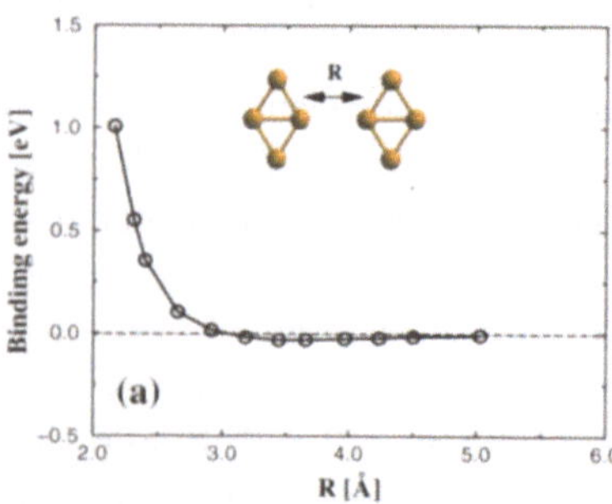

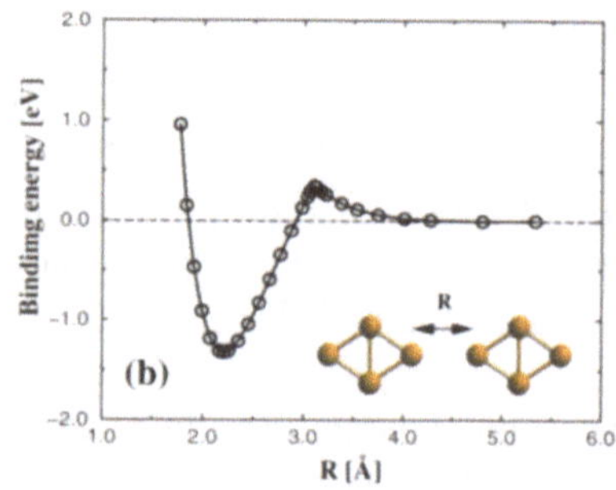

Fig. 1. Calculated potential energy curves for two neutral interacting Si_4 clusters. Two different reaction channels have been calculated: the two tetramers approaching each other with the flat (a) and sharp (b) corners ahead. Case (a) is repulsive, while in case (b) a bond is formed. In (b) an energy barrier is observed at a distance of 3.1A and a height of 0.3 eV.

to the situation of the clusters in the gas phase. However, since the interaction with the Van-der-Waals surface is small the results obtained can be considered a good approximation to the case of deposited clusters.

Figure 1 displays the calculated dependencies of potential energy of two interacting Si_4 clusters as a function of distance. Neutral Si_4 in its electronic ground state is a planar rhombus and there are several geometries possible for two tetramers approaching each other. We assume the Si_4 clusters lie flat on the surface and, therefore, we are restricted to planar geometry of two Si_4 approaching each other with the two obtuse (a) or sharp (b) corners encountering. For the geometry displayed in Fig.1a the potential energy increases monotonously with decreasing distance corresponding to a repulsive interaction. The two clusters do not fuse. If the two Si_4 approach with the sharp corners ahead a bond is formed (Fig. 1b). A minimum with a binding energy of 1.3 eV is calculated corresponding to the formation of a Si_8 cluster. Important in Fig.1b is the small increase of the potential energy at a distance of 3.1 Å. This barrier is 0.3 eV high and, therefore, it might not be overcome at kinetic energies corresponding to room temperature. Accordingly, for both reaction channels the calculation predict a repulsive interaction at low temperatures.

These theoretical findings support the results of an experimental [7] study of Si_4 clusters deposited on HOPG at room temperature (AG Ganteför). Both findings support the idea that this magic silicon cluster is suitable as a building block for a new clusters material consisting of pure silicon. If it will finally turn out to be really possible to synthesize such a material this will open a door to a whole new world of material science based on the many magic clusters already found in the gas phase.

3 Electronic and structural properties of nano wires

In this part of the project the structural and electronic properties of atomic gold wires have been computed. Such systems were studied recently by experimental methods [30, 31, 32], where wires have been stretched down to single atom contacts. In this context nano contacts under stress have been simulated [33] using three different interactions: the "surface embedded atom"-interaction with a (up to 70 %) reduced electron density turned out to be not sufficiently stable, with Lennard-Jones interactions shifts of planes and single-atom contacts have been observed, however no atom chains, s. Fig. 2. In case of the stretching of a nano contact with interactions according to the "effective medium theory" (EMT) [34] single atom contacts as well as atom chains have been found, s. Fig. 3. In order to prevent the heating of the wire due to the stretching work, a Nosé-Hoover thermostat has been implemented in the molecular dynamics simulation.

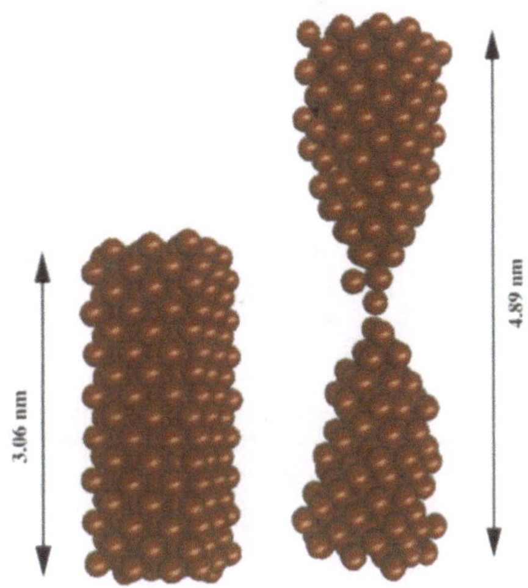

Fig. 2: Typical atom configurations during a stretching process of a Lennard-Jones nano contact at T = 273 K. In the left picture the stretching factor is 1, in the right 1.6.

In cooperation with Jan Heurich and Carlos Cuevas, who developed [35]- by using a tight-binding-model and Greens function techniques- a program for the current through a nano contact (in different channels), conductivity curves have been determined [33]. The qualitative agreement with the experiment is good, the conductivity fluctuations seem to be slightly higher compared to the experiment. The current through the contact depends on the atom type and the atomic configurations at the thinnest part of the chain. However it turns out that the atomic configuration in the surrounding of this position plays an important role as well.

In the experimental studies of a single stretching process a first plateau is found at a conductivity value slightly smaller than G_0, s. Fig. 4. A histogram in Fig. 4 shows the result of averaging over many stretching processes at different temperatures. In order to compute such histograms and to be able to analyze the experimentally observed effects, a large number of single stretching processes is required. A detailed numerical analysis is scheduled as well as a comparative study for different materials. In parallel an improved treatment of the electronic components of the system at the single atom contact is planned by use of the Car-Parrinello- method and the results obtained in the present project (see previous section) at the SSC.

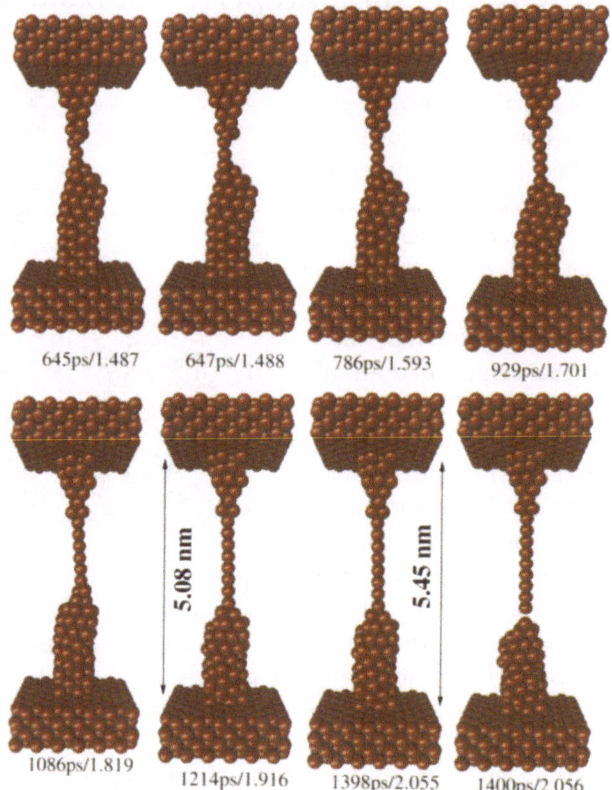

Fig. 3. Typical configuration of an EMT-interacting Au-nano contact at T=4.2 K. During the stretching process more and more atoms from the lower electrode go to the chain, finally resulting in an eight-atom chain. This nano contact can still be stretched considerably, until it breaks finally.

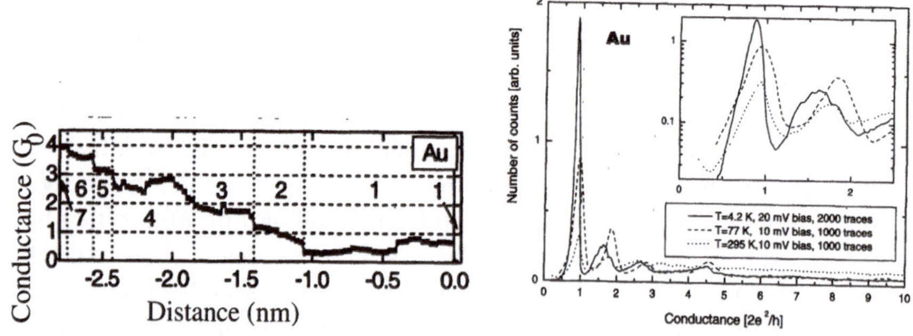

Fig. 4. Left: Experimentally obtained conductance (in units of the conductivity quantum $G_0 = 2e^2/h$) as function of distance (zero corresponds to the break of the nano contact) [From Ref. [31]]. Right: Experimental conductance histogram for gold wires. [From Ref. [36]].

4 Phase transitions in nano-pores

Another research topic in our HLRS-project were structures, phase transitions and quantum effects in pore condensates, which we studied by a combination of finite-size-scaling methods and PIMC [37, 38, 39, 40]. The effect of the finite pore diameter and the strength of the interaction between the particle and the cylinder wall on the structures and phase diagrams has been computed, in particular in the solid phase. Besides a reduction of the critical temperature of the adsorbate-condensate phase transition with decreasing pore diameter several interesting effects on the crystallization scenario have been found. In case of strong particle-wall interactions the system freezes at low temperatures in layers from the cylinder wall to the axis, in case of weak particle-wall interactions solid structures appear locally, which are known from bulk-materials. The crystallization process can consist of two stages, in which parts of the system close to the cylinder wall freeze at a higher temperature than the rest of the system at the pore axis. In experiments [41] two stages crystallization phenomena have been observed as well. Besides this solid-liquid- interfaces with meniscus shape have been found in our studies. Quantum effects on phase diagrams have been quantified by the path integral Monte Carlo method, which is briefly described below:

Canonical averages $< A >$ of an observable A in a system defined by the Hamiltonian $\mathcal{H} = E_{kin} + V_{pot}$ of N particles in a volume V are given by:

$$\langle A \rangle = Z^{-1} \quad \mathrm{tr} \quad [A \exp(-\beta\mathcal{H})] \quad . \tag{1}$$

Here $Z = \mathrm{tr} \quad [\exp(-\beta\mathcal{H})]$ is the partition function and $\beta = 1/k_B T$ is the inverse temperature. Utilizing the Trotter–product formula,

$$\exp(\beta\mathcal{H}) = \lim_{P\to\infty} (\exp(-\beta E_{kin}/P)\exp(-\beta V_{pot}/P))^P \quad , \tag{2}$$

we obtain the path integral expression for the partition function:

$$Z(N,V,T) = \lim_{P\to\infty} \left(\frac{mP}{2\pi\beta\hbar^2}\right)^{3NP/2} \prod_{s=1}^{P} \int d\{\mathbf{r}^{(s)}\} \cdot \tag{3}$$
$$\cdot \exp\left\{-\frac{\beta}{P}\left[\sum_{k=1}^{N} \frac{mP^2}{2\hbar^2\beta^2}(\mathbf{r}_k^{(s)} - \mathbf{r}_k^{(s+1)})^2 + V_{pot}(\{\mathbf{r}^{(s)}\})\right]\right\}$$

Here, m is the particle mass, integer P is the Trotter number and $\mathbf{r}_k^{(s)}$ denotes the coordinate of particle k at Trotter-index s, and periodic boundary conditions apply, $P + 1 = 1$. This formulation of the partition function allows us to perform Monte Carlo simulations [42] for increasing values of P approaching the true quantum limit for $P \to \infty$. Thermal averages in the ensemble with constant pressure p are given via the corresponding partition function $\Delta(N,p,T) = \int_0^\infty dV \exp[-\beta pV] Z(N,V,T)$.

For light (Ne) particles the critical temperature and the condensate density is reduced by about 10 % compared to the classical case. The quantum mechanical ground state oscillations destabilize the system and result in smaller phase transition temperatures. Structural changes due to ground state oscillations have been found as well, in particular the occupancy of positions on the cylinder axis due to packing effects have been analyzed for light and heavy particles. As an important conclusion of our studies it turns out, that a complete overview on the fluid and solid structures and phase diagrams in pore condensates at low temperatures can only be obtained by taking the quantum effects into account- which can be done efficiently by PIMC. Based on these results, we investigate [37, 43] the properties of molecular pore condensates (N_2, CO). Our studies [37, 43] show that at low temperatures interesting solid structures appear, in which particles locally have FCC- or HCP- neighborhood, s. Fig. 5. The influence of the quantum mechanics and the strength of the particle-wall

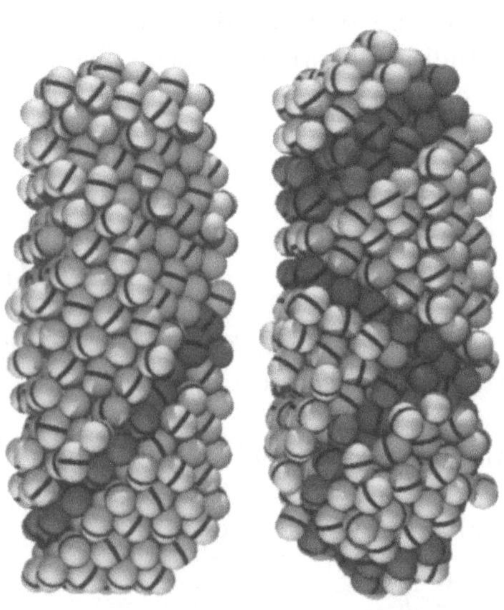

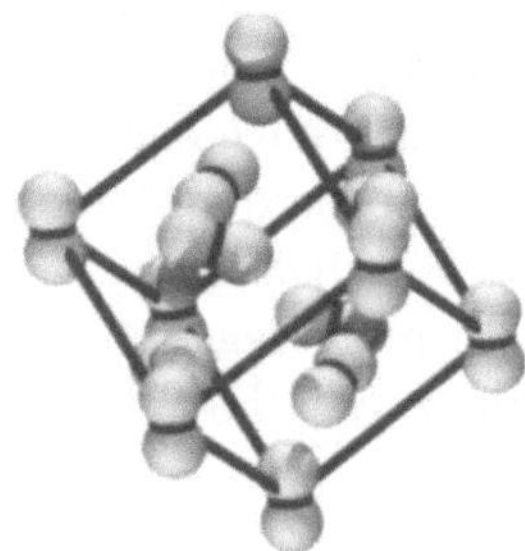

Fig. 5. Solid CO- (left) and N_2- pore condensate in a hard R = 14 Å-pore at 2 K. Light: Particles with FCC-type neighborhood. Dark: HCP-type neighborhood. Particles, which cannot assigned be to any symmetry, are not shown (border particles). Top: a FCC-cell taken from the CO- pore condensate.

interaction and the pore diameter as well as the new phase transitions due to the molecular orientational degrees of freedom will be analyzed.

5 PIMC – studies of hard Bose fluids

Exploiting path integral Monte Carlo methods we studied [44] the behavior of the fluid phase of quantum mechanical hard sphere- and disk- systems including the superfluid phase transition. The partition function for a bosonic system is obtained by - in addition to Eq. (3) - taking into account the connectivity of the permutations $\mathcal{P}$ of the Trotter-chain end points of Bose particles with the Trotter-chain starting points of other Bose particles. By sampling the winding number $\mathbf{W}$ we analyzed the superfluid fraction [45]:

$$\frac{\rho_S}{\rho} = \frac{m}{\hbar^2} \frac{\langle \mathbf{W}^2 \rangle L^2}{2\beta N} + \mathcal{O}(\mathbf{v}^4) \ , \tag{4}$$

where L is the side length of the simulation box, ρ is the density, ρ_S is the superfluid density, and the winding number is the number of times we have to invoke periodic boundary conditions when following a path: $\sum_{i=1}^{N} (\mathbf{r}_{\mathcal{P}_i} - \mathbf{r}_i) = \mathbf{W}L$. In Fig. 6 we show a typical connected Trotter chain spreading over the box and the superfluid fraction of a two dimensional hard Bose fluid as a function of temperature for P=8. The transition is rounded due to the finite

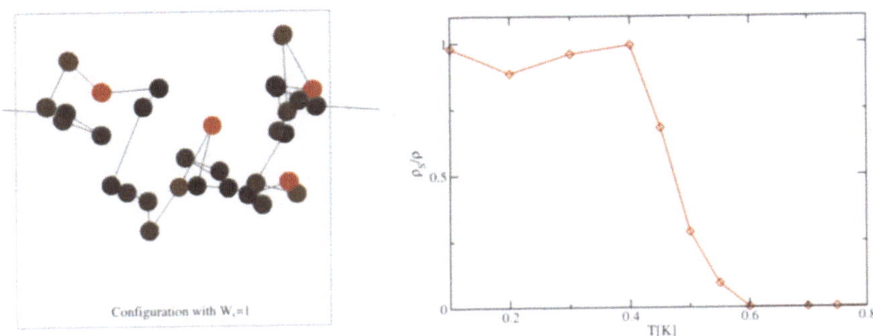

Fig. 6. Left side: Connected Trotter chains for two dimensional hard Bose fluids (P=8) in a configuration with $W_x = 1$. Right side: Superfluid fraction of a two dimensional hard Bose fluid with particles of diameter $\sigma = 2.56$ Å and mass $m = 4u$ at a density $\rho^* = 0.282$ (N=23, P=8).

size of the system (N=23). Further studies of the effect of external potentials on the Bose condensation are scheduled.

6 Elastic and structural properties of model colloids

Other research topics which have been studied by us aim at a better understanding of the properties of colloidal systems in two dimensions. Since about 40 years the nature of the melting transition in the system of hard disks with translational degrees of freedom in two dimensions is under debate in the literature. By a new finite-size-scaling procedure [46] the elastic constants in this system have been analyzed [48, 49]. According to these results the behavior of the system is consistent with the predictions of the KTHNY theory [47, 50]. By application of such a method to configurations obtained experimentally by video microscopy methods it was possible to analyze precisely experimental results on the elasticity of colloidal systems [61].

In future studies we plan to analyze in detail colloidal mixtures with different diameters in two and three dimensions, and to study the composition dependency of their phase behavior and the elastic properties. A priori it is not obvious if such systems are softer or harder compared to the corresponding monodisperse systems, and a systematic study is required in order to design materials with well defined elastic properties at a later stage. Besides this, already in two spatial dimensions interesting structures have been found which significantly deviate from the traditional triangular lattice for certain diameter ratios, s. Fig. 7.

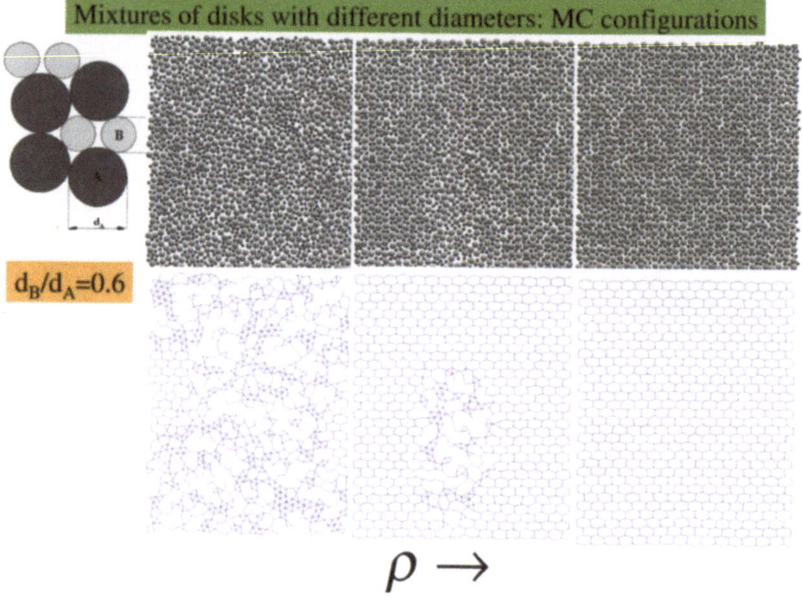

Fig. 7. Typical configurations [60] of a binary system of hard disks with a diameter ratio of 0.6 for small, medium and large densities (from left to right). In the upper part of the picture the particles are shown, in the lower part the connection lines between the nearest neighbors of the large particles for the enhancement of the solid structures at high densities.

The nature of the phase transitions in systems with elliptical particle shapes is of great interest as well, since in this case soft and hard "directions" exist in the solid (parallel and perpendicular to the main particle axes) and deviations from the KTHNY scenario can be expected. Such systems shall be studied in the future as well as systems with antiferromagnetic interactions.

7 Quantum effects on phase diagrams of nano-systems in external potentials

Hard and soft disks in external periodic potentials show rich phase diagrams with freezing and melting phase transitions when the density of the system is varied [51]. In the HLRS-project large scale Monte Carlo simulations have been done in order to determine the phase diagram by detailed finite-size-scaling analyses of different thermodynamic quantities like the order parameter, its cumulants and other quantities for various values of the density and amplitude of the external potential [52, 56, 53, 54, 57, 58]. In case of hard disks we found clear evidence for the appearance of a reentrant fluid phase over a significant region of parameter values [53]. We thus have shown by our simulations, that the hard disk system has a phase behavior as it is known from experiments with charge stabilized colloids [51], a system which undergoes a phase transition from a fluid to a solid phase with increasing amplitude of a periodic laser field, and for larger amplitudes another phase transition to a reentrant fluid phase. Our data are in partial agreement with the results of a theory of laser induced melting. The differences and similarities of the systems with weak potentials (DLVO, $1/r^{12}$) and the relation to the experiment have been analyzed [54, 57, 58].

These results shall be completed by further comparative studies. Besides these classical studies we explore the validity of our results on atomic length scales. In this context we were able to investigate the properties of quantum hard disks with a finite particle mass m and interaction diameter σ in an external periodic potential by PIMC [52, 59]. Due to the quantum delocalization effect a larger effective particle diameter results, and in the external potential this delocalization is asymmetrical: in the direction perpendicular to the potential valleys we obtain a stronger particle localization than parallel to the valleys, s. Fig. 8. As a result the reentrance region in the phase diagram is significantly modified in comparison to the classical case, s. Fig. 9.

without ext. potential

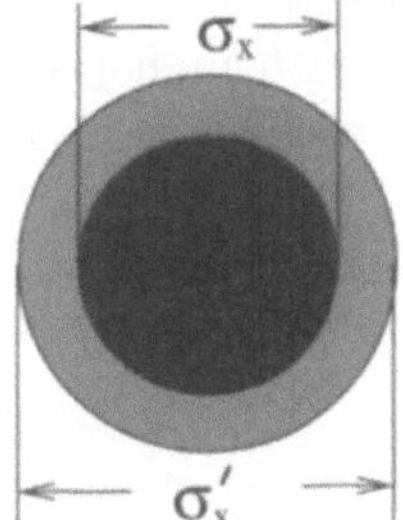

with ext. potential

Fig. 8. Schematic picture of the effect of an external periodic potential of the form $V(x, y) = V_0 \sin(x/a)$ on the "effective" diameter of quantum hard disks.

Due to the larger quantum "diameter" the transition densities at small potential amplitudes are reduced in comparison to the classical values. At large amplitudes the classical and quantum transition densities merge. This effect is due to the approach of the effective quantum disk size to the classical value in the direction perpendicular to the potential valleys and leads to the surprizing prediction, that the quantum crystal in a certain density region has a direct transition to the phase of the modulated liquid by an increase of the potential amplitude. This scenario is not known in the classical case. We plan to explore this interesting topic for systems with different particle masses by PIMC studies and finite-size-scaling methods.

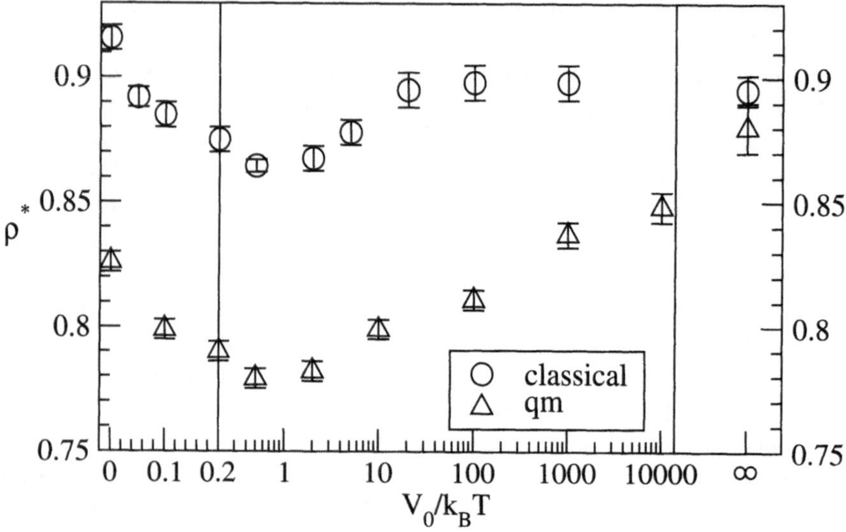

Fig. 9. Phase diagram in the density ($\rho^* = \rho\sigma^2$)- potential amplitude (V_0/k_BT)-plane for a system with N=400 particles, $m^* = mT\sigma^2 = 10.000$ ("qm") and $m^* = \infty$ (classical) and Trotter order $P = 64$.

Besides this we plan to analyze the order of the phase transition at high potential amplitudes by an application of our new method for the computation of elastic constants. The influence of a potential with higher symmetry shall be explored as well as the effect of the potential amplitude on the Bose condensation in case of systems with Bose statistics.

Acknowledgements:

We grateful acknowledge useful discussions with W. Andreoni, C. Bechinger, K. Binder, C. Cuevas, J. Heurich, E. Scheer and S. Sengupta, support by the

SFB 513 and the SFB-TR6 and granting of computer time from the HLRS and the SSC.

References

1. P. Nielaba, in *Annual Reviews of Computational Physics V*, edited by D. Stauffer, p. 137-199 (1997).
2. P. Nielaba, J.L.Lebowitz, H. Spohn, J.L.Valles, J. Stat. Phys. **55**, 745 (1989); P. de Smedt, P. Nielaba, J.L. Lebowitz, J. Talbot, L. Dooms, Phys. Rev. **A38**, 1381 (1988); P. Nielaba, in *Quantum Simulations of Condensed Matter Phenomena*, (Doll, Gubernatis (eds.)) (World Scientific, Singapur 1990); D. Marx, P. Nielaba, und K. Binder, Phys. Rev. Lett. **67**, 3124 (1991); D. Marx, P. Nielaba, und K. Binder, Int. J. Mod. Phys. C **3**, 337 (1992); D. Marx und P. Nielaba, Phys. Rev. A **45**, 8968 (1992); D. Marx, O. Opitz, P. Nielaba, und K. Binder, Phys. Rev. Lett. **70**, 2908 (1993); S. Sengupta, O. Opitz, D. Marx, P. Nielaba, Europhys. Lett. **24**, 13 (1993); D. Marx, P. Nielaba, K. Binder, Phys. Rev. **B 47**, 7788 (1993); S. Sengupta, D. Marx, P. Nielaba, Europhys. Lett. **20**, 383 (1992); D. Marx, S. Sengupta, P. Nielaba, K. Binder, Phys. Rev. Lett. **72**, 262 (1994); O. Opitz, D. Marx, S. Sengupta, P. Nielaba, K. Binder, Surf. Sci. Lett. **297**, L122 (1993); D. Marx, S. Sengupta, O. Opitz, P. Nielaba, K. Binder, Mol.Phys. **83**, 31 (1994); D. Marx, S. Sengupta, P. Nielaba, J. Chem. Phys. **99**, 6031 (1993); D. Marx, S. Sengupta, P. Nielaba, K. Binder, J. Phys.: Condensed Matter **6**, A175 (1994); A.C. Mitus, D. Marx, S. Sengupta, P. Nielaba, A.Z. Patashinskii, H. Hahn, J. Phys.: Condensed Matter **5**, 8509 (1993).
3. P. Nielaba, in: *Computational Methods in Surface and Colloid Science*, M. Borowko (Ed.), Marcel Dekker Inc., New York (2000), pp.77-134.
4. *Bridging Time Scales: Molecular Simulations for the Next Decade*, edited by P. Nielaba, M. Mareschal, G. Ciccotti, Springer, Berlin (2002).
5. M. Dreher, D. Fischer, K. Franzrahe, P. Henseler, J. Hoffmann, W. Strepp, P. Nielaba, in *High Performance Computing in Science and Engineering 02*, edited by E. Krause and W. Jäger, Springer, Berlin, 2003, pp.168.
6. D. Fischer, W. Andreoni, A. Curioni, H. Gröbeck, S. Burkart, G. Ganteför, Chem. Phys. Lett. **361**, 389 (2002).
7. M. Grass, D. Fischer, M. Mathes, G. Ganteför, P. Nielaba, Applied Physics Letters **81**, 3810 (2002).
8. D. Fischer, A. Curioni, W. Andreoni, Langmuir (2003, in press).
9. H.W.Kroto et al, Science **242**, 1139 (1988).
10. M.S.Dresselhaus, G.Dresselhaus, and P.C.Eklund, "Science of Fullerenes and Carbon Nanotubes", Academic Press, San Diego, 1995.
11. H.Kietzmann, R.Rochow, G.Ganteför, W.Eberhardt, K.Vietze, G.Seifert, P.W.Fowler, Phys.Rev.Lett. **81**, 5378 (1998).
12. O.Cheshnovsky, S.H.Yang, C.L.Pettiette, M.J.Craycraft, Y.Liu, and R.E.Smalley, Chem.Phys.Lett. **138**, 119 (1987).
13. J.Müller, Bei Liu, A.A.Shvartsburg, S.Ogut, J.R.Chelikowsky, K.W.M.Siu, Kai-Ming Ho, and G.Ganteför, Phys.Rev.Lett. **85**, 1666 (2000).
14. S.N.Khanna, and P.Jena, Chem.Phys.Lett. **218**, 383 (1993).
15. S.Burkart, N.Blessing, B.Klipp, J.Müller, G.Ganteför, and G.Seifert, Chem.Phys.Lett. **301**, 546 (1999).

16. U.Röthlisberger, W.Andreoni, and M.Parrinello, PRL **72**, 665 (1994).
17. B.Klipp. M.Grass, J.Müller, D.Stolcic, U.Lutz, G.Ganteför, T.Schlenker, J.Boneberg, and P.Leiderer, Appl.Phys.A **73**, 547 (2001).
18. P.Scheier, B.Maersen, M.Lonfat, W.D.Schneider, K.Sattler, Surf.Sci. **458**, 113(2000).
19. W.Yamaguchi, K.Yoshimura, Y.Maruyama, K.Igaraschi, S.Tanemura, and J.Murakami, Chem.Phys.Lett. **311**, 415 (1999).
20. S.J. Caroll, S.G. Hall, R.E. Palmer, R. Smith, Phys. Rev. Lett. **81**, 3715 (1998).
21. J.E.Bower, and M.Jarrold, J.Chem.Phys. **97**, 8312 (1992).
22. P. Hohenberg, W. Kohn, Phys. Rev. **136**, B864 (1964); W. Kohn, L.J. Sham, Phys. Rev. **140**, A1133 (1965).
23. J. P. Perdew, K. Burke, M. Ernzerhof, Phys. Rev. Lett. **77**, 3865 (1996).
24. Calculations used the CPMD code by J.Hutter: CPMD 3.0 Copyright IBM Corporation (1990-1997) and MPI Festkörperforschung Stuttgart, 1997.
25. H. Grönbeck, W. Andreoni, Chem. Phys. **262**, 1 (2000).
26. H. Grönbeck, A. Curioni, W. Andreoni, J. Am. Chem. Soc. **122**, 3839 (2000); W. Andreoni, A. Curioni, H. Grönbeck, Int. J. Quant. Chem. **80**, 598 (2000).
27. R. Car, M. Parrinello, Phys. Rev. Lett. **55**, 2471 (1985).
28. Si PBE pseudopotentials and cluster structures from W. Andreoni (upublished).
29. P. Fenter, A. Eberhardt, P. Eisenberger, Science **266**, 1216 (1994).
30. E. Scheer, P. Joyez, D. Esteve, C. Urbina, M. Devoret; PRL **78**, 3535 (1997).
31. E. Scheer, N. Agrait, J. Cuevas, A. Yeyati, B. Ludolph, A. Rodero, G. Bollinger, J. Ruitenbeck, C. Urbina; Nature **394**, 154 (1998).
32. E. Scheer, W. Belzig, Y. Naveh, M. Devoret, D. Esteve, C. Urbina; Phys. Rev. Lett. **86**, 284 (2000).
33. M. Dreher, Diplomarbeit, Konstanz (2002).
34. K.W. Jacobsen, P. Stoltze, J.K. Norskov; Surf. Sci. **366**, 394 (1996).
35. J. Cuevas et al.; Phys. Rev. Lett. **81**, 2990 (1998).
36. A.I. Yanson, Ph.D. thesis, U. Leiden (2001).
37. J. Hoffmann, Ph.D.-thesis, University of Konstanz (2002).
38. J. Hoffmann, P. Nielaba, Phys. Rev **E67**, 036115 (2003).
39. J. Hoffmann, P. Nielaba, in *High Performance Computing in Science and Engineering 01*, edited by E. Krause and W. Jäger, Springer, Berlin (2002) pp.92.
40. J. Hoffmann, P. Nielaba, in *Computer Simulation Studies in Condensed Matter Physics XV*, ed. by D.P. Landau, S.P. Lewis, H.-B. Schüttler, Springer, Berlin (2003).
41. D. Wallacher, K. Knorr, Phys. Rev. **B63**, 104202 (2001).
42. D.P. Landau and K. Binder, *A Guide to Monte Carlo Simulations in Statistical Physics*, Cambridge University Press (2000).
43. J. Hoffmann, P. Nielaba (in preparation).
44. G. Günther, Ph.D. thesis (in work).
45. E.L. Pollock, D.M. Ceperley; Phys. Rev. **B 36**, 8343 (1987)
46. S. Sengupta, P. Nielaba, M. Rao, K. Binder, Phys. Rev. **E 61**, 1072 (2000).
47. J. M. Kosterlitz, D. J. Thouless, J. Phys. **C 6**, 1181 (1973); B.I.Halperin and D.R.Nelson, PRL **41**,121(1978); D.R.Nelson and B.I.Halperin, PR **B19**, 2457 (1979); A.P. Young, PR **B 19**, 1855 (1979).
48. S.Sengupta,P.Nielaba,K.Binder, PR**E61**,6294(2000).
49. K. Binder, S. Sengupta, P. Nielaba, J. Phys.: Cond. Mat. **14**, 2323 (2002).
50. E.Frey,D.R.Nelson,L.Radzihovsky,PRL**83**,2977(1999).

51. C. Bechinger, M. Brunner, P. Leiderer, Phys. Rev. Lett. **86**, 930 (2001).
52. W. Strepp, Ph.D-thesis, Univ. of Konstanz (2003).
53. W. Strepp, S. Sengupta, P. Nielaba, Phys. Rev. **E63**, 046106 (2001).
54. W. Strepp, S. Sengupta, P. Nielaba, Phys. Rev. **E66**, 056109 (2002).
55. K. Binder, Z. Phys. **B43**, 119 (1981); K. Binder, Phys. Rev. Lett. **47**, 693 (1981).
56. M. Lohrer, Diplom-thesis, Univ. of Konstanz (2001).
57. W. Strepp, S. Sengupta, M. Lohrer, P. Nielaba, Mathematics and Computers in Simulation **62**, 519 (2003).
58. W. Strepp, S. Sengupta, M. Lohrer, P. Nielaba, Comp. Phys. Commun. **147**, 370 (2002).
59. W. Strepp, P. Nielaba, in preparation.
60. G. Schafranek, Diplomarbeit, Konstanz (2002).
61. K. Zahn, A. Wille, G. Maret, S. Sengupta, P. Nielaba; Phys. Rev. Lett. **90**, 155506 (2003).

Gas-Phase Epitaxy Grown InP(001) Surfaces From Real-Space Finite-Difference Calculations

W.G. Schmidt, P.H. Hahn, K. Seino, M. Preuß, and F. Bechstedt

Computational Materials Science Group
Institut für Festkörpertheorie und Theoretische Optik
Friedrich-Schiller-Universität, Max-Wien-Platz 1, 07743 Jena
http://www.ifto.uni-jena.de/en/begin.html

Summary. Density-functional calculations based on finite-difference discretization and multigrid acceleration are used to explore the atomic and spectroscopic properties of P-rich InP(001)(2×1) surfaces grown in gas-phase epitaxy. These surfaces have been reported to consist of a semiconducting monolayer of buckled phosphorus dimers. This apparent violation of the electron counting principle was explained by effects of strong electron correlation. Our calculations show that the (2×1) reconstruction is not at all a clean surface: it is induced by hydrogen adsorbed in an alternating sequence on the buckled P-dimers. Thus, the microscopic structure of the InP growth plane relevant to standard gas-phase epitaxy conditions is resolved and shown to obey the electron counting rule.

1 Introduction

The most widely used numerical discretization methods to solve the equations of density-functional theory (DFT) are linear combination of atomic orbitals (LCAO), plane-waves (PW) and finite differences (FD). Of these three, FD is the most recent method. Fully three-dimensional grid-based electronic structure representations using finite differences as approximate numerical scheme for partial differential operators have started being widely used in the last ten years only. Traditionally, chemists have mostly used LCAO methods. Atomic orbitals expressed as linear combinations of Gaussian functions allow for a numerically very efficient description of the electronic structure of localized systems such as molecules. This is exploited in the wide spread commercial code GAUSSIAN. Plane waves, on the other hand, efficiently describe systems characterized by periodic boundary conditions [1]. Because plane waves are eigenfunctions of the momentum operator, the PW approach has been most successful at describing systems with almost free electrons, such as metals. By its nature, the PW method has mainly been developed and used by condensed matter physicists. In this approach, known as pseudo-spectral method in the mathematics community, the numerical basis set is completely independent of the positions of the atoms present in the simulation. It can be made as accurate as desired by systematically increasing the number of basis functions. Similarly, the finite-difference method is an alternative to the linear combination of atomic orbitals when highly accurate electronic wave functions are required. No assumption on where the atoms and electrons are located is made. In recent years, large parallel supercomputers have become an essential tool in *first-principles* molecular dynamics simulations. They not only allow for calculations to be completed within hours or days that would take months or years to be completed on single processor machines, but also for calculations that would just not fit into the memory of a workstation or personal computer. In a parallel environment, the electronic wave functions described in a PW approach are usually distributed between the processors. Every time a matrix element between two wave functions is required, a considerable data traffic through the whole machine is necessary. In a real-space FD approach, all the expensive operations can be done locally, thanks to the real-space nature of the DFT Hamiltonian. To compute matrix elements between wave functions, local contributions are computed on every processor before being summed up globally at the end. This is one of the main advantages of FD over PW for nowadays simulations. It has been exploited extensively by our group in the context of developing codes which allow for the efficient calculation of many-body effects in excitation spectra [2, 3, 4]. Furthermore, large-scale real-space calculations involve large sparse matrices, and multigrid methods [5], either as solvers or preconditioners, allow to design very efficient scalable algorithms [6, 7, 8]. More recently, in the context of the search for linear scaling algorithms, real-space methods have appeared to be appropriate for imposing natural localization constraints on the orbitals [9]. Such an approach leads to a dramatic reduction in computer time and memory requirements for very large systems. Other advantages of FD over standard PW approaches include the possibility of introducing local mesh refinements [10] and Dirichlet boundary conditions in a natural fashion.

Advanced numerical methods such as discussed above are useful only if they allow for treating real-world problems. One such problem is the accurate determination of the atomic and electronic structure of semiconductor growth planes. While the properties of clean surfaces under ultrahigh vacuum (UHV) conditions are reasonably well understood, very little is known about the structure of surfaces under growth conditions. In the present article we clarify the hitherto open structure of the gas-phase grown InP(001)(2×1) surface by means of accurate total-energy and electronic structure calculations based on a FD scheme [8]. To our knowledge, this is the first example of an atomically resolved gas-phase epitaxy growth structure.

The ideal III-V(001) surface is polar, being terminated either entirely by cations or anions. However, such an ideal termination is usually not observed [11, 12, 13]. InP(001) seems to be a remarkable exception in that respect. For specific P-rich preparation conditions it is apparently terminated by a complete phosphorus monolayer, forming a nominal (2×1) reconstructed surface [14, 15]. Such a structure clearly violates the electron counting rule [16] and should be metallic. However, in this case an energy gap between valence and conduction states of more than one eV was measured. The formation of this unusual surface structure and the band gap were explained by strong electron correlation effects between the P dangling bonds across the dimer rows [14]. As a result, the P dimers buckle and form zig-zag chains along the [110] direction. Zig-zag chains are indeed clearly resolved by scanning tunneling microscopy (STM) [14, 15]. Depending on whether two adjacent chains are in or out of phase, (2×2) and c(4×2) reconstructed domains, reminiscent of the Si(001) surface, are observed.

Such observations, however, have exclusively been made by groups using metal-organic vapor phase epitaxy (MOVPE) and chemical beam epitaxy (CBE) to prepare their InP samples [14, 15, 17, 18, 19]. Samples prepared by molecular beam epitaxy (MBE) show only two reconstructions with long range periodicity, c(4×4) and (2×4) [20, 21]. On the other hand, to our knowledge the c(4×4) surface has never been observed on gas-source grown samples. The MBE results agree with total-energy calculations [13], which identified several stable surface geometries with (2×4) and c(4×4) periodicity. An extensive computational search for geometries able to explain the peculiar (2×1) surface ordering failed [22]. Symmetric, rather than asymmetric P dimers were found to be energetically favored.

A solution of this puzzle is highly desirable: gas-phase epitaxy is an essential technology for manufacturing compound semiconductor devices, in particular those containing phosphide-based materials. The microscopic understanding of surfaces prepared by MOVPE or CBE is the prerequisite for any reliable computational modeling of the complex transport and reaction phenomena during the epitaxial growth. In the present study we explore the possibility that the apparent discrepancies between the outcome of gas-phase epitaxy and MBE experiments can be explained by the presence of hydrogen in the former case: While hydrogen is present in MOVPE and CBE, it is absent under MBE conditions. Furthermore, hydrogen is difficult to detect. It may thus be the key to resolve the puzzle of the (2×1) surface.

2 Computational Method

Our calculations are based on density-functional theory [23] which states that the electronic ground state of a physical system can be described by a system of orthogonal one-particle electronic wave functions $\psi_j, j = 1, ..., N$ that minimizes the Kohn-Sham total-energy functional

$$E_{KS}[\{\psi_i\}_{i=1}^N, \{\mathbf{R}\}_{\alpha=1}^{\mathbf{N}_\alpha}] = \sum_{i=1}^N f_i \int_\Omega \psi_i^*(\mathbf{r})(-\frac{1}{2}\nabla^2)\psi_i(\mathbf{r})d\mathbf{r} \tag{1}$$
$$+\frac{1}{2}\int_\Omega\int_\Omega \frac{\rho_e(\mathbf{r}')\rho_e(\mathbf{r}'')}{|\mathbf{r}'-\mathbf{r}''|}d\mathbf{r}'d\mathbf{r}'' + E_{XC}[\rho_e],$$
$$+\int_\Omega \psi_i^*(\mathbf{r})(V_{ext}\psi_i)(\mathbf{r})d\mathbf{r}$$

where N_α atoms are located at $\{\mathbf{R}\}_{\alpha=1}^{\mathbf{N}_\alpha}$, and the electronic density is given by

$$\rho_e(\mathbf{r}) = \sum_{i=1}^N f_i|\psi_i(\mathbf{r})|^2 , \tag{2}$$

with $0 \leq f_i \leq 2$ being the occupation numbers. E_{XC} models the exchange and correlation between electrons. Here we use the local density approximation (LDA) in the parameterization of Ref. [24]. The minimum of the energy functional (1) can be found by solving the associated Euler-Lagrange (or Kohn-Sham) equations [25]

$$H\psi_j = \left\{-\frac{1}{2}\nabla^2 + v_H(\rho_e) + \mu_{XC}(\rho_e) + V_{ext}\right\}\psi_j = \epsilon_j\psi_j \tag{3}$$

self-consistently for the N lowest eigenvalues ϵ_j, while imposing the orthonormality constraints $\langle\psi_j|\psi_i\rangle = \delta_{ji}$. The Hartree potential v_H represents the Coulomb potential due to the electronic charge density ρ_e, and $\mu_{XC} = \delta E_{XC}[\rho_e]/\delta\rho_e$ is the exchange and correlation potential. In order to discretize the Kohn-Sham equations, a real-space rectangular grid Ω_h of mesh spacing h_x, h_y, h_z in the x, y, and z directions is introduced, covering the computation domain Ω. Let M be the number of grid points. The wave functions, potentials and the electronic density are represented by their values at the grid points $\mathbf{r}_{ijk}$. Integrals over Ω are performed using the discrete summation rule

$$\int_\Omega u(\mathbf{r})d\mathbf{r} \approx h_xh_yh_z \sum_{i,j,k\in\Omega_h} u(\mathbf{r}_{ijk}) . \tag{4}$$

Given the values of a function $u(\mathbf{r})$ on a set of nodes $\mathbf{r}_{ijk}$, the traditional FD approximation $w_{i,j,k}$ to the Laplacian of the function at a given node is expressed as a linear combination of values of the function at the neighboring nodes

$$w_{i,j,k} = \sum_{n=-p}^{p} c_n(u(x_i+nh_x, y_j, z_k) + u(x_i, y_j+nh_y, z_k) + u(x_i, y_j, z_k+nh_z)) \tag{5}$$

where the coefficients $\{c_n\}$ can be computed from the Taylor expansion of u near $\mathbf{r}_{ijk}$. Such an approximation has an order of accuracy $2p$, that is for a sufficiently smooth function u, $w_{i,j,k}$ will converge at the rate $\mathcal{O}(h^{2p})$ as the mesh spacing $h \longrightarrow 0$. For the second order approximation for example ($p = 1$), we have $c_0 = 2/h^2$ and $c_1 = -1/h^2$. High order versions of this scheme were first introduced in electronic structure calculations by Chelikowski and co-workers [26]. In the present work we use an alternative, compact FD scheme called *Mehrstellenverfahren* [27]. In contrast to the central difference approximation, this discretization uses more *local* information. The fourth-order *Mehrstellen* discretization leads to an eigenvalue problem of the form

$$\left\{\frac{1}{2}L_h + B_h V_h\right\} \psi_i = \epsilon_i B_h \psi_i \tag{6}$$

for the Kohn-Sham equations (3) with operators defined by

$$L_h u(\mathbf{r}) = \frac{4}{h^2}u(\mathbf{r}_0) - \frac{1}{3h^2}\sum_{\substack{r\in\Omega_h \\ |\mathbf{r}-\mathbf{r}_0|=h}} u(\mathbf{r}) - \frac{1}{6h^2}\sum_{\substack{r\in\Omega_h \\ |\mathbf{r}-\mathbf{r}_0|=\sqrt{2}h}} u(\mathbf{r}) \tag{7}$$

$$B_h u(\mathbf{r}) = \frac{1}{2}u(\mathbf{r}_0) + \frac{1}{12}\sum_{\substack{r\in\Omega_h \\ |\mathbf{r}-\mathbf{r}_0|=h}} u(\mathbf{r}). \tag{8}$$

For simplicity, we have assumed here that $h_x = h_y = h_z$, but this expression is easily generalized [7]. This scheme requires only values at grid points within a sphere of radius $\sqrt{2}h$. Besides its good numerical properties (see, e.g. Ref. [7]), its compactness reduces the amount of communication in a domain-decomposition based parallel implementation compared to traditional FD approximations.

This scheme was introduced in electronic structure calculations by Briggs *et al.* [6] and Fattebert [28]. To efficiently solve (6), we use multigrid iteration techniques that accelerate convergence by employing a sequence of grids of varying resolutions. The final solution is obtained on a grid fine enough to accurately represent the potential and wave functions. This procedure provides excellent preconditioning for all length scales present in the system and leads to very rapid convergence rates. The operation count to converge one wave function with a fixed potential is $\mathcal{O}(M)$, compared to $\mathcal{O}(M \log M)$ for PW approaches.

The InP(001) surface structures studied in the following were modeled using periodic supercells. They contain material slabs of 16 InP(001) layers, separated by vacuum regions large enough to decouple the surfaces. This corresponds to a grid of $M = 32 \times 32 \times 144$ points. Between 170 – 270 wave functions were included in the calculations. The real-space grid was mapped on 64 – 256 Cray T3E processors. Parallelizing is done using Message Passing Interface (MPI). Earlier we have shown the excellent scaling behaviour of this method using up to 512 Cray T3E processors (Fig. in 1 in Ref. [29]). Typical job run times amount to 4 – 12 wall clock hours.

The possibility of hydrogen adsorption on gas-phase grown InP surfaces leads to a large number of conceivable structures. We investigate more than

50 models, which differ with respect to their geometries, their In/P ratio, and the number of adsorbed hydrogen atoms. The energetically favored hydrogen-induced surface reconstructions are shown in Fig. 1. The notation of surface structures is chosen such that a leading P indicates adsorption on top of a P-terminated substrate and a hyphen followed by P, D or MD denotes adsorption of P atoms, P dimers or mixed In-P dimers, respectively. The number of hydrogens per surface unit cell concludes the notation.

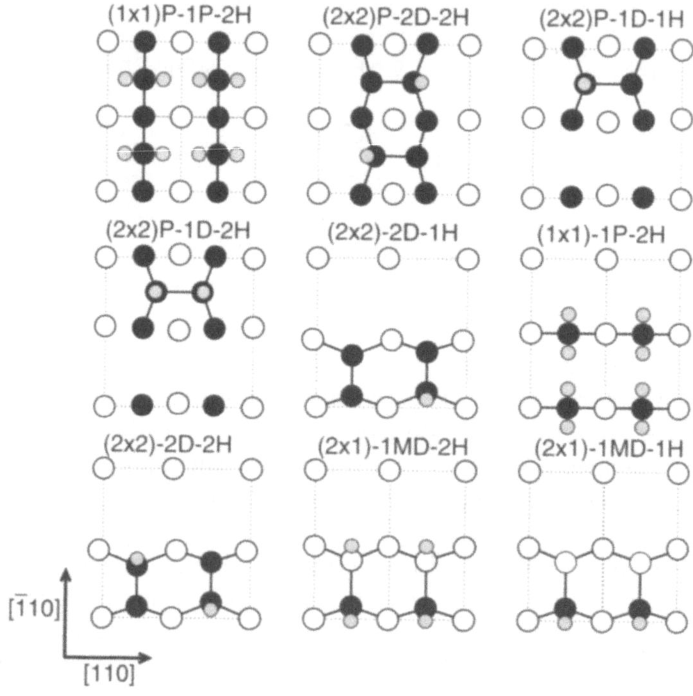

Fig. 1. Top view of relaxed InP(001):H surface structures. Empty (filled,grey) circles represent In (P, H) atoms. The surface unit cells are indicated.

Due to the varying surface stoichiometry, the total energies of the studied structures cannot directly be used to determine the surface ground state. Rather, the thermodynamic grandcanonical potential Ω in dependence on the chemical potentials μ of In, P and H needs to be considered. Since the surface is in equilibrium with the bulk compound, μ_{In} and μ_P are related to each other: their sum equals the chemical potential of bulk InP. Consequently, the surface formation energy may be written as a function of only two variables, which we take to be the relative chemical potential of indium with respect to its bulk phase and the chemical potential of hydrogen with respect to its molecular phase. The computational accuracy in determining the chemical potentials is of the order of 0.1 eV [30]. From convergence tests, the uncertainty of the calculated surface energies is estimated to be smaller than 0.01 eV per surface atom.

3 Results

The calculated surface phase diagram in terms of its dependence on the chemical potentials of In and H is shown in Fig. 2. Here zero hydrogen chemical potential corresponds to the situation where the surface is exposed to molecular hydrogen at $T = 0$ K. For higher temperatures the surface phase diagram will change, due to vibrational contributions to the energy and entropy of the surface structures, and due to the temperature (and pressure) dependence of the chemical potentials of the surface constituents. By far the largest change of the surface energetics, however, is related to the hydrogen chemical potential. The temperature and pressure dependence of $\Delta\mu_H$ can be approximated by that of a two-atomic ideal gas [31].

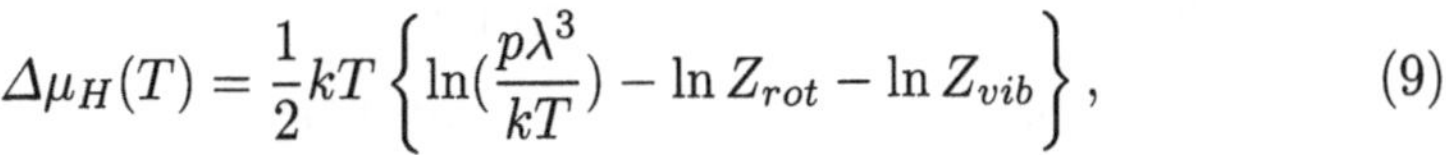

$$\Delta\mu_H(T) = \frac{1}{2}kT\left\{\ln(\frac{p\lambda^3}{kT}) - \ln Z_{rot} - \ln Z_{vib}\right\}, \tag{9}$$

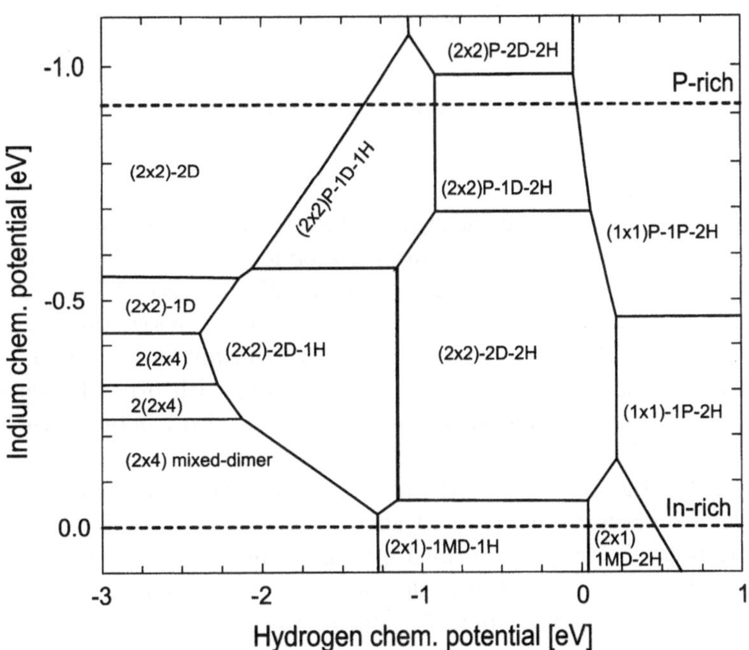

Fig. 2. Calculated phase diagram of the hydrogen-exposed InP(001) surface. Dashed lines indicate the approximate range of the thermodynamically allowed values of the In chemical potential. The chemical potential of hydrogen is given with respect to molecular hydrogen at $T = 0$ K.

where k is the Boltzmann constant, T is the temperature, p the pressure, $\lambda = (\frac{h^2}{2\pi mkT})^{1/2}$ the de Broglie thermal wavelength of the H_2 molecule, and Z_{rot} and Z_{vib} its rotational and vibrational partition functions, respectively. By increasing the temperature, the hydrogen chemical potential is lowered, i.e., less energy is gained by taking a H atom out of the reservoir and attaching it to the InP surface. In Fig. 3 we show the temperature and pressure values corresponding to specific values for $\Delta\mu_H(T)$. A hydrogen chemical potential $\Delta\mu(\mathrm{H})$ of about -1 eV can be estimated to correspond to typical MOVPE growth conditions.

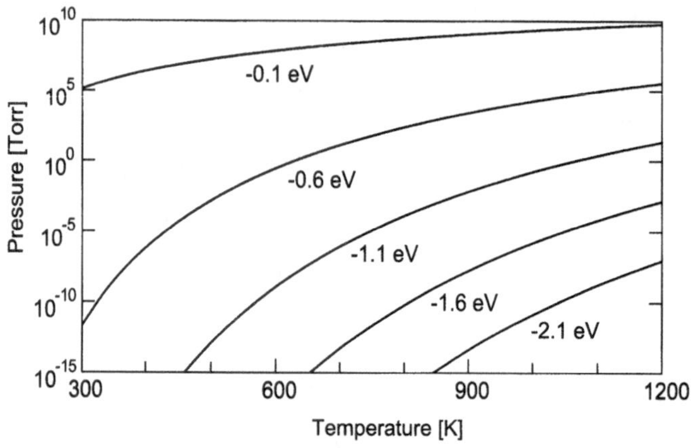

Fig. 3. Temperature and pressure values corresponding to given relatives change of the hydrogen chemical potential with respect to its molecular value at T = 0 K.

The structures indicated for $\Delta\mu_H > 0$ in the phase diagram of Fig. 2 may form when atomic hydrogen becomes available for the surface reaction. In that case the reaction barrier is also substantially lowered. If no hydrogen is present, i.e. for $\Delta\mu_H \ll 0$, InP forms the surface reconstructions typical for the clean surface [13]. The $c(4{\times}4)$ reconstruction, not visible in the present phase diagram, is energetically nearly degenerate with other P-rich surface reconstructions. It is stable for a narrow range of $\Delta\mu_{In}$ between the (2×2)-2D and (2×2)-1D surfaces.

The (2×2)-2D-2H surface is the most dominant structure in the calculated surface phase diagram in Fig. 2. It should occur for a wide range of surface preparation conditions. It is formed by a periodic arrangement of oppositely buckled ($\Delta z = 0.30$ Å) P dimers on top of an In-terminated substrate. One H atom is bonded to the "down" atom of the P dimer (cf. Fig. 1). While this structure seems to represent the surface ground state for annealed gas-phase epitaxy grown InP(001) surfaces, energetical arguments alone are only an indication that this structure indeed corresponds to the surface observed experimentally [14, 15], because its formation may be kinetically hindered.

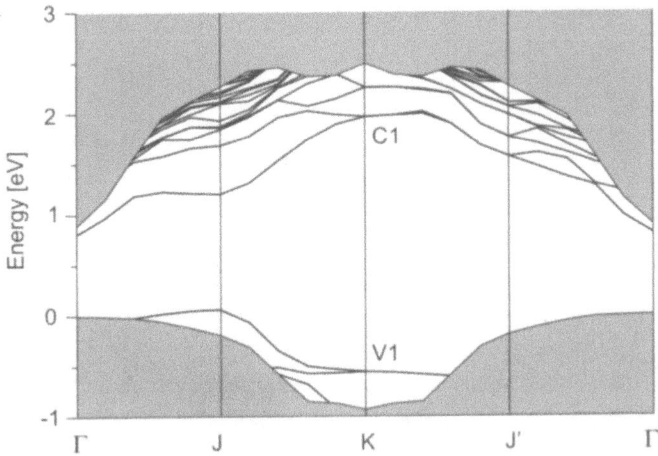

Fig. 4. Surface band structure for the (2x2)-2D-2H surface. Grey regions indicate the projected bulk band structure.

In order to clarify whether the 2D-2H structure indeed corresponds to the experimental observations, we investigate its electronic structure. The model obeys the electron counting rule. Its surface bands calculated along the high-symmetry lines of the (2×2) surface Brillouin zone are shown in Fig. 4. The surface band gap calculated in DFT-LDA amounts to 0.75 eV. This value is lower than the one measured for the zig-zag chain structure of 1.2 ± 0.2 eV [14]. This discrepancy, however, can be explained by the insufficient description of the electronic self-energy within the DFT-LDA [32]. The inclusion of quasiparticle effects results in an opening of the InP(001) surface band gap by about 0.4 – 0.5 eV [33].

The highest occupied surface band, V1, is due to the dangling bonds on the "up" atoms of the Phosphorus dimers. The lowest unoccupied surface band, C1, is related to an antibonding σ^* combination of in-plane p orbitals localized at the P dimer atoms (cf. Fig. 5).

In Fig. 6 we show a STM image of the (2x2)-2D-2H surface, calculated according to the Tersoff-Hamann approach [34]. A bias voltage of -5 eV was used, to allow for a meaningful comparison with the experiments of Refs. [14, 15]. Clearly, the simulated STM image is in very good agreement with the measured data (cf., e.g., Fig. 2 of Ref. [14]). The bright spots visible in the image are due mainly to the dangling bonds at the "up" atom of the P dimer, i.e., the V1 surface state. The alternating arrangement of the dangling bonds causes the appearance of zig-zag chains in the STM. By varying the phase shift between adjacent zig-zag chains, the observation of (2×2) and $c(4\times2)$ reconstructed domains can be explained.

We mention that experimental evidence for the existence of hydrogen at annealed, MOVPE-grown InP(001)(2×1) surfaces has recently been obtained

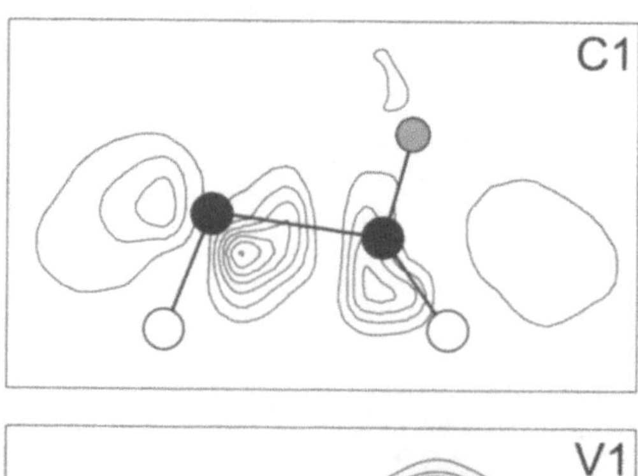

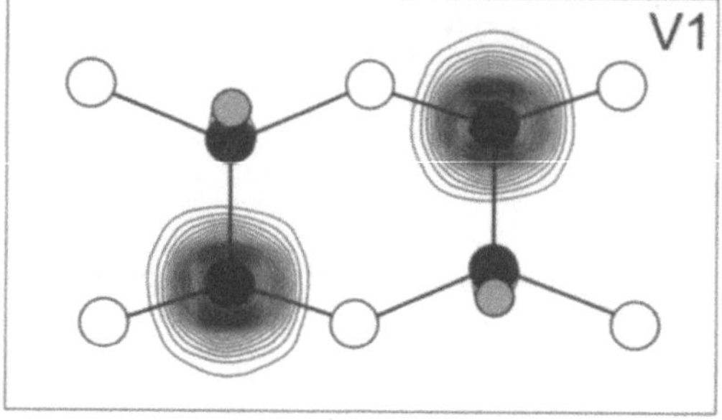

Fig. 5. Contour plots of the squared wave functions of the (2x2)-2D-2H surface at the K point of the (2×2) surface Brillouin zone. Top and side views are given for V1 and C1, respectively.

by optical spectroscopy. The comparison between the calculated and measured surface optical response [35] lends further credibility to the 2D-2H structure predicted here.

4 Conclusions

In conclusion, we performed large-scale real-space finite-difference calculations for a large variety of clean and H-covered InP(001) surfaces. Oppositely buckled P dimers with one hydrogen adsorbed are energetically favored for a wide range of the surface chemical potentials. The simulated STM image as well as the calculated surface band gap for this structure agree very well with the experimental findings obtained for P monolayer-terminated InP(001) surfaces prepared by the annealing of gas-phase epitaxy grown samples. The adsorption of hydrogen thus provides a simple explanation, as opposed to electron-correlation effects, for the experimental observation of a Si(001)-like surface ordering and the existence of a band gap. Our results underline the strong influence of the surface preparation procedures on the geometries finally obtained. In particular the importance of adsorbates resulting from decomposition products of precursors on the microscopic structure of gas-phase epitaxy grown surfaces has been demonstrated.

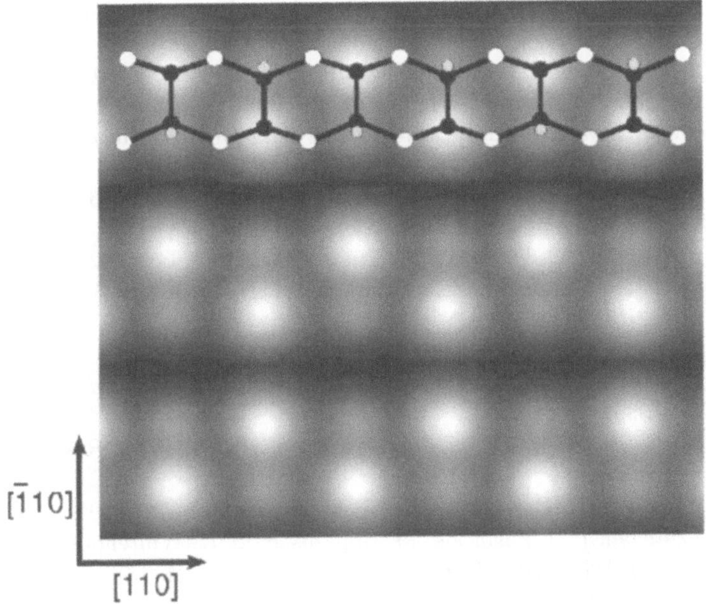

Fig. 6. STM image calculated for the (2x2)-2D-2H surface.

Acknowledgments

We thank Norbert Esser, Patrick Vogt, Wolfgang Richter, Lars Töben and Frank Willig for stimulating discussions on the growth of III-V compound semiconductors. Jean-Luc Fattebert, Emil Briggs and Jerry Bernholc are acknowledged for their help concerning the numerical and computational aspects of this work. Generous grants of computer time from the the Höchstleistungsrechenzentrum Stuttgart and the Leibniz-Rechenzentrum München made the calculations possible.

References

1. M. C. Payne, M. P. Teter, D. C. Allan, T. A. Arias, and J. D. Joannopoulos, Rev. Mod. Phys. **64**, 1045 (1992).
2. W. G. Schmidt, J. L. Fattebert, J. Bernholc, and F. Bechstedt, Surf. Rev. Lett. **6**, 1159 (1999).
3. P. H. Hahn, W. G. Schmidt, and F. Bechstedt, Phys. Rev. Lett. **88**, 016402 (2002).
4. W. G. Schmidt, S. Glutsch, P. H. Hahn, and F. Bechstedt, Phys. Rev. B **67**, 085307 (2003).
5. A. Brandt, Math. Comp. **31**, 333 (1977).
6. E. L. Briggs, D. J. Sullivan, and J. Bernholc, Phys. Rev. B **52**, R5471 (1995).
7. E. L. Briggs, D. J. Sullivan, and J. Bernholc, Phys. Rev. B **54**, 14362 (1996).

8. J. Bernholc, E. L. Briggs, C. Bungaro, M. B. Nardelli, J. L. Fattebert, K. Rapcewicz, C. Roland, W. G. Schmidt, and Q. Zhao, phys. stat. sol. (b) **217**, 685 (2000).
9. J. L. Fattebert and J. Bernholc, Phys. Rev. B **62**, 1713 (2000).
10. J. L. Fattebert, J. Comput. Phys. **149**, 75 (1999).
11. W. Mönch, *Semiconductor Surfaces and Interfaces* (Springer-Verlag, Berlin, 1995).
12. Q.-K. Xue, T. Hashizume, and T. Sakurai, Prog. Surf. Sci. **56**, 1 (1997).
13. W. G. Schmidt, Appl. Phys. A **75**, 89 (2002).
14. L. Li, B.-K. Han, Q. Fu, and R. F. Hicks, Phys. Rev. Lett. **82**, 1879 (1999).
15. P. Vogt, T. Hannappel, S. Visbeck, K. Knorr, N. Esser, and W. Richter, Phys. Rev. B **60**, R5117 (1999).
16. M. D. Pashley, Phys. Rev. B **40**, 10481 (1989).
17. L. Li, B.-K. Han, D. Law, C. H. Li, Q. Fu, and R. F. Hicks, Appl. Phys. Lett. **683**, 75 (1999).
18. L. Li, Q. Fu, C. H. Li, B.-K. Han, and R. F. Hicks, Phys. Rev. B **61**, 10223 (2000).
19. B. X. Yang and H. Hasegawa, Jpn. J. Appl. Phys. **33**, 742 (1994).
20. K. B. Ozanyan, P. J. Parbrook, M. Hopkinson, C. R. Whitehouse, Z. Sobiesierski, and D. I. Westwood, J. Appl. Phys. **82**, 474 (1997).
21. V. P. LaBella, Z. Ding, D. W. Bullock, C. Emery, and P. M. Thibado, J. Vac. Sci. Technol. A **18**, 1492 (2000).
22. O. Pulci, K. Lüdge, W. G. Schmidt, and F. Bechstedt, Surf. Sci. **464**, 272 (2000).
23. P. Hohenberg and W. Kohn, Phys. Rev. **136**, B864 (1964).
24. J. P. Perdew and A. Zunger, Phys. Rev. B **23**, 5048 (1981).
25. W. Kohn and L. J. Sham, Phys. Rev. **140**, A1133 (1965).
26. J. R. Chelikowski, N. Troullier, and Y. Saad, Phys. Rev. Lett. **72**, 1240 (1994).
27. L. Collatz, *The Numerical Treatment of Differential Equations* (Springer-Verlag, Berlin, 1966).
28. J. L. Fattebert, BIT **36**, 509 (1996).
29. W. G. Schmidt, P. H. Hahn, and F. Bechstedt, in *High Performance Computing in Science and Engineering 2001* (Springer-Verlag, Berlin, 2002), Chap. GaAs and InAs(001) Surface Structures from Large-scale Real-space Multigrid Calculations, p. 178.
30. W. G. Schmidt, Appl. Phys. A **65**, 581 (1997).
31. L. D. Landau and E. M. Lifshitz, *Lehrbuch der Theoretischen Physik* (Akademie-Verlag, Berlin, 1987), Vol. 5.
32. F. Bechstedt, in *Festköperprobleme / Advances in Solid State Physics*, edited by U. Rössler (Vieweg, Braunschweig/Wiesbaden, 1992), Vol. 32, p. 161.
33. W. G. Schmidt, N. Esser, A. M. Frisch, P. Vogt, J. Bernholc, F. Bechstedt, M. Zorn, T. Hannappel, S. Visbeck, F. Willig, and W. Richter, Phys. Rev. B. **61**, R16335 (2000).
34. J. Tersoff and D. R. Hamann, Phys. Rev. B **31**, 805 (1985).
35. W. G. Schmidt, P. H. Hahn, F. Bechstedt, N. Esser, P. Vogt, A. Wange, and W. Richter, Phys. Rev. Lett. **90**, 126101 (2003).

Amorphous Silica at Surfaces and Interfaces: Simulation Studies

J. Horbach[1], T. Stühn[1], C. Mischler[1], W. Kob[2], and K. Binder[1]

[1] Institut für Physik, Johannes Gutenberg–Universität, Staudinger Weg 7, D–55099 Mainz, Germany
[2] Laboratoire des Verres, Université Montpellier II, Place E. Bataillon, F–34095 Montpellier, France

Summary. The structure of surfaces and interfaces of silica (SiO_2) is investigated by large scale molecular dynamics computer simulations. In the case of a free silica surface, the results of a classical molecular dynamics simulation are compared to those of an *ab initio* method, the Car–Parrinello molecular dynamics. This comparative study allows to check the accuracy of the model potential that underlies the classical simulation. By means of a pure classical MD, the interface between amorphous and crystalline SiO_2 is investigated, and as a third example the structure of a silica melt between walls is studied in equilibrium and under shear. We show that in the latter three examples important structural information such as ring size distributions can be gained from the computer simulation that is not accessible in experiments.

1 Introduction

The understanding of the structural and dynamic properties of silica at surfaces and interfaces is of special importance for technologically interesting systems such as semiconductor devices and nanoporous materials [1, 2, 3]. Furthermore, since amorphous silica surfaces are very reactive, they are used in catalysis and chromatography. Although there are different experimental methods such as infrared and Raman spectroscopy, atomic force microscopy and NMR to study surface properties, at least in the case of silica it is still difficult to make conclusive statements about the microscopic surface structure of SiO_2 from experiments that use the latter methods [4, 5, 6, 7].

In order to shed more light on these issues, a molecular dynamics (MD) computer simulation is a very useful tool. In this method Newton's equations of motion are solved and thus one obtains the trajectories of the particles in the system from which one can calculate all the relevant quantities to characterize the microscopic structure (and dynamics). However, the predictive power of a MD simulation depends strongly on the quality of the potential with which one models the interactions between the particles. One possibility to check the accuracy of the model potential is a comparison to experiment. But, especially in the case of the surface structure of amorphous silica, there is a lack of experimental data. An alternative method to check the potential is presented in the next section: We use a combination of classical MD with a model potential and an *ab initio* simulation technique, the so–called Car–Parrinello MD. The use of classical MD for the investigation of the non–bulk behavior of amorphous SiO_2 is then presented for two other subjects, namely the structure of an interface between amorphous and crystalline SiO_2 (Sec. 3) and the behavior of amorphous silica between two walls in equilibrium and under shear (Sec. 4). Finally we summarize the results in Sec. 5.

2 Free silica surfaces: *Ab initio* and classical molecular dynamics

As a model to describe the interactions between the atoms in silica we use a pair potential which has been proposed by van Beest *et al.* [8]. It contains a Coulomb part and a short–ranged part,

$$\phi(r) = \frac{q_\alpha q_\beta e^2}{r} + A_{\alpha\beta} \exp\left(-B_{\alpha\beta} r\right) - \frac{C_{\alpha\beta}}{r^6} \quad \alpha, \beta \in [\mathrm{Si}, \mathrm{O}] \, , \tag{1}$$

where r is the distance between an atom of type α and an atom of type β. The effective charges are $q_{\mathrm{O}} = -1.2$ and $q_{\mathrm{Si}} = 2.4$, and the parameters $A_{\alpha\beta}$, $B_{\alpha\beta}$, and $C_{\alpha\beta}$ can be found in the original publication. They were fixed by using a mixture of *ab initio* calculations and classical lattice dynamics simulations. The long–ranged Coulomb forces (and the potential) were evaluated by means of the Ewald summation technique. As an integrator for the simulation we used a velocity Verlet algorithm [9] with a time step of 1.6 fs. More details on the potential and the calculation of the forces can be found in Ref. [10].

Although the BKS potential has turned out to be very good in the description of *bulk properties* of amorphous silica (see Ref. [11] and references therein), it is much less obvious that this potential is also reliable to model silica surfaces. The reason is that it was optimized to reproduce bulk properties such as the experimental elastic constants of α quartz. A similar fitting of the potential parameters to surface properties of real silica is difficult because of the lack of experimental data in this case. Although it is straightforward to investigate the properties of free surfaces of SiO_2 by means of a MD using the BKS potential, it is not clear how one could test whether the results have anything in common with real silica.

One possibility to circumvent this problem is to use Car–Parrinello molecular dynamics (CPMD) [12] in which, different from a classical MD, the electronic degrees of freedom are taken into account via a density functional theory. Therefore, in contrast to a classical MD an effective potential between the ions is calculated self–consistently on the fly, i.e. the instantaneous geometry

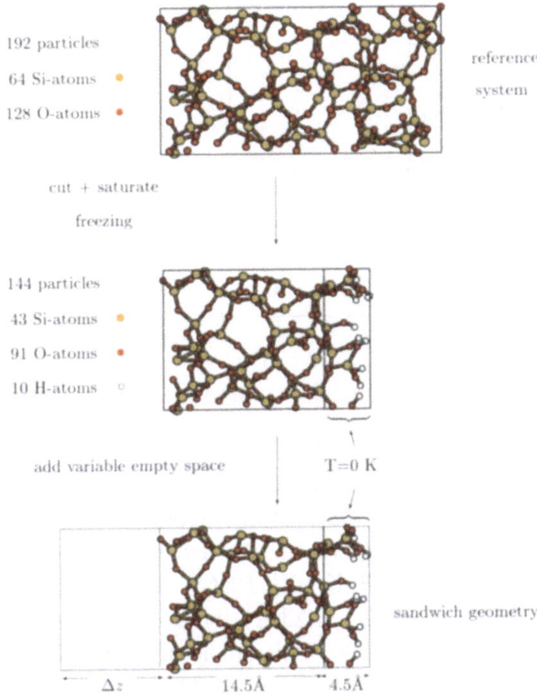

Fig. 1. Procedure to create the used sandwich geometry.

of the ions is always taken into account and one does not have to model the potential energy between the ions by a given function such as Eq. (1). However, it is not yet possible to replace the classical MD method by the CPMD, since due to the huge computational burden only relatively short time scales, a few ps, as well as small systems, a few hundred particles, can be simulated. In contrast to that classical MD allows one to simulate thousands of particles over several ns [9].

Since the time scale which is accessible in a CPMD is very restricted the idea of our approach is to combine a classical MD and a CPMD (see also Ref. [13]). To this end, we first equilibrate the system with a classical MD in which we again model the interactions between the atoms by the BKS potential, Eq. (1). Then, we use these configurations as the starting point of a CPMD. The goal of this investigation is twofold: Firstly, we want to see whether the classical configurations are stable in CPMD. If this is the case we can subsequently compare the structural differences as obtained by the two methods. So we get an idea of how accurate the BKS potential is able to describe silica surfaces, and we have hints of how this potential energy model could be improved.

In order to investigate a free silica surface one could consider a film geometry, i.e. periodic boundary conditions (PBC) in two directions and an infinite free space above and below the remaining third direction. Unfortunately, this

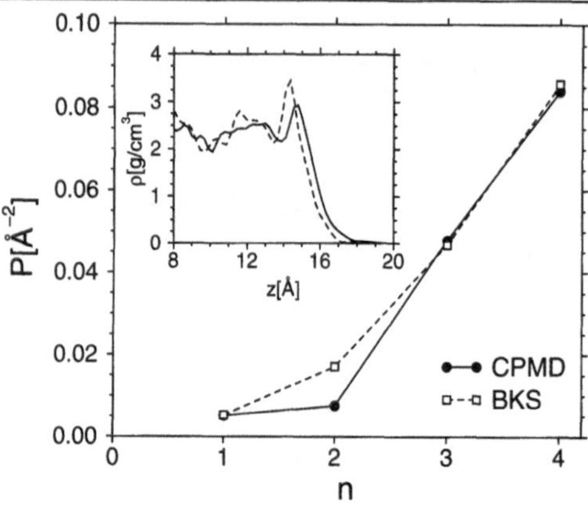

Fig. 2. Probability to find a ring of size n. Inset: z–dependence of the mass density.

is not a very good solution since the Ewald summation technique for the long ranged Coulomb interactions becomes inefficient in this case [14]. Therefore, we have adopted the following strategy which is illustrated in Fig. 1: i) We start with a system at $T = 3400$ K with PBC in three dimensions (box dimensions: $L_x = L_y = 11.51$ Å and $L_z' = 23$ Å). ii) We cut the system perpendicular to the z–direction into two pieces. This cut is done in such a way that we get only free oxygen atoms at this interface. iii) These free oxygen atoms are now saturated by hydrogen atoms. The place of these hydrogen atoms is chosen such that each of the new oxygen–hydrogen bonds is in the same direction as the oxygen–silicon bonds which were cut and have a length of approximately 1 Å. The interaction between the hydrogen and the oxygen atoms as well as the silicon atoms are described only by a Coulombic term. The value of the effective charge of the hydrogen atoms is set to 0.6 which ensures that the system is still (charge) neutral. iv) Atoms for which the distance from the interface that is less than 4.5 Å are made immobile whereas atoms that have a larger distance can propagate subject to the force field. v) We add in z–direction an empty space $\Delta z = 6.0$ Å and thus generate a free surface at around 14.5 Å. With this sandwich geometry we now can use periodic boundary conditions in all three directions. We have made sure that the value of Δz is sufficiently large that the results do not depend on it anymore [10]. Eventually, we have a system of 91 oxygen, 43 silicon, and 10 hydrogen atoms in a simulation box with $L_x = L_y = 11.51$ Å and $L_z \approx 25$ Å.

We have fully equilibrated 100 independent configurations for about 1 ns. Using a subset of these configurations as starting points we subsequently started CPMD simulations [15]. We used conventional pseudopotentials for silicon and oxygen and the BLYP exchange functions [16, 17]. The electronic wave–functions were expanded in a plane wave basis set with an energy cutoff of 60 Ry and the equations of motion were integrated with a time step of 0.085 fs for 0.2 ps. More details on the CPMD simulations can be found in Ref. [10].

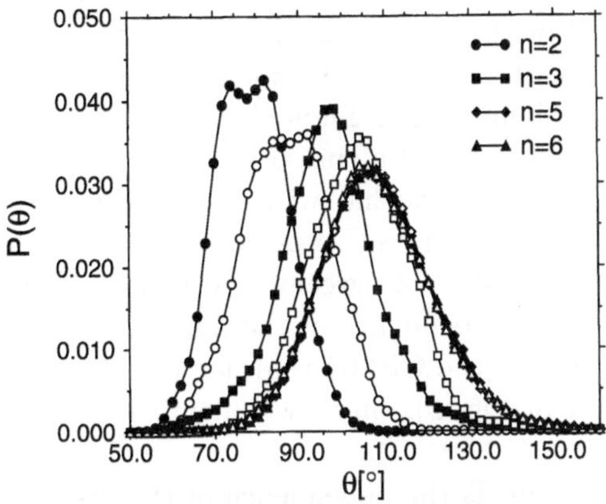

Fig. 3. Distribution of O–Si–O angles for different ring sizes (BKS: filled symbols, CPMD: open symbols).

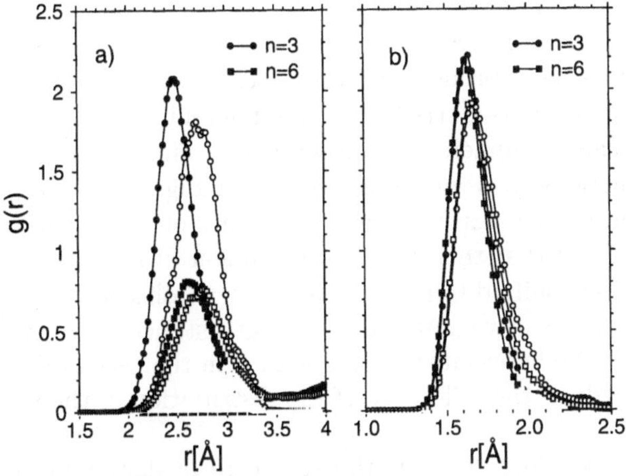

Fig. 4. Radial distribution function for different ring sizes, (a) O–O and (b) Si–O (BKS: filled symbols, CPMD: open symbols).

In the analysis of the CPMD run only those configurations were taken into account that were produced later than 5 fs after the start of the CPMD run in order to allow the system to equilibrate at least locally [13]. From the 100 BKS configurations we have picked up those for which one of the three following cases holds: i) The surface exhibits no defects, i.e. all Si and O atoms are four and two–fold coordinated, respectively. ii) The surface has an undercoordinated oxygen atom and an undercoordinated silicon atom. iii) The surface has an overcoordinated oxygen atom and an undercoordinated silicon atom. We used two BKS configurations for each of the cases i)–iii) and started the CPMD runs. The computational effort for the latter was very large: To simulate 1 ps one needs 11000 CPU hours of single processor time (or about one month on 16 processors of a Cray T3E).

A quantity which is appropriate to characterize the network structure is the distribution of the ring size. A ring is defined as a closed loop of n consecutive Si–O segments. The largest differences between the results of the classical and that of the CPMD simulation is found for the short rings, i.e. $n < 5$. Fig. 2 shows the probability to find a ring of size n for the case of the BKS simulation and the CPMD. We see that with the BKS model the frequency with which a ring of size two occurs is overestimated by about a factor of two as compared to the CPMD result. In agreement with this observation we find that the overshoot in the z–dependence of the mass density profile near the surface is less pronounced for the CPMD than for the classical MD with the BKS potential (see inset of Fig. 2), since two–membered rings are relatively dense.

Another interesting result is the dependence of the distribution of angles O–Si–O on the ring size n (Fig. 3). For large n, $n > 4$, i.e. for rings which are normally found in the bulk, the results of the two different methods are in good agreement [13]. For smaller n, however, the mean O–Si–O angle from CPMD is shifted to larger values in comparison to that of the classical MD. This shift becomes more pronounced with decreasing n. Furthermore also the shape of the distributions starts to become different if n is small.

This effect can be understood better if one analyzes the partial radial distribution functions $g(r)$ which are shown in Fig. 4. We see that for the Si–O correlation the curves from the CPMD are shifted to larger distances by about 0.04 Å and that this shift is independent of n. Also the $g(r)$ for the O–O correlation are shifted to larger r, but now we observe different shifts for different n. In particular we note that the O–O distance is nearly independent of n for the CPMD whereas it increases with n in the case of the BKS result. These effects result in the difference of the distribution of the O–Si–O angles if n is small.

In conclusion we find that for the structure on larger length scales the BKS simulation and the CPMD yield similar results whereas the details of the structural elements on short scales (for instance, the short rings) are different in both methods. This shows that it is probably necessary to use *ab initio* methods like CPMD if one wants to reproduce quantitatively the properties of silica surfaces on short length scales.

3 Interfaces between liquid and crystalline silica

This section is devoted to a classical MD of an interface between liquid and crystalline silica where we have also used the BKS potential to model the interactions between the Si and O atoms. A straightforward method to produce the latter interface would be to wait for the formation of a crystalline nucleus in a supercooled melt. But presently this is an impossible task since the viscosity of a silica melt at the melting temperature, $T_{\mathrm{m}} \approx 2000$ K, is around 10^7 Poise and thus the time scale that can be covered by a MD simulation

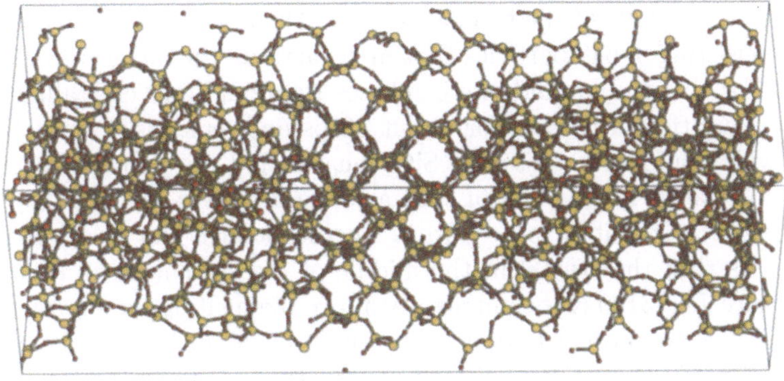

Fig. 5. Snapshot of the system at $T = 3100$ K.

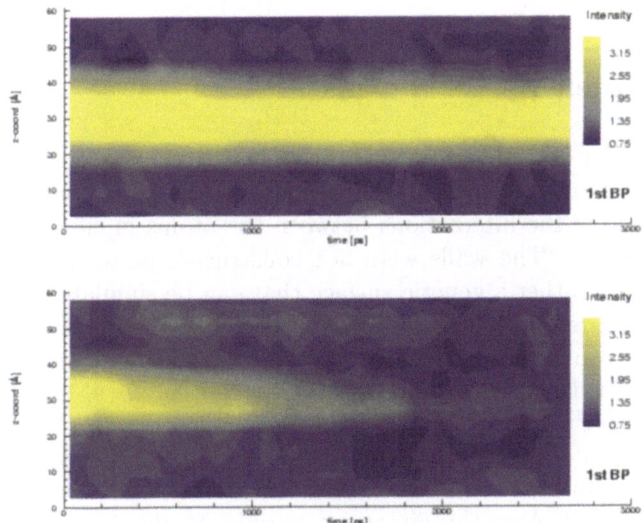

Fig. 6. Time evolution of the intensity of the first Bragg peak in the static structure factor for two different samples at the temperature $T = 3100$ K.

is by far too short to observe the growth of a nucleus in a supercooled silica melt.

Our strategy to prepare a liquid–crystalline interface is as follows: In a first step a pure melt and a pure crystal (β cristobalite), both containing 1944 particles, are relaxed at $T = 3100$ K for 1.6 ns and 160 ps, respectively. Thereby, the box lengths are $L_x = L_y = 21.375$ Å in x and y direction and $L_z = 61.465$ Å for the melt and $L_z = 64.125$ Å for the crystal in z direction. The corresponding densities are 2.32 g/cm^3 in the case of the melt and 2.22 g/cm^3 in the case of the crystal. In a second step 648 particles are removed from the equilibrated liquid configuration such that a free space is created in the middle of the simulation box. Then one cuts out a part of the crystal consisting of 648 particles that fits exactly in the latter free space (see Ref. [18] for the details) and one combines the liquid and crystalline pieces. This configuration is relaxed for about 30 ps and as a result one obtains a system with liquid and crystalline SiO_2 phases that are joined to each other via two interfaces. A snapshot of such a configuration is shown in Fig. 5.

After the preparation procedure a microcanonical run is started where one expects that the crystal in the middle of the system melts because the temperature $T = 3100$ K is significantly above the melting temperature of our system [19]. A good order parameter to quantify the melting of the crystal is the intensity of the first Bragg peak in the static structure factor [18]. The latter is shown in Fig. 6 for two different samples. Whereas in the first sample (upper plot of Fig. 6) the crystal does not melt at all even after about 2.6 ns, in the same time span the second sample the crystal is completely melted. More details on that can be found in Ref. [18].

4 A silica melt between walls

We turn now our attention to a silica melt between walls to study the wall–fluid interface structure and to see how the melt behaves under shear. As a model to describe the interactions between the atoms in silica we use again the BKS potential. The walls were not constructed as to model a particular material but rather a generic surface that can be simulated conveniently. Each wall consisted of 563 point particles forming a rigid face–centered cubic lattice with a nearest–neighbor distance of 2.33 Å. These point particles interact with the atoms in the fluid according to a 12–10 potential, $v(r) = 4\epsilon\left[(\sigma/r)^{12} - (\sigma/r)^{10}\right]$ with $\sigma = 2.1$ Å, $\epsilon = 1.25$ eV, r being the distance between a wall particle and a Si or O atom. The main details of the MD simulations are as follows: The simulation box had linear dimensions $L_x = L_y = 23.066$ Å in the directions parallel to the walls (in which also periodic boundary conditions were applied), and $L_z = 31.5$ Å in the direction perpendicular to the walls. Thus, $N = 1152$ atoms (384 Si atoms and 768 O atoms) were contained in the system to maintain a density around 2.3 g/cm^3 which is close to the experimental one at ambient pressure. All runs were done in the NVT ensemble whereby the temperature was kept constant by coupling the fluid to a Nosé–Hoover thermostat [9]. We investigated the temperatures $T = 5200$ K, $T = 4300$ K, and $T = 3760$ K at which we first fully equilibrated the system for 29, 65, and 122 ps, respectively. At $T = 5200$ K and $T = 3760$ K we continued with additional runs over 164 and 490 ps, respectively, from which we analyzed the equilibrium structure. Then we switched on a "gravitational" field of strength $a_e = 9.6$ Å/ps^2 that was coupled to the mass of the particles. With the acceleration field, runs were made over 736 ps, 1.23 ns, and 3.27 ns at $T = 5200$ K, 4300 K and 3760 K, respectively. In addition at $T = 5200$ K we did a run over 1.72 ns with field strength

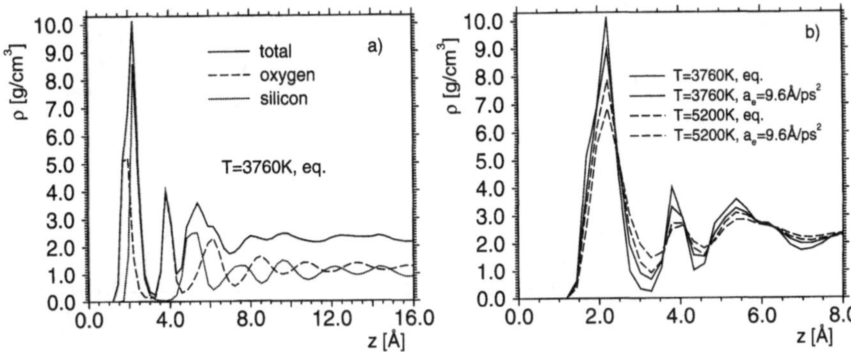

Fig. 7. (a) Total density profile and partial density profiles for oxygen and silicon in a system at equilibrium at $T = 3760$ K. (b) Total density profiles at equilibrium and with acceleration field $a_e = 9.6$ Å/ps^2 for the indicated temperatures.

$a_e = 3.8$ Å/ps^2. The total amount of computer time spent for these simulations was 16 years of single processor time on a Cray T3E. More details on the simulation can be found in Ref. [20].

Fig. 7a shows the density profiles across half of the film for all atoms and for the oxygen and silicon atoms only. In contrast to typical density profiles in simple monoatomic liquids, the oscillations of the total profile, which indicate a layering near the walls, do not have a regular character. This behavior can be explained by the partial density profiles for oxygen and silicon. Obviously, the tetrahedra adjacent to the walls prefer to be aligned such that a two–dimensional plane forms which contains three out of the four oxygen atoms of a tetrahedron, as well as the silicon atom at their center (slightly further away from the wall), while the fourth oxygen atom of the tetrahedron has to be further away from the wall for geometrical reasons: this causes the second peak of the oxygen distribution. Thus, the walls have a tendency to "orient" the network of coupled SiO_4 tetrahedra in the fluid locally. It is evident that the oscillations in the local density of both silicon and oxygen are rather regular, like a damped cosine function, but the wavelength and phase of both cosine functions are different: their superposition causes then the rather irregular layering structure of the SiO_2 total density. We expect that similar effects also occur in many other associating molecular fluids confined between walls if the wall–fluid interaction is weak enough that it does not affect the chemical ordering in the fluid as it is the case in our system.

For $z > 8$ Å the total density profile shows only small oscillations around a constant value of 2.3 g/cm^3 which is an indication for bulk behavior. Thus, keeping in mind that the density profile is symmetric, the bulk in our system seems to extend in z direction from about 8 Å to 23.5 Å, i.e. it has a width of about 15.5 Å.

The behavior of the total density profiles at the temperatures $T = 5200$ K and $T = 3760$ K in equilibrium and with an external force with an acceleration of 9.6 Å/ps^2 can be seen in Fig. 7b. In equilibrium the effect of decreasing

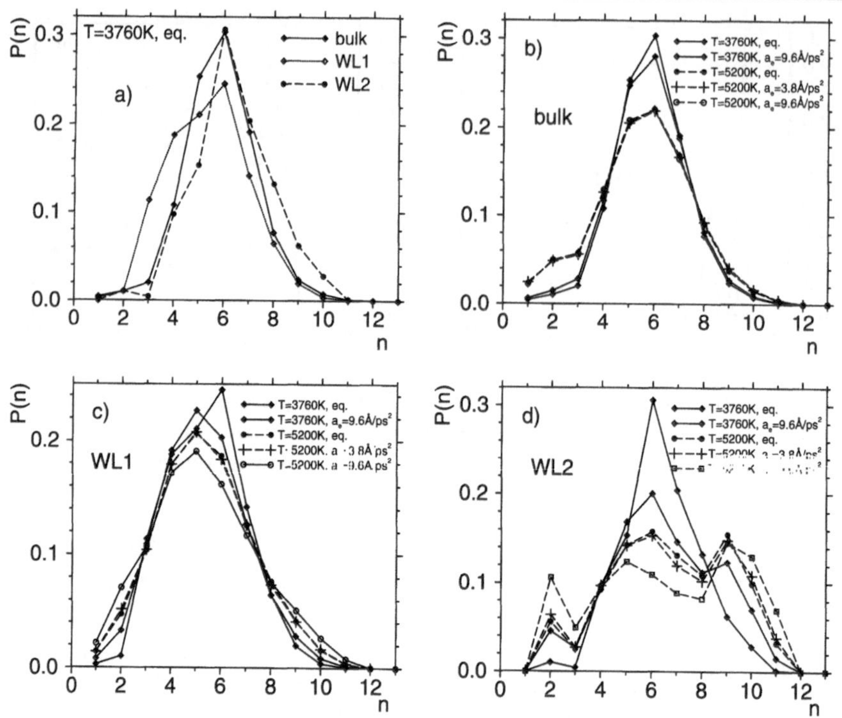

Fig. 8. Probability $P(n)$ that a ring has a length n, a) at $T = 3760$ K in equilibrium for the bulk, WL1, and WL2, b) at $T = 3760$ K and $T = 5200$ K in equilibrium and under the indicated acceleration fields in the bulk, c) the same as in b) but for WL1, and d) also the same as in b) but for WL2. See text for the definition of WL1 and WL2.

temperature on the oscillations near the wall is an increase of the peak heights that is accompanied with a smaller value of $\rho(z)$ at the minima between the peaks. Thus, the layering becomes more pronounced if one decreases the temperature. If one switches on the gravitational–like field the effect is similar to an increase of the temperature. In the bulk region the density profiles are not very sensitive to a variation of temperature and/or the presence of the external force. Within the accuracy of our data the same value of about 2.3 g/cm^3 is reached for all four cases under consideration.

Fig. 8a shows the ring distribution function $P(n)$ for $T = 3760$ K in the bulk and in two different wall layers denoted by WL1 and WL2. WL1 and WL2 are defined as the regions which are respectively within a distance of 6.25 Å and 3.0 Å away from the wall corresponding to the second minimum of the density profile for silicon and the first minimum of the total density profile, respectively (see Fig. 7a). In each region, i.e. bulk, WL1, and WL2, we took only those rings into account that fit completely into it. Thus, in WL2 those rings are counted that are formed at each case by the first and the second O and Si layers (with respect to the distance from the wall), whereas with WL2 only those rings are taken into account that are formed by the first O and the first Si layer. This is justified because the first two oxygen and the first two silicon layers are well–defined in that the minima in the corresponding density profiles are close to zero density in the case of WL2 and around the small value $\rho = 0.5$ g/cm^3 in the case of WL1. Furthermore, we can infer from Fig. 7a that in contrast to the second layers the first oxygen layer overlaps strongly with the first silicon layer and the overall thickness of both layers is only about 2 Å. Thus, the first oxygen and the first silicon layer form a quasi–two–dimensional plane and $P(n)$ for WL2 gives a distribution of rings that have an orientation parallel to the walls whereas in $P(n)$ for WL1 also the rings perpendicular to the walls are included.

In bulk simulations of SiO_2 one finds a maximum around $n = 6$ [21]. This is plausible since in silica the high–temperature crystalline phase at zero pressure, β–cristobalite, exhibits only six–membered rings. In WL1 the probability for $n \geq 6$ is smaller than in the bulk in favor of a relatively high probability of $n = 3$ and $n = 4$. In WL2 it is the other way round: $n = 4$ and even more $n = 5$ are less frequent than in the bulk in favor of $n = 8, 9, 10$. The ring structure near the walls that corresponds to these findings is as follows: Perpendicular to the walls small rings with $n = 3, 4$ are seen such that, e.g., $n = 3$ is formed by two silicon atoms from the first silicon layer with a third one from the second silicon layer. In contrast to that, parallel to the walls (considering the first oxygen and the first silicon layer) an open structure with relatively large rings is observed which compensates somewhat the dense packing of SiO_4 tetrahedra perpendicular to the walls.

Figure 8b shows the behavior of $P(n)$ in the bulk at the two temperatures $T = 3760$ K and $T = 5200$ K in equilibrium and under shear. We can immediately infer from the figure that the considered shear fields have only a small effect on the structure. At $T = 5200$ K one has a relatively large amount of two– and three–membered rings and their frequency is more than a factor of two smaller at $T = 3760$ K. In Ref. [21] it was shown that their frequency of occurrence decreases further with decreasing temperature such that the amount of two–membered rings falls far below 1% for systems that have typical structural relaxation times of the order of 1 ns.

In contrast to the bulk in WL1 significant changes in the ring distribution take place if the system is sheared (see Fig. 8c), and the external force field affects the ring structure such that small and large rings are formed, while at the same time especially the amount of six–membered rings decreases. Only for the smaller field strength $a_{\rm e} = 3.8$ Å/ps^2 there are no significant changes in the ring distribution. As we have shown elsewhere [20] the latter is accompanied with a very small slip motion at the walls whereas a large slip velocity is correlated with strong rearrangements in the ring structure. One can also infer the remarkable fact from Fig. 8c that the probability to find rings with $n = 3, 4$ does not change very much when an external acceleration field is switched on. This is reasonable because these small–membered rings, as we have seen before, are located perpendicular to the walls and thus they are very stable to shear forces that are imposed parallel to the walls.

The strongest rearrangements in the ring structure due to a shear field are found when we consider the region WL2. The corresponding curves are shown in Fig. 8d. Again, there are only minor changes in $P(n)$ at $T = 5200$ K and $a_e = 3.8$ Å/ps^2 as compared to the corresponding equilibrium case which is, as mentioned before, related to the presence of only a very small slip velocity. For the higher acceleration field, $a_e = 9.6$ Å/ps^2, the ring structure becomes more heterogeneous and the effect of the external field is if one would locally increase the temperature. The rearrangements in the ring structure can be summarized as follows: Rings mainly of size $n = 6, 7, 8$ are broken under the influence of the shear force and instead small rings with $n < 4$ and very large rings with $n \geq 9$ are formed.

5 Summary

We have shown that large scale MD simulations are able to give a lot of insight into the microscopic surface and interface structure of silica. In all three examples that we have presented in this report it is necessary to simulate relatively large systems on a relatively large time scale and thus the use of parallel supercomputers is indispensable. In our study of the free silica surface the Car–Parrinello MD was used for which length and time scales that can be covered are even very restricted on a parallel computer (CPMD simulations are typically on a ps time scale for systems of about 100 particles). Thus, the development and application of more powerful parallel computers is required to gain further insight from atomistic simulations.

Acknowledgments: We thank the HLRZ Stuttgart for a generous grant of computer time on the CRAY T3E. C. M. and T. S. are grateful to SCHOTT Glas for partial financial support.

References

1. A. P. Legrand, *The Surface Properties of Silica* (Wiley, New York, 1998).
2. R. K. Iler, *The Chemistry of Silica* (Wiley, New York, 1979).
3. C. R. Helms and B. D. Deal, *The Physics and Chemistry of SiO_2 and the Si–SiO_2 Interface* (Plenum, New York, 1993).
4. A. Grabbe, T. A. Michalske, and W. L. Smith, J. Phys. Chem. **99**, 4648 (1995).
5. P. K. Gupta, D. Inniss, C. R. Kurkjian, and Q. Zhong, J. Non–Cryst. Sol. **262**, 200 (2000).
6. J.-F. Poggemann, A. Go, G. Heide, E. Rädlein, and G. H. Frischat, J. Non–Cryst. Sol. **281**, 221 (2001).
7. I.-S. Chuang and G. F. Maciel, J. Am. Chem. Soc. **118**, 401 (1996).

8. B. W. van Beest, G. J. Kramer, and R. A. van Santen, Phys. Rev. Lett. **64** 1955 (1990).
9. K. Binder, J. Horbach, W. Kob, W. Paul, and F. Varnik, J. Phys.: Condens. Matter **15** (2003).
10. C. Mischler, *Molekulardynamik-Simulation zur Struktur von* SiO_2*-Oberflächen mit adsorbiertem Wasser*, Ph.D. thesis, Mainz 2002; C. Mischler, W. Kob, and K. Binder, Comp. Phys. Comm. **147**, 222 (2002).
11. J. Horbach and W. Kob, Phys. Rev. B **60**, 3169 (1999).
12. R. Car and M. Parrinello, Phys. Rev. Lett. **55**, 2471 (1985).
13. M. Benoit, S. Ispas, P. Jund, and R. Jullien, Eur. Phys. J. B **13**, 631 (2000).
14. A. Arnold, J. de Joannis, and C. Holm, J. Chem. Phys. **117**, 2496 (2002); J. de Joannis, A. Arnold, and C. Holm, J. Chem. Phys. **117**, 2503 (2002).
15. CPMD Version 3.3a, J. Hutter, P. Ballone, M. Bernasconi, P. Focher, E. Fois, S. Goedecker, M. Parrinello, and M. Tuckermann, MPI für Festkörperforschung and IBM Research 1990–2000.
16. N. Trouiller and J. L. Martins, Phys. Rev. B **43**, 1993 (1991).
17. C. Lee, W. Yang, and R. G. Parr, Phys. Rev. B **37**, 785 (1988).
18. T. Stühn, Ph.D. thesis, Mainz, 2003.
19. A. Roder, W. Kob, and K. Binder, J. Chem. Phys. **114**, 7602 (2001).
20. J. Horbach and K. Binder, J. Chem. Phys. **117**, 10798 (2002).
21. K. Vollmayr, W. Kob, and K. Binder, Phys. Rev. B **54**, 15808 (1996).

Quantum Monte-Carlo Simulations of Correlated Bosonic and Fermionic Systems

C. Lavalle[1], M. Rigol[1], M. Feldbacher[1], M. Arikawa[1], F. F. Assaad[1,2], and A. Muramatsu[1]

[1] Institut für Theoretische Physik III, Universität Stuttgart, Pfaffenwaldring 57, D-70550 Stuttgart, Germany
[2] Max Planck institute for solid state research, Heisenbergstr. 1, D-70569, Stuttgart.

Summary. We review recent results of quantum Monte Carlo simulations applied to correlated electronic and bosonic systems. We concentrate on three subjects. 1) Using a recently developed hybrid quantum Monte-Carlo algorithm we investigate the excitation spectra of the one-dimensional $t-J$ model. Our results give strong numerical support for the existence of antiholons, which along with spinons and holons correspond to the elementary excitations of this model. 2) Very recently, it was experimentally demonstrated, that it is possible to attain temperatures low enough, such that degenerate quantum gases can be studied in magneto-optical traps, the most prominent example being Bose-Einstein condensation of alkali atoms. Under the action of a periodic potential created by interfering laser beams, such systems can be brought to a strongly correlated state. We present numerical simulations in one-dimension in order to understand theses new states of matter. 3) Taking the step from one to two and three dimensions poses a formidable numerical challenge. In particular for fermionic models the quantum Monte Carlo method suffers from the so-called sign problem which renders simulations exponentially expensive in CPU time as a function of inverse temperature at lattice size. We show that by considering multi-flavored models this problem is reduced and in some special cases altogether removed.

1 Introduction

We review here the application of several quantum Monte Carlo (QMC) methods to investigate the physics of correlated electronic and bosonic systems. Correlation effects are at the heart of modern condensed-matter physics and pose a formidable challenge in the sense that standard many body physics approximations – such as the random phase approximation – are in general not justified. In particular in the one-dimensional (1D) case, quantum fluctuations lead to the breakdown of Fermi liquid theory. That is, elementary excitations are not quasiparticles with charge e (the elementary electronic charge) and spin 1/2 but of solitonic nature. In the Sec. 2, we will concentrate on simulations of the 1D t-J model. It is generally believed that elementary excitations of 1D systems are spinons carrying charge 0 and spin 1/2 and holons with charge -e and spin 0 [VOIT94]. Progress in the understanding of the excitation spectra of the 1D models with long-range interactions, such as the supersymmetric t-J model with $1/r^2$ interactions, demonstrate the existence of another elementary excitation which carries charge $2e$ and spin 0: the antiholon [HAHA94, ARIK01]. In the next section we present numerical simulations giving support for the existence of antiholons in the generic t-J model with short range interactions [LAVA03].

Recently, and after the experimental achievement of Bose-Einstein condensation (BEC) with confined ultra-cold atomic gases [ANDE95, DAVI95], a number of perspectives opened for the realization of new states of matter. Of particular interest to condensed matter physics is the possibility of loading confined degenerated quantum gases on optical lattices generated by interfering laser beams. These systems offer the closest realization of the Hubbard model for bosons and fermions with repulsive or attractive interactions. Recently such an experimental set-up was used to study the superfluid-Mott insulator transition in the bosonic case [GREI02].

In Sec. 3 we discuss 1D systems of bosonic and fermionic atoms confined in harmonic traps with an underlying lattice. We show that the confining potential introduces new features not present in the periodic case, like coexistence of different local phases and a more complicated phase diagram. Surprisingly, features common to critical phenomena like critical exponents and universality are found on passing from one phase to another, giving clear signals of local quantum criticality [RIGO03].

From the numerical point of view, lattice fermions in dimensions larger than unity pose a formidable challenge. In particular the stochastic approach is in general plagued by the so-called sign problem which leads to an exponential increase of the CPU time as a function of inverse temperature and lattice size. In Sec. 4 we will show that by enhancing the number of flavors (i.e. fermion species) in Hubbard-type models, the sign problem is reduced and is some special cases altogether removed [ASSA03].

In the last section we draw conclusions.

2 The one-dimensional nearest-neighbor $t-J$ model at finite doping

The t-J model in one dimension reads:

$$\begin{aligned} H_{t-J} &= \sum_i H_i \\ &= -t\sum_{i,\sigma}\left(\tilde{c}^\dagger_{i,\sigma}\tilde{c}_{i+1,\sigma} + h.c.\right) + J\sum_i\left(\mathbf{S}_i\cdot\mathbf{S}_{i+1} - \frac{1}{4}\tilde{n}_i\tilde{n}_{i+1}\right). \quad (1) \end{aligned}$$

Here, $\tilde{c}^\dagger_{i,\sigma}$ are projected fermion operators $\tilde{c}^\dagger_{i,\sigma} = (1 - c^\dagger_{i,-\sigma} c_{i,-\sigma}) c^\dagger_{i,\sigma}$, $\tilde{n}_i = \sum_\alpha \tilde{c}^\dagger_{i,\alpha} \tilde{c}_{i,\alpha}$, $\mathbf{S}_i = (1/2) \sum_{\alpha,\beta} c^\dagger_{i,\alpha} \boldsymbol{\sigma}_{\alpha,\beta} c_{i,\beta}$, and the sum runs over nearest neighbors only. In Sec. 2.1, we will describe shortly a new algorithm, the hybrid-loop QMC algorithm, that we developed, and that allows us to study static and dynamic properties of the $t-J$ model for arbitrary doping. We will then give a description of the excitation content of the system with finite doping.

As a first step to set up the new algorithm for the $t-J$ model at arbitrary doping, we perform a canonical transformation [KHAL90]. This canonical transformation leads to a Hamiltonian bilinear in the fermion fields, a form that is very adequate for the hybrid-loop algorithm described below.

In this alternative formulation the H_i become

$$\tilde{H}_i = +t\, P_{i,i+1} \left(f_i^\dagger f_{i+1} + h.c. \right) + \frac{J}{2} \Delta_{i,i+1} (P_{i,i+1} - 1), \tag{2}$$

where $P_{i,j} = (1 + \boldsymbol{\sigma}_i \cdot \boldsymbol{\sigma}_j)/2$, $\Delta_{i,j} = (1 - n_i - n_j)$, and $n_i = f_i^\dagger f_i$.

Starting from a system of interacting spinfull fermions, after the canonical transformation, we obtain a system describing free spinless fermion interacting with a quantum pseudospin background. The constraint to avoid doubly occupied states transforms in a conserved quantity.

2.1 The hybrid-loop algorithm

Once the canonical transformation is performed, we obtain a Hamiltonian bilinear in the fermion field (Eq. 2). Our aim is now to use the separation of degrees of freedom of the model while performing a $T = 0$ projection that filters the ground state out of a trial wave function.

The projection is performed via a QMC evolution treating the fermionic degrees of freedom with the determinantal algorithm [SUGI86],[BLAN81] and the spin degrees of freedom with the loop algorithm[EVER97]. The combination of those two algorithms is the hybrid loop algorithm.

We consider now the following definition of an expectation value:

$$\langle \hat{\mathcal{O}} \rangle = \lim_{\Theta \to \infty} \frac{\sum_{\{\sigma_0\}} \langle \sigma_0 | \otimes \langle \Psi_T | \mathcal{P}\, \mathrm{e}^{-\frac{\Theta}{2}\mathcal{H}}\, \hat{\mathcal{O}}\, \mathrm{e}^{-\frac{\Theta}{2}\mathcal{H}}\, \mathcal{P} | \Psi_T \rangle \otimes | \sigma_0 \rangle}{\sum_{\{\sigma_0\}} \langle \sigma_0 | \otimes \langle \Psi_T | \mathcal{P}\, \mathrm{e}^{-\Theta\mathcal{H}}\, \mathcal{P} | \Psi_T \rangle \otimes | \sigma_0 \rangle} \tag{3}$$

where $\{| \sigma_0 \rangle\}$ is a complete set of spin states and $| \Psi_T \rangle$ a trial wavefunction for the spinless fermions. $\mathcal{P}$ is a projector ensuring the constraint against double occupancy. Taking the limit $\Theta \to \infty$ leads each state $\mathcal{P} | \Psi_T \rangle \otimes | \sigma_0 \rangle$ to converge to the GS as long as the GS has a finite overlap with it. The multiplicity is corrected by the normalization factor.

The first step to perform, in order to do a QMC evaluation of this expectation value, is a so called checkerboard decomposition, where the Hamiltonian is split in non-commuting terms H_{even} and H_{odd}, where the subscript even or

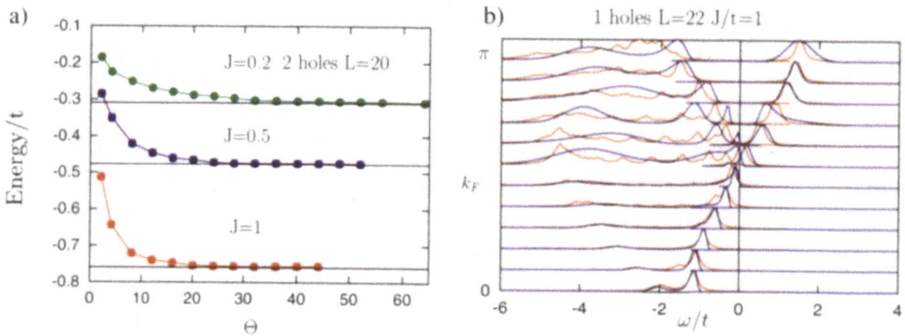

Fig. 1. a) Ground-state energies *vs.* exact diagonalization results as a function of the projection parameter Θ. **b)** Single particle spectral function calculated with hybrid loop algorithm (blue lines) and exact diagonalization (red lines) .

odd refers to the index i in (2). Finally, a Trotter decomposition of the partition sum maps the d-dimensional quantum system to a $d+1$-dimensional classical system.

Since the Hamiltonian is already quadratic in the fermionic degrees of freedom we can directly apply the determinantal algorithm without the need of introducing auxiliary fields. In order to integrate the fermions, a complete set of pseudospin states has to be inserted for every time slice.

The fermionic dynamics is fixed by the spin background and the fermions contribute to the weight of the spin field realization with $D_f(\boldsymbol{\sigma})$, i.e. with the fermionic determinant for the fixed spin field $\{\boldsymbol{\sigma}\}$, where $\boldsymbol{\sigma}$ is a vector containing all intermediate states $(\sigma_0, \ldots \sigma_n, \ldots, \sigma_0)$. On the other hand the pseudospins are quantum degrees of freedom with their own dynamics and they are simulated via a loop algorithm that leads to a weight for every spin field realization $P(\boldsymbol{\sigma})$ corresponding to the probability distribution of a Heisenberg antiferromagnet for the configuration $\boldsymbol{\sigma}$.

From this follows that the weight of each configuration is given by $P(\boldsymbol{\sigma})\ D_f(\boldsymbol{\sigma})$ so that Eq. 3 becomes:

$$\langle \hat{\mathcal{O}} \rangle = \lim_{\Theta \to \infty} \frac{\sum_{\{\sigma_0,\ldots,\sigma_L\}} P(\sigma)\ D_f(\sigma)\ \langle \mathcal{O} \rangle(\sigma)}{\sum_{\{\sigma_0,\ldots,\sigma_L\}} P(\sigma)\ D_f(\sigma)} \,. \tag{4}$$

The simulation is therefore divided in two steps. First, a pseudospin configuration is proposed via the loop-algorithm. This fixes the fermion evolution that is performed via the determinantal algorithm in the frame of which all the observables can be calculated.

The hybrid-loop MC incorporates the advantages of the projector- and of the loop-algorithm. It is very efficient since the loop-algorithm allows for global up-dates that lead to extremely short autocorrelation times (< 2 sweeps), i.e. critical slowing down is avoided. On the other hand, the fermionic evolution is calculated exactly for each spin-background and both static and dynamic

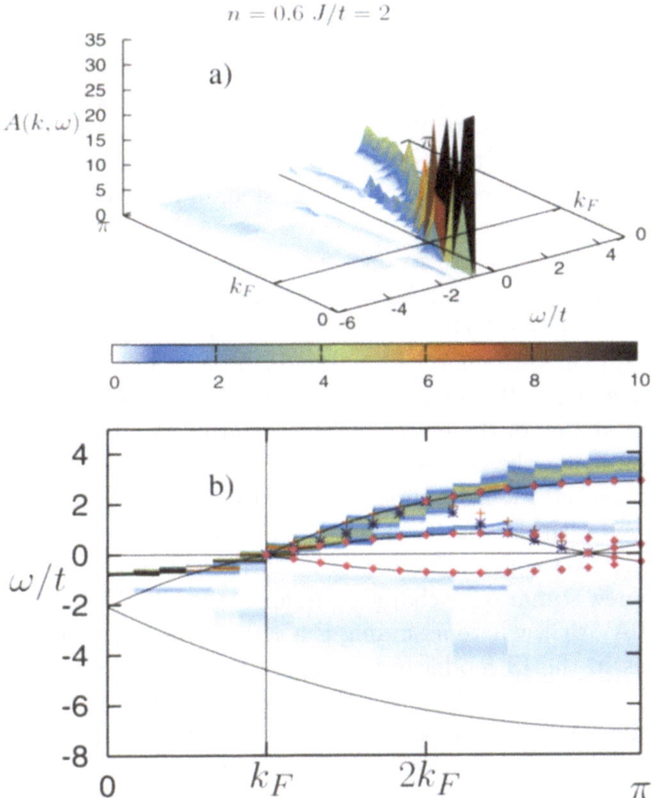

Fig. 2. a) $A(k,\omega)$ for $J = 2t$ at a density $n = 0.6$. b) Projection of intensities on the (ω, k) plane. Solid lines: compact support of the IS SuSy t-J model. Red crosses: spinons, blue asterisks: holons, magenta diamonds: antiholons. .

observables are measurable. Dynamical data are obtained from the imaginary time Green's function and analytically continued using the maximum entropy method [JARR96].

2.2 Antiholons in one-dimensional t-J models

The 1D n.n. $t - J$ model is highly interesting for the understanding of correlated electron systems. In fact due to its rich phase diagram it is possible to study in a lower dimension all the phases that are suspected in the two-dimensional high-temperature superconductors. The issue is still largely unresolved especially for what concerns the dynamical properties of the system for finite values of the parameters and of doping.

Using the hybrid-loop algorithm we are able for the first time to study in detail the spectral function of the model for very large systems (L=80,100) at finite doping and without limitations in parameter space.

Figure 2 a) shows the spectral function $A(k,\omega)$ for $J/t = 2$ at electron density $n = 0.6$ for both photoemission ($\omega < 0$) and inverse photoemission ($\omega > 0$) processes. A splitting of the spectral weight into two branches can be readily seen on the inverse photoemission side, in contradiction with what is expected for an ordinary metal.

In order to interpret the results, we compared them with those analytically obtained from a $t-J$ model with both hopping and exchange interaction decaying as $1/r^2$ [HAHA94], where the electron separates into three free excitations: a spinon with charge $Q = 0$ and spin $S = 1/2$, a holon with charge $Q = -e$ and spin $S = 0$, and an antiholon with charge $Q = 2e$ and spin $S = 0$. Here e is the charge of the electron. The quantum numbers above show that the antiholon is not the charge conjugate excitation of a holon, and hence, it is a new excitation. Figure 2 b) shows that the excitations of the $1/r^2$ model agree very well with the results of the simulation for the n.n. model [LAVA03], strongly indicating that antiholons are also present in the n.n. $t-J$ model, leading to a new scenario for charge spin separation. This feature is still visible at $J \neq 2t$ [LAVA03], where assuming the same dispersions for the holon, and antiholon, as in the IS model but changing the scale of energy to J for the spinon, a fairly good description of the spectrum can be given.

The results above strongly indicate, that antiholons, that are not charge conjugate of holons, are generic excitations in the nearest neighbor t-J model.

3 Bosons and fermions confined in optical lattices

Here, we discuss 1D systems of bosonic and fermionic atoms confined in harmonic traps with an underlying lattice, in order to shed light on recent experiments that produced a strongly correlated state starting from a Bose-Einstein condensate [GREI02].

In the case of the Hubbard model for confined bosons, we study the Hamiltonian [BATR02]

$$H = -t\sum_i \left(a_i^\dagger a_{i+1} + h.c.\right) + V_0 \sum_i n_i(n_i - 1) + \left(\frac{2}{N}\right)^2 V \sum_i (i - N/2)^2\, n_i, \tag{5}$$

where t measures the boson kinetic energy, V_0 the on-site repulsive interaction ($V_0 > 0$) and the last term models the potential of the harmonic trap. N is the number of lattice sites considered. The simulations were performed with the world-line quantum Monte Carlo (QMC) algorithm [BATR95, HIRS81].

In Fig. 3 we show the evolution of the local boson density $n_i = \langle a_i^\dagger a_i \rangle$ with increasing total occupancy of the trap. This quantity indicates the formation

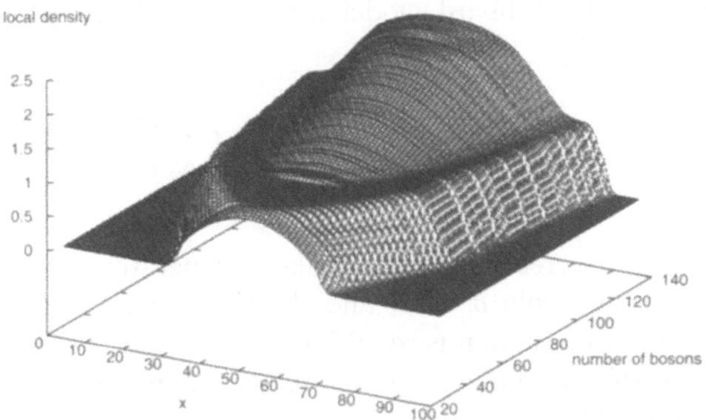

Fig. 3. Evolution of the local density as a function of the position x and increasing total number of bosons. The parameters are $V_0/t = 4.0$, $V/t = 20$ and $N = 100$.

of Mott insulator regions for integer values of the local density and also that there is coexistence of Mott insulating and superfluid phases in the trap.

Different runs like the ones presented in Fig. 3 allowed us to determine the phase diagram as a function of the total boson filling and the interaction strength, for a given trap curvature. Results are shown in Fig. 4. Region A admits locally incompressible Mott insulating domains with $n_i = 1$ in the center of the trap. Region B has $n_i > 1$ in the trap center surrounded by incompressible $n_i = 1$ regions. In region C the central part of the trap has an $n_i = 2$ Mott insulating phase inside superfluid and Mott ($n_i = 1$) regions respectively. Region D is where the center of the system is compressible with $n_i > 2$ surrounded by Mott ($n_i = 2$), superfluid and Mott ($n_i = 1$) regions. Region E has no locally incompressible regions.

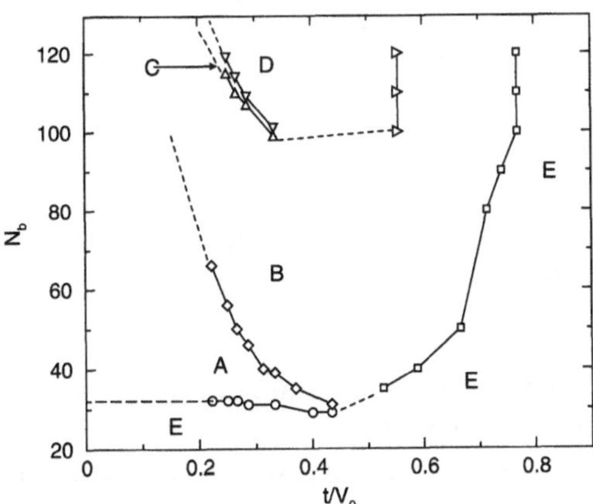

Fig. 4. Phase diagram of confined bosons in optical lattices. Solid lines are to guide the eye and dashed lines are extrapolations.

In the case of the Hubbard model for confined fermions, we study the Hamiltonian [RIGO03]

$$H = -t \sum_{i,\sigma} \left(c^{\dagger}_{i\sigma} c_{i+1\sigma} + h.c. \right) + U \sum_{i} n_{i\uparrow} n_{i\downarrow} + \left(\frac{2}{N} \right)^2 V \sum_{i\sigma} \left(i - \frac{N}{2} \right)^2 n_{i\sigma}, \tag{6}$$

where $c^{\dagger}_{i\sigma}$ and $c_{i\sigma}$ are creation and annihilation operators, respectively, for a fermion on site i with spin $\sigma = \uparrow, \downarrow$ and the local density per spin is $n_{i\sigma} = c^{\dagger}_{i\sigma} c_{i\sigma}$. The contact interaction is repulsive ($U > 0$) and the last term models the potential of the parabolic trap. The QMC simulations were performed using a projector QMC algorithm [MURA99].

Figure 5 shows density profiles along a harmonic trap for different fillings such that the system goes from an entirely metallic phase to a phase with insulating regions due to full occupancy of the sites, coexisting with metallic regions. Intervening phases with coexisting metallic and Mott-insulating domains can be seen (as in the bosonic case local phases develop in the trap).

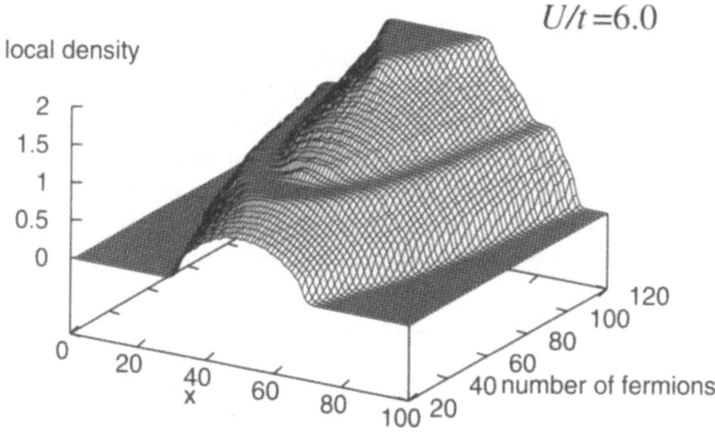

Fig. 5. Density profiles along the trap for different fillings. Flat terraces are the Mott insulating regions.

Finally we consider the phase diagram of the system. As in the unconfined case, we would expect to be able to relate systems with different number of particles and/or sizes by their density. Given the harmonic potential, a characteristic length (in units of the lattice constant) is given by $N\,(4V/t)^{-1/2}$, such that a characteristic density can be defined. Figure 6 shows that the characteristic density $\tilde{\rho} = \rho\sqrt{4V/t}$ ($\rho = N_f/N$) is a meaningful quantity to characterize the phase diagram. There, the phase diagrams for two systems

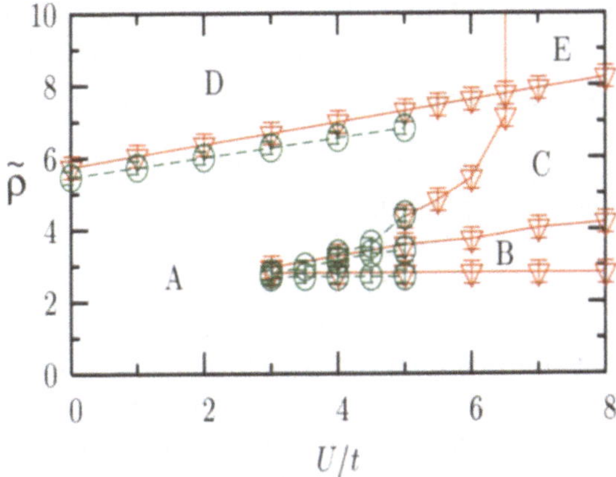

Fig. 6. Phase diagram for a system with $N = 100$ ($\bigtriangledown$) and $N = 150$ ($\bigcirc$) sites. The different phases are explained in the text.

with different sizes ($N = 100$ and $N = 150$) and different strength of the harmonic potential ($V = 15t$ and $V = 11.25t$ respectively) are depicted showing that such a scaling allows to compare systems with different sizes, different number of particles, and different strength of the potential. This makes possible to relate the results of numerical simulations to much larger experimental systems. The different phases obtained are: A pure metal without insulating regions (A), a Mott-insulator at the center of the trap (B), a metallic intrusion at the center of a Mott-insulator (C), a "band insulator" (i.e. with $n = 2$) at the center of the trap surrounded by a metal (D), and finally a "band insulator" surrounded by a metal, surrounded by a Mott-insulator with the outermost region being again a metal (E).

4 Multiflavored Hubbard models

In this section, we will show with a specific example that by enhancing the number of flavors in the Hubbard model we can suppress the sign problem and in some special cases remove it altogether [ASSA03].

Our starting point is the Hamiltonian:

$$H = -t \sum_{\langle \mathbf{i},\mathbf{j} \rangle} \mathbf{c}_\mathbf{i}^\dagger \mathbf{c}_\mathbf{j} - \frac{U}{N} \sum_{\mathbf{i}} \left(\mathbf{c}_\mathbf{i}^\dagger \lambda \mathbf{c}_\mathbf{i} \right)^2 \tag{7}$$

where $\mathbf{i}$ labels the sites of a square lattice and the first sum runs over nearest neighbors. The spinors $\mathbf{c}_\mathbf{i}^\dagger = \left(c_{\mathbf{i},1}^\dagger \cdots c_{\mathbf{i},N}^\dagger \right)$ correspond to fermions with N flavors. For even values of N, $\lambda_{\alpha,\gamma} = \delta_{\alpha,\gamma} f(\alpha)$ with $f(\alpha) = 1$ for $\alpha \leq N/2$ and -1 otherwise. At $N = 2$ the model reduces to the standard $SU(2)$-spin invariant

Hubbard model since the interaction is the square of the magnetization. Away from $N = 2$ the model has an $SU(N/2) \otimes SU(N/2)$ symmetry which becomes clear when writing the Hamiltonian in terms of the spinors: $\left(c^\dagger_{\mathbf{i},1} \cdots c^\dagger_{\mathbf{i},N/2}\right)$ and $\left(c^\dagger_{\mathbf{i},N/2+1} \cdots c^\dagger_{\mathbf{i},N}\right)$. For $N \to \infty$ we will show below that the exact solution of the model is a spin-density wave mean-field approximation of the $N = 2$ model based on the Ansatz: $\langle n_{\mathbf{i},1} - n_{\mathbf{i},2} \rangle = m_i$.

The central observation is that at $N = 4n$ the minus sign problem in the QMC approach is never present regardless of the lattice topology and band-filling. To illustrate this and to keep the notation simple we consider the finite temperature auxiliary field QMC approach. After the usual decoupling of the interaction term with a Hubbard-Stratonovich transformation, the partition function becomes $Z = \int \mathcal{D}\phi e^{-NS(\phi)}$, where

$$S(\phi) = U \frac{\sum_{\mathbf{i}} \int_0^\beta \mathrm{d}\tau \phi^2_{\mathbf{i}}(\tau)}{4} - \frac{1}{2} \ln \mathrm{Tr}\left[T e^{-\int_0^\beta \mathrm{d}\tau H(\tau)} \right] \tag{8}$$
$$H(\tau) = -t \sum_{(\mathbf{i},\mathbf{j}),\sigma=\pm} c^\dagger_{\mathbf{i},\sigma} c_{\mathbf{j},\sigma} - U \sum_i \phi_{\mathbf{i}}(\tau) \left(n_{\mathbf{i},+} - n_{\mathbf{i},-} \right)$$

Here T corresponds to time ordering and in terms of our original fermions we can identify $c_{\mathbf{i},+} = c_{\mathbf{i},1}$ and $c_{\mathbf{i},-} = c_{\mathbf{i},N/2+1}$ and $n_{\mathbf{i},+} = c^\dagger_{\mathbf{i},+} c_{\mathbf{i},+}$, $n_{\mathbf{i},-} = c^\dagger_{\mathbf{i},+} c_{\mathbf{i},-}$. As $N \to \infty$ the integral over the fields ϕ is dominated by the saddle point configuration, $\phi^\star$ satisfying $\partial S(\phi)/\partial\phi|_{\phi=\phi^\star} = 0$. A time independent solution is $\phi^\star_i = \langle n_{\mathbf{i},+} - n_{\mathbf{i},-} \rangle$ where the expectation value is taken with respect to the Hamiltonian in Eq. 8. This is precisely the set of self-consistent equations obtained from the mean-field decoupling: $n_{i,+} - n_{i,-} = \phi^\star_i + [(n_{i,+} - n_{i,-}) - \phi^\star_i]$ for the Hamiltonian of Eq. 7 at $N = 2$.

Since $H(\tau)$ in Eq. 8 is quadratic in the fermion variables we can carry out the trace analytically to obtain:

$$e^{-NS(\phi)} = e^{-NU \frac{\sum_{\mathbf{i}} \int_0^\beta \mathrm{d}\tau \phi^2_{\mathbf{i}}(\tau)}{4}} \left[\mathrm{det} M^+(\phi) \right]^{N/2} \left[\mathrm{det} M^-(\phi) \right]^{N/2}. \tag{9}$$

For even values of $N/2$ ($N = 4n$) the Boltzmann weight, $e^{-NS(\phi)}$, is positive and the sign problem never occurs regardless of doping and lattice topology. Here, we study the ground state properties of the Hamiltonian 7 with the related projector auxiliary field QMC algorithm. The detail of how to apply this algorithm to our Hamiltonian, may be found in Ref. [CAPP00] where very similar model – at least from the technical point of view – is considered.

To test and interpret our approach, we show in Fig. 7 the static spin and charge structure factors,

$$S_{\substack{c\\s}}(q) = \sum_{\mathbf{r}} e^{i\mathbf{r}\mathbf{q}} \langle (n_{0,+} \pm n_{0,-})(n_{r,+} \pm n_{r,-}) \rangle, \tag{10}$$

as well as the ground state energy as a function of $1/N$ on a 4×4 lattice, $U/t = 4$ and two holes doped away from half-filling. The electron density is defined as: $\langle n \rangle = \frac{1}{V} \sum_{\mathbf{i}} \langle n_{\mathbf{i},+} + n_{\mathbf{i},-} \rangle$ where V is the number unit cells $\mathbf{i}$. In spite of sign problems for $N \neq 4n$, the QMC data interpolate between the mean-field solution at $N = \infty$ and the exact diagonalization results at $N = 2$. It is interesting to note that the sign problem becomes rapidly less and less severe for growing values of $N = 4n+2$. In the context of the large-N approach [AUER94] Gaussian fluctuations around the mean-field saddle point (i.e. the random phase approximation (RPA)) correspond to $1/N$ corrections. Figure 7 shows that for $N \geq 16$ we can understand our results in the framework of this approximation. The model at $N = 4$ may be seen as a model in it's own right or in the framework of an approximation to the $N = 2$ Hubbard model which goes beyond the RPA approximation.

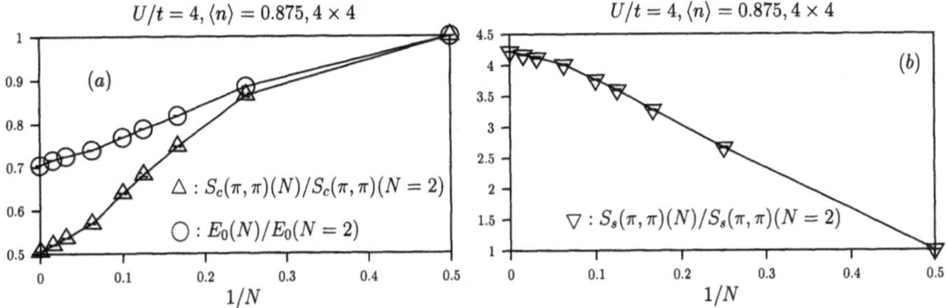

Fig. 7. Ground state energy as well as spin and charge structure factors at $\mathbf{q} = (\pi, \pi)$ as a function of $1/N$. The data points at $N = 2$ stem from exact diagonalization studies [PARO91].

The above approach has been used to investigate the metallic phase in the proximity of the Mott insulating phase. It is shown [ASSA03] that the above model at $N = 4$ has a striped phase. In particular it appears that the single particle spectra is gapped but that the systems remains metallic due to long-wave length *gapless* collective charge excitations related to the stripe structure.

The generalization of the Hubbard model to a $SU(N/2) \otimes SU(N/2)$ model is only one of many choices. One can consider $SU(N)$ Hubbard-Heisenberg models. In this case the large-N saddle point has been discussed by Affleck and Martson [AFFL88, KOTL88], and shows a d-density wave or staggered flux phase state. This state has recently been conjectured to describe the pseudo-gap phase of cuprates [CHAK01]. Simulations at half-band filling show that this state survives up to $N = 4$ [ASSAUN].

5 Conclusions

In summary, we have used and developed quantum Monte Carlo algorithms to investigate various aspects of correlated bosonic and fermionic systems.

In the one-dimensional case without confining potential, we have shown that the antiholon is required to understand the excitation spectra of the doped t-J model. We then considered one-dimensional models in traps. In this case the confining potential introduces new features not present in the periodic case, like coexistence of different *local* phases within the trap. Finally, we have presented ideas on how to reduce the sign problem present in higher dimensional interacting fermion systems by enhancing the number of fermionic flavors. A result of this approach are microscopic four flavored models which describe complex quantum ordered states such as d-density wave states and striped phases.

We thank G. G. Batrouini, S. Capponi, P. J. H. Denteneer, F. Hébert, V. Rousseau, R. T. Scalettar, and M. Troyer for collaborations. This work was supported by the Deutsche Forschungsgemeinschaft (grant numbers AS 120/1-3 , AS 120/1-1), Sonderforschungsbereich 382, as well as a joint Franco-German cooperative grant (PROCOPE). The numerical calculations were performed at HLR-Stuttgart. We thank the above institutions for their support.

References

[VOIT94] J. Voit, Rep. Prog. Phy. **57**, 977 (94).
[ARIK01] M. Arikawa, Y. Saiga, and Y. Kuramoto, Phys. Rev. Lett. **86**, 3096 (2001).
[LAVA03] C. Lavalle, M. Arikawa, S. Capponi, F. Assaad, and A. Muramatsu, Phys. Rev. Lett. **90**, 216401 (2003).
[ANDE95] M. H. Anderson, J. R. Ensher, M. R. Matthews, C. E., Wieman, and E. A. Cornell, Science **269**, 198 (1995).
[DAVI95] K. B. Davis, M.-O. Mewes, M. R. Andrews, N. J. van Druten, D. S. Durfee, D. M. Kurn, and W. Ketterle, Phys. Rev. Lett. **75**, 3969 (1995).
[GREI02] M. Greiner, O. Mandel, T. Esslinger, T. W. Hänsch, and I. Bloch, Nature (London) **415**, 39 (2002).
[RIGO03] M. Rigol, A. Muramatsu, G. Batrouni, and R. Scalettar, cond-mat/0304028, submitted to Phys. Rev. Lett.
[ASSA03] F. Assaad, V. Rousseau, F. Hébert, M. Feldbacher, and G. G. Batrouni, submitted to EPL.
[KHAL90] G. Khaliullin, JETP Lett. **52**, 389 (1990).
[SUGI86] G. Sugiyama and S. E. Koonin, Anals of Phys. **168**, 1 (1986).
[BLAN81] R. Blankenbecler, R. L. Sugar, and D. J. Scalapino, Phys. Rev. D **24**, 2278 (1981).
[EVER97] H. G. Evertz, Adv. Phys. **52**, 1 (2003).
[JARR96] M. Jarrell and J. Gubernatis, Phys. Rep. **269**, 133 (1996).
[HAHA94] Z. N. C. Ha and F. D. M. Haldane, Phys. Rev. Lett. **73**, 2887 (1994).
[BATR02] G. G. Batrouni, V. Rousseau, R. T. Scalettar, M. Rigol, A. Muramatsu, P. J. H. Denteneer, and M. Troyer, Phys. Rev. Lett. **89**, 117203 (2002).
[BATR95] G. G. Batrouni, R. T. Scalettar, and G. T. Zimanyi, Phys. Rev. Lett. **65**, 1765 (1995).
[HIRS81] J. E. Hirsch, D. J. Scalapino, R. L. Sugar, and R. Blankenbecler, Phys. Rev. B **26**, 5033 (1981).

[MURA99] A. Muramatsu, in *Quantum Monte Carlo Methods in Physics and Chemistry*, edited by M. P. Nightingale and C. J. Umrigar (NATO Science Series, Kluwer Academic Press, Dordrecht, 1999), pp. 343–373.

[CAPP00] S. Capponi and F. F. Assaad, Phs. Rev. B **63**, 155113 (2001).

[AUER94] A. Auerbach, *Interacting electrons and quantum magnetism.*, *Graduate texts in contemporary physics* (Springer, New York, Berlin, Heidelberg, 1994).

[PARO91] A. Parola, S. Sorella, M. Parrinello, and E. Tosatti, Phys. Rev. B **43**, 6190 (1991).

[AFFL88] I. Affleck and J. B. Martson, Phys. Rev. B **37**, 3774 (1988).

[KOTL88] G. Kotliar, Phys. Rev. B **37**, 3664 (1988).

[CHAK01] S. Chakravarty, R. B. Laughlin, D. K. Morr, and C. Nayak, Phys. Rev. B **63**, 094503 (2001).

[ASSAUN] F. Assaad, unpublished.

Ab initio Simulation of Clusters: Modeling the Deposition Dynamics and the Catalytic Properties of Pd_N on MgO Surface F-Centers

M. Moseler[1,2], B. Huber[1], H. Häkkinen[3], and U. Landman[3]

[1]Faculty of Physics, University of Freiburg, Herrmann-Herder-Str. 3, D-79104 Freiburg
[2]Fraunhofer-Institute for Mechanics of Materials, Wöhlerstr. 11, 79108 Freiburg
[3]School of Physics, Georgia Institute of Technology

Summary. Nano-catalysts are studied in an ab inito framework by solving the Kohn-Sham equations of density functional theory for the supported clusters and a finite zone of the underlying surface. An efficient and accurate numerical parallel implementation of the Kohn-Sham solver using plane waves for the kinetic energy calculations and a real space grid for the potential energy evaluations permits first principle molecular dynamics simulations of the nano-catalyst formation process namely the low-energy deposition of neutral Pd_N clusters (N = 2–7 and 13) on a MgO(001) surface with oxygen vacancies (so called F-centers, FC). The main findings of this simulations are a steering effect by an attractive "funnel" due to the polarizing F-center. This results in strong adsorption of the cluster, with one of its atoms pinned atop of the FC confirming that corresponding experiments are performed with supported size-selected nano-clusters and not with larger structures grown by coalescence. Interestingly, the deposited Pd_2-Pd_6 clusters retain their gas-phase geometries, while for N>6 the clusters adopt structures which maximize the contact area with the surface. Furthermore, we show that a large number of NO molecules can adsorbe on the low coordinated sites of the supported Pd clusters. For instance, the Pd_4 was able to capture up to 5 NO in our simulations (4 on Pd-Pd bridges and one molecule on top of the tetrahedral cluster). In order to demonstrate the accuracy of our method, we report on an additional study of finite temperature photoelectron spectra for sodium cluster anions.

1 Introduction

The scientific and industrial importance of transition metal particles on ceramics supports is based on their usage as catalysists in many applications like the cleaning of exhaust gases (i.e. the CO-NO reaction in car catalytic converters) or cyclomerization reactions in C-H chemistry [1]. Despite the broad interest in these systems little details are known concerning their structure and the pathways of the products in the catalised reactions. In this context a deeper theoretical understanding is surely usefull and could assist chemists in their attempts to improove existing catalytic components or to design new catalysts.

Of course, quantum mechanics plays a crucial role in the description of these complicated d-electron systems and therefore a numerical solution of the electronic Schrödinger equation is required in order to compute accurate forces in the ionic system allowing a faithfull description of structure and dynamics in the catalyst and the reactants. Unfortunatelly, the reliable calculation of industrially used catalytic transition metal particles still exceeds the capacities of modern super computers by several orders of magnitude and therefore one might loose the hope to learn anything about these fascinating systems in the near future. However, the last few years have seen major advances in experimental surface and ion beam techniques initiating a new branch in catalysis research namely the creation and study of nano-catalyists consisting of only a small and precisely determined number of atoms [1] (mainly Pd, Pt and Au) on well defined metal oxide surfaces like MgO(001). Temperature programmed reaction studies under very clean (i.e. ultra high vacuum) conditions already provided many usefull experimental results and ask for theoretical explanations of the fundamental chemical processes on nano-catalysts. Of course, an understanding of this model devices would also accelerate the knowledge gain in macroscopic catalysis.

One major finding in nano catalyst research consists of a strong variation of the observed reactivity with nano particle (cluster) size. For instance, the reaction of NO and CO to N_2 and CO_2 is strongly reduced on palladium clusters smaller than 6 atoms [2] whereas this process takes place on larger particles with a very high efficiency suggesting that cluster structure plays an important role in nano catalysis.

Fortunatelly, the progress in computer hardware and computational chemistry allows the numerical investigation of supported transition metal clusters constisting of more than a dozend atoms on surface models composed by almost 100 atoms [3]. In this context the work horse of large scale quantum chemistry namely Kohn's density functional theory [4] plays a dominant role since it reduces the complicated many electron system to a more tracktable picture of a single electron in the mean field of the other electrons resulting in a three-dimensional eigenvalue equation, the so called Kohn-Sham equation. For large systems, this equation can be solved with great accuracy and efficiency using a plane wave basis for the single electron wave functions [5]. However, the required memory and CPU speed still exceeds modern serial hardware and thus massive parallel computing is the only way to solve the Kohn-Sham equations for a large number of atoms. For instance, for the description of a Pd_{13} cluster on an MgO(001) surface [3], we utilized 3D grids of the order of 200 grid points in each dimension for the Fourier and real space representation of approximatelly 200 electronic wave functions resulting in a memory consumption of the order of 10 Gbyte! This demanding computation could only be done after reception of a generous computer grant for the Cray

T3E at the HLRS Stuttgart. The optimization of the electronic and ionic structure took roughly 160 hours on 128 processors and had to be performed for 4 different spin isomers. Thus, several tenthousands of CPU hours were already required for the electronic structure calculation of a single cluster size. Needless to say that this kind of computations are by no means a simple task. The selfconsistent nature of the Kohn-Sham approach often effects convergence problems (at least in d-metals and systems with spin flip behaviour) and sometimes it's more an art than science to find ways into selfconsistency. Therefore, it's not surprising that this contribution represent the first ab initio cluster softlanding simulation reported in the literature.

This paper is organised as follows. First, we give some theoretical and numerical details of the special method used for the density functional calculations [5]. Than, its usefullness and explanatory power is demonstrated for a simple application namely the finite temperature dynamics of sodium cluster anions [6]. These sytems were studied in the context of a diploma thesis (BH) as an introduction into the methodology. Meanwhile BH started his doctoral thesis on the more complicated catalytic systems[1]. In section 4, the main result of our work performed on the Cray T3E of the HLRS Stuttgart is presented: the deposition dynamics and the structure of Palladium clusters in the size range from two to thirteen atoms interacting with a typical MgO(001) surface defect (an F-center). We close with very recent results namely the determination of the adsorption sites of NO on the supported Pd clusters and an outlook on our future work.

2 The Born-Oppenheimer-Spin-Density-Molecular-Dynamics-Method

As mentioned in the introduction, we look for the solution of the Kohn-Sham equation for the electrons

$$\Big(-\frac{1}{2}\nabla^2 + v_{eff,\sigma}(\mathbf{r})\Big)\phi_{i,\sigma}(\mathbf{r}) = \epsilon_{i,\sigma}\phi_{i,\sigma}(\mathbf{r}). \tag{1}$$

Here the $\phi_{i,\sigma}$ are a set of a single particle electronic wave function, $\epsilon_{i,\sigma}$ their energies and the effective potential is given by

$$v_{eff,\sigma}(\mathbf{r}) = v(\mathbf{r}) + \int d^3r' \frac{n(\mathbf{r}')}{|\mathbf{r}-\mathbf{r}'|} + v_{xc,\sigma}(\mathbf{r}). \tag{2}$$

The electron density n of the system as the central quantity of density functional theory derives from the occupied Kohn-Sham orbitals

[1] Our nano-catalyst project is embedded and financed in the framework of the recently founded SPP 1153 of the Deutsche Forschungsgemeinschaft "Clusters on surfaces: electronic structure and magnetism".

$$n(\mathbf{r}) = \sum_{i,\sigma}^{occ} |\phi_{i,\sigma}(\mathbf{r})|^2. \tag{3}$$

In order to make the computations less expensive only chemical active electrons are considered and therefore a pseudo potential v is used for the confinement of the valence electrons representing the influence of the naked ions and the core electrons [7]. The so called exchange-correlation potential v_{xc} takes into account many body effects that are not included in the classical Coulomb field $\int d^3r' \frac{n(\mathbf{r}')}{|\mathbf{r}-\mathbf{r}'|}$ in the above equation. It is treated in the frame work of the generalized gradient approximation [8]. The spin of the system is explicitely taken into account by calculating the wave functions of both spin manifolds $\sigma = \uparrow, \downarrow$ and thus the description of magnetism is possible within this formalism. For more details on spin density functional theory, the reader is refered to standart text books [9].

The method for the numerical solution of eq. (1) utilizes the Born-Oppenheimer-local-spin-density-molecular-dynamics (BO-LSD-MD) approach of Barnett and Landman [5] and benefits from the fact that the differential operator $-\frac{1}{2}\nabla^2$ is a simple multiplication by $-\frac{1}{2}k^2$ for the Fourier transform $\phi_{\mathbf{k}}$ of the wave function. An iterative Block-Davidson eigenvalue solver only needs the action of the hamiltonean $-\frac{1}{2}\nabla^2 + v_{eff}$ onto a wave function and therefore a dual space technique treating the kinetic energy in Fourier and the potential energy part in real space provides a very efficient scheme to solve eq. (1). A domain decomposition of both spaces and an efficient parallelisation of the fast fourier transform (FFT) conecting k- and real space results in a very good parallel efficiency on massive parallel machines like a Cray T3E. The FFT is also used to calculate the Coulomb field $\int d^3r' \frac{n(\mathbf{r}')}{|\mathbf{r}-\mathbf{r}'|}$ since it satisfies Poissons equation which is algebraic and thus easily solvable in k-space. For more details on the numerical aspects of the method see [5].

After the solution of the Kohn-Sham equations, the forces on the ions are calculated employing the Hellmann-Feynman-Theorem [9] and the molecular dynamics of the respective system can be studied with very high accuracy. In certain cases where finite temperature infomation are needed a Langevin thermostat is used in order to simulate a canonical ensemble.

3 Thermal effects in the photo electron spectra of sodium cluster anions

In order to demonstrate the usefullness and accuracy of the BO-LSD-MD method, we report in this section on a theoretical study of room temperature Na_N^- clusters (N=4−19). Photoelectron spectroscopy is a frequently used experimental method to extract electronic binding energies from atomic, molecular and condensed matter systems. For molecules and clusters at low temperatures, a comparison of the measured photoelectron spectra (PES)

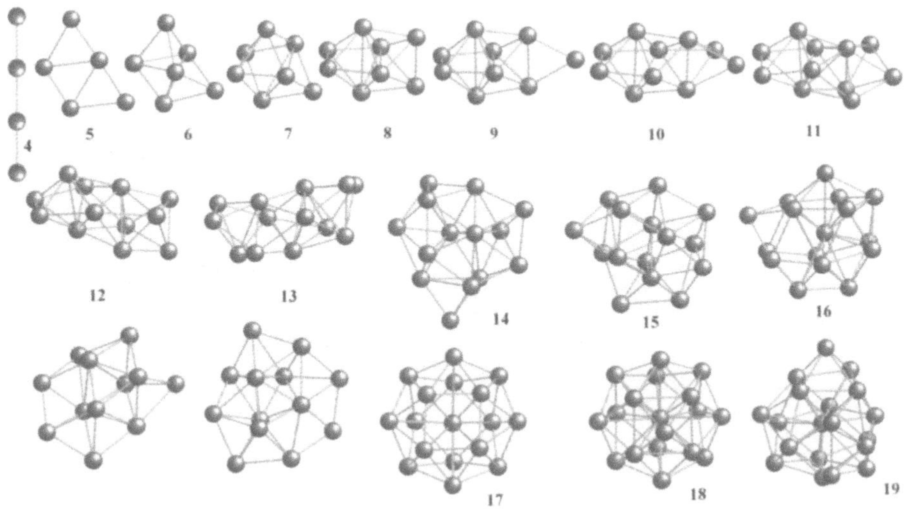

Fig. 1. Ground state structures of Na_N^- clusters (N=4−19). For most sizes energetically close lying isomers were found. For Na_{12}^- and Na_{13}^-, two degenerate isomers with different shapes occur - those in the middle row are prolate, and those in the bottom row are oblate.

with the electronic density of states (DOS) obtained from ab-initio calculations for T=0 optimal structures, provides informations about the underlying electronic structure and may lead to assignments of pertinent structures. At finite temperatures floppy systems like sodium clusters undergo large amplitude vibrations or even isomerisation (in fact medium sized sodium clusters actually melt already significantly below room temperature [10]). In such a case the PES represent superpositions of the spectra of more or less different structures and therefore yield information about these structures as well as on the isomerization dynamics.

Sodium is one of the best representatives of a free-electron-metal, which is the reason why many of the properties of sodium clusters have been successfully described by jellium [11] and Clemenger-Nilson [13] models with a total neglect of the ionic backbone. Of course, these models are quite crude approximations, and it is therefore desirable to check their predictions by theoretical and experimental work which yields detailed information about the geometrical and electronic structure of the clusters.

Fugure 1 displays the ground state (GS) structures of sodium cluster anions with less than 20 atoms found by us employing BO-LSD-MD. For a discussion of the apparent growth patterns, the reader is refered to ref. [6]. In addition to these simple structural optimizations, we also performed extensive

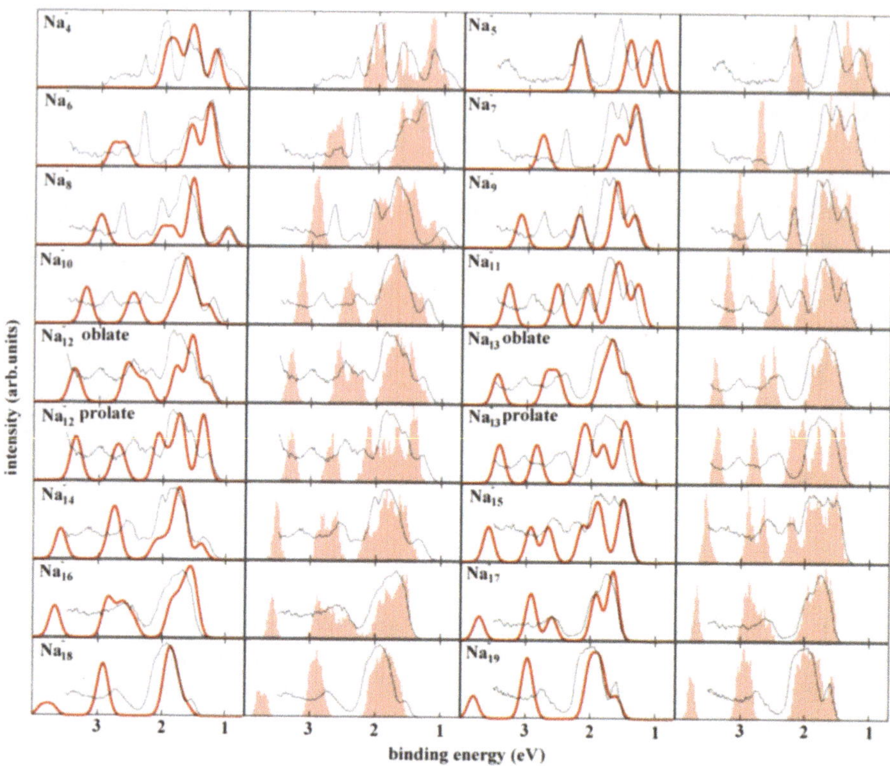

Fig. 2. Experimental photoelectron spectra of sodium cluster anions (thin solid curves) [6] are compared to the Gaussian broadened (σ= 0.08 eV) ground state DOS (thick solid curves in the 1st and 3rd columns), and to the thermally broadened DOS obtained from room temperature BO-LSD-MD trajectories (histograms in the 2nd and 4th columns).

ab initio finite temperature (T=300 K) MD simulations extending over tens of picoseconds.

A comparison of recently measured photoelectron spectra with the electronic density of states (obtained by histogramming the Kohn-Sham energies ϵ_i over the entire trajectory) allows us to extract detailed information about static and dynamic shape properties of simple metal clusters (see Fig. 2). Note, the overall good agreement between the simulated finite temperature spectra and the experiment justifying theoretical conclusion about the (experimentally not observable) cluster shapes (see ref. [6] for a detailed discussion of the spectra).

The cluster shapes were analyzed by calculating the radii of gyration ($R_{\text{min}} \leq R_{\text{middle}} \leq R_{\text{max}}$) from an analysis of the principal moments of inertia for the ionic coordinates. The radii of the ground state clusters and the average radii from our finite temperature simulations are plotted in Figs. 3a

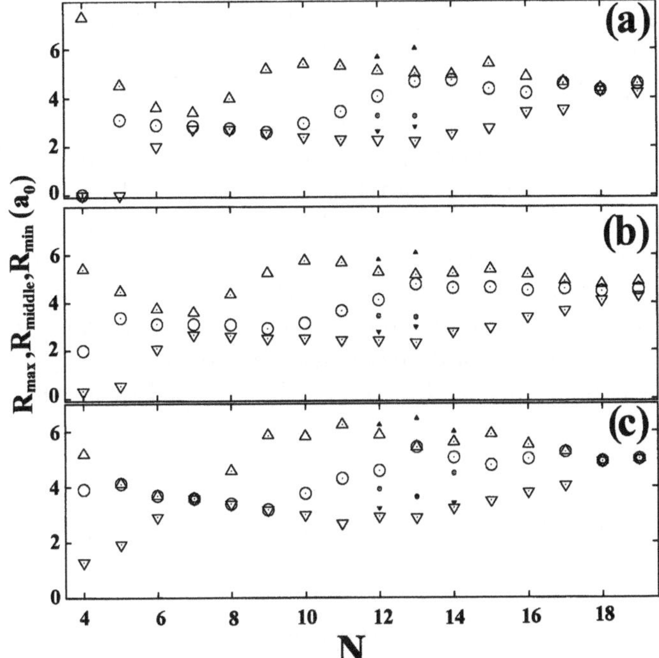

Fig. 3. The three radii (R_{max}, R_{middle}, R_{min}) of the anionic sodium clusters along the principal axis, plotted versus the number of atoms in the cluster N. Down-triangles, circles, and up-triangles correspond to R_{min}, R_{middle} and R_{max}, respectively. The radii were obtained from the moments of inertia of the ionic background. (a) Radii pertaining to the GS structure, see Fig. 1. (b) Thermally averaged radii from room temperature BO-LSD-MD trajectories. (c) Radii calculated from the reported Hill-Wheeler parameters in the ultimate jellium results [12]. Small filled symbols in (a) and (c) represents other energetically degenerate GS isomers and in (b) averages over the trajectories in the shape basins of this isomers.

and 3b, respectively. They exhibit allmost the same size-evolutionary pattern as already found for a variety of jellium models, see for instance the ultimate jellium results [12] plotted in Fig. 3c.

However, there are also subtle differences. For instance, the Na_7^- ground state from our BO-LSD-MD calculations exhibits an elipsoidal distortion (reflected in the appearance of two different radii in Fig. 3a). On the other hand, the jellium picture relates this cluster with an electronic shell closing and consequently predicts it to be spherical (i.e. it has three equal radii in Fig. 3c). Remarkably, at room temperature the Na_7^- breaks even the elipsoidal symmetry resulting in an triaxial cluster as can be seen in Fig. 3b. This symmetry breaking directly affects the photoelectron spectrum in Fig. 2 where three close lying lines at low electron binding energies are observed. These three features can be directly related with the three different radii of gyration [6].

As another interesting detail we mention the shape isomerism occuring in Na_{12}^- as well as in Na_{13}^- (see Fig. 1). While for Na_{12}^- both isomers exhibit strong triaxial shape deformation (one more oblate, the other more prolate), the Na_{13}^- ground state structures are almost ideally oblate and prolate spheroidal. These findings are in excellent agreement with the predictions of the jellium model [12] (compare Fig. 3a with Fig. 3c). Interestingly, the experimental photoelectron spectra of both Na_{12}^- and Na_{13}^- can only be interpreted by the simulated *oblate* DOS spectra (Fig. 2) which is a strong indication that only one shape isomer is favored when those clusters are formed in the experiment. A likely explanation for the preference of oblate Na_{12}^- and Na_{13}^- is that these clusters are formed by evaporation of monomers from the slighlty larger parent clusters (which are *oblate*, see Fig. 3) and thus still reflect the shapes of their parents.

4 Softlanding of palladium cluster on a MgO(001) surface

An understanding of the catalytic properties of supported Pd clusters requires the knowledge of the structures of the supported particles. Even the deposition dynamics might be of interest in order to clarify the presence or absence of coalescence and fragmentation processes during and after cluster deposition. Furthermore, the important role of spin in a previous gas-phase Pd cluster study [14] suggests that magnetism might also remain after deposition.

This was our motivation for extensive BO-LSD-MD simulations treating for the first time soft-landing on an ab initio level. We calculated the structures of Pd_2 to Pd_7 and Pd_{13} deposited on a realistic surface namely a MgO (001) with one oxygen vacancy. The MgO substrate with the F-center was modeled with a two-layer *ab-initio* cluster $Mg_{13}O_{12}$, embedded into a lattice of point-charges in order to simulate the long-range Madelung potential. The lattice parameter of the embedding part of the substrate was fixed to the experimental value (4.21 Å) of bulk MgO. The Pd_N cluster and the F-center's 4 nearest-neighbor Mg atoms and 4 nearest-neighbor O atoms of the first layer were treated dynamically during the deposition. To model the heat conductivity of the extended MgO surface the equations of motion of the dynamic surface atoms included an added damping term with a damping constant $\pi\omega_D/6$, where ω_D is the Debye frequency of bulk MgO.

The initial spin states (triplet for N=2−7 and nonet for N=13) and geometrical structures of the Pd_N clusters were taken from our recent gas-phase study [14]. The clusters were placed with a random orientation 4 Å above the FC (measured from the cluster atom closest to the surface) and an initial velocity directed perpendicular to the MgO surface, corresponding to a kinetic energy of 0.1 eV per atom to simulate softlanding conditions. The spin of the cluster-substrate system was dynamically evaluated at each MD time step. Subsequent to the dynamical evaluation of the deposition process for about 1 ps the simulation was stopped, and starting from the last recorded configuration a corresponding potential energy minimum was located by an energy-gradient optimization with variable spin; other spin-isomers (SPIs) were optimized (starting from the aforementioned optimal configuration), in order to explore the thermal stability of the lowest-energy SPI.

The adsorption of a single Pd atom on top of the FC (*t*FC site) is characterized by a strong binding energy (3.31 eV) and a short equilibrium adsorption distance (1.65 Å), compared to adsorption on-top of an oxygen (*t*O) atom at the ideal MgO surface (1.16 eV and 2.17 Å). The bonding between the Pd atom and the FC involves the localized FC electronic orbital, located in the band gap of MgO (separated from the top of the valence band by 2.3 eV), and (mainly) the $d(m = 0)$ orbital of the Pd atom. The attractive interaction to the F-center is rather long-ranged extending up to about 5 Å above the surface; e.g., the interaction energy of a Pd atom placed 5.2 Å above the FC is 0.1 eV. This weak attraction is due to polarization of the d(m=0) valence orbital of Pd by the FC. Surprisingly, we found that none of the other adsorption sites for the Pd atom, lying in the vicinity of the F-center (e.g., on-top of the neighboring oxygen (*t*O), on-top of the neighboring Mg atom (*t*Mg), the Mg-Mg bridge (*b*MgMg), the Mg-O bridge (*b*MgO) and the Mg-Mg-O hollow site (*h*MgMgO)), are stable; i.e., optimization starting from any of these sites leads to a spontaneous (barrierless) transition to the aforementioned *t*FC configuration.

We conclude that the F-center acts as a rather wide attractive "funnel" for the Pd atom, extending several Å both laterally and vertically[15]. This funneling effect steers the incident cluster and dominates the dynamics of the initial phases of the deposition process, as illustrated in the following for the representative case of a Pd_5 cluster.

When the Pd_5 cluster is placed 4 Å above the oxygen vacancy, the FC electronic state (located just below E_F) combines with d-orbitals of the closest Pd atom to form two bonding molecular orbitals (see the up-spin HOMO-1 and the down-spin HOMO in Fig. 4a). All other orbitals (for example the lowest unoccupied orbital (LUMO) of Pd_5 shown in Fig. 4a) remain to a large degree eigenstates of the separated systems, and consequently the corresponding density of states (DOS in Fig. 4b) may be represented as a superposition of those of the bare surface and the gas-phase cluster.

The long-range attraction between the cluster and the FC accelerates the lowermost Pd atom towards the *t*FC site (note the strong deformation of the Pd_5 cluster at 0.2 ps in Fig. 4c and the increase of the kinetic energy in Fig. 4d). Subsequently, other Pd atoms are attracted to neighboring *b*MgO positions (Fig. 4c, t=0.5 ps) accompanied by additional release of kinetic energy. Consequently, the center of mass (CM) velocity toward the surface increases to almost twice its initial value (Fig. 4d) leading to a strong flattening of the cluster at t=0.5 ps (see the minimum in the z component of the cluster CM in Fig. 4e). The cluster shape deformation causes a transient reordering of the molecular orbitals, i.e. it raises the energy of the up-spin HOMO-1 level

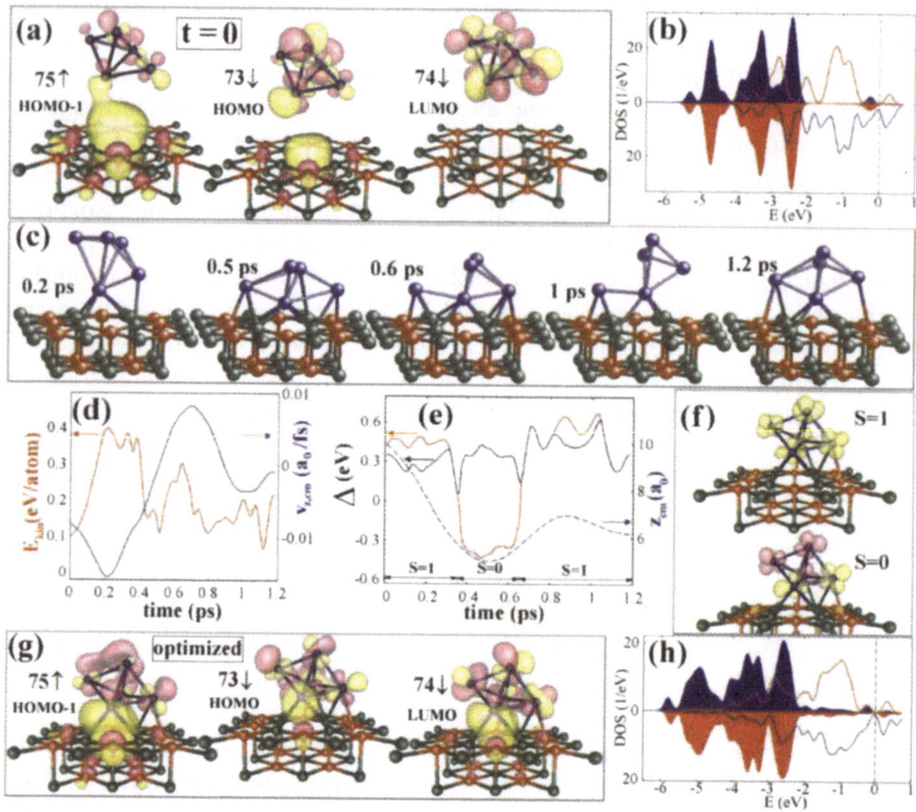

Fig. 4. A Pd_5 cluster impinges with 0.1 eV/atom kinetic energy on an F-center in a MgO (001) surface. Pd atoms are depicted as blue, Mg as green and O as red spheres. (a) Isosurfaces of the highest occupied up-spin molecular orbital (HOMO-1), the highest occupied down-spin orbital (HOMO) and the lowest unoccupied down-spin orbital (LUMO) of the initial configuration at t=0 (color coding distinguishes the sign of the wave function). Note, that the 75 up-spin and 73 down-spin orbitals are both occupied. (b) The corresponding local densities of states of the surface (blue area for up- and red area for the down-spin DOS) and of the cluster (red line for up- and blue line for down-spin DOS). The Fermi level, $E_F = 0$. We note that the DOS of the isolated surface and free cluster are essentially identical to that shown here, except for the first peak below E_F that corresponds to the long-range interaction discussed in the text. (c) Snapshots from the MD simulation recorded at the indicated times. (d) Time-evolution of the kinetic energy (red line) and the z-component of the CM velocity (blue line) of the cluster. (e) Evolution of the HOMO-LUMO gap (black line), of the eigenvalue energy difference $\epsilon_{74\downarrow} - \epsilon_{75\uparrow}$ (red line, see panel (a) for explanation of the orbital numbers), and of the z-component of the cluster's CM coordinate (blue dashed line). The triplet-singlet-triplet transition is indicated by the black arrows drawn on the time axis. (f) Isosurfaces of spin polarization density for the optimized triplet (S=1) and singlet (S=0) states. Yellow and purple denote excess of up and down spins, respectively. (g) Isosurfaces of the orbitals in (a) for the optimized adsorbed cluster. (h) The local DOS corresponding to the optimized cluster (color coding as in (b)).

(marked 75↑ in Fig. 4a) and turns it into a HOMO at t=0.32 ps, and even into a LUMO state at t=0.36 ps; this sequence is portrayed in Fig. 4e by closing of the HOMO-LUMO gap (black curve) and the minimum in the eigenvalue energy difference $\epsilon_{75\uparrow}-\epsilon_{74\downarrow}$ (red curve). Since the down-spin LUMO (74↓ in Fig. 4a) temporarily becomes the HOMO state, the total spin flips for a short time period from S=1 to S=0 (Fig. 4e). After 0.5 ps the cluster recoils the reverse process drives the cluster back into the triplet spin state at t=0.63 ps (Fig. 4e).

Optimization of the adsorbed cluster after a 1.2 ps MD simulation (see Fig. 4c for the last MD configuration), resulted in a trigonal bipyramide structure (which coincides with the gas-phase optimal configuration) lying with a triangular facet against a *t*FC-*b*MgO-*b*MgO surface triangle (Fig. 4f). The spin polarization of the triplet ground state (see $S = 1$ isosurface in Fig. 4f) resembles that of the free cluster with a minor additional contribution from four surface oxygen atoms closest to the FC. As expected from our gas-phase calculations [14] the slightly higher-lying singlet state (ΔE=24 meV) consists of an anti-ferromagnetic ordering of the local magnetic moments (see $S = 0$ isosurface in Fig. 4f). The spatial character of the orbitals close to E_F and the surface and cluster contributions to the DOS of the triplet ground-state of Pd_5/MgO(FC) are shown in Figs. 4g and 4h, respectively.

Using the above-mentioned methodology, we have determined the ground-states for the other deposited Pd_N clusters. For $3 \leq N \leq 6$ we observed a regular structural size evolution (Fig. 5a) where the gas-phase ground state structures are anchored to the MgO surface with one Pd atom on the *t*FC, another Pd on *h*MgMgO site (for $N = 2$) or 2 additional Pd atoms on *b*MgO sites close to the *t*O position (for $3 \leq N \leq 6$). However, for Pd_7 and Pd_{13} the free clusters transform to structures that exhibit a higher degree of commensurability with the underlying surface, incorporating a Pd_6 and Pd_7 subunit, respectively (Fig. 5a). In this case, the loss in the intracluster cohesion is counterbalanced by a considerable gain of adhesion energy E_{ad} [defined as $E_{ad} = E(\mathrm{MgO(FC)}) + E(\mathrm{Pd}_N) - E(\mathrm{Pd}_N/\mathrm{MgO(FC)})$, see the red curve in Fig. 5b]. Consequently, the cohesive energy E_c per Pd atom [defined as $E_c = (E(\mathrm{MgO(FC)}) + NE(\mathrm{Pd}) - E(\mathrm{Pd}_N/\mathrm{MgO(FC)}))/N$, see the blue curve in Fig. 5b] continues to increase after Pd_6 and remains well above the gas-phase E_c values.

The HOMO-LUMO gap (Fig. 5b) of the combined Pd_N/MgO(FC) system is governed mainly by the metal cluster since the top part of its density of states lies in the MgO band gap. Most interestingly, the deposited Pd_N clusters with $N \geq 4$ remain magnetic: S=1 for $4 \leq N \leq 7$ and S=3 for N=13. The crossover from nonmagnetic to magnetic states between Pd_3 and Pd_4 correlates with an increased "thickness" of the cluster (Fig. 5c), corroborating our finding that flattening of the cluster on the surface tends to be accompanied by quenching of the spin (see discussion in the context of Fig. 4e).

In general, the deposition of the cluster reduces the energy separation between SPIs, thus lowering the threshold temperature for their coexistence.

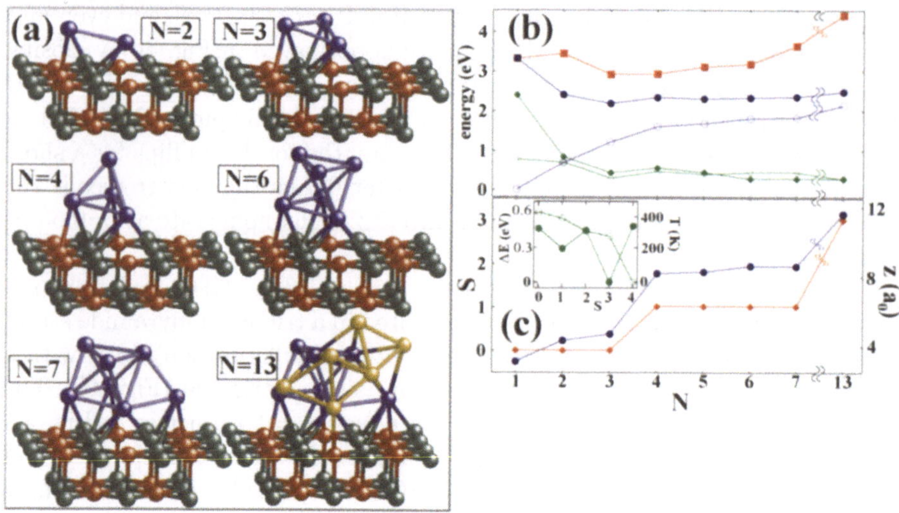

Fig. 5. Structural and magnetic size-evolution of supported Pd_N clusters. (a) GS structures of Pd_N (N=2,3,4,6,7 and 13). Color coding as in Fig. 4 except for Pd_{13} where a subset of the Pd atoms is colored in yellow in order to highlight the Pd_7 subunit (blue). (b) Size-evolution of the adhesion energy E_{ad} (red filled squares), the binding energy per atom E_b for the supported (blue solid dots) and free (blue circles) clusters, and the HOMO-LUMO gap of the supported (green solid diamonds) and free (green open diamonds) clusters. (c) Size evolution of GS spin S (red diamonds) and the distance of the highest cluster atom to the surface (blue solid dots). The inset in (c) shows the SPI energies ΔE (with reference to the GS configuration, S = 3 for the adsorbed cluster and S = 4 for the free one) and corresponding activation temperatures $T = 2\Delta E/k/(3N - 6)$, of supported (green solid dots) and free Pd_{13} (green circles) clusters.

For instance, the triplet-singlet energy difference of supported Pd_4 is ΔE =65 meV compared to the gas-phase values of ΔE =136 meV; for Pd_{13} five SPIs can be found within a 0.5 eV range, which expressed in terms of temperature corresponds to about 350 K (Fig. 5c, inset). This result indicates that experiments aiming at distinguishing magnetic states of the adsorbed clusters could be carried out at room temperature.

5 NO adsorption sites on Pd_4

A first step into an understanding of nano catalysis can be performed by studying the most likely initial configurations of reactant molecules. For example, prior to an investigation of the NO-CO reaction the prefered adsorption sites of NO and CO have to be determined. We report here on first results for NO molecules adsorbed on Pd_4. Figure 6 displays the consecutive occupation of possible adsorption sites on the cluster. NO preferes undercoordinated

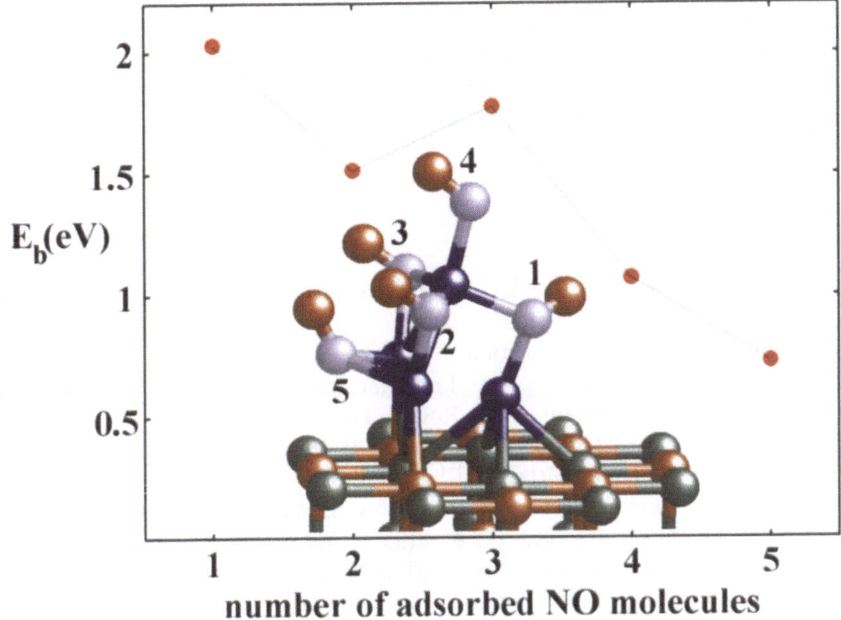

Fig. 6. Binding energy E_b of an additional NO molecule on a Pd_4 cluster supported by a MgO(001) F-center. The inset displays the adsorption sites. The numbers in the inset correspond to the numbers on the abscisa and represent addition order. Color coding for Mg, O, and Pd is the same as in Fig. 4. Light blue spheres represent nitrogen atoms.

sites like bridges (locations 1-3 and 5 in inset of Fig. 6) and on top of the tetrahedron defined by the cluster (location 4). Threefold coordinated NO on the tetrahedral faces is energetically less favorable. The average binding energy per NO is of the order of 1.5 eV and the energy gain by adding the 5th NO to an already 4fold occupied Pd_4 cluster is still roughly 1 eV expressing the strong tendency of Pd-clusters to adsorb NO molecules.

6 Outlook

The next step on the route to a NO-CO reaction model is the determination of the CO reaction sites (calculations are on the way) and than the investigation of clusters with mixed NO-CO occupation. This will form a suitable starting point to study the actual NO-CO reaction by constrained dynamics methods.

7 Acknowlegdement

We thank Ueli Heiz and Bernd v. Issendorff for fruitfull discussions. This work is supported by the Deutsche Forschungsgemeinschaft (MM,BH), the Academy of Finland (HH) and the US Department of Energy (UL).

References

1. U. Heiz and W-D. Schneider, in *Metal Clusters at Surfaces*, edited by K.H. Meiwes-Broer (Springer, Berlin, 2000)
2. U. Heiz, private communication
3. M. Moseler, H. Häkkinen and U. Landman, Phys. Rev. Lett. **89**, 176103 (2002)
4. W. Kohn and L.J. Sham, Phys. Rev. A **140**, 1133, (1965)
5. R. Barnett and U. Landman, Phys. Rev. B **48**, 2081 (1993).
6. M. Moseler et al., submitted to Phys. Rev. B
7. N. Troullier, and J.L. Martins, Phys. Rev. B **43**, 1993 (1991). The core radii (in a_0) are: Pd $s(2.45)$ local, $p(2.6)$, $d(2.45)$; Mg $s(2.5)$, $p(2.75)$ local; O $s(1.45)$, $p(1.45)$ local. A plane-wave basis with a 62 Ry cutoff was used.
8. J.P. Perdew et al., Phys. Rev. Lett. 77, 3865 (1996)
9. R. Parr and W. Yang, *Density functional theory of atoms and molecules* (Oxford university press, 1989)
10. M. Schmidt, R. Kusche, B. v. Issendorff and H. Haberland, Nature **393**, 6682 (1998).
11. W. Ekardt, Phys. Rev. Lett. **52**, 1925 (1984)
12. M. Koskinen, PO Lipas, and M. Manninen, Z. Phys. D **35**, 285 (1995).
13. W. deHeer, Rev. Mod. Phys. **65**, 611 (1993).
14. M. Moseler, H. Häkkinen, R. Barnett and U. Landman , Phys. Rev. Lett, **86**, 2545 (2001). The optimal gas-phase geometries of the Pd_N clusters are as follows: $N = 3$, equilateral triangle; $N = 4$, tetrahedron; $N = 5$, trigonal bipyramide; $N = 6$, octahedron; $N = 7$, pentagonal bipyramide; $N = 13$, icosahedron.
15. Our study is relevant to a series of experiments[1] where nanocatalysts are prepared by softlanding metal clusters on thin MgO films containing typically a few % ML coverage of FCs. In this case, an average FC-FC distance is a few lattice constants. Therefore, most metal clusters either experience directly the "funnel effect" of the nearest FCs while approaching the MgO surface, or become trapped at FCs after a rather short surface diffusion path.

Reactive Flows

Prof. Dr. Dietmar Kröner

Institut für Angewandte Mathematik, Universität Freiburg
Hermann-Herder-Str. 10, 79104 Freiburg

In this section two highly advanced contributions concerning reactive flow and dissipative solitons are presented.

The first contribution shows that the direct numerical simulations of turbulent premixed flames on a distributed memory machine has now become a standard task. This is important since it opens the door for new investigations on fundamental mechanisms in turbulence-chemistry-interactions and in the refinement of turbulent combustions models. Therefore the numerical simulation has become a reliable tool to study turbulent flames at detailed levels.

This contribution of Lange is also important because he uses the code for the direct numerical simulation of premixed flames as a benchmark problem for different distributed memory platforms. The performance of the code on the most important current microprocessor architectures like Intel Xeon, Itanium 1 and Itanium 2 and IBM Power 4 are presented. Therefore by these results important and reliable information of the power of different computer systems for real applications are available.

The second contribution concerns a slightly different application. The mathematical model for pattern formation phenomena in semiconductor gas discharge systems consists of a three component reaction-diffusion system in two spatial dimensions. The authors have performed several numerical experiments on the parallel computer to demonstrate a new type of bifurcation of localized structures without rotational symmetry, possible in non-gradient dissipative systems with more than one spatial dimensions.

DNS of Turbulent Premixed CO/H_2/Air Flames

Marc Lange

High-Performance Computing Center Stuttgart (HLRS), Stuttgart University
Allmandring 30, D-70550 Stuttgart, Germany
lange@hlrs.de

Summary. A program for the direct numerical simulation (DNS) of reactive flows is presented. In favor of using detailed models for chemical kinetics and molecular transport only spatially two-dimensional simulations are performed. This scientific application code has been used as a benchmark on several platforms, including Intel Itanium 2 systems, IBM's SMP server pSeries 690, and a Fujitsu-Siemens hpc-Line cluster with Intel Xeon CPUs. The overhead for parallelization on these systems is discussed separately from the single processor performance. Up to 512 processor elements of a Cray T3E have been used and it is shown that even more processors could be used while maintaining a high parallel efficiency. Direct simulations of flames evolving after induced ignition of a premixed CO/H_2/air mixture under turbulent conditions have been carried out. Results of this investigation are presented in the application part of the paper.

1 DNS of Turbulent Reactive Flows

Energy conversion in numerous industrial power devices like automotive engines or gas turbines is still based on the combustion of fossil fuels. In most applications, the reactive system is turbulent and the reaction progress is influenced by turbulent fluctuations and mixing in the flow. The understanding and modeling of turbulent combustion is thus vital in the conception and optimization of these systems in order to achieve higher performance levels while decreasing the amount of pollutant emission.

During the last few years, direct numerical simulations (DNS), i. e. the computation of time-dependent solutions of the Navier-Stokes equations for reacting ideal gas mixtures (as given in Sect. 2), have become an important tool to study turbulent combustion at a detailed level. As DNS does not make use of any turbulence or turbulent combustion models, this technique may be interpreted as high-resolution (numerical) experiments, enabling new investigations on fundamental mechanisms in turbulence-chemistry-interaction and aiding in the refinement of turbulent combustion models.

Due to the broad spectrum of length and time scales apparent in turbulent reactive flows, DNS require a very high resolution in space and time. To be able to perform DNS of reactive flows including detailed chemical reaction mechanisms and a detailed description of molecular transport, it is therefore necessary to make efficient use of HPC-systems. The computation of the chemical source-terms and the multicomponent diffusion velocities are the most time-consuming parts in such DNS. Therefore, almost all DNS carried out so far have been (at least) restricted to the use of simplified models (e. g. one global reaction and equal diffusivities) or to two-dimensional simulations. Even with these restrictions it is crucial to make efficient use of the computational power provided by parallel supercomputers to be able to carry out DNS of reactive flows in acceptable time.

2 Governing Equations

In the context of DNS, chemically reacting flows are described by a set of coupled partial differential equations expressing the conservation of total mass, chemical species masses, momentum and energy [1]. By using summation convention, which shall be applied to roman italic indices (i, j, k) only whereas no summation is carried out over greek indices (α, λ), these equations can be written in the form

$$\frac{\partial \varrho}{\partial t} + \frac{\partial(\varrho u_j)}{\partial x_j} = 0 \quad , \tag{1}$$

$$\frac{\partial(\varrho Y_\alpha)}{\partial t} + \frac{\partial(\varrho Y_\alpha u_j)}{\partial x_j} = -\frac{\partial(\varrho Y_\alpha V_{\alpha,j})}{\partial x_j} + M_\alpha \dot{\omega}_\alpha \qquad (\alpha = 1, \ldots, N_\mathrm{S}) \quad , \tag{2}$$

$$\frac{\partial(\varrho u_i)}{\partial t} + \frac{\partial(\varrho u_i u_j)}{\partial x_j} = \frac{\partial \tau_{ij}}{\partial x_j} - \frac{\partial p}{\partial x_i} \quad , \tag{3}$$

$$\frac{\partial e_\mathrm{t}}{\partial t} + \frac{\partial((e_\mathrm{t} + p) u_j)}{\partial x_j} = \frac{\partial(u_j \tau_{kj})}{\partial x_k} - \frac{\partial q_j}{\partial x_j} \quad , \tag{4}$$

where ϱ is the density and u_i the velocity component in ith coordinate direction. Y_α, $\boldsymbol{V}_\alpha$ and M_α are the mass fraction, diffusion velocity and molar mass of the species α. N_S is the number of chemical species occuring in the flow, τ_{ij} denotes the viscous stress tensor and p the pressure, $\boldsymbol{q}$ is the heat flux and e_t is the total energy given by

$$e_\mathrm{t} = \varrho \left(\frac{u_i u_i}{2} + \sum_{\alpha=1}^{N_\mathrm{S}} h_\alpha Y_\alpha \right) - p \quad . \tag{5}$$

where h_α is the specific enthalpy of the species α.

The chemical production rate $\dot{\omega}_\alpha$ of the species α, which appears in the source-term on the right-hand sides of the N_S species mass equations (2), is given as the sum over the formation rate equations for all N_R elementary reactions,

$$\dot{\omega}_\alpha = \sum_{\lambda=1}^{N_\mathrm{R}} k_\lambda (\nu_{\alpha\lambda}^{(\mathrm{p})} - \nu_{\alpha\lambda}^{(\mathrm{r})}) \prod_{\alpha=1}^{N_\mathrm{S}} c_\alpha^{\nu_{\alpha\lambda}^{(\mathrm{r})}} , \qquad (6)$$

where $\nu_{\alpha\lambda}^{(\mathrm{r})}$ and $\nu_{\alpha\lambda}^{(\mathrm{p})}$ denote the stoichiometric coefficients of reactants and products respectively, and c_α is the concentration of the species α. The rate coefficient k_λ of the elementary reaction λ is given by a modified Arrhenius law

$$k_\lambda = A_\lambda T^{b_\lambda} \exp\left(-\frac{E_{\mathrm{a}\lambda}}{RT}\right) \quad , \qquad (7)$$

where the parameters A_λ, b_λ of the pre-exponential factor and the activation energy $E_{\mathrm{a}\lambda}$ are determined by a comparison with experimental data [2].

The viscosity and multicomponent diffusion velocities are computed using standard formulae [3, 4], thermodynamical properties are computed using fifth-order fits of experimental measurements [2]. This set of equations is complemented by the ideal-gas state-equation

$$p = \frac{\varrho}{\overline{M}} RT \quad . \qquad (8)$$

with R being the gas constant and $\overline{M}$ the mean molar mass of the mixture.

3 Structure of the Parallel DNS Code

A code has been developed for the DNS of reactive flows on parallel computers with distributed memory using message-passing communication [5, 6, 7]. In favor of being able to include detailed models for the chemical reaction kinetics and the molecular transport as outlined in Sect. 2, only two-dimensional simulations are performed.

The spatial discretization in DNS codes is typically done by spectral methods or high-order finite-difference schemes. The main advantages of the latter ones are a greater flexibility with respect to boundary-conditions and the possibility of a very efficient parallelization, as in a fully explicit formulation no global data-dependencies occur. We chose a finite-difference scheme with sixth-order central-derivatives, avoiding numerical dissipation and leading to very high accuracy. Depending on the type of boundaries — periodic and symmetry conditions, different inflow conditions, non-reflecting outflow conditions and non-reacting walls are implemented — lower order schemes may be used on the global domain boundaries. The integration in time is carried out using a fourth-order fully explicit Runge-Kutta method with adaptive time-stepping. The control of the timestep is based on (up to) three independent criteria: A Courant-Friedrichs-Lewy (CFL) criterion and a Fourier criterion for the diffusion terms are checked to ensure the stability of the integration and an additional accuracy-control of the result for one or more selectable variables is obtained through step doubling [8].

The fully explicit formulation leads to a parallelization strategy, which is based on a regular two-dimensional domain-decomposition with halo boundaries. The global computational grid is split up into non-overlapping rectangular subdomains. After this initial decomposition each of these subdomains is controlled by one processor element (PE). In addition to the points belonging to its subdomain, on each PE the values of a three points wide surrounding region are stored in so called ghost cells. By using the locally stored values of this halo region, an integration step in the subdomain can be carried out independently from the other nodes. After each integration step, the values in the ghost cells are updated by point-to-point communications with the neighboring nodes. MPI is used for the communication.

4 Single Processor Performance

A high single processor performance is obviously crucial for any parallel code to achieve a high overall performance. Besides the performance on the Cray T3E, in our last report already performance data for the Alpha EV68 and the AMD AthlonMP have been presented [9]. New benchmarks have been performed on systems using some of the most important current microprocessor architectures:

hpcLine Fujitsu-Siemens hpcLine cluster: Dual-Nodes with 2.0 GHz Intel Xeon CPUs (Prestonia) with 512 KB L2 Cache, Myrinet 2000 and Fast Ethernet node-to-node interconnects

AzusA NEC AzusA ccNUMA server with 16 800 MHz Intel Itanium CPUs (C0 stepping)

zx6000 IA64-Cluster, Nodes: HP zx6000 with two 900 MHz Itanium 2 CPUs (1.5 MB L3 Cache) and HP's ZX1 chipset, Interconnect: Myrinet 2000

Tiger Intel's IA64 evaluation system with four 1.0 GHz Itanium 2 CPUs each having 3 MB L3 Cache

p690 IBM pSeries 690 "Regatta" with 1.3 GHz Power 4 CPUs

Table 1 lists the performance of one processor of each of these systems normalized with the problem size and given as a value relative to the performance of a Cray T3E-1200 PE for the 50^2 grid points problem. The 1.3 GHz Power 4, Intel's 1.0 GHz Itanium 2, and the 2.0 GHz Xeon all perform at a very similar level for our application. In contrast to the AthlonMP and especially to the Alpha 21264B (EV68) no significant decrease for increasing problem size is observed.

Table 1. Monoprocessor performance for different problem sizes related to the performance of the Cray T3E-1200 solving the 50 × 50 problem

Problem Size	50 × 50	100 × 100	200 × 200	400 × 400
T3E-1200, Alpha EV5, 600 MHz	1.00	1.11	1.19	1.23
API-M2K, Alpha EV68, 833 MHz	4.40	4.18	3.67	2.69
HELICS, AMD AthlonMP, 1.4 GHz	4.22	3.87	3.84	3.69
hpcLine, Intel Xeon, 2.0 GHz	5.57	5.77	5.50	5.52
AzusA, Intel Itanium (C0), 800 MHz	2.25	2.52	2.57	2.52
zx6000, Intel Itanium 2, 900 MHz	4.53	5.17	5.28	5.37
Tiger, Intel Itanium 2, 1.0 GHz	5.07	5.61	5.45	5.51
p690, IBM Power 4, 1.3 GHz	5.41	5.79	5.84	5.49

5 Parallelization Overhead and Scaling Behavior

In this section, the parallelization overhead time t_c, i. e. the time spent for communication and synchronization, which is needed for the computation on the different systems listed above when using two or more CPUs is compared for different problem sizes and numbers of processors. In our benchmark, symmetric boundary conditions are used in the x-direction and periodic boundary conditions in the y-direction. In the cases using two processors with a load of $N \times N$ grid points per processor, the global grid has $2N$ points in the y-direction and N points in the x-direction. Hence the boundary between the two subdomains is parallel to the x-axis and due to the periodicity in the y-direction each of the two processors has to exchange the values at two sides of its subdomain-boundary with the other processor. In cases using four or eight processors every processor has to exchange values at three of the four sides of its subdomain-boundary and in the cases with 16 and more processors all "inner" processors, i. e. all except for those controlling subdomains that are located at the left or right boundary of the global domain, have to exchange the values at all four sides of their subdomain-boundaries with the neighboring processors.

In Table 2 the parallelization overhead time t_c is listed for cases with a constant load per processor of 50 × 50 and 100 × 100 grid points computed using either 4 or 16 processors. In the case of the clusters with dual-nodes, the suffix "-single" means that only one processor per node has been used whereas "-dual" denotes the use of two processors per node. By the separate measurements for these two cases it is possible to distinguish between the influence of the intra-node parallelization and the parallelization over the node-to-node interconnect. For the problem with the smaller messages the intra-node communication of the p690 SMP server is the fastest, while for the larger problem the Myrinet 2000 of HELICS performs best. Although the API-M2K cluster also uses a Myrinet 2000 node-to-node interconnect it does not achieve the communication performance of HELICS. Possible reasons are a superior MPI implementation on HELICS and differences in the data transfer to and from the Myrinet hardware inside the nodes. As there are already some Myrinet 2000 clusters in this comparison, the Fast Ethernet interconnect has been used on the hpcLine, which is slower by a factor of about 5.9 compared to the Myrinet 2000 of HELICS for the smaller and about 4.3 times slower for the larger messages. In the 100×100 grid-points per processor case the intra-node parallelization on the PC clusters (HELICS-dual, hpcLine-dual) is even worse than the pure Ethernet solution (hpcLine-single). This is caused by memory conflicts in these simple bus-based systems and the severity of this performance problem strongly depends on the problem size [9]. This dependency can clearly be seen from Table 3, in which the parallelization overhead times on two CPUs for a constant load per processors of 50×50 and 400×400 grid points are listed respectively. However, in most cases in which the increase of t_c with the problem size is much larger than the increase of the amount of data to be communicated, the time for computation increases even stronger

Table 2. Parallelization overhead t_c/s for 10 timesteps

Problem Size per Processor	50 × 50	50 × 50	100 × 100	100 × 100
Number of Processors	4	16	4	16
T3E-1200	-	0.80	-	2.52
API-M2K-single	-	0.81	-	1.33
API-M2K-dual	-	0.56	-	1.68
HELICS-single	-	0.33	-	0.85
HELICS-dual	-	0.75	-	4.36
hpcLine-single	-	1.94	-	3.65
hpcLine-dual	-	1.90	-	5.67
AzusA	0.84	0.96	2.47	3.59
Tiger	0.33	-	1.34	-
p690	-	0.29	-	1.54

Table 3. Parallelization overhead t_c and parallel efficiency for two scaled problems with different size using two processors

	t_c for 10 timesteps (s)		Parallel Efficiency (%)	
Problem Size per Processor	50 × 50	400 × 400	50 × 50	400 × 400
HELICS-single	0.19	4.11	91.0	97.0
hpcLine-single	0.86	7.79	62.7	92.3
zx6000-single	0.30	11.51	85.6	89.3
API-M2K-dual	0.35	21.43	83.9	89.9
HELICS-dual	0.56	36.66	77.4	79.1
hpcLine-dual	0.33	36.77	81.4	71.7
zx6000-dual	0.30	16.15	85.5	85.6
p690	0.07	5.19	95.3	94.7

than t_c leading to a still higher parallel efficiency for the larger problem. The only case in which the parallel efficiency is below 70 % is the hpcLine using the Fast Ethernet for the small problem, in which the latency of the interconnect becomes more important. For current 2D production runs the load per processor typically is between 50×50 and 100×100 grid points. If the same methods as implemented in our code would be used in a 3D DNS, the CPU-time required to carry out one timestep would increase by less than 50 % per grid point. The increase in the memory needed per grid point for such an extension from 2D to 3D would also be less than 50 %. The number of timesteps to be carried out as well as the stepsize is the same for 2D and 3D DNS. The number of grid points in the 2D simulations presented in Sect. 6 is about 10^6. While a corresponding 3D DNS with 10^9 grid points is currently out of reach, a computational grid with $200 \times 200 \times 400 = 16 \cdot 10^6$ points would be a reasonable minimum for a 3D DNS. Computational loads varying between $40000 = 200^2$ and $160000 = 400^2$ grid points per processor are expected to be typical for 3D detailed chemistry DNS to be carried out in the near future. As tests on current state-of-the-art vector processors with versions of our code that are optimized for these systems show, the performance of these processors for our application is far higher than that of all microprocessors discussed above. Therefore simulations with more grid points per processor could be run on these systems. To perform detailed chemistry DNS in 3D, a very good scaling behavior of the code for a large number of processors becomes a necessity, especially in the case in which a system based on microprocessors shall be used.

Figure 1 shows the parallel efficiency measured on a Cray T3E-900 using up to 512 PEs with a constant load per PE of 28×28 and 42×42 grid points respectively. The main observation is that the parallel efficiency remains nearly constant from 16 PEs, which is the first configuration in which the "inner" nodes have to exchange the values of all four sides of their subdomain's boundary with the neighboring nodes, up to 512 PEs. The parallel efficiency which is achieved depends on the ratio of the amount of data to be communicated to the amount of computation to be performed. The number of grid points to be exchanged by an "inner" node after each timestep divided by the number of grid points in the subdomain can act as a simple indicator for

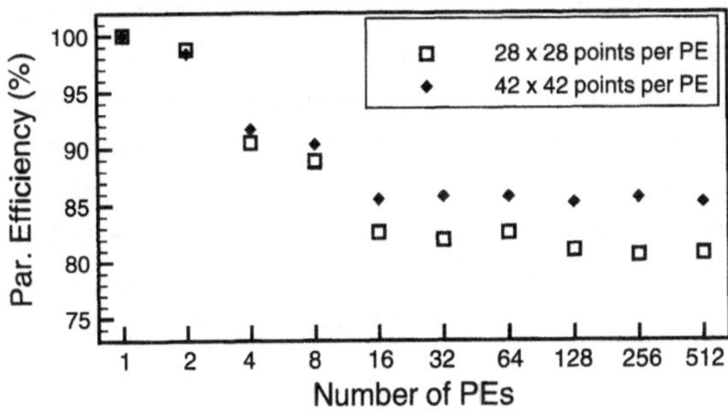

Fig. 1. Parallel efficiency on the Cray T3E-900 for a constant load per PE of 28^2 and 42^2 grid points respectively

this ratio. It is about 0.31 in the case with 42×42 grid points per subdomain and about 0.47 for the 28×28 points subdomains. In a 3D simulation with cubic subdomains this quotient would be about 0.55 for an $40 \times 40 \times 40$ points subdomain and about 0.42 for a subdomain with $50 \times 50 \times 50$ grid points.

6 Flame Evolution in Turbulent CO/H_2/Air Mixtures

Induced ignition and the subsequent evolution of premixed turbulent flames is a phenomenon of large practical importance, e. g. in Otto engine combustion and safety considerations. This process has been studied using DNS in a model configuration of an initially uniform premixed gas under turbulent conditions which is ignited by an energy source in a small region at the center of the domain. DNS of this configuration have been performed earlier with simple one-step chemistry [10] and using a detailed reaction mechanism for the hydrogen/air system (9 species, 37 elementary reactions) as well as a detailed transport model including thermodiffusion [9]. In this section, results are presented of an investigation of flames evolving after induced ignition of a turbulent mixture of syngas (i. e. a mixture of carbon monoxide and hydrogen) and air. A detailed reaction mechanism has been used in this investigation that consists of 67 elementary reactions among the 13 chemical species CO, HCO, CH_2O, CO_2, H_2O, O_2, O, H, OH, HO_2, H_2O_2, H_2, and N_2 [11].

The initial conditions for the DNS of this study are a superposition of a cold ($T = 300\,\mathrm{K}$) uniform mixture consisting of 14.9 % CO, 11.2 % H_2, 13.1 % O_2, and 60.8 % N_2 (mole fractions) with a turbulent flow field computed by inverse FFT from a von-Kármán-Pao-spectrum with randomly chosen phases. The mixture is ignited by an energy source in a small round region ($r = 0.2\,\mathrm{mm}$) at the center of the domain, which is active during the first 15 µs of the simulation. During this time, the mixture at the center of the domain heats up and radicals are formed. A shockwave is observed which propagates in outward direction towards the boundaries of the domain. Non-reflecting outflow conditions based on characteristic wave relations [12] are imposed on all boundaries, which allow the shock wave to leave the domain without disturbing the solution [13]. Above a minimum ignition energy an expanding flame kernel is observed.

Two DNS of such turbulent flame kernels with differently sized computational domains have been carried out. In the first simulation (Kernel 1), a 800×800 points grid has been used to discretize a quadratic domain with a side length of 15.98 mm. In the second case the size of the domain was 22.38 mm $\times$ 22.38 mm and the computational grid had 1120×1120 points. A second difference between the two simulations are the turbulent flow fields. While both have the same statistical properties, i. e. a root mean square of the velocity fluctuations of 3 m/s and an integral length scale of 2 mm, the individual realizations are different.

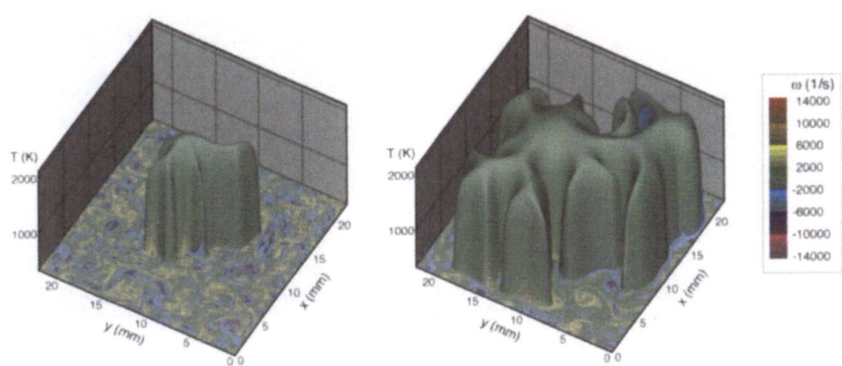

Fig. 2. Temperature (z-coordinate) and vorticity (contour colors) at $t = 0.7$ ms (left) and $t = 1.4$ ms (right) in a flame (Kernel 2) evolving after induced ignition of a turbulent mixture

Both simulations have been carried out up to the time at which parts of the flame left the computational domain, which was after a physical time about $t = 1.2$ ms for Kernel 1 (173270 timesteps) and after about $t = 1.55$ ms for Kernel 2 (226353 timesteps). Both DNS have been performed on the HEidelberger LInux Cluster System (HELICS, see http://www.helics.de/), a cluster with 256 Dual AthlonMP nodes (1.4 GHz) and Myrinet 2000 interconnect. The author hereby gratefully acknowledges the Interdisciplinary Center for Scientific Computing (IWR) at Heidelberg for granting him access to this system. Each whole simulation has been carried out with several restarts and a changing number of nodes has been used for different parts of the computation. E. g. 7460 timesteps were needed to advance the DNS of Kernel 2 from $t = 1.075$ ms to $t = 1.125$ ms which took 7950 seconds using 256 processors (128 nodes).

Figure 2 shows two snapshots of the temporal evolution of Kernel 2 at $t = 0.7$ ms (left) and $t = 1.4$ ms (right). The temperature is shown as the height while the contour colors represent the vorticity. The lack of a vortex-stretching mechanism in two-dimensional simulations of decaying turbulence leads to an inverse cascade with growing structures. There is a very strong damping of the turbulence in the hot region of the burnt gas due to the high viscosity. A comparison shows a good qualitative agreement of the DNS presented with the results of an experimental investigation of turbulent flames performed under similar conditions [14].

Figure 3 shows the spatial distribution of the mass fraction of CO, HCO, H_2O_2, and H at a physical time of 1.4 ms. The highest concentrations of H_2O_2 occur at regions where flame extinction due to mutual annihilation of different parts of the flame front which propagate towards each other takes place. This process leads to a flame surface reduction and is frequently found at regressing cusps like the one that can be seen in Fig. 3 in the region around

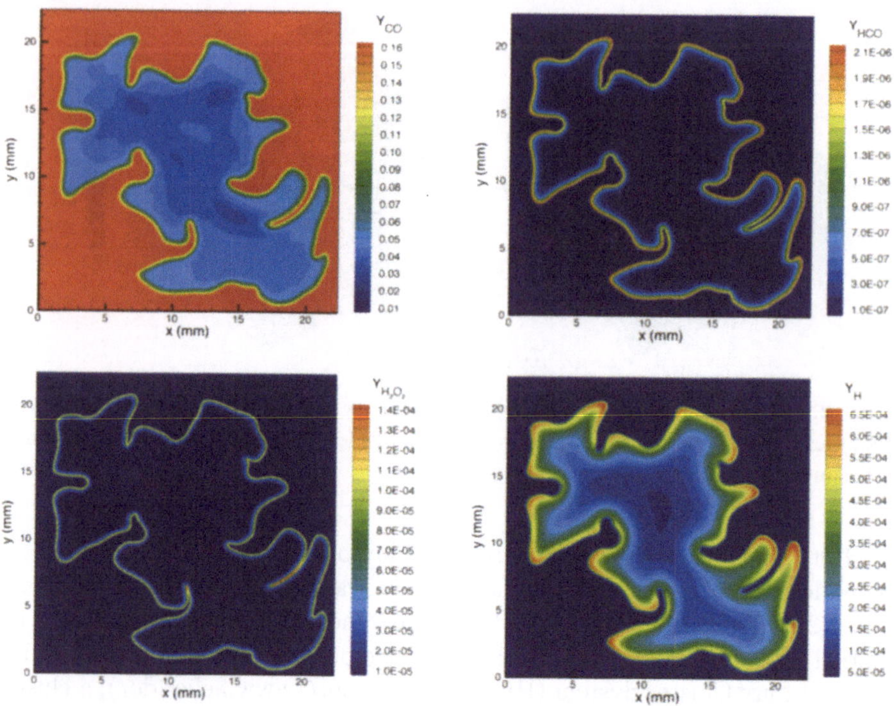

Fig. 3. Distribution of mass fraction of CO, HCO, H_2O_2, and H in a turbulent premixed syngas flame (Kernel 2)

($x \approx 19\,\mathrm{mm}, y \approx 7\,\mathrm{mm}$). If this mutual annihilation of flame elements occurs more behind than directly at the cusp, some fresh gas is enclosed on the burnt side of the main flame front. The latest stage of the burnout of these pockets of fresh gas is the extinction of the flame front around the pocket by mutual annihilation in a roughly cylindrical geometry. The formation and burnout of one such pocket can be seen in Fig. 4 in which the H_2O_2 mass fraction in a part of the computational domain of Kernel 1 is shown at $t = 0.65\,\mathrm{ms}, 0.70\,\mathrm{ms}, \ldots, 0.85\,\mathrm{ms}$. This mechanism of pocket formation and burnout has also been observed in earlier DNS of premixed methane/air [15, 7] and hydrogen/air flames [16] as well as in experiments [17].

The variation of the chemical composition along the flame front is strongly influenced by the flame front curvature. A curvature of the flame front that focusses the mass flow through the front leads to an increase of the concentration of the faster diffusing species. This effect can especially well be seen by looking at the spatial distribution of the H mass fraction (lower left part of Fig. 3). Such dependencies of the local chemical composition on other variables like e. g. strainrate or curvature constitute a central point in many combustion models. As the changes of the chemical composition in curved flame fronts are

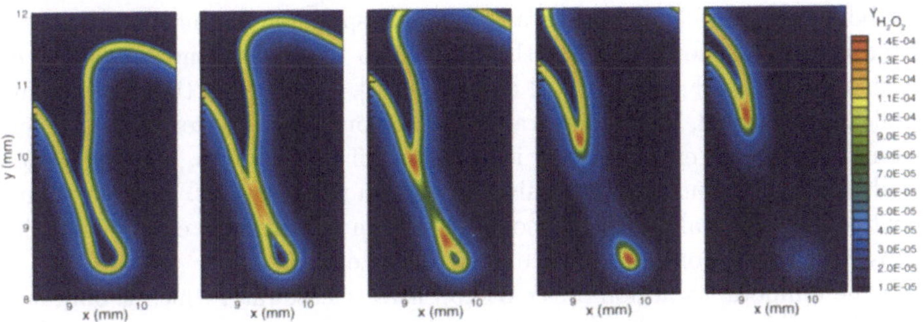

Fig. 4. Temporal evolution of H_2O_2 mass fraction in a part of the computational domain of Kernel 1 (from left to right: $t = 0.65\,\text{ms}$, $t = 0.70\,\text{ms}$, $t = 0.75\,\text{ms}$, $t = 0.80\,\text{ms}$, and $t = 0.85\,\text{ms}$)

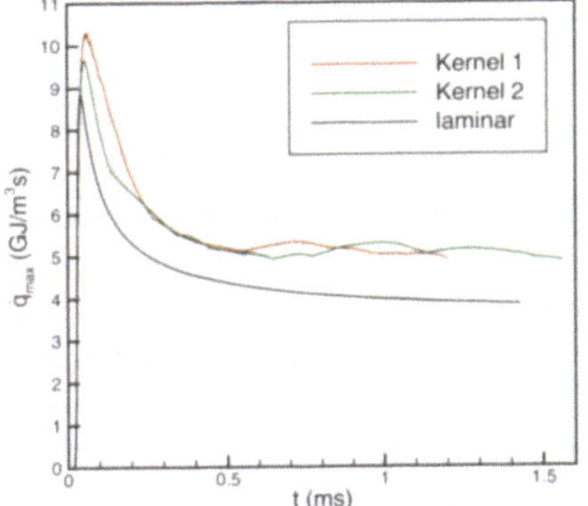

Fig. 5. Temporal evolution of maximum heat-release in flames evolving after induced ignition of turbulent and laminar CO/H$_2$/air mixtures

Fig. 6. Temporal evolution of overall heat-release in flame kernels evolving in turbulent and laminar CO/H$_2$/air mixtures after induced ignition

caused by the combined effects of curvature and preferential diffusion, these dependencies can only be found from the analysis of data obtained by high-resolution experiments or simulations with a sufficiently accurate description of molecular diffusion and chemical kinetics.

An important quantity in turbulent combustion modelling is the reaction intensity. There are several possibilities to define the reaction intensity, e. g. the local heat release rate

$$\dot{q} = \sum_{\alpha=1}^{N_S} Y_\alpha \dot{\omega}_\alpha h_\alpha \tag{9}$$

can be used for this purpose. Figure 5 shows the temporal evolution of the spatial maximum of the local heat-release

$$\dot{q}_{\text{max}}(t) = \max_{x,y} \dot{q}(x, y, t) \tag{10}$$

for both turbulent flame kernels and the corresponding laminar one. The existence of an ignition-delay, i. e. the time up to the first strong increase of $\dot{q}$, is characteristic for chain-radical explosions [2]. During this time, important radicals are formed, which cause rapid ignition once their concentration is high enough. After the energy source is switched off at $t = 15\,\mu\mathrm{s}$, the turbulence leads to a faster and more intensive ignition in some regions. The maximum heat-release stays also clearly above the one in the laminar case during the subsequent evolution of both turbulent flame kernels.

The temporal evolution of the overall heat-release rate, which is obtained by integrating $\dot{q}$ over the computational domain, is shown in Fig. 6. While there are almost no differences during the first $25\,\mu\mathrm{s}$, there is a much stronger increase of the overall heat-release in the subsequent evolution of the turbulent flame kernels. Figure 7 shows the change of the overall heat-release in time,

$$\frac{\mathrm{d}}{\mathrm{d}t} \int \dot{q}(x, y, t) \mathrm{d}A \quad , \tag{11}$$

as a function of time. As could already be seen from Fig. 6, the increase of the total heat-release is nearly constant in the laminar case. After the ignition phase, the laminar flame propagates circularly outwards with approximately constant velocity causing an approximately constant increase of flame surface. In the turbulent case, there is an additional flame surface increase due to the wrinkling of the flame caused by the turbulent eddies. On the other hand, mutual annihilation of parts of the flame front, like found in regressing cusps or during pocket formation and burnout, leads to a decrease of flame surface. The decline of the increase of overall heat release of Kernel 1 around $t \approx 0.7\,\mathrm{ms}$ is caused by the mutual annihilation of parts of the flame front behind the pocket shown in Fig. 4. The decline at the end of the curves for both turbulent flames is caused by parts of the flame leaving the computational domain.

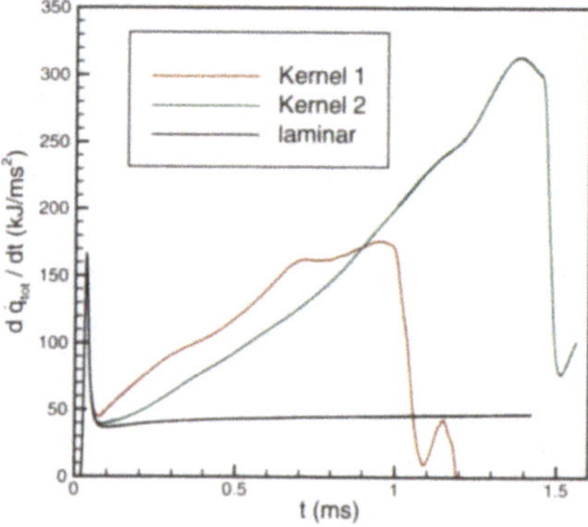

Fig. 7. Temporal evolution of the change rate of total heat-release

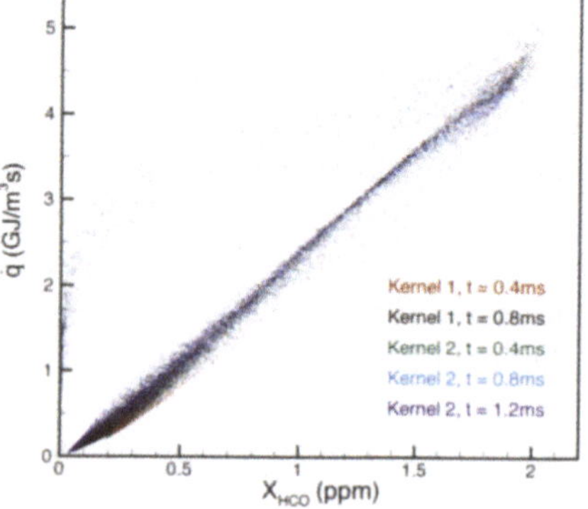

Fig. 8. Correlation of local heat-release with HCO mole fraction

In experimental investigations of turbulent flames the reaction intensity is typically indirectly measured by using some tracer species that is well correlated to it. It has been found in a simulation of a premixed methane flame interacting with a vortex-pair that HCO is well suited as a tracer for reaction intensity [18]. This has been confirmed in DNS of turbulent premixed methane/air flames [7]. Figure 8 shows the instantaneous correlation of heat-release with the concentration of HCO in both turbulent premixed syngas flames at some points in time. It can clearly be seen that HCO is also well suited as a tracer for the reaction intensity in the type of flames under investigation here.

7 Conclusions

A detailed chemistry DNS code has been used as a benchmark of computer systems using a variety of modern microprocessors (AMD AthlonMP, Intel Xeon, Alpha 21264B (EV68), Intel Itanium 1 and Itanium 2, IBM POWER4). The overhead for the parallelization on these systems has been measured for several node-to-node interconnects and shared memory implementations (T3E, Fast Ethernet, Myrinet 2000, Dual Nodes, NEC AzusA, Intel Tiger, IBM p690). The scaling behavior of the code has been analyzed using measurements on the Cray T3E for different loads per PE and up to 512 PEs. Some implications for future 3D DNS with detailed chemistry and transport on forthcoming HPC systems have been presented.

DNS of flames kernels evolving after induced ignition of mixtures of carbon-monoxide, hydrogen, and air under turbulent conditions have been performed. The combined effect of flame front curvature and preferential diffusion on the chemical composition in the flame has been discussed. The total heat-release is increased compared to the corresponding laminar flame due to a surface increase caused by the turbulence. Mutual annihilation of distinct elements of the flame front leads to the regression of flame cusps and the formation of pockets of fresh gas enclosed by hot products. It has been shown that the HCO radical could serve as a tracer for the reaction intensity in experiments.

References

1. Bird, R. B., Stewart, W. E., and Lightfoot, E. N., *Transport Phenomena*, Wiley, New York, 1960.
2. Warnatz, J., Maas, U., and Dibble, R. W., *Combustion*, Springer, Berlin, Heidelberg, New York, 2nd ed., 1999.
3. Hirschfelder, J. O., Curtiss, C. F., and Bird, R. B., *Molecular Theory of Gases and Liquids*, John Wiley & Sons, 1954, Corrected Printing 1964.

4. Kee, R. J., Dixon-Lewis, G., Warnatz, J., Coltrin, M. E., and Miller, J. A., "A Fortran Computer Code Package for the Evaluation of Gas-Phase Multicomponent Transport Properties," Tech. Rep. SAND86-8246, 1986.
5. Thévenin, D., Behrendt, F., Maas, U., Przywara, B., and Warnatz, J., "Development of a Parallel Direct Simulation Code to Investigate Reactive Flows," *Computers and Fluids*, Vol. 25, No. 5, 1996, pp. 485–496.
6. Lange, M., Thévenin, D., Riedel, U., and Warnatz, J., "Direct Numerical Simulation of Turbulent Reactive Flows Using Massively Parallel Computers," *Parallel Computing: Fundamentals, Applications and New Directions*, edited by E. D'Hollander, G. Joubert, F. Peters, and U. Trottenberg, No. 12 in Advances in Parallel Computing, Elsevier Science, Amsterdam, 1998, pp. 287–296.
7. Lange, M. and Warnatz, J., "Investigation of Chemistry-Turbulence Interactions Using DNS on the Cray T3E," *High Performance Computing in Science and Engineering '99*, edited by E. Krause and W. Jäger, Springer, Berlin, Heidelberg, New York, 2000, pp. 333–343.
8. Press, W. H., Teukolsky, S. A., Vetterling, W. T., and Flannery, B. P., *Numerical Recipes in C*, Cambridge University Press, 2nd ed., 1992.
9. Lange, M., "Massively Parallel DNS of Flame Kernel Evolution in Spark-Ignited Turbulent Mixtures," *High Performance Computing in Science and Engineering '02*, edited by E. Krause and W. Jäger, Springer, Berlin, Heidelberg, New York, 2002, pp. 425–438.
10. Echekki, T., Poinsot, T. J., Baritaud, T. A., and Baum, M., "Modeling and Simulation of Turbulent Flame Kernel Evolution," *Transport Phenomena in Combustion*, edited by S. H. Chan, Vol. 2, Taylor & Francis, 1995, pp. 951–962.
11. Maas, U. and Pope, S. B., "Simplifying Chemical Kinetics: Intrinsic Low Dimensional Manifolds in Composition Space," *Combustion and Flame*, Vol. 88, 1992, pp. 239–264.
12. Baum, M., Poinsot, T. J., and Thévenin, D., "Accurate Boundary Conditions for Multicomponent Reactive Flows," *Journal of Computational Physics*, Vol. 116, 1995, pp. 247–261.
13. Lange, M., "Direct Numerical Simulation of Turbulent Flame Kernels Using HPC," *High Performance Computing in Science and Engineering '01*, edited by E. Krause and W. Jäger, Springer, Berlin, Heidelberg, New York, 2002, pp. 418–432.
14. Renou, B., Boukhalfa, A., Puechberty, D., and Trinité, M., "Local Scalar Flame Properties of Freely Propagating Premixed Turbulent Flames at Various Lewis Numbers," *Combustion and Flame*, Vol. 123, 2000, pp. 507–521.
15. Chen, J. H., Echekki, T., and Kollmann, W., "The Mechanism of Two-Dimensional Pocket Formation in Lean Premixed Methane-Air Flames with Implications to Turbulent Combustion," *Combustion and Flame*, Vol. 116, 1999, pp. 15–48.
16. Lange, M., Riedel, U., and Warnatz, J., "Parallel DNS of Turbulent Flames with Detailed Reaction Schemes," AIAA Paper 98-2979, 1998.
17. Baillot, F. and Bourhela, A., "Burning Velocity of Pockets from a Vibrating Flame Experiment," *Combustion Science and Technology*, Vol. 126, 1997, pp. 201–224.
18. Najm, H. N., Paul, P. H., Mueller, C. J., and Wyckhoff, P. S., "On the Adequacy of Certain Experimental Observables as Measurements of Flame Burning Rate," *Combustion and Flame*, Vol. 113, 1998, pp. 312–332.

Transition from Stationary to Rotating Bound States of Dissipative Solitons

A. W. Liehr, A. S. Moskalenko, and H.-G. Purwins

Institut für Angewandte Physik, Correnssstr. 2/4, D-48149 Münster, Germany

Summary. By the example of a cluster of two dissipative solitons, which are well localized solitary solutions of a 3-component reaction-diffusion system in 2-dimensional space, we demonstrate that in dissipative systems a bifurcation of stationary well localized structures to uniform rotating structures is possible. The underlying mechanism is similar to the mechanism of the drift (traveling) bifurcation. For appropriate choice of the path in parameter space of the considered reaction-diffusion system the rotational bifurcation precedes the drift bifurcation. The theoretically predicted velocities are compared to solutions of the reaction-diffusion system.

In the last decades much attention has been devoted to localized particle-like solutions of conservative nonlinear spatially extended systems, so-called solitary structures or solitons. They became especially interesting from the theoretical point of view as exact analytical solutions of Korteweg-de Vries, sin-Gordon and nonlinear Schrödinger equations in one spatial dimension (see e.g. [1], and references therein). At the same time, the modeling of many real systems leads to continuous nonlinear equation systems, which are essentially dissipative [2]. In addition, frequently two or three equivalent spatial dimensions come into play in a natural manner. This leads to new physical phenomena, which are not possible in one dimensional systems. In this paper we report on the transition from stationary to rotating or traveling bound states of dissipative solitons and compare the analytically predicted velocities with numerical results [3].

The mathematical description of pattern formation in spatially extended systems mostly is based on partial differential equations. Many of them take the form

$$\partial_t U(\mathbf{r}, t) = G[U, \partial_\mathbf{r} U, ...; R], \quad U = (u_1, ..., u_n), \tag{1}$$

thus defining the time evolution of the considered system by a vector function G that is generally nonlinear and that contains the unknown field quantity U, its derivatives and one or more control parameters R [2]. Bounded nonlinear dissipative systems are characterized by the property that in the course of time their infinitely dimensional phase space generally contracts to several attractors. As a consequence, in many circumstances the various solutions being possible in nonlinear dissipative systems described by Eq. (1) can be classified according to the set of attractors.

Suppose system (1) has a stationary well localized solution $\bar{U}(\mathbf{r}; R)$, which continuously depends on R. The dynamics of the system in the neighborhood of $\bar{U}$ is governed by the time evolution of the amplitudes of modes, which are eigenfunctions of the linearization $\mathcal{D}[\bar{U}; R]$ of $G[U, \partial_{\mathbf{r}} U, ...; R]$ about $\bar{U}$:

$$\partial_t \tilde{U} = \mathcal{D}[\bar{U}; R]\tilde{U}, \quad \mathcal{D}_{ij} = \delta G_i / \delta u_j \big|_{U=\bar{U}}, \tag{2}$$

where $U = \bar{U} + \tilde{U}$.

If the system (1) is invariant under shift in a spatial coordinate ξ, which corresponds to translation or rotation, and $\bar{U}(\mathbf{r}; R)$ is not invariant then for any R there exists a mode $\mathcal{G} = \bar{U}_\xi$ with zero eigenvalue, such that

$$\mathcal{D}\mathcal{G} = 0. \tag{3}$$

$\mathcal{G}$ is called Goldstone mode. The existence of Goldstone modes leads to neutral stability of the stationary solution $\bar{U}$ if the eigenvalues of all other modes have negative real parts. One of the possible destabilizations of $\bar{U}$ occurs when one of the modes exactly coincides with $\mathcal{G}$ while changing R [4, 5]. Because of this degeneracy a generalized eigenfunction $\mathcal{P}$ of $\mathcal{D}[\bar{U}; R]$, the so-called propagator mode, appears:

$$\mathcal{D}\mathcal{P} = \mathcal{G}. \tag{4}$$

In the case that $\mathcal{G}$ is a translational mode such situation corresponds to a drift (traveling) bifurcation [6], which causes translational motion of the localized structure and is accompanied by its symmetry breaking in the direction of motion [7]. Concerning two-dimensional dissipative solitons this destabilisation has recently been detected experimentally [8, 9]. The solvability condition for equation (4) leads to the following equation defining the bifurcation value R_c of the control parameter R:

$$< \mathcal{G}^\dagger | \mathcal{G} > = 0, \tag{5}$$

where $< \cdot | \cdot >$ denotes a scalar product and $\mathcal{G}^\dagger$ is the eigenfunction of the operator $\mathcal{D}^\dagger$, adjoint to $\mathcal{D}$, with zero eigenvalue and the same symmetry properties as $\mathcal{G}$.

Here we are investigating the case where $\mathcal{D}[\bar{U}; R]$ can be represented as

$$\mathcal{D}[\bar{U}; R] = M_R \mathcal{L}_{\bar{U}}, \tag{6}$$

where $\mathcal{L}_{\bar{U}}$ is a self-adjoint operator: $\mathcal{L}_{\bar{U}}^\dagger = \mathcal{L}_{\bar{U}}$, and M_R is a reversible self-adjoint $n \times n$-matrix depending on R. Using Eq. (6), neutral modes $\mathcal{G}^\dagger$ of the operator $\mathcal{D}^\dagger$ can be expressed analytically through $\mathcal{G}$:

$$\mathcal{G}^{\dagger} = M_R^{-1}\mathcal{G}. \tag{7}$$

Thus, the numerical calculation of the modes $\mathcal{G}^{\dagger}$, which is generally supposed to be necessary [5, 10], can be avoided if a stationary solution $\bar{U}$ is known so that the Goldstone modes can be easily found by differentiation. As a consequence, from Eq. (5) we get an algebraic equation for R, whose real solution, if it exists, gives R_c.

In order to determine the dependence of the velocity of motion Ω in the direction given by ξ on the control parameter R near to the bifurcation point $R = R_c$ one should search for stationary solutions $\breve{U}$ of the equation

$$\partial_t U = \Omega U_{\xi} + G[U, \partial_{\mathbf{r}} U, ...; R], \tag{8}$$

using the fact that $R \approx R_c$ and $\breve{U} \approx \bar{U} - \Omega \mathcal{P}$. Stationary solutions $\breve{U}$ of Eq. (8) correspond to uniform moving solutions $U = \breve{U}(\xi - \Omega t, \eta)$ of Eq. (1), where η denotes all remaining space variables. It should be mentioned that, in general, it can be a problem to extract the dependence $\Omega(R - R_c)$ from Eq. (8) analytically, even if perturbation methods are applied. In such a case numerical methods should be used.

As an example system we have chosen a 3-component reaction-diffusion system investigated as a phenomenological model in the context of pattern formation phenomena in semiconductor-gas discharge systems [8, 11]:

$$\begin{aligned} u_t &= D_u \Delta u + \lambda u - u^3 - \kappa_3 v - \kappa_4 w + \kappa_1, \\ \tau v_t &= D_v \Delta v + u - v, \\ \theta w_t &= D_w \Delta w + u - w, \end{aligned} \tag{9}$$

where $u = u(\mathbf{r}, t)$, $v = v(\mathbf{r}, t)$, $w = w(\mathbf{r}, t)$ with $\mathbf{r} \in \mathbb{R}^2$, and $D_u, D_v, D_w, \tau, \theta, \kappa_3, \kappa_4 \geq 0$.

There are solutions of Eq. (9) in the form of stationary well localized structures with rotational symmetry. By changing system parameters they can undergo a drift bifurcation loosing their rotational symmetry [4, 6]. In both cases we use the notion *dissipative solitons* (DSs) (cf. [11, 12, 13]). In the russian literature such solutions of reaction-diffusion systems are called *autosolitons* [14]. Other nomenclatures like *spots* [4, 6], *pulses* [15] etc. are also used.

In a parameter range near to the Turing bifurcation of the homogeneous background state stationary DSs possess spatially oscillating tails, which enable the formation of stationary bound states (BSs) of DSs (see Fig. 1 and [16]).

In what follows, we use the time constant τ as control parameter. Rewriting Eq. (9) in the form of Eq. (1), linearizing it about $\bar{U}(\mathbf{r}) = (\overline{u}, \overline{v}, \overline{w})$ (which does not depend on τ) and decomposing the right hand side according to Eq. (6) we get:

$$M_{\tau} = \begin{pmatrix} 1 & 0 & 0 \\ 0 & -\frac{1}{\kappa_3 \tau} & 0 \\ 0 & 0 & -\frac{1}{\kappa_4 \theta} \end{pmatrix}, \tag{10}$$

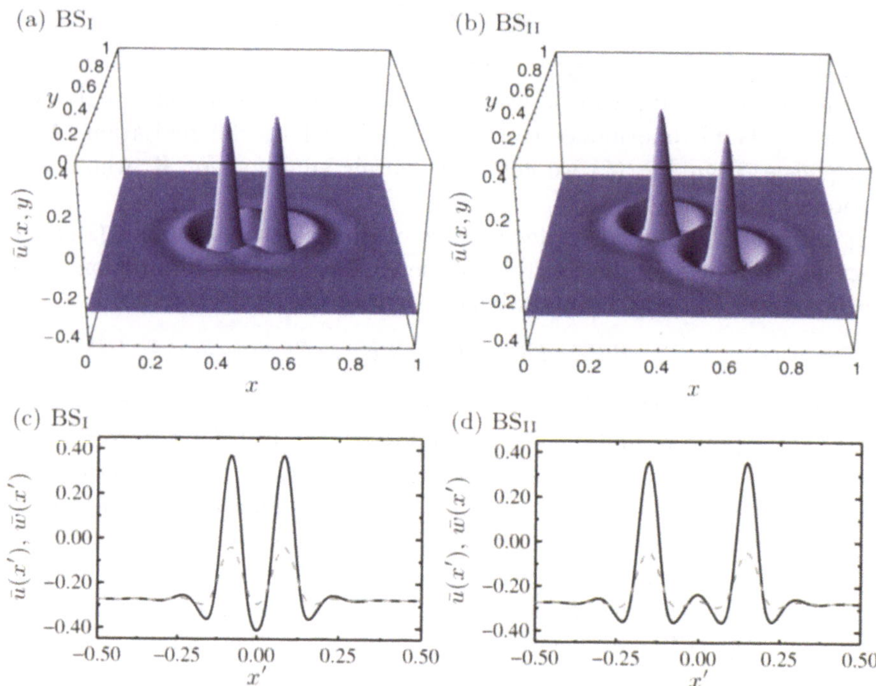

Fig. 1. Two-dimensional stationary BSs of two DSs. The upper row shows three-dimensional plots of the activator distribution $\bar{u}(x, y)$ for configurations with first and second shortest binding distances. The lower row shows intersections of the activator $\bar{u}(x')$ (solid curve) and the fast inhibitor $\bar{w}(x')$ (dashed curve) at the longitudinal axis x' of the BSs. The slow inhibitor $\bar{v}$ is identical with the activator $\bar{u} = \bar{v}$ for the chosen parameters. Parameter: $D_u = 1.1 \times 10^{-4}$, $D_v = 0$, $D_w = 9.64 \times 10^{-4}$, $\lambda = 1.01$, $\kappa_1 = -0.1$, $\kappa_3 = 0.3$, $\kappa_4 = 1$, $\theta = 0$, $\Omega = [0,1] \times [0,1]$, no-flux boundary condition, $\Delta x = 5 \cdot 10^{-3}$ und $\Delta t = 0.1$.

$$\mathcal{L}_{\bar{U}} = \begin{pmatrix} D_u \Delta + \lambda - 3\bar{u}^2 & -\kappa_3 & -\kappa_4 \\ -\kappa_3 & -\kappa_3 D_v \Delta + \kappa_3 & 0 \\ -\kappa_4 & 0 & -\kappa_4 D_w \Delta + \kappa_4 \end{pmatrix}. \tag{11}$$

Using Eq. (10) in Eq. (7) and inserting the latter in Eq. (5), we arrive at the formula for the critical values of τ (see also [4], where another method was used):

$$\tau_{c\xi} = \frac{< \bar{u}_\xi^2 > -\kappa_4 \theta < \bar{w}_\xi^2 >}{\kappa_3 < \bar{v}_\xi^2 >}. \tag{12}$$

Here $\xi = x, y$ relates to the onset of translational motion in the x- and y-directions, respectively, and $\xi = \phi$ relates to the onset of rotation of the structure, where ϕ is the angle of a polar coordinate system.

For a single DS (see e.g. Fig. 1a) all directions in (x, y) space are equivalent. Therefore, the bifurcation point is degenerate: in such an ideal situation the

DS can move uniformly in any spatial direction. In real systems the direction of motion will be determined by small inhomogeneties or fluctuations that are always present. Because of the rotational symmetry in the case of a single stationary DS, the onset of rotation does not exist.

The situation is different when the BS of two DSs is considered (see Fig. 1). In this case we choose a coordinate system with its center at the geometrical center of the structure, and the x- and y-axes parallel and perpendicular to the longitudinal axis of the BS, respectively, as shown in Fig. 1a. Then, because of the reflection symmetry with respect to the x- and y axes, we obtain

$$< \overline{u}_x \overline{u}_y > = < \overline{v}_x \overline{v}_y > = < \overline{w}_x \overline{w}_y > = 0 \, .$$

Using these properties the critical value τ_{ce} of the bifurcation parameter τ for the onset of uniform motion in an arbitrary spatial direction $\mathbf{e} = a\mathbf{e}_x + b\mathbf{e}_y$ can be expressed through τ_{cx} and τ_{cy}:

$$\tau_{ce} = \frac{a^2 < \overline{v}_x^2 > \tau_{cx} + b^2 < \overline{v}_y^2 > \tau_{cy}}{a^2 < \overline{v}_x^2 > + b^2 < \overline{v}_y^2 >} \, . \tag{13}$$

From Eq. (13) it can be seen that τ_{ce} lies between τ_{cx} und τ_{cy}. Thus, the smallest value from τ_{cx}, τ_{cy} and $\tau_{c\phi}$ defines which kind of destabilization occurs first: the onset of motion in x-direction, the onset of motion in y-direction or the onset of rotation. In the limit case $D_v = 0$, $\theta = 0$ of Eq. (9) the values of τ_{cx}, τ_{cy}, and $\tau_{c\phi}$ coincide and are equal to $1/\kappa_3$ [4], and the dynamics of the dissipative structure beyond the bifurcation point is determined by the translational and rotational modes as well as by their interaction.

For the sake of simplicity we choose $D_v = 0$ but consider the case of $\theta > 0$. In this situation, Eq. (12) reads as:

$$\tau_{c\xi} = \frac{1}{\kappa_3} - \theta \frac{\kappa_4}{\kappa_3} \frac{<\bar{w}_\xi^2>}{<\bar{u}_\xi^2>}, \tag{14}$$

and for the BS of two DSs the values of τ_{cx}, τ_{cy}, and $\tau_{c\phi}$ are different. For example, for $\theta = 0.8$ and the remaining parameters from Fig. 1 we get $\tau_{cx} = 3.0359$, $\tau_{cy} = 3.0364$, and $\tau_{c\phi} = 3.0314$. Note, that for increasing binding distance the two-body character of the BS vanishes and the bifurcation points of the BS can not be discriminated from the bifurcation point of a single DS with respect to the numerical error. The dependence of τ_{ce} on the motion direction of the cluster is depicted in Fig. 2. We see that, in such a case, by increase of the control parameter τ the destabilization of the stationary dissipative structure should occur due to the onset of rotation.

In order to determine the dependence of the angular velocity ω of uniform rotation of the BS on the control parameter τ in the vicinity of the bifurcation point $\tau = \tau_{c\phi}$, we rewrite Eq. (9) using a rotating coordinate system with its origin in the center of the cluster:

Table 1. Bifurcation points $\tau_{c_{xi}}$ of BSs calculated with a numerical error of $\pm 10^{-3}$. For the investigated parameter range the bifurcation point of a single dissipative soliton is $\tau_c = 3.0399$. Parameters like in Fig. 1.

	τ_{cx}	τ_{cy}	$\tau_{c\phi}$
BS_I (Fig. 1a)	3.0359	3.0364	3.0314
BS_{II} (Fig. 1b)	3.0397	3.0398	3.0393
BS_{III}	3.0408	3.0409	3.0409

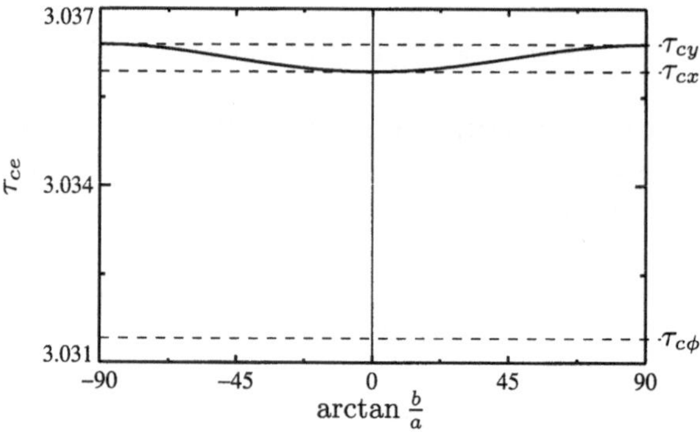

Fig. 2. Dependence of τ_{ce} on the motion direction of the BS of two DSs (Fig. 1a) after Eq. (13) in comparison with $\tau_{c\phi}$. Parameter like in Fig. 1 with $\Delta x = 1.25 \cdot 10^{-3}$, $\langle(\frac{\partial}{\partial x}\bar{v})^2\rangle = 2.236$, $\langle(\frac{\partial}{\partial x}\bar{v})^2\rangle = 2.062$ and $\tau_{c_{xi}}$ from Tab. 1.

$$\begin{aligned} u_t &= \omega u_\phi + D_u \Delta u + \lambda u - u^3 - \kappa_3 v - \kappa_4 w + \kappa_1, \\ v_t &= \omega v_\phi + (u - v)/\tau, \\ w_t &= \omega w_\phi + (D_w \Delta w + u - w)/\theta. \end{aligned} \tag{15}$$

We consider a stationary solution $\breve{U}(r,\phi) = (\breve{u}(r,\phi), \breve{v}(r,\phi), \breve{w}(r,\phi))$ of this equation in polar coordinates, thus the left hand side of Eq. (15) vanishes. The right hand side is then projected onto the vector function $\breve{\mathcal{G}}^\dagger = M_R^{-1}\breve{U}_\phi$ (cf. [4], [12]). This leads after several integrations by parts to the following equation:

$$S(\omega) := \omega(<\breve{u}_\phi^2> -\tau\kappa_3 <\breve{v}_\phi^2> -\theta\kappa_4 <\breve{w}_\phi^2>) = 0. \tag{16}$$

In addition, from the second equation of Eq. (15) we obtain

$$\breve{u} = \breve{v} - \omega\tau\breve{v}_\phi. \tag{17}$$

Inserting Eq. (17) in Eq. (16), we get after some transformations

$$\begin{aligned} S(\omega) =& \omega\, \tau^2 < \breve{v}_{\phi\phi}^2 > \\ &\times \left[\omega^2 - \frac{\kappa_3}{\tau^2} \frac{<\breve{v}_\phi^2>}{<\breve{v}_{\phi\phi}^2>} \left(\tau - \left(\frac{1}{\kappa_3} - \theta \frac{\kappa_4 <\breve{w}_\phi^2>}{\kappa_3 <\breve{v}_\phi^2>}\right)\right)\right]. \end{aligned} \tag{18}$$

From Eq. (18) it can be seen that equation $S(\omega) = 0$ always has a trivial solution $\omega = 0$. In the vicinity of the bifurcation point we have $\tau \approx \tau_{c\phi}$ and $\breve{u} \approx \breve{v} \approx \overline{u}$, $\breve{w} \approx \overline{w}$, so that $S(\omega)$ can be rewritten as

$$S(\omega) = \omega\, \tau_{c\phi}^2 < \overline{u}_{\phi\phi}^2 > \left[\omega^2 - \frac{\kappa_3}{\tau_{c\phi}^2} \frac{< \overline{u}_\phi^2 >}{< \overline{u}_{\phi\phi}^2 >} (\tau - \tau_{c\phi}) \right] . \tag{19}$$

For $\tau > \tau_{c\phi}$ two new real solutions of $S(\omega) = 0$, which correspond to a uniform rotating localized structure, branch off:

$$\omega^2 = \frac{\kappa_3}{\tau_{c\phi}^2} \frac{<\overline{u}_\phi^2>}{<\overline{u}_{\phi\phi}^2>} (\tau - \tau_{c\phi}). \tag{20}$$

In an analogous manner the velocity c_x of the uniform motion in x-direction can be found:

$$c_x^2 = \frac{\kappa_3}{\tau_{cx}^2} \frac{< \overline{u}_x^2 >}{< \overline{u}_{xx}^2 >} (\tau - \tau_{cx}). \tag{21}$$

In order to compare the theoretically predicted velocities with solutions of the three-component reaction-diffusion-system (9) the latter is solved on two-dimensional domains with no-flux boundary conditions. The time-dependent partial differential equations are approximated by finite differences in space and a Crank-Nicholson time stepping scheme. Both discretisation lengths in space and time are constant. The resulting set of equations is solved with a successive over-relaxation algorithm whereby the discretisation points are taken account of in red-black order. For problems considered in this paper an individual simulation runs approximately twelve hours on 32 nodes of the Cray T3E in order to calculate $120 \cdot 10^3$ time steps. The solution of the problem is parallelized by dividing the domain in sub-domains with minimal internal boundaries. Each sub-domain is assigned to one node, whereby communication between the nodes is realized via the Message Parsing Interface (MPI) [17]. Speed-up and scale-up of the parallel program have been measured and show good results [18].

Numerical simulations of Eq. (9) confirm that stationary BSs of two DSs become unstable if τ exceeds $\tau_{c\phi}$ and start to rotate. The equilibrium angular velocity is in good agreement with the result of Eq. (20) (see Fig. 3). For $\tau > \tau_{cx}$ the uniform moving solution is also possible. However, to get translational motion, the corresponding initial conditions or an appropriate perturbation of the rotating localized structure should be prepared. As it can be seen from Fig. 3, the equlibrium velocity of motion in x-direction is also in good agreement with the analytical prediction of Eq. (21). It turns out that only the motion in the direction parallel to the longitudinal axis of the BS is stable [19].

In the investigated case we have shown that a stable stationary localized structure is destabilized due to the onset of rotation by a continuous increase of the control parameter, whereby the angular velocity increases as a square

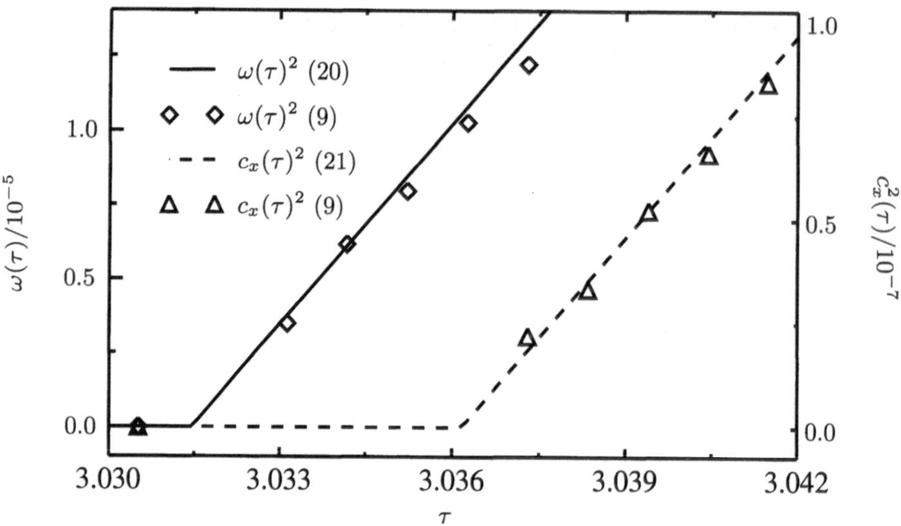

Fig. 3. Square of the angular velocity ω of a uniform rotation and square of the velocity c_x of a uniform drift parallel to the longitudinal axis of the cluster as functions of the control parameter τ. A uniform drift in other directions is unstable. Parameter of Eq. (9) are that from Fig. 1 except for τ and $\theta = 0.8$.

root of the distance from the bifurcation point. We classify such a qualitative change in the behaviour of a localized structure as *rotational bifurcation.*

We suppose that this type of bifurcation could be found in gas-discharge systems, where stationary and moving bound states of DSs have already been reported [20, 21], and in other two-dimensional dissipative systems, too. As it was pointed out in Ref. [5] for fronts, stable localized dissipative structures can undergo transition to motion only if the corresponding system is non-gradient. In the opposite case the operator $\mathcal{D}$ is self-adjoint and Eq. (5) can never be fulfilled. In this context, driven optical systems [22, 23, 24] and vibrated granular layers [25, 26], exhibiting stable BSs of well localized structures, seem to be promising. In contrast, the recently reported rotating clusters of localized states in the complex Ginzburg-Landau equation (CGLE) [10] and in the vector CGLE [27] can not appear as a result of the rotational bifurcation of a stationary cluster because they are stabilized by the rotation and do not exist as stationary structure.

In conclusion, we have analytically and numerically demonstrated a new type of bifurcation of localized structures without rotational symmetry, possible in non-gradient dissipative systems with more then one spatial dimensions: the rotational bifurcation. In the case of the 3-component reaction-diffusion system it is supercritical, occurs due to the coincidence of one mode of the structure with the rotational Goldstone mode at the bifurcation point, and can be separated from the drift bifurcation.

We thank the Deutsche Forschungsgemeinschaft (DFG) and the High-Performance Computing-Center Stuttgart (HLRS) for their support. We also like to thank M. C. Röttger and M. Bode for fruitful discussions.

References

[1] REMOISSENET, M.: *Waves Called Solitons: Concepts and Experiments*. 3. Berlin : Springer, 1999

[2] CROSS, M. C. ; HOHENBERG, P. C.: Pattern formation outside of equilibrium. In: *Reviews of Modern Physics* 65 (1993), Nr. 3, S. 851–1112

[3] MOSKALENKO, A. S. ; LIEHR, A. W. ; PURWINS, H.-G.: Rotational bifurcation of localized dissipative structures. In: *Europhysics Letters* 63 (2003), Nr. 3, S. 361–367

[4] OR-GUIL, M. ; BODE, M. ; SCHENK, C. P. ; PURWINS, H.-G.: Spot bifurcations in three-component reaction-diffusion systems: The onset of propagation. In: *Physical Review E* 57 (1998), Nr. 6, S. 6432–6437

[5] MICHAELIS, D. ; PESCHEL, U. ; LEDERER, F. ; SKRYABIN, D. V. ; FIRTH, W. J.: Universal criterion and amplitude equation for a nonequilibrium Ising-Bloch transition. In: *Physical Review E* 63 (2001), Nr. 6, S. 066602

[6] KRISCHER, K. ; MIKHAILOV, A.: Bifurcation to Traveling Spots in Reaction-Diffusion Systems. In: *Physical Review Letters* 73 (1994), Nr. 23, S. 3165–3168

[7] COULLET, P. ; LEGA, J. ; HOUCHMANZADEH, B. ; LAJZEROWICZ, J.: Breaking Chirality in Nonequilibrium Systems. In: *Physical Review Letters* 65 (1990), Nr. 11, S. 1352–1355

[8] BÖDEKER, H. ; RÖTTGER, M. C. ; LIEHR, A. W. ; FRANK, T. ; FRIEDRICH, R. ; PURWINS, H.-G.: Noise-covered drift bifurcation of dissipative solitons in a planar gas-discharge system. In: *Physical Review E* 67 (2003), Nr. 056220, S. 1–12

[9] LIEHR, A. W. ; BÖDEKER, H. U. ; RÖTTGER, M. C. ; FRANK, T. ; FRIEDRICH, R. ; PURWINS, H.-G.: Drift Bifurcation Detection for Dissipative Solitons. In: *New Journal of Physics* 5 (2003), Nr. 89, S. 1–9. – URL: `http://stacks.iop.org/1367-2630/5/89`

[10] SKRYABIN, D. V. ; VLADIMIROV, A. G.: Vortex induced rotation of clusters of well localized states in the complex Ginzburg-Landau equation. In: *Physical Review Letters* 89 (2002), Nr. 4, S. 044101

[11] BODE, M. ; LIEHR, A. W. ; SCHENK, C. P. ; PURWINS, H.-G.: Interaction of dissipative solitons: particle-like behaviour of localized structures in a three-component reaction-diffusion system. In: *Physica D* 161 (2002), Nr. 1-2, S. 45–66

[12] BODE, Mathias ; PURWINS, Hans-Georg: Pattern formation in reaction-diffusion systems – dissipative solitons in physical systems. In: *Physica D* 86 (1995), S. 53–63

[13] CHRISTOV, C. I. ; VELARDE, M. G.: Dissipative Solitons. In: *Physica D* 86 (1995), S. 323–347

[14] KERNER, B. S. ; OSIPOV, V. V.: *Fundamental Theories of Physics*. Bd. 61: *Autosolitons. A New Approach to Problems of Self-Organization and Turbulence*. Dordrecht : Kluwer Academic Publishers, 1994

[15] OHTA, T.: Pulse dynamics in a reaction-diffusion system. In: *Physica D* 151 (2001), Nr. 1, S. 61–72

[16] SCHENK, C. P. ; SCHÜTZ, P. ; BODE, M. ; PURWINS, H.-G.: Interaction of self-organized quasiparticles in a two-dimensional reaction-diffusion-system: The formation of molecules. In: *Physical Review E* 57 (1998), Nr. 6, S. 6480–6486

[17] Message Passing Interface Forum: *MPI: A Message-Passing Interface Standard.* 1995. – URL: `http://www.hlrs.de/organization/par/services/models/mpi/mpi-11.ps.gz`

[18] LIEHR, A. W. ; MOSKALENKO, A. S. ; RÖTTGER, M. C. ; BERKEMEIER, J. ; PURWINS, H.-G.: Replication of Dissipative Solitons by Many-Particle Interaction. In: KRAUSE, E. (Hrsg.) ; JÄGER, W. (Hrsg.): *High Performance Computing in Science and Engineering '02. Transactions of the High Performance Computing Center Stuttgart (HLRS) 2002*, Springer, 2003, S. 48–61

[19] MOSKALENKO, A.: *Dynamische gebundene Zustände und Drift-Rotations-Dynamik von dissipativen Solitonen*, Institut für Angewandte Physik, Westfälische Wilhelms-Universität Münster, Diplomarbeit, 2002

[20] AMMELT, E. ; ASTROV, Yu. A. ; PURWINS, H.-G.: Stripe Turing Structures in a Two-Dimensional Gas Discharge System. In: *Physical Review E* 55 (1997), Nr. 6, S. 6731–6740

[21] ASTROV, Yuri A. ; PURWINS, Hans-Georg: Plasma Spots in a Gas Discharge System: Birth, Scattering and Formation of Molecules. In: *Physics Letters A* 283 (2001), S. 349–354

[22] ACKEMANN, T. ; LANGE, W.: Optical pattern formation in alkali metal vapors: Mechanisms, phenomena and use. In: *Applied Physics B* 72 (2001), S. 21–34

[23] VLADIMIROV, A. G. ; MCSLOY, J. M. ; SKRYABIN, D. V. ; FIRTH, W. J.: Two-dimensional clusters of solitary structures in driven optical cavities. In: *Physical Review E* 65 (2002), Nr. 046606, S. 1–11

[24] TARANENKO, V. B. ; O.WEISS, C.: *Spatial semiconductor-resonant solitons.* – nlin.PS/0206029

[25] UMBANHOWAR, P. B. ; MELO, F. ; SWINNEY, H. L.: Localized excitations in a vertically vibrated granular layer. In: *Nature* 382 (1996), Nr. 6594, S. 793–796

[26] CRAWFORD, C.; RIECKE, H.: Oscillon-type structures and their interaction in a Swift-Hohenberg model. In: *Physica D* (1999), Nr. 129, S. 83–92

[27] ARANSON, I. S. ; PISMEN, L. M.: Interaction of Vortices in a Complex Vector Field and Stability of a "Vortex Molecule". In: *Physical Review Letters* 84 (2000), Nr. 4, S. 634–637

Computational Fluid Dynamics

Prof. Dr.-Ing. Siegfried Wagner

Institut für Aerodynamik und Gasdynamik, Universität Stuttgart
Pfaffenwaldring 21, 70550 Stuttgart

The Sixth Results and Review Workshop on High Performance Computing in Science and Engineering 2003 shows again that big progress was achieved in the numerical simulation of complex flows in various fields of computational fluid dynamics. However, there remains still a big demand in increasing the performance and storage capacity of High Performance Computers (HPC) as the following paragraphs will show. It should be emphasized that only a small selection of progress reports could be included in the publication. They represent examples of the big variety of flow problems. Although most of the reports revealed a very high scientific standard, those papers were preferably selected for publication that clearly demonstrated the unalterable usage of HPC for the solution of the problem.

The following paragraph starts with a basic research problem, the laminar-turbulent transition of boundary layers, that, however, has a big impact on technical applications. Meyer et al. investigate the flow randomization process in a transitional boundary layer by direct numerical simulation, DNS. The simulation of transition and turbulence is a typical example that the capacity of current HPCs is still too small. Therefore, the investigations have to be restricted to Reynolds numbers that are orders of magnitude smaller than those in reality. Despite that fact DNS still allows many insights into the complicated mechanisms of transition which cannot be gained without HPC. Due to the extraordinary performance requirements of spatial direct numerical simulations Meyer et al. depend on a high processing speed on a single CPU that can usually be provided only by vector-type processors. To decrease the turn around-time further they applied two fundamentally different parallelization techniques at the same time. Their hybrid approach, using directive-based shared-memory parallelization with OpenMP and explicitly coded distributed memory parallelization with MPI takes advantage of the efficient use of state-of-the-art multi-node supercomputers with multiple CPUs per node. This smart usage of HPC allowed them for the first time the investigation of the late stages of the laminar-turbulent transition process in a flat plate boundary layer that was confirmed by experiments.

Hase and Weigand use numerical simulation to investigate three-dimensional unsteady heat transfer at strongly deformed droplets at high Reynolds numbers. They find a strong difference in the time dependent Nusselt number and thus in the heat transfer for the different deformation cycles. They use 32 CPUs of the CRAY T3E/512-900. The computational time is about 2,08 h with 6310 time steps and a cycle time of 1,22 s per time for the computation of the initially cylindrical droplet using a grid resolution of $128 \times 64 \times 64$. They also demonstrate the grid independency on a $256 \times 128 \times 128$ grid and the behaviour on a longer time scale ($128 \times 64 \times 64$ grad, 64 CPUs and 11 h computation time). The parallel efficiency of the code is very good.

Schneider et al. investigate by 3D numerical simulation the mixing and combustion of hydrogen in a model scramjet (supersonic combustion ramjet) engine. Computational grids of up to 3,5 million volume grids are necessary due to the complex physical phenomena in high speed flows. The use of finite rate chemistry makes the partial differential equations very stiff and causes long computation times. For the proper representation of chemistry reaction mechanisms they choose 20 reactions and 9 different species. Thus, very efficient numerical solvers and high performance computers are required. In this case 208 nodes of the CRAY T3E/512-900 are applied using MPI for the parallelization. 900 to 1200 hours in total CPU-time is required for one simulation in order to optimize the shape of a lobed strat injector.

Khalifa and Laurien investigate numerically the semi-turbulent pipe flow and validate it by experiments. They use the NEC SX-5 and need on the basis of 129 720 grid points approximately 10 hours. They observe varicose instabilities and can explain surface deformation of the pipe by their simulations.

Albina and Peric study the forced brake-up of a liqued jet by numerical simulations that were run on the CRAY T3E/512-90 using up to 64 CPUs. The simulation requires between 144 and 108 CPU-hours.

Michelassi et al. use Large Eddy Simulations (LES) to study the flow in a highly loaded low-pressure cascade, with and without oncoming wakes. They use 64 processors on the Hitachi-SR800 F1 and a grid of $614 \times 294 \times 64 =$ 11.553.024 nodes resulting in 4.26×107 unknowns. They need approximately 8.000 CPUs per run. The simulations show that increasing the strength of the impinging wake successfully inhibits separation of the boundary layer at the suction side of the blade and reduces the total pressure loss.

Melber-Wilkending et al. perform numerical simulations of the viscous flow around high-lift configurations and of the wake vortices of transport aircraft including an optimization task for selected 3-D high-lift flow problems. For a simulation of a 4-vortex configuration behind the wing of a transport aircraft they needed a mesh of 12 million grid points. The total simulation time of one configuration is 200 h in a single processor mode on the NEC-SX5.

Braun et al. perform a direct numerical simulation of fluid-structure interactions on aircraft wings. The simulation takes 15 hours and 2 GB memory on a single NEC-SX5 processor for a computational rush of 4 million grid points. 99 per cent of the computation time is needed to slove the Reynolds averaged

Navier-Stokes equations (RANS). A more frequent use of the FLOWER code will require the parallel option.

Hartleb and Tilgner study the Rayleigh-Bénard convection in a plane layer with periodic boundary conditions in the horizontal directions. A spectral method allows them to reach Rayleigh numbers up to 107 even for an aspect ratio of 10. They use LES methods and logically shared, distributed access (SHMEM) routines available on the CRAY T3E 512/900 supercomputer. Computation time for a 16 processor run was 6.9 times larger than for a job with 128 processors in parallel compared to factor 8 in the increase of the number of processors.

Investigation of the Flow Randomization Process in a Transitional Boundary Layer

Daniel Meyer, Ulrich Rist, and Markus Kloker

Institut für Aerodynamik und Gasdynamik, Universität Stuttgart, Pfaffenwaldring 21, D-70550 Stuttgart, Germany, e-mail: [name]@iag.uni-stuttgart.de

This work is devoted to the investigation of the late stages of the laminar-turbulent transition process in a flat-plate boundary layer without pressure gradient. Similar to a turbulent boundary layer these stages are dominated by nonlinear flow dynamics and by the occurrence of coherent vortex motions in the boundary layer, which are far from being well understood, up to now. The work consists of two parts: a description of some salient features of the numerical method, and an account of some specific results of a numerical study of the flow randomization process in K-type transition that has been performed conjointly with a wind-tunnel investigation at the Technical University of Berlin.

1 Introduction

The different aspects of laminar-turbulent transition have been subject of intensive research for the past few years, especially in Germany where a national research program has been established in 1996 by the DFG [12]. From a better understanding of transition, improvements in its prediction, new means for flow control, and basic contributions to turbulence research can be expected. In addition to performing various kinds of laboratory or free-flight experiments, direct numerical simulation has become a reliable tool to perform "virtual experiments" under carefully controlled boundary conditions which allow to trace down several observations to their physical origins. The focus of the present work is on the late stages of the transition process, which are dominated by nonlinear flow dynamics and by the occurrence of coherent vortex motion in the boundary layer. In contrast to the earlier stages of the transition process that can be adequately described by linear (primary) and secondary stability theory, these are not as well understood and their numerical simulation demands large computer resources, despite the use of adapted and optimized numerical algorithms.

2 Numerical method

The well-proven numerical method is based on a velocity-vorticity formulation of the Navier-Stokes equations for an incompressible fluid. The so-called spatial model is applied, where all three spatial dimensions and time are discretized without any modeling assumptions for a spatially growing boundary layer. For the present simulations, the flow is split into a steady two-dimensional baseflow and an unsteady three-dimensional disturbance flow. The baseflow has to be calculated separately before the equations for the disturbance flow field can be solved. In this model the nonlinear meanflow deformation is obtained as the temporal mean of the fluctuating disturbance quantities at any given point in the flow field. Since the flow can be considered periodic in spanwise direction, a Fourier ansatz in this direction is used, and the equations and boundary conditions are transformed accordingly. The nonlinear terms are solved using a pseudo-spectral technique introduced by [10] which ensures an aliasing-free computation of all spanwise modes. High-order compact finite differences are used for discretization of the wall-normal and the downstream direction. Explicit time integration is performed by a standard Runge-Kutta scheme of 4th-order accuracy. A scheme of alternating high-order up- and downwind-biased compact finite differences is used in subsequent intermediate Runge-Kutta steps for computation of the streamwise derivatives of the nonlinear terms. This method leads to a finite-difference scheme with the dispersion characteristics of a central scheme, but, in addition, provides the appropriate amount of numerical damping for possibly occuring high-wavenumber modes on the used grid.

The disturbances in the DNS are generated by suction and blowing within a disturbance strip at the wall. Shortly upstream of the outflow boundary the disturbance vorticity vector is forced to zero in a buffer domain to prevent undue reflections induced otherwise by large-amplitude disturbances passing through. The numerical method has been extensively verified (checked for consistent discretization and convergence) and validated (compared with linear stability theory and experimental data) and has proven to be a useful research tool (cf. [12]).

Due to the extraordinary performance requirements of spatial direct numerical simulations we strongly depend on high processing speed on a single CPU. This can usually only be provided by vector-type processors. Therefore, our code is especially prepared to take maximum advantage of such CPUs by providing simple, well vectorizable, high work-load algorithmic loops. However, to decrease the turn around-time further we apply two fundamentally different parallelization techniques at the same time. A hybrid approach, using directive-based shared-memory parallelization with OpenMP and explicitly coded distributed-memory parallelization with MPI, allows for efficient use of state-of-the-art multi-node supercomputers with multiple CPUs per node. We combine classical domain decomposition in physical space with parallel computation of the temporally decoupled spanwise Fourier modes in spectral space in our code. A sketch of our parallelization strategy is given in figure 1, showing, as a simple example, a run with two MPI processes and three OpenMP threads running in each MPI process, i.e. using a total of 6 CPUs. As long as we are in physical space (upper part of figure 1), we use domain decomposion for the MPI processes by dividing the downstream direction into as many equally sized regions as MPI processes are available. The wall-normal direction will be worked on in parallel by the different OpenMP threads in each MPI process, because all computations in that direction can be performed independently of each other. When in spectral space (lower part of figure 1), the number of Fourier modes is divided by the number of MPI processes and each MPI process is responsible for his share of consecutive Fourier modes. Within the MPI process the OpenMP threads work on different Fourier modes in parallel, because the computations for different modes are independent of each other. During the computation data have to be exchanged between both arrangements in order to compute the non-linear product terms of the Navier-Stokes equations in physical space. These terms provide the coupling between the different Fourier modes. The data exchange between the two different arrangements of data in memory is done by a call to the MPI subroutine MPI_ALLTOALL because every MPI process must exchange data with all other MPI processes. Since MPI only provides facilities to exchange data stored in linear buffers in memory (consecutive regions in memory), a quite elaborate data sorting is required in order to write into or read from these linear buffers. The involved three-dimensional arrays used for data storage in memory have different dimension sizes for both data arrangements, which also contributes to a tedious sorting process.

The performance on the NEC-SX/5 is about 2 GFlops per CPU for the whole code, which corresponds to roughly 50% of the theoretical peak performance. In single subroutines a performance of up to 3.8 GFlops is achieved. Our code seems to be bandwidth limited, because on the NEC-SX/4 we achieve about 60% of the theoretical peak performance, which is most likely due to the fact that the NEC-SX/4 has a faster memory interface (relative to the available CPU performance on that machine). The code scales very well in shared-memory environments up to about 12 CPUs using OpenMP directives. For larger numbers of parallel CPUs the computations should be distributed on more than one node using the MPI parallelization additionally. Using MPI within the same node is not desirable, because of the overhead generated by the data exchange required between the MPI processes. Up to now, we have used the code on the Cray-T3E with up to 230 CPUs in parallel (using MPI), and we did successful test runs on two NEC-SX/5 nodes during maintenance times with our hybrid approach as described above. However, up to now production runs have always been done on a single NEC-SX/5 node using a maximum of 14 CPUs.

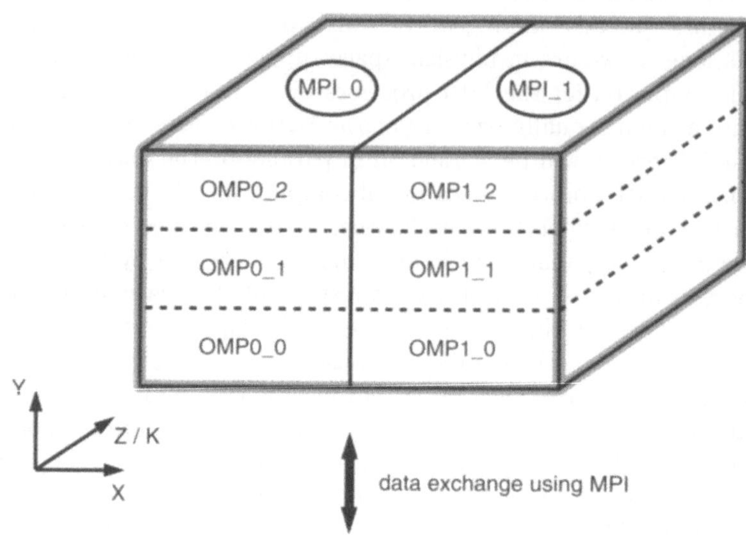

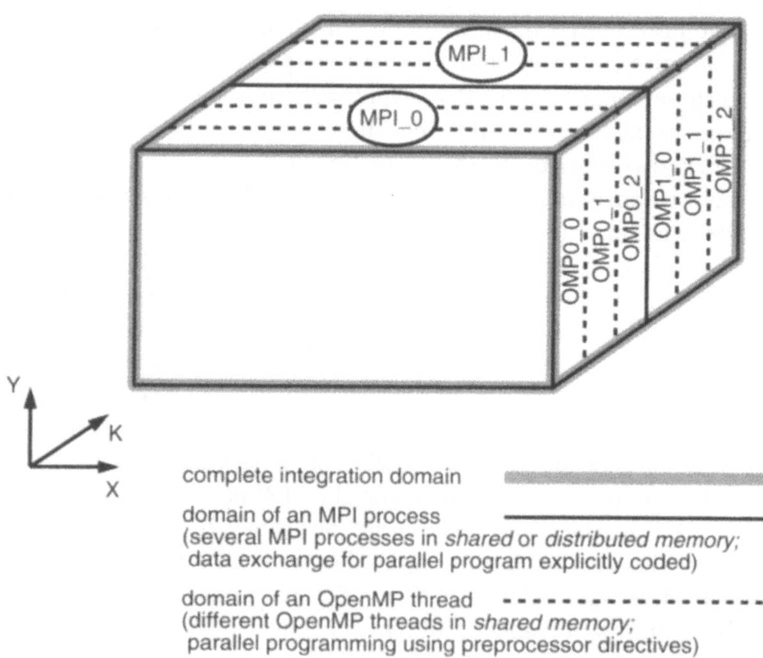

Fig. 1. Sketch of strategy for the hybrid parallelization approach using OpenMP and MPI at the same time

3 Results

In the HLRS-project "LAMTUR" a considerable number of investigations related to laminar-turbulent transition have been performed recently, ranging from the reception of initial disturbances from the free-stream flow into the boundary layer [15] to investigations of the late stages of the transition process [3, 7, 13]. Only the last of these problems will be presented here. For more references see [4, 6, 11, 12, 14]).

For a mutual verification and for gaining additional insights wind-tunnel experiments have been performed at the Hermann-Föttinger-Institute of the Technical University of Berlin in parallel to the present numerical simulations for a corresponding set-up. More information about the experiment can be found in [1] together with several direct quantitative comparisons between simulations and experiment. A sketch of the numerical integration domain related to the experimental set-up is given in figure 2 and the parameters of the simulation are summarized in table 1. All downstream positions are given in experimental coordinates.

Time-wise periodic suction and blowing with a disturbance frequency of 62.5 Hz is applied within a disturbance strip at the wall located at $x = 550$ mm which corresponds to a local Reynolds number (based on the displacement thickness) of $Re_{\delta 1} = 730$. The disturbance input consists of a large-amplitude two-dimensional disturbance (a so-called Tollmien-Schlichting wave) and a number of spanwise periodic disturbances which sum up to a peak amplitude in the spanwise centre of the integration domain at $z = 0$ that mimics the disturbances produced in the experiment by a series of loudspeakers (see Bake *et al.* [1]. However, it turned out that the amplitudes of the timewise periodic disturbance had to be quite large right from the very beginning in order to trigger the transition process within a short distance downstream of their source. Therefore, nonlinear wave-interactions are already important at the first data acquisition point of the experiment which is located at $x = 640$ mm. Iterative adjustments of the amplitudes of the periodic disturbances at the disturbance strip in the DNS were necessary in order to get a good overall agreement with the experimental measurements. This did not only lead to a close agreement

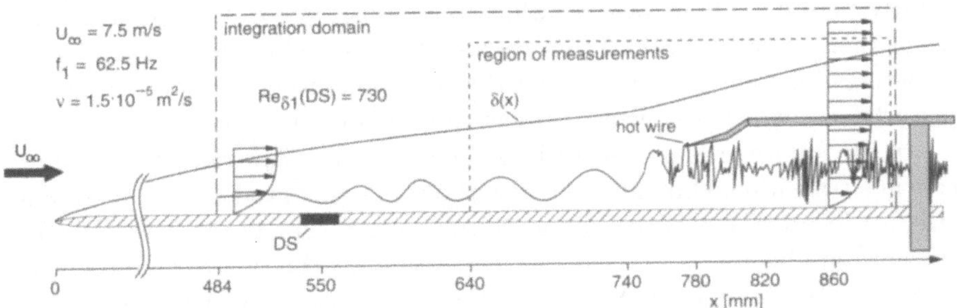

Fig. 2. Sketch of the integration domain

Table 1. Simulation parameters

integration domain in mm ($X \times Y \times Z$)	$541.53 \times 18.48 \times 80.00$
grid points in x- and y-direction ($N \times M$)	2266×361
grid points within the wall zone	33
spectral modes $(K+1)$ (de-aliased)	154
grid points per wave length in z-direction (for the computation of non-linear terms)	512
number of unknowns (de-aliased)	$\approx 125 \cdot 10^6$
approximate number of grid points per disturbance wave length	180
resolution in mm ($\Delta x \times \Delta y$ (wall zone) $\times \Delta z$)	$0.2391 \times 0.0537(0.0269) \times 0.2614$
resolution in wall units at $x = 900$ mm ($\Delta x^+ \times \Delta y^+ \times \Delta z^+$)	$\approx 7.2 \times 0.8 \times 7.8$
non-dimensional circular disturbance frequency	$\beta = 10.472$
time step	$\Delta t = 1.067 \cdot 10^{-5}$ s
time steps per disturbance cycle	1500
non-dimensional wave number in x-direction	$\alpha = 29.2$
non-dimensional wave number in z-direction	$\gamma = 15.708$
reference values for normalization	$U_\infty = 7.5\ \frac{m}{s}$, $L = 200$ mm $\nu = 1.5 \cdot 10^{-5}\ \frac{m^2}{s}$
number of disturbed spectral modes	21

for the disturbance velocity-profiles in wall-normal and spanwise direction at the first measurement position, but also for the development of the disturbance amplitudes over the whole transition process (see [1]). All numerical results have been checked by grid refinement tests using partially a higher resolution than shown in table 1. During these tests, the wall-normal direction turned out to be the most critical one in terms of resolution requirements.

Since turbulent flow is by definition connected with random fluctuations with respect to the time- or (in our case) the phase-averaged flow, the question arises where these random fluctuations occur for the first time in a predominantly periodical flow, and by which mechanism they are amplified. Obviously, there must be a connection between the instability mechanisms and the structures inherent to the flow and the small-amplitude random background disturbances which are always present in real flow situations.

In the following, our observations concerning the flow randomization process in a transitional boundary layer without pressure gradient will be summarized. Additional details can be found in [1, 7, 8]. For the case with adverse pressure gradient see [5]. Here, we will focus on the flow structures which become only available via DNS, an appropriate post-processing, and flow visualizations. It will be shown that the flow randomization process can be understood by taking into account the development of the vortical structures in the transitional boundary layer. However, apart from becoming increasingly smaller and more and more entangled with each other during the transition process, these structures are always closely related to high-shear layers inside the boundary layer. This is illustrated in figure 3 which is obtained in the so-called "two-spike stage", i.e. at a spatial position where two subsequent highly characteristic velocity defects in the time signal at the edge of the boundary layer can be detected per disturbance cycle. These low-velocity regions mark the passage of small Ω-shaped vortex loops which can be seen in figure 3 between $x = 780$ and 800 mm. It turns out that high-shear layers are a by-product of the activity of the vortices embedded in the boundary layer. For instance, the legs of the characteristic so-called Λ-vortex in the left part of the figure rotate in such a way that they move low-speed fluid away from the wall in between them, while high-speed fluid is transported towards the wall on their outward sides. These two effects cause the high-shear layer on the inner top of the vortex and the outer bottom, respectively. At the 'neck' of the Λ-vortex at $x \approx 770$ mm, we find a strong upward movement of fluid in between the vortex' legs due to induction by their counter rotation. This transport of low speed fluid away from the wall generates a high-shear layer above the vortex. Interestingly, the high-shear layer located on top of the Λ-structure from the previous disturbance cycle (i.e. the one further downstream) is always connected to the wall shear layer at the end of the next one. By this, structures from different disturbance cycles interact with each other. This is the beginning of a complicated process whose continuation will be shown and discussed below.

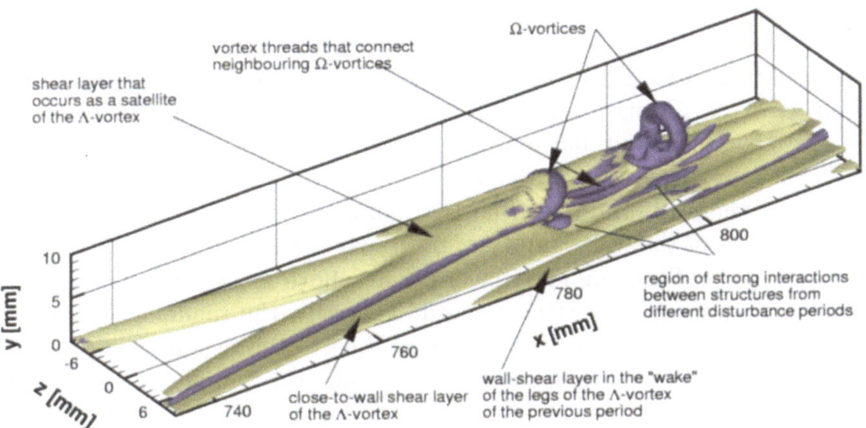

Fig. 3. Comparison of shear layers (yellow) and vortices (blue) inside the transitional boundary layer

The first random fluctuations occur at the tip of the Λ-vortex, where the ring-like Ω-vortices are formed. This formation process is very sensitive to background perturbations, as was also observed in turbulent channel flow by [16]. The formation and detachment of the Ω-vortex is accompanied by strong vortex stretching when the developing vortex lifts up in the boundary layer where it is accelerated. The stretching leads to increased fluid motion by induction, and, therefore, to a region of extremely high shear slightly above the stretched vortex legs. This shear layer disappears when the Ω-vortex detaches from the tip of the Λ-vortex and the legs, still connecting it with the Λ's tip, start moving towards the wall. The sensitivity of the process of formation of the Ω-vortices to small random background perturbations is responsible for a slightly varying occurrence of the Ω-vortices in space and time from one cycle to another. These variations are perceived in the experiment as phase jitter and small amplitude variations of the spike signals. Since the fluid region around the Ω-vortices is a region of strong local velocity gradients, small-amplitude background perturbations will be amplified at these locations. Therefore, the amplified random motions will still be connected to the periodically occurring events in the flow, i.e. they will be strongest in regions where the local flow structures generate the largest gradients; they are no longer completely independent background perturbations. Such a correlation is actually found in the DNS as can be seen in figure 4 which shows a superposition of the non-periodic (i.e. random) fluctuations versus time together with the phase-averaged disturbance field in a crosscut at constant x. The same distribution of the random fluctuations has also been observed in the experiment (see [1]). Note: In the experiment it is easier to measure time signals for constant x versus time than to measure at several x-position at the same time (or phase) instant. The amplitude level of these random perturbations strongly depends on the amplitude level of the background disturbances at the inflow.

Once formed, the Ω-vortices have a strong effect on the surrounding fluid, especially down to the region close to the wall as already presented by Borodulin *et al.* [2]. Positive velocity fluctuations that move with the same speed as the Ω-vortices but much closer to the wall are observed in planes at $z \approx \pm 2$ mm. This influence can also be clearly observed in animations of $\partial U/\partial y$ at the wall produced from DNS data (not shown here). The modulation patterns of the wall values can be clearly attributed to the Ω-vortices in the outer part of the boundary layer because both move with the same velocity, which is almost free-stream speed. This way, the random motion in the outer part of the boundary layer is transfered to the region close to the wall where strong mean velocity gradients occur and the non-periodic fluctuations are amplified.

Another important effect of the vortices on the surrounding flow becomes apparent when we consider the instantaneous pressure field generated by the Λ-vortex in the boundary layer. This is shown in figure 5 with the help of two cuts through the pressure field in the plane $z = 0$ mm, and at the wall, respec-

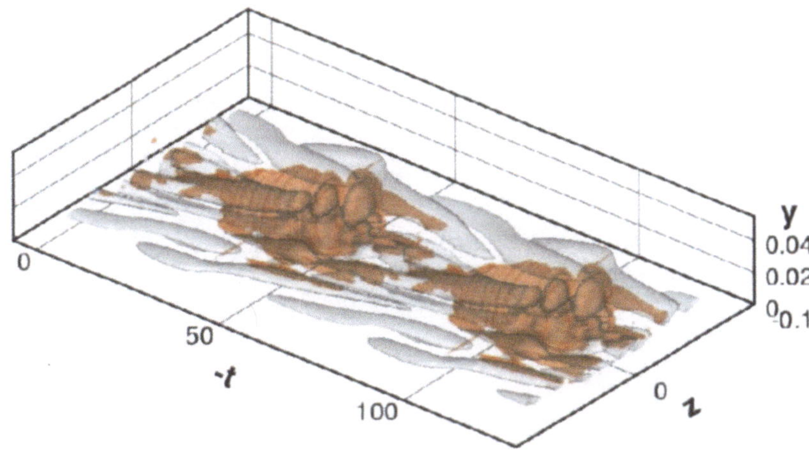

Fig. 4. Comparison of non-periodic fluctuations (red) with iso-surfaces of instantaneous disturbance velocity (grey) at $x = 780$ mm. Data computed from 6 disturbance cycles of a DNS with additional random background perturbations

tively. Because of the low pressure regions present within the vortex core, and at the locations where very strong induction occurs, pressure gradients are also generated at the wall. Such pressure gradients influence the production process of new vorticity at the wall. Therefore, the evolution of the coherent structures in the outer part of the boundary layer will directly influence the vorticity production at the surface of the flat plate. The pressure gradients may not be as pronounced in the ensuing turbulent boundary layer because of the huge number of structures in the flow, but for this stage of the flow development they seem to be relevant.

Thus, it turns out that the vortices and their residing high-shear layers are the main elements for a qualitative (and quantitative) description of the transition process. Their downstream development and their evolution towards a fully developed turbulent flow will be discussed now in connection with figures 6 and 7 at five phases of one disturbance cycle.

In figure 6, we can observe many Ω-vortices which evolve continuously at the tip of the Λ-structures. Despite the durability of these vortices in the outer part of the boundary layer, their initial formation seems to be very sensitive to small perturbations as already said above. Consequently, their streamwise distance with respect to each other varies slightly from one disturbance cycle to the next. The re-connections that are generated between the legs of the Λ-vortex, out of which the Ω-vortices emerge, must obviously be created by a viscous process because vortex lines are reconnecting at this stage of the development which cannot be explained by a purely inviscid mechanism. The subsequent detachment of the vortex loops from the tip of the Λ-vortex is a predominantly inviscid process, basically due to self-induction of the deformed

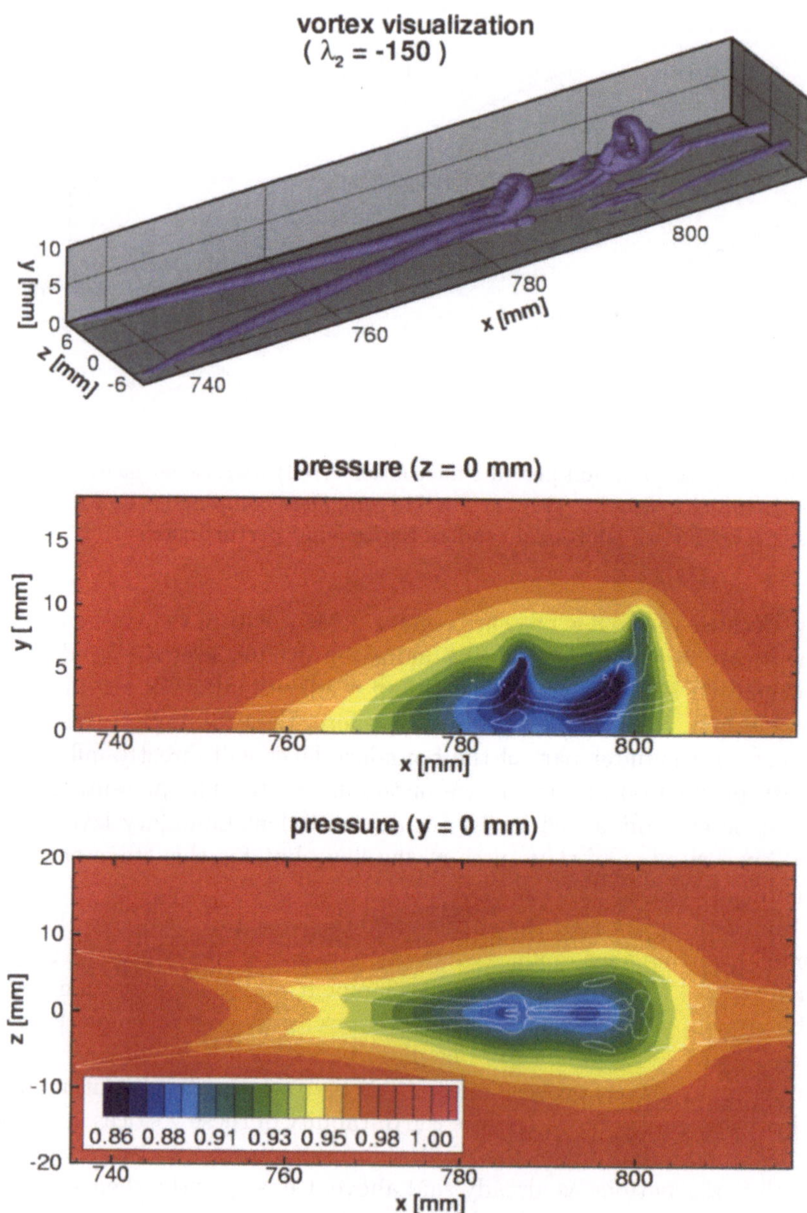

Fig. 5. Comparison of vortex visualization with the instantaneous pressure in the symmetry plane $z = 0$ and at the wall $y = 0$

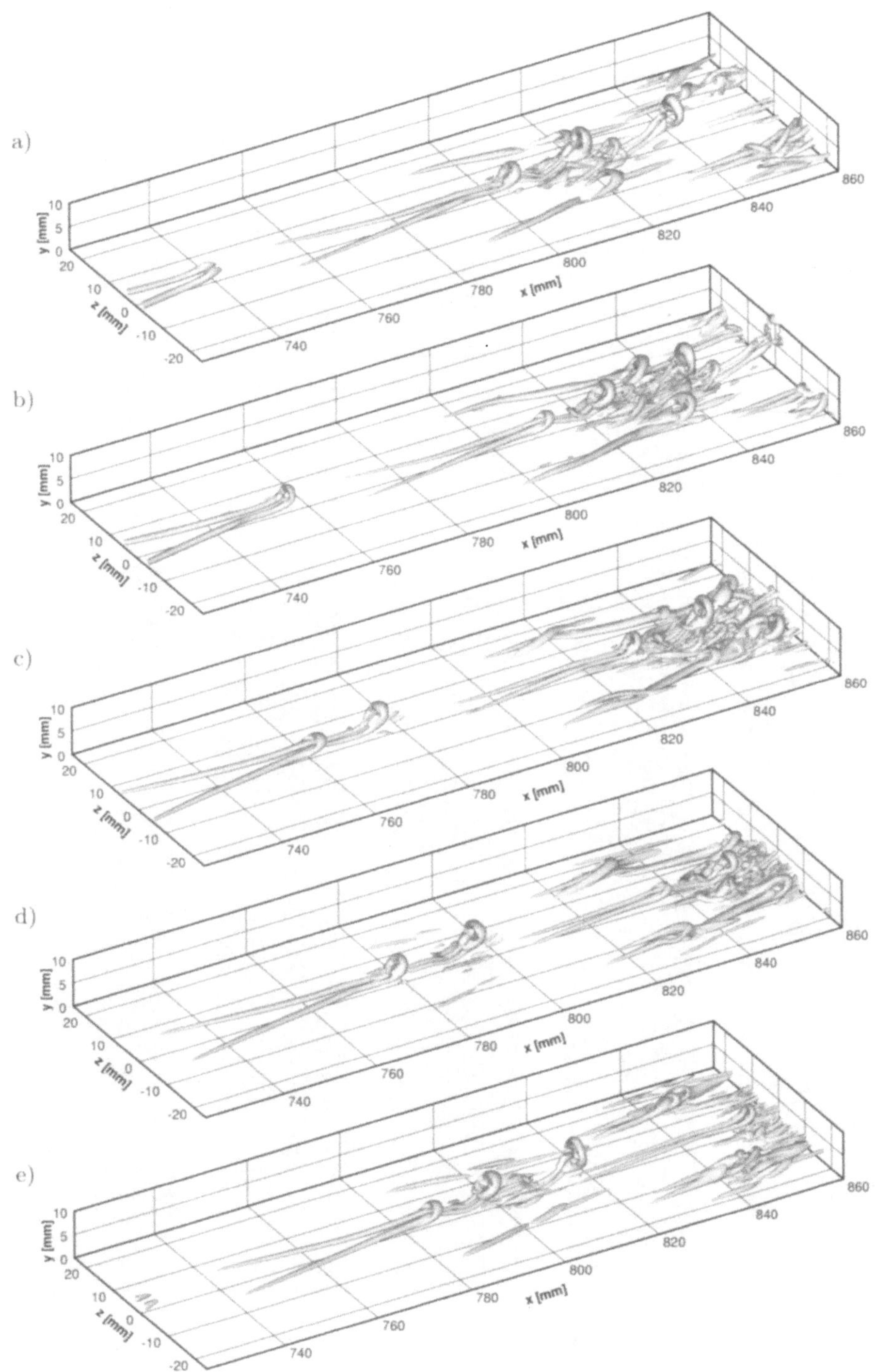

Fig. 6. Vortex visualization at five phases within one disturbance cycle; a) $t/T = 0.2$; b) $t/T = 0.4$; c) $t/T = 0.6$; d) $t/T = 0.8$; e) $t/T = 1$). Iso value $\lambda_2 = -150$

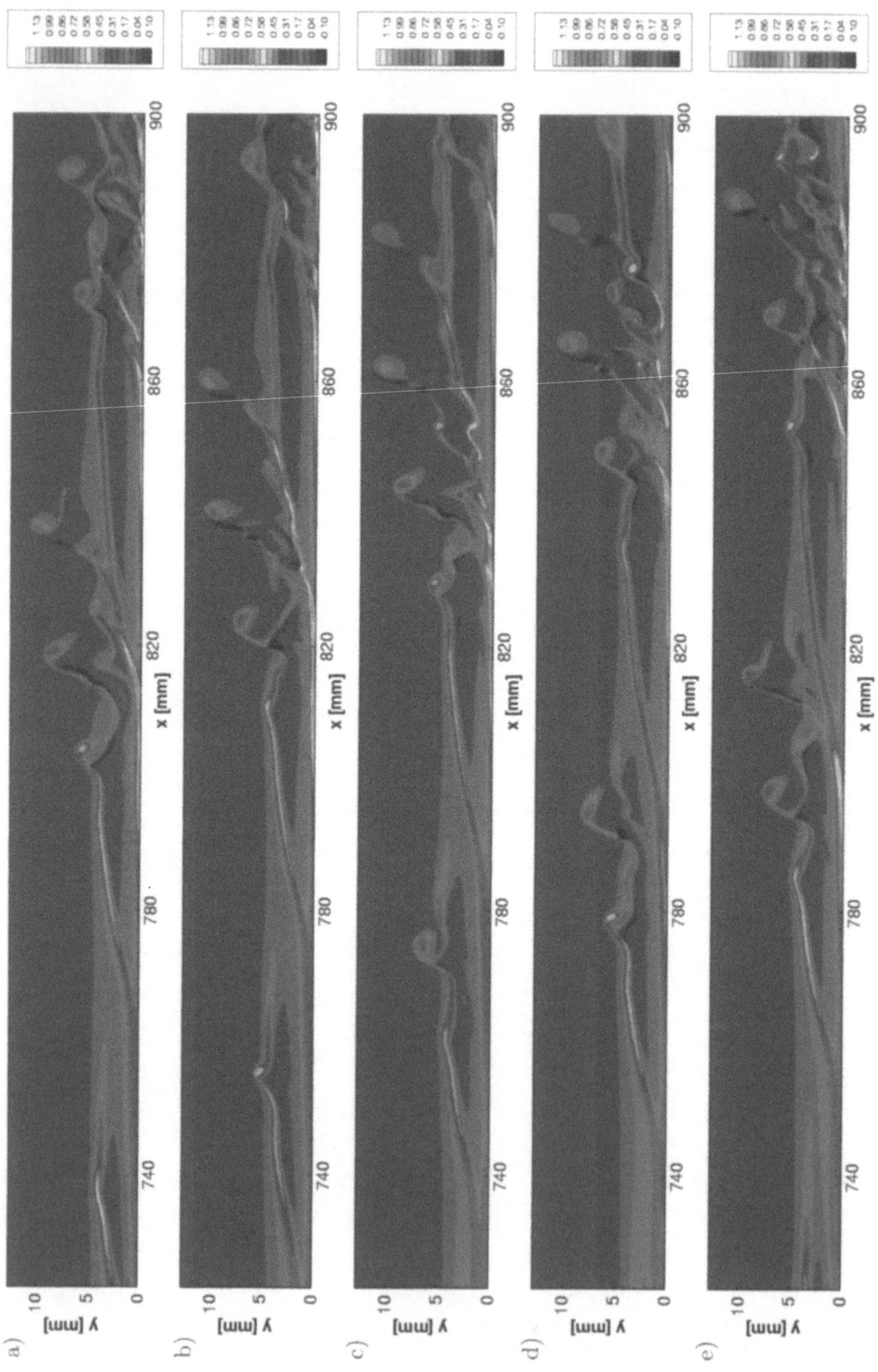

Fig. 7. Vorticity Ω_z at $z = 0$ mm for five consecutive instants from $t/T = 0.2$ to $t/T = 1$ (from left to right) as in fig. 6. $\Omega_z = \frac{\partial U}{\partial y} - \frac{\partial V}{\partial x}$

vortex line moving in the mean shear, as suggested in the work of Moin, Leonard und Kim [9].

Between any two Ω-vortices that were originally generated at the tip of the Λ-vortex and already have left the tip, there is only a relatively weak shear layer visible in figure 7. But this is not necessarily due to the fact that the high-shear layer "rolls up" into the currently developing Ω and therefore disappears [1]. Thus, the nonlinear breakdown process to turbulence is not caused by a mere inflectional instability of the high-shear layer on top of the Λ-vortex, even if the shear layer visualization in figure 7 might suggest this, but is a much more complicated process involving several different mechanisms.

Further downstream, the amplification process of background disturbances becomes even more complex when the vortical structures and high-shear layers generated in different disturbance cycles start to interact. Starting at about $x = 800$ mm in figure 7(d), the Ω-vortices of a given disturbance cycle catch up with the structures generated one cycle before. The interaction of the Ω-vortices with the high-shear layer generated by the tail of the preceding Λ-vortex close to the wall, and with its legs, generates very strong shear that moves very quickly towards the wall (starting at $x \approx 810$ mm in figure 7(a)). The close-to-the-wall structures are modulated by the structures moving at the outer edge of the boundary layer. These modulations initiate the breakdown of the former into smaller ones. This accelerates the development towards fully turbulent flow. As an additional effect, the legs that connect the accelerated Ω-vortices with the tip of the Λ-vortex, or with the Ω next to them, are strongly stretched as they propel themselves towards the wall where the already existing high-shear layer is intensified (compare $x \approx 810$ mm in figure 6(a) and 7(a)).

Many new small vortical structures and high-shear layers appear in the following development through these interactions, which can be well observed comparing figures 6, 7 and 8. Looking at the latest figure, we can easily see why we need such a high resolution in our DNS: the boundary layer is filled with a lot of small-scale vortices compared to the wavelength of the original Tollmien-Schlichting disturbance which is $\lambda_{TS} \approx 43$ mm. This conglomeration of new structures is also very sensitive to background disturbances because of the complex interactions that occur between them. Most likely, the non-periodic motion from the upstream development as described earlier will be amplified and spread to the sides of the center plane ($z = 0$ mm) in this environment. Figure 8 also shows that the structures at the off-center position break down independently from those in the spanwise center. The generation and early development of these secondary structures can be observed in figure 6 at $z = \pm 12$ mm. Finally, a lot of symmetric, as well as asymmetric, structures develop during the late-stage transition process. The influence of asymmetric background perturbations at these very late stages of the transition process has been discussed in detail in [7] and [8].

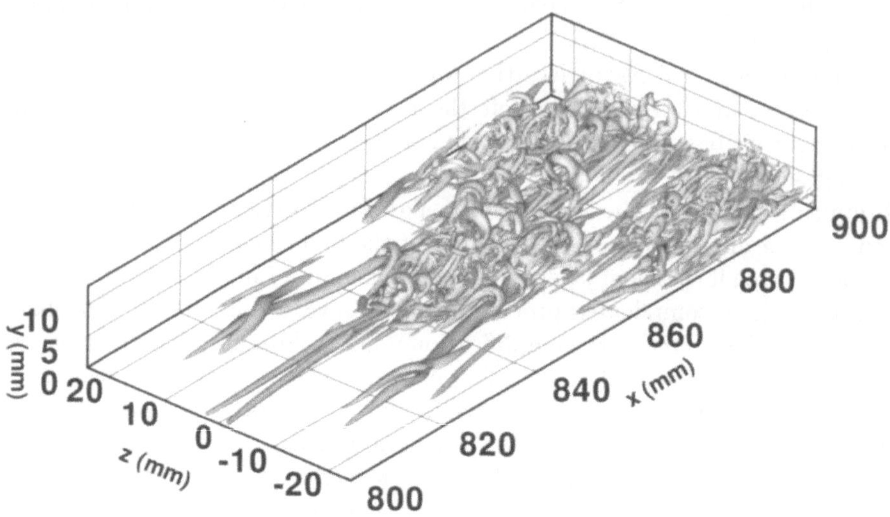

Fig. 8. Vortices in the later stage of the transition process at $t/T = 0.7$ downstream of the region shown in figure 6

4 Conclusions

The present picture of the randomization process seems to be fully confirmed by the experimental data in [1]. The inclined high-shear layer between the legs of the Λ-vortex exhibits increasing phase jitter (i.e. randomization) starting from its tip towards the wall region. In this process, the strongest instantaneous wall-normal shear $\partial U/\partial y$ occurs close to the wall below the Ω-loops and not in the high-shear layer that appears as a satellite on top of the Λ-vortex. Eventually, the regions of strong random velocity fluctuations in the layer closest to the wall enlarge, spread sideways and merge.

The direct numerical simulations provide the full unsteady, instantaneous flow field and are therefore extremely helpful for interpreting and understanding the results obtained by the measurements.

References

1. S. Bake, D. G. W. Meyer, and U. Rist. Turbulence mechanism in Klebanoff-transition. A quantitative comparison of experiment and direct numerical simulation. *J. Fluid Mech.*, 459:217–243, 2002.
2. V. I. Borodulin, V. R. Gaponenko, Y. S. Kachanov, D. G. W. Meyer, U. Rist, Q. X. Lian, and C. B. Lee. Late-stage transitional boundary-layer structures. Direct numerical simulation and experiment. *Theor. Comput. Fluid Dynamics*, 15(5):317–337, 2002.

3. C. Gmelin. *Numerische Untersuchungen zur aktiven Dämpfung von Störungen im Vorfeld der laminar-turbulenten Transition.* Dissertation, Universität Stuttgart, 2003.
4. C. Gmelin, U. Rist, M. Kloker, and S. Wagner. DNS of active control of disturbances in a Blasius boundary layer. In E. Krause and W. Jäger, editors, *High Performance Computing in Science and Engineering '01*, pages 273–285, Heidelberg, 2002. Springer.
5. M. Kloker. A robust high-resolution split-type compact FD-scheme for spatial direct numerical simulation of boundary-layer transition. *Appl. Sci. Research*, 59(4):353–377, 1998.
6. U. Maucher, U. Rist, M. Kloker, and S. Wagner. DNS of laminar-turbulent transition in separation bubbles. In E. Krause and W. Jäger, editors, *High Performance Computing in Science and Engineering '99*, pages 279–294, Heidelberg, 2000. Springer.
7. D. G. W. Meyer. *Direkte numerische Simulation nichtlinearer Transitionsmechanismen in der Strömungsgrenzschicht einer ebenen Platte.* Dissertation, Universität Stuttgart, 2003.
8. D. G. W. Meyer, U. Rist, and S. Wagner. Direct numerical simulation of the development of asymmetric perturbations at very late stages of the transition process. In S. Wagner, M. Kloker, and U. Rist, editors, *Recent Results in Laminar-Turbulent Transition – Selected Numerical and Experimental Contributions from the DFG-Verbundschwerpunktprogramm "Transition" in Germany*, NNFM, pages 213–222, Heidelberg, 2003. Springer.
9. P. Moin, A. Leonard, and J. Kim. Evolution of a curved vortex filament into a vortex ring. *Phys. Fluids*, 29(4):955–963, 1986.
10. S. A. Orszag. Numerical simulation of incompressible flows within simple boundaries. I. Galerkin (spectral) representations. *SIAM*, 1(4):293–327, 1971.
11. C. Stemmer, M. Kloker, U. Rist, and S. Wagner. DNS of point-source induced transition in an airfoil boundary-layer flow. In E. Krause and W. Jäger, editors, *High Performance Computing in Science and Engineering '98*, pages 213–222, Heidelberg, 1999. Springer.
12. S. Wagner, M. Kloker, and U. Rist, editors. *Recent Results in Laminar-Turbulent Transition – Selected Numerical and Experimental Contributions from the DFG-Verbundschwerpunktprogramm "Transition" in Germany.* NNFM, Springer, Heidelberg, 2003.
13. P. Wassermann and M. Kloker. Mechanisms and passive control of crossflow-vortex-induced transition in a three-dimensional boundary layer. *J. Fluid Mech.*, 456:49–84, 2002.
14. P. Wassermann, M. Kloker, U. Rist, and S. Wagner. DNS of laminar-turbulent transition in a 3-d aerodynamics boundary-layer flow. In E. Krause and W. Jäger, editors, *High Performance Computing in Science and Engineering 2000*, pages 275–289, Heidelberg, 2001. Springer.
15. A. Wörner. *Numerische Untersuchungen zum Entstehungsprozess von Grenzschichtstörungen durch die Interaktion von Schallwellen mit Oberflächenrauigkeiten.* Dissertation, Universität Stuttgart, 2003.
16. J. Zhou, R. J. Adrian, S. Balachandar, and T. M. Kendall. Mechanisms for generating coherent packets of hairpin vortices in channel flow. *J. Fluid Mech.*, 387:353–396, 1999.

Numerical Simulation of 3D Unsteady Heat Transfer at Strongly Deformed Droplets at High Reynolds Numbers

Matthias Hase and Bernhard Weigand

Institute of Aerospace Thermodynamics, University of Stuttgart, Pfaffenwaldring 31, 70569 Stuttgart, Germany, matthias.hase@itlr.uni-stuttgart.de

Summary. The dependency of the heat transfer on an initial deformation of droplets has been investigated at high droplet Reynolds numbers. The two-phase flow has been computed with an inhouse 3D DNS program (FS3D) using the Volume-of-Fluid method. For the droplets initial prolate and oblate shapes with an axial approaching flow has been studied. In addition, a spherical shape has been used as reference. The initial droplet Reynolds number for the present study has been $Re_0 = 660$ for all investigated cases. Due to the fact that the steady droplet velocity for the considered droplets has been much lower than the initial velocity of the droplets, the droplet velocity is decreased during the simulation. To gain more knowledge about the influence of deformation on the heat transfer, the time dependent, spatial averaged Nusselt number Nu_t and the time and spatial averaged Nusselt number Nu_m has been matched by the temperature and velocity field around a deformed droplet. By this comparison the oscillation phase with the largest heat transfer has been observed. The simulations have been performed on the Cray T3E/512-900 at the HLRS with 32 processors. The parallel performance in dependency of the number of processors has been investigated.

1 Introduction

The rate of heat transfer from the surrounding gas to droplets in sprays is a critical design parameter of many technical spray systems as for instance in automotive engines or gas turbines. In these processes the considered droplets respectively liquid ligaments are in many cases strongly deformed and the droplet velocity is heavily unsteady. Additionally due to the high velocities the flow around the droplet is transient and fully 3D. Because of this difficulties strongly deformed droplets have been studied rarely in the past, neither numerically nor experimentally.

In the present study it has been assumed that the droplets are deformed initially due to the primary breakup. During this breakup process strongly deformed liquid ligaments emerge which are approximated by two idealized droplet shapes in this study as a first step. The numerical investigation has been performed at high droplet Reynolds numbers ($Re > 270$), which means that the flow is fully 3D and time dependent. The initial droplet Reynolds number $Re_0 = D_0 U_0/\nu_L$ of the considered droplets is $Re_0 = 660$, where D_0 is the diameter of a spherical droplet with the same volume, U_0 the initial droplet velocity and ν_L the kinematic viscosity of the droplet fluid. To take the flow character into account the investigation has been performed fully 3D. Additionally no restrictions on the deformation of the droplets are assumed. The programs efficiency and reliability for the computation of strongly deformed two-phase flow has been presented already in [1].

2 Analysis and numerical method

The inhouse 3D CFD program FS3D (Free Surface 3D) has been developed to compute the Navier-Stokes equations for incompressible flows with free surfaces. The equations are solved without using a turbulence model by *direct numerical simulation* (DNS). The governing conservation equations for momentum and mass are

$$\frac{\partial(\rho\mathbf{u})}{\partial t} + \nabla\cdot[(\rho\mathbf{u})\otimes\mathbf{u}] = -\nabla p + \nabla\cdot\mu\,[\nabla\mathbf{u} + (\nabla\mathbf{u})^T] + \nabla\cdot\mathbf{T} \qquad (1)$$

$$\nabla\cdot\mathbf{u} = 0 \quad , \qquad (2)$$

where $\mathbf{T}$ is the capillary stress tensor which adds the surface tensor force to the momentum equation. Furthermore $\mathbf{u}$, ρ, μ and p are the velocity vector, the density, the dynamic viscosity and the pressure, respectively. In addition, the energy equation is solved. For the above mentioned incompressible flow and for a fluid with constant fluid properties in each phase the energy equation is decoupled from the equations of motion. Therefore, the energy equation

$$\frac{\partial}{\partial t}(\rho c_p T) + \nabla\cdot(\rho c_p \mathbf{u}T) = \nabla\cdot(k\nabla T) + \Phi \quad . \qquad (3)$$

can be solved after the computation of the flow field, where T is the temperature, c_p the specific heat and k the heat conductivity. The dissipation term Φ can be neglected for all mentioned flows due to the low Eckert number. The implementation and validation of the energy equation has been described in [2].

In two phase flows additional information about the interface position between the disperse and the continuous phase are needed. In FS3D a Volume-Tracking method, well known as the *Volume-of-Fluid* (VOF) method, is used [3]. In the VOF-method an additional transport equation

$$\frac{\partial f}{\partial t} + \nabla\cdot(\mathbf{u}f) = 0 \qquad (4)$$

for the volume fraction of the disperse phase is solved.

The variable f is called the VOF-variable. The VOF-variable is

$$f(\mathbf{x},t) = \begin{cases} 0 & \text{in the continuous phase} \\ 0 < f < 1 & \text{at the interface} \\ 1 & \text{in the disperse phase} \end{cases} \tag{5}$$

With the VOF-variable the change of the fluid properties across the interface can be computed by using the equations

$$\rho(\mathbf{x},t) = \rho_d + (\rho_c - \rho_d)\, f(\mathbf{x},t) \tag{6}$$

$$\mu(\mathbf{x},t) = \mu_d + (\mu_c - \mu_d)\, f(\mathbf{x},t) \tag{7}$$

where the subscript c indicates the continuous phase and d the disperse phase. Additional fluid properties like the specific heat c_p and the heat conductivity k are obtained in the same way. To ensure a sharp interface and to suppress numerical dissipation of the disperse phase in each time step the interface is reconstructed by the PLIC-method (*Piecewise linear interface reconstruction computation*) [4]. After the reconstruction, the disperse phase is transported on the basis of the reconstructed distribution of the disperse phase. The spatial discretization is realized by a structured Finite Volume scheme on a staggered grid. In each phase the discretization is second-order accurate. Due to the high gradients across the interface a limiter is used to prevent the generation of oscillations and spurious solutions. The program is parallelized with domain decomposition using the communication library *MPI.* A multigrid solver is included to solve the Poisson equation for the pressure. Additionally a coordinate transformation from the inertial system to a coordinate system moving with the droplet is implemented [5] to track the droplet for a longer time without generating extremely large computational domains.

3 Results

3.1 Computational domain and fluid properties

In the presented simulations, the liquid has been assumed to have the properties of water at $20°C$ except for the dynamic viscosity. The dynamic viscosity is $\mu_L = 10 \cdot \mu_{H_20} = 1 \cdot 10^{-3}\,\mathrm{kg/(ms)}$. The higher viscosity has been chosen to avoid *parasitic currents* [6]. For the surrounding gas the properties of air have been chosen. The initial temperature of the liquid is $T_L = 350\,K$ and $T_G = 293.15\,K$ for the gaseous phase.

The computational domain is displayed in Fig. 1. The 3D channel geometry for a spherical droplet with the diameter $D = 1 \cdot 10^{-3}\,m$ is $x = 12 \cdot 10^{-3}\,m$, $y = 6 \cdot 10^{-3}\,m$ and $z = 6 \cdot 10^{-3}\,m$. The gravitational force acts in the negative x-direction. The inlet boundary is on the left with an inflow velocity which is equal to the droplet velocity at this time. The temperature at the inlet is $T_{inflow} = T_G = T_\infty$. On the right an outlet boundary with a damping

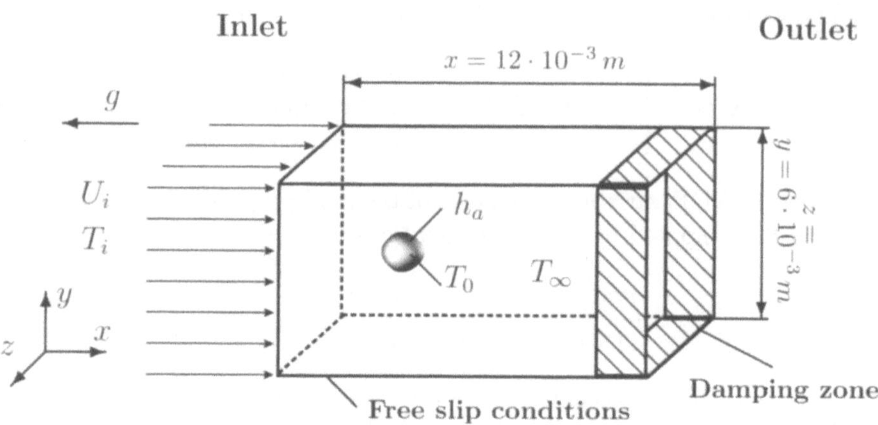

Fig. 1. Channel geometry and boundary conditions for a computation of a droplet diameter $D = 1 \cdot 10^{-3}\,m$.

zone is placed. The damping zone avoids back-flow into the computational domain. At the other boundaries Dirichlet boundary conditions for the y- and z- velocities $v = w = 0$ and Neumann boundary conditions for the x-velocity $(du/dx)_W = 0$ are used.

To perform the study concerning the heat transfer of deformed droplets, two initially strongly deformed droplets have been computed. One is a cylinder with diameter $D = 0.52 \cdot 10^{-3}\,m$ and the cylinders axis parallel to the flow direction. The other initial shape is a disk with diameter $D = 1.38 \cdot 10^{-3}\,m$ with the flow direction perpendicular to the disk surface. The volume of this droplets are the same as for a spherical droplet with $D = 1 \cdot 10^{-3}\,m$.

3.2 Droplet deformation

In Fig. 2 the surface area of the three different initial droplet shapes in dependency of the Fourier number $Fo = at/R_0^2$ is displayed, where a, t, R_0 are the thermal diffusivity, the time and the radius of the spherical droplet with the same volume, respectively. The marks indicate the times used in Figs. 3 and 4. For both deformed droplets oscillations can be seen, which were damped over the time. At $Fo = 0.01$ nearly the same surface area as a spherical droplet has been reached for the deformed droplets. From this time a spherical droplet can be assumed for the computation of the heat transfer. For the oscillation periods a different time range for the prolate and the oblate oscillation cycle can be seen. This asymmetry of the period parts is well known for strong deformations and depends on the oscillation amplitude [7]. For the first oscillation period with the highest oscillation amplitude the difference in the oscillation time is very pronounced. So the time needed by the cylindrical droplet to reach the spherical shape for the first time is twice as high than for the discoidal droplet. However, when the oscillations reach the spherical

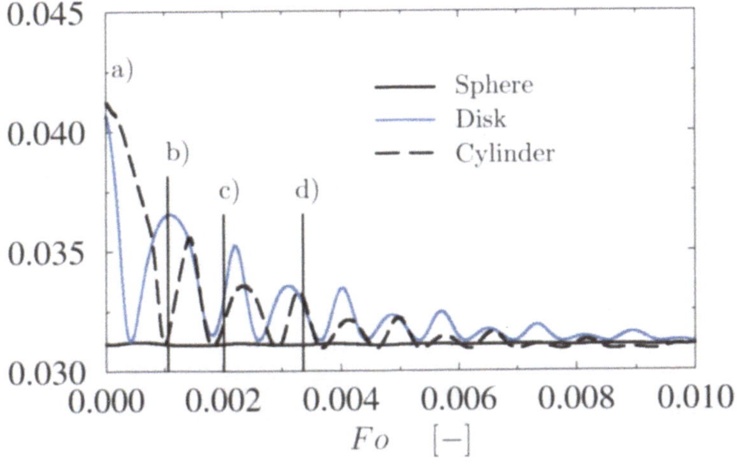

Fig. 2. Surface area A for three different, initial deformed droplets with $\mu_L = 10 \cdot \mu_{H_20}$ as a function of Fo.

shape for the second time the time for both droplets is nearly the same. In the further progression the difference between the time needed for the oblate and the prolate oscillation cycle become smaller due to the lower deformation rate. At about $Fo = 0.006$ the difference between oscillation times has vanished and the oscillation can be described by the linear theory for small deformation.

3.3 Temperature and velocity field

In Fig. 3 the temporal evolution of the droplet shape and the temperature field for the initial cylindrical droplet is shown at different times. Additionally the velocity field around the droplet is displayed. The droplet shape and the temperature field for an initial disk shaped droplet is displayed in Fig. 4. The figures give an impression of the evolution of the temperature field due to the deformation of the droplet and due to the flow field. At $Fo = 0$ the initial conditions for both shapes can be seen. The disk shaped droplet passes through a half oscillation period at the next displayed time $Fo = 0.00108$ to a prolate shape as depicted in Fig. 4b. The initial cylindrical droplet passes only through one quarter of an oscillation period and is just about spherical at this time (3b).

At $Fo = 0.00108$ a very interesting difference in the temperature field can be seen. In Fig. 3b a significant increase in the gas temperature occurs behind the droplet. But for the initial disk shape droplet (Fig. 4b) the temperature increase is weak. At $Fo = 0.00202$ in Fig. 3c the initial cylindrical droplet is oscillating from an oblate to a prolate shape short-time after the largest oblate deformation (see Fig. 2). The temperature distribution behind this droplet is similar to the initial disk shape in Fig. 4c at the same time, but

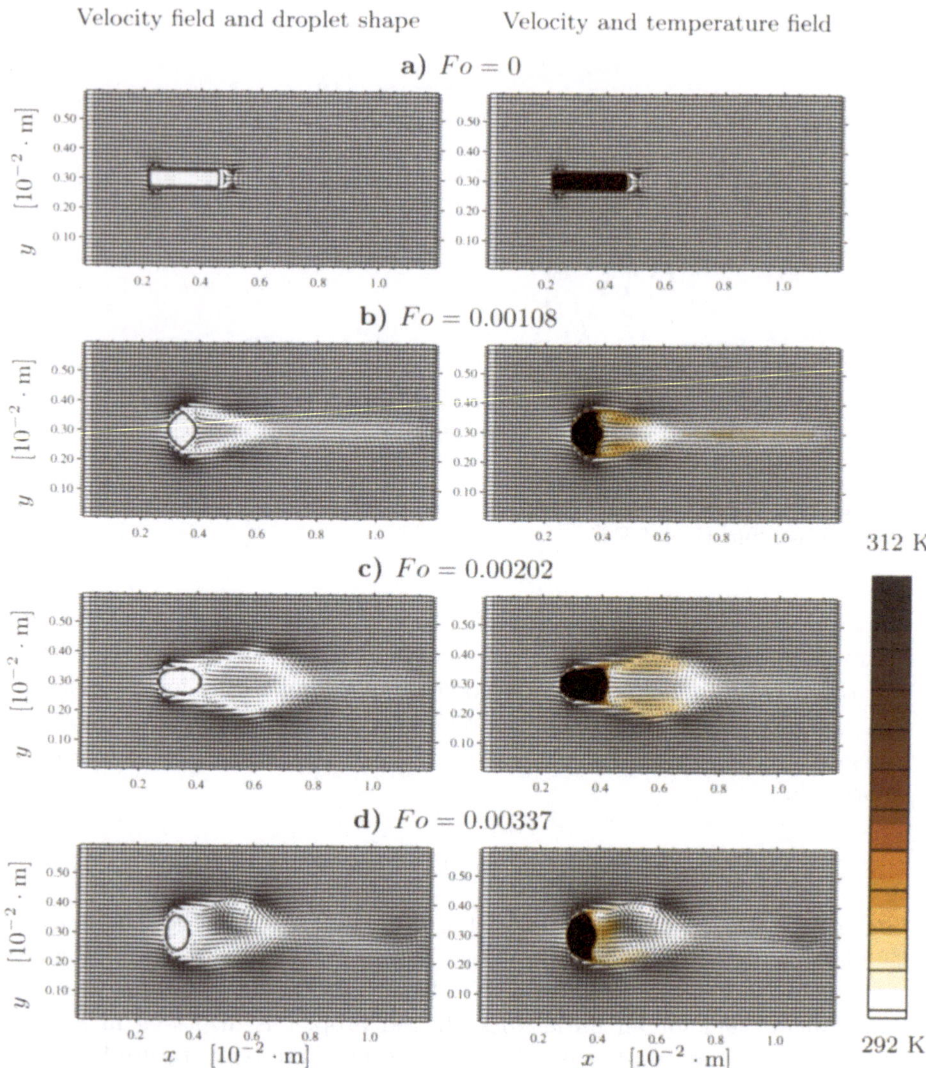

Fig. 3. Velocity field and temperature field around an initial cylindrical droplet with the diameter $D = 1 \cdot 10^{-3}\, m$ in dependency of Fo.

the temperature is higher for the cylindrical case. The droplet in Fig. 4c at $Fo = 0.00202$ is oscillating from prolate to oblate shape (see Fig. 2). The displayed droplet will immediately reach the largest oblate deformation. As pointed out later, the highest heat transfer occurs during the oscillation to the oblate shape. Due to the time delay by the heat transport in the gas phase the effect of the higher heat transfer on the gas temperature is more significant after the droplet passed through the largest oblate deformation. In Fig. 3c it

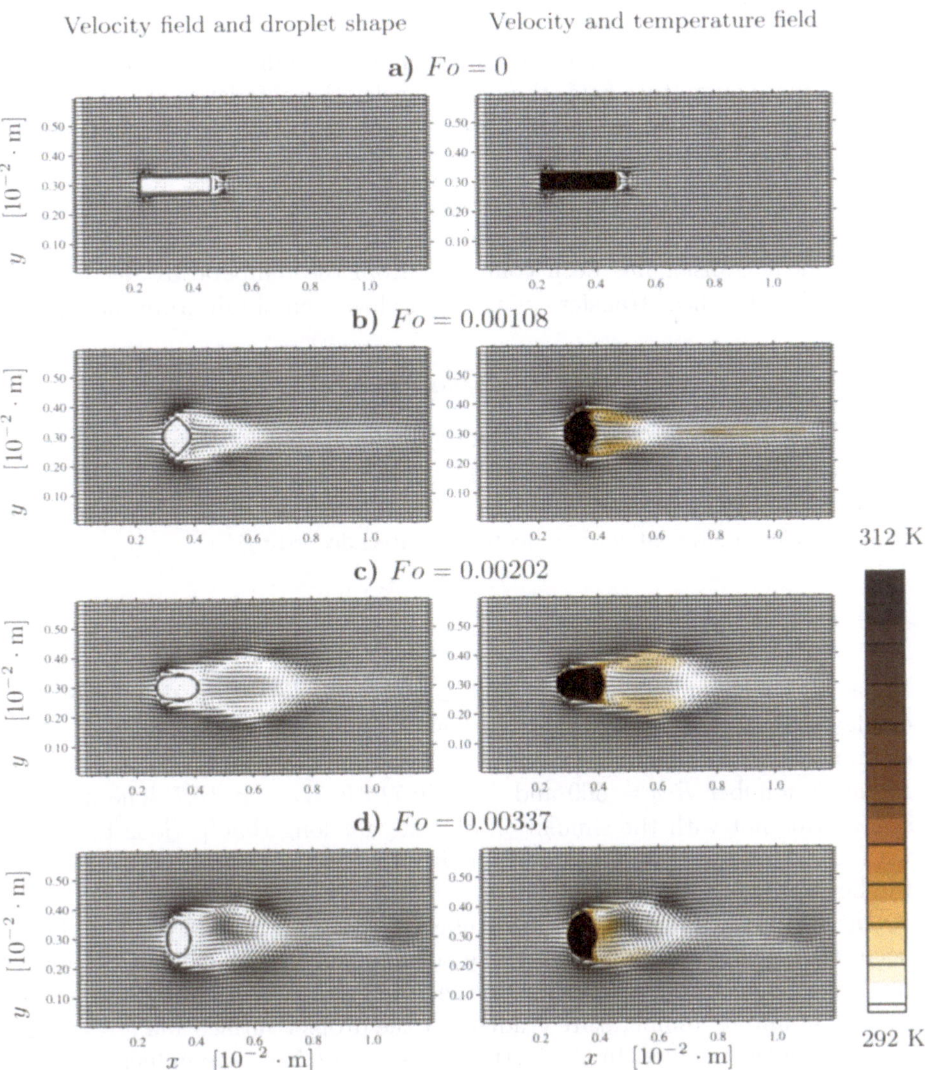

Fig. 4. Velocity field and temperature field around an initial discoidal droplet with the diameter $D = 1 \cdot 10^{-3}\,m$ in dependency of Fo.

can also be seen, that directly behind the droplet the temperature is lower than in the vortices in the recirculation zone. This clarifies a period of high heat transfer is followed by a period of lower heat transfer. Figs. 3d and 4d at $Fo = 0.00337$ show nicely the fully 3D character of the flow and temperature field.

3.4 Heat transfer computations

After investigating the droplet deformation and the influence of the deformation on the temperature field, the heat transfer characteristics of the process will now be explained in detail. To carry out this study a time and space averaged Nusselt number

$$Nu_m = \frac{Dh_a}{k_G} \tag{8}$$

for the heat transfer has been computed from the temperature decrease of the droplet. The heat transfer coefficient h_a has been obtain from the energy balance

$$-\rho_L V c_{v,L} \frac{\partial T_m}{\partial t} = -A h_a (T_{W,m} - T_\infty) \quad , \tag{9}$$

where c_v is the specific heat at constant volume, T_m the averaged droplet temperature, $T_{W,m}$ a time and space averaged wall temperature and T_∞ denotes the temperature in the undisturbed gas flow. Explained in discrete values the heat transfer coefficient h_a has been computed according to

$$h_a = \frac{\rho_L V c_{v,L}}{A(T_{W,m} - T_\infty)} \frac{\Delta T_m}{\Delta t} \quad . \tag{10}$$

It has to be noted that because of the short time scales in this study the displayed Nu_m is far away from the thermal fully developed state Nu_∞. For example the Nusselt number from the correlation given by [8] for the initial Reynolds number $Re_0 = 660$ and $Pr = 0.714$ is $Nu_\infty = 15.7$. The present study deals not with the simulation of Nu_∞ for long time periods but only with the strongly transient period for droplet dynamic and heat transfer at the beginning of this process.

In addition averaged Nusselt numbers Nu_m of deformed droplets have been studied. The steep gradient at the beginning of the mean Nusselt number Nu_m evolution in Fig. 5 occurs due to the initial temperature conditions which lead to an infinite temperature gradient at the droplet surface and a resulting infinite Nusselt number. In the further evolution of Nu_m the influence of the deformation can be seen from the peaks in the curves of the initial deformed droplets. The displayed peaks only occur due to the enhanced heat transfer by the motion of the droplets, because the larger surface area of the deformed droplets has taken into account in the computation of the Nusselt number (see the calculation of the heat transfer coefficient, Eq. (10)). A comparison between the Nusselt number computed with the real surface area and with the surface area of a spherical droplet, is given in [9].

For $Fo = 0.01$ the mean Nusselt number for both deformed droplets is $Nu_m \approx 20.8$ and $Nu_m \approx 20$ for the spherical droplet. At this time the influence of the initial deformation is nearly vanished and must not longer taken into account. These results show that the influence of the deformation at the mentioned Reynolds number of $Re_0 = 660$ is not so strong if the deformation

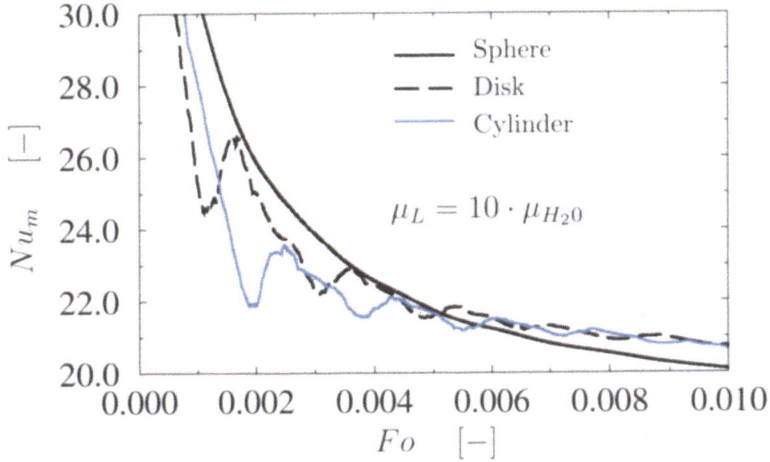

Fig. 5. Comparison of time and space averaged Nusselt numbers Nu_m for three initial droplet shapes in dependency of Fo.

occurs only initially. If additional effects like deformation due to the aerodynamic forces are important the heat transfer could be much more effected [9].

In Fig. 6 the dimensionless surface areas A/A_0 (upper diagrams) and the time dependent, spatial averaged Nusselt numbers Nu_t (lower diagrams) for both deformed droplets are displayed in dependency of Fo. As a reference Nu_t and A/A_0 for a spherical droplet (thick black line in all diagrams) is also plotted in Fig. 6. The values for Nu_t are scattered a lot due to the complexity of the flow and temperature field. Therefore, the values of Nu_t have been averaged over fifty time steps (small black line) to get a more unique evolution of Nu_t. Additional the droplet shapes of the maximum deformation are displayed for the respective oscillation period.

As described before the first steep gradient in the Nusselt number evolution has its reason in the thermal initial conditions. After this period for the initially discoidal droplet (left side) a region of lower heat transfer ($Nu_t \approx 20$) is following, which matches with the prolate oscillation cycle. When the droplet reaches after this cycle the spherical shape and move on into the oblate deformation regime the Nusselt number increases up to $Nu_t \approx 40$. This period of increasing Nusselt number continues up to approximately 2/3 of the maximum deformation in the oblate direction. After reaching this maximum Nusselt number $Nu_{t,max}$ the Nusselt number decreases and for the prolate oscillation period a nearly constant value of the Nusselt number Nu_t is found again. The same behavior is shown on the right side for the initially cylindrical droplet. The period of the highest heat transfer is the oscillation from the spherical to the oblate droplet shape and for the prolate oscillation no significant increase of the Nusselt number can be seen.

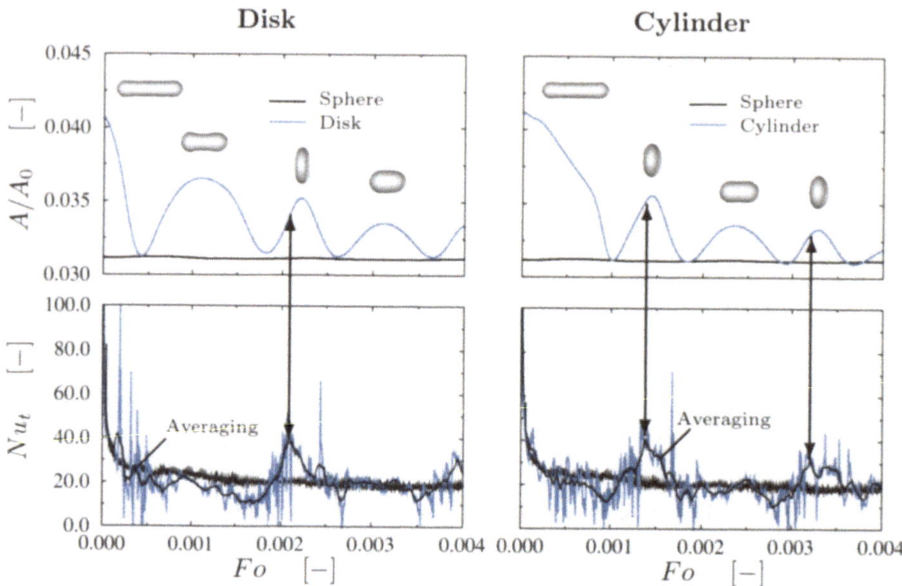

Fig. 6. Deformation A/A_0 with droplet shapes (upper diagram) and time dependent, space averaged Nusselt numbers Nu_t (lower diagram) of an initial discoidal and cylindrical droplet with $\mu_L = 10 \cdot \mu_{H_2O}$ in dependency of Fo .

The reason for the higher Nusselt numbers Nu_t during the oblate oscillation period are the axial motion of the droplet in direction to the boundary layers. Due to this motion the boundary layers are very thin which leads to steep gradients in the temperature profile at the droplet surface and therefore to higher Nusselt numbers. The maximum heat transfer matches not with the highest deformation of the droplet. This is because of the dependency of the enhanced heat transfer on the axial deformation velocity of the droplet and not on the absolute deformation.

4 Computational Resources

All results presented in this paper have been computed on the CRAY T3E/512-900 at the HLRS on 32 CPU's. The grid resolution was $128 \times 64 \times 64$ (524288 cells) with $32 \times 32 \times 32$ cells on each CPU. Due to the explicit time discretization the governing restriction for the time steps is the CFL-condition. The computational time is approximately 2.08 h with 6310 time steps and a cycle time of 1.22 s per time step for the computation of the initially cylindrical droplet. For the other computations the needed resources are very similar.

Additional simulations have been made to investigate the grid independency on a 256 × 128 × 128 grid and the behavior on a longer time scale (128 × 64 × 64 grid, 64 CPU's and 11 h computational time) but the results

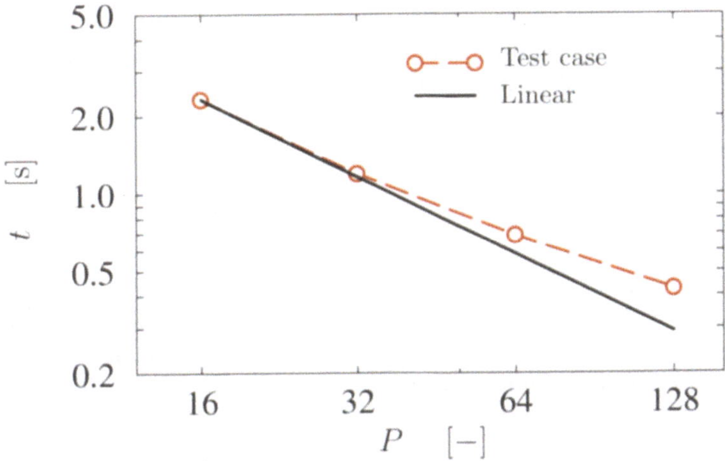

Fig. 7. Comparison of the computational time per cycle in dependency of different processor numbers P.

are not presented here, because they have been very similar to the ones carried out on the coarse grid.

In Fig. 7 the computational time per time step (without the setup time) in dependency of the number of CPU's P is displayed. As test case a heat transfer computation of a spherical droplet on a $128 \times 64 \times 64$ grid and 400 time steps is chosen. Additional a linear progression on the basis of the computational time needed with 16 CPU's is plotted in the diagram. The comparison between the linear progression and the needed computational time shows that the parallel efficiency of the program is very well.

5 Concluding Remarks

A 3D CFD program has been used to compute the heat transfer on initial deformed droplets at high Reynolds numbers. For the different deformation cycles – prolate/oblate droplet shape – a strong difference in the time dependent Nusselt number Nu_t has been found. The highest heat transfer has been found during the oscillation period for spherical to oblate droplet shape. For the prolate oscillation period no significant higher heat transfer due to the deformation has been recognized. Consequently for a modeling of the heat transfer of deformed droplet not the absolute deformation but the deformation period is the governing parameter.

The mean Nusselt number Nu_m for the deformed droplets shows that the absolute increase in the Nusselt number for the mentioned properties are not very high. A wider range of Reynolds numbers and Weber numbers will be investigated in the future.

Acknowledgments

The authors would like to thank the *Deutsche Forschungsgemeinschaft* (DFG) for the financial support of this project and the *High Performance Computing Center Stuttgart* (HLRS) for support and computational time on the high performance computers.

References

1. M. Rieber and A. Frohn. A numerical study on the mechanism of splashing. *International Journal of Heat and Fluid Flow*, Vol. 20, No. 5, pp. 455–461, 1999.
2. M. Hase and B. Weigand. Numerical study of the temperature field of unsteady moving droplets and of the surrounding gas. In *Proceedings ILASS-Europe 2001*, Zuerich, 2001.
3. C.W. Hirt and B.D. Nichols. Volume of fluid (VOF) method for the dynamics of free boundaries. *Journal of Computational Physics*, Vol. 39, pp. 201–225, 1981.
4. W.J. Rider and D.B. Kothe. Reconstructing volume tracking. *Journal of Computational Physics*, Vol. 141, pp. 112–152, 1998.
5. M. Hase, M. Rieber, F. Graf, N. Roth, and B. Weigand. Parallel computation of the time dependent velocity evolution for strongly deformed droplets. In *High-Performance Computing in Science and Engineering 2001: Transactions of the High Performance Computing Center Stuttgart (HLRS)*, pp. 342–351. Springer-Verlag, Februar 2002.
6. R. Scardovelli and S. Zaleski. Direct numerical simulation of free-surface and interfacial flow. *Annual Review Fluid Mechanics*, Vol. 31, pp. 567–603, 1999.
7. S.A. Kowalewski and D. Bruhn. Nonlinear oscillations of viscous droplets,. In *Proc. of Japanese-Centr. European Workshop on Adv. Comp. in Eng.*, pp. 63.68, 1994.
8. W.E. Ranz and W.R. Marshall. Evaporation from drops, Part II. *Chemical Engineering Progress*, Vol. 48, No. 4, pp. 173–180, 1952.
9. M. Hase and B. Weigand. Transient heat transfer of deforming droplets at high Reynolds numbers. *Int. Journal of Numerical Methods for Heat & Fluid Flow*, 2003. Accepted for publication.

3D Simulations of Supersonic Chemically Reacting Flows

Fernando Schneider, Peter Gerlinger, and Manfred Aigner

Institut für Verbrennungstechnik, DLR Stuttgart
Pfaffenwaldring. 38-40, 70569 Stuttgart, Germany

Summary. The mixing and combustion of hydrogen in a model scramjet (supersonic combustion ramjet) engine is investigated numerically. For an improved mixing a lobed strut injector is employed. It will be shown that due to the chosen shape of the injector strong streamwise vortices are induced which improve the mixing and therefore shorten the necessary combustor length. The hydrogen is injected with Mach 1.4 into a Mach 2 supersonic air flow. Computational grids with 0.43 and 3.2 million volumes have been used for these combustion simulations. Due to the complex physical phenomena in high speed flows extremely fine grids are required in the vicinity of walls. The use of finite-rate chemistry additionally causes long computational times. The chosen chemistry reaction mechanism employs 20 reactions and 9 different species. Thus efficient numerical solvers are required as well as facilities that allow high performance computing. The numerical code is parallelized by domain decomposition using MPI and the simulations shown are performed on a Cray T3E using up to 208 nodes.

1 Introduction

The understanding and modelling of supersonic combustion is an important step for the developement of future air breathing space transportation systems. Due to the difficulties in realizing scramjet combustor conditions in ground test facilities numerical simulations are of great importance. In contrast to rocket propulsion systems scramjets utilize the oxygen of the atmosphere for combustion and therefore may carry more pay load. By keeping the flow supersonic even in the combustion chamber losses in total pressure usually caused by normal shocks are avoided. On the other hand due to the high flow velocities the available time for achieving a perfect mixing between fuel (mostly hydrogen) and air is extremely short. Techniques for mixing enhancement in supersonic flow are required. A large number of different injector concepts have been investigated during the last decades [1, 2, 3, 4]. In this paper a lobed strut injector is investigated [5, 6] that allows the hydrogen to be injected directly into the core region of the combustion chamber. Additionally streamwise vortices are induced by the chosed strut geometry. The design of the fuel injector and the combustion chamber geometries significantly depends on accurate multidimensional prediction capabilities.

2 Numerical Method

The fluid mechanical and chemical processes in supersonic combustors are characterized by very small time scales. Thus combustion has to be described by finite-rate chemistry. The consequence is a set of governing equations which is numerically stiff. Usually explicit numerical solvers are not able to deal with this stiffness. For the present investigation an implicit LU-SGS (Lower-Upper Symmetric Gauss-Seidel) method [7] is used for time integration which solves the chemistry fully coupled with the fluid motion [8, 9]. Moreover low-Reynolds-number turbulence closures are needed for an accurate simulation of the turbulent near wall behaviour. This type of turbulence modeling requires an accurate resolution of all near wall layers and therefore extremely fine meshes. In addition the computational structured grid is refined in mixing and combustion zones. Consequently the total number of grid nodes even for simple combustor geometries exceeds one million in most cases. For any grid point conservation equations for mass, momentum, energy, turbulence and species masses have to be solved.

2.1 Governing Equations

The numerical simulation of high-speed reactive flows is based on the solution of the unsteady compressible Reynolds-averaged Navier-Stokes equations and a detailed chemical reaction mechanism. These equations are discretized in physical space using a cell centred finite-volume method. The governing equations for the vector of conservatives variables $\mathbf{Q_c} = [\rho,\ \rho u,\ \rho v,\ \rho w,\ \rho E,\ \rho q,\ \rho\omega,\ \rho Y_i]^T$ are given by

$$\frac{\partial \mathbf{Q_c}}{\partial t} = -\frac{\partial \mathbf{E}}{\partial x} - \frac{\partial \mathbf{F}}{\partial y} - \frac{\partial \mathbf{G}}{\partial z} + \mathbf{S}\ . \tag{1}$$

The variables in eq. (1) are density $\bar{\rho}$, velocity components $\widetilde{u}$, $\widetilde{v}$ and $\widetilde{w}$, the total specific energy $\widetilde{E}$, the turbulence variables $q = \sqrt{k}$ (k = turbulent kinetic energy) and $\omega = \epsilon/k$ (ϵ = dissipation rate of k) and the species mass fractions, $\widetilde{Y}_i$. N_k is the number of different species. The chemical reaction scheme involves 9 species (N_2, O_2, H_2, H_2O, OH, O, H, HO_2, and H_2O_2) and 20 reactions [10]. The fluxes are divided into inviscid (index c) and viscous (index ν) parts $\mathbf{E} = \mathbf{E}_c - \mathbf{E}_\nu$, $\mathbf{F} = \mathbf{F}_c - \mathbf{F}_\nu$, and $\mathbf{G} = \mathbf{G}_c - \mathbf{G}_\nu$. The source vector $\mathbf{S}$ is given as $\mathbf{S} = [0, 0, 0, 0, 0, S_q, S_\omega, S_i]^T$ where S_q and S_ω are source terms of the $q - \omega$ turbulence closure [11] and S_i are species source due to chemistry.

2.2 Time Integration

For high speed flows compressible solvers are required. The disadvantage of such schemes is that they usually fail for simulations where significant parts of the field are in the low-Mach number limit. All-Mach number preconditioning techniques may be used to circumvent this problem. Using such a technique the compressible flow solver can be used over the total Mach number range [12]. Equation (1) is preconditioned by a joint preconditioning and transformation matrix

$$\mathbf{P}\frac{\partial \mathbf{Q_c}}{\partial \mathbf{Q_v}}\frac{\partial \mathbf{Q_v}}{\partial t} := \mathbf{PT}\frac{\partial \mathbf{Q_v}}{\partial t} := \mathbf{\Gamma}\frac{\partial \mathbf{Q_v}}{\partial t} = -\frac{\partial \mathbf{E}}{\partial x} - \frac{\partial \mathbf{F}}{\partial y} - \frac{\partial \mathbf{G}}{\partial z} + \mathbf{S} \tag{2}$$

with $\mathbf{\Gamma} = \mathbf{PT}$ and $\mathbf{Q_v} = [\, p, u, v, w, E, q, \omega, Y_i \,]^T$ being the vector of primitive variables. The transfer from conservative to primitive variables allows the implementation of an artificial compressibility term in the partial derivative of density with respect to pressure. For steady flows the time derivative $\frac{\partial \mathbf{Q_v}}{\partial t}$ disappears. An artificial time step is introduced and the dual-time-stepping scheme

$$\mathbf{\Gamma}\frac{\partial \mathbf{Q_v}}{\partial \tau} + \frac{\partial \mathbf{Q_c}}{\partial t} = -\frac{\partial \mathbf{E}}{\partial x} - \frac{\partial \mathbf{F}}{\partial y} - \frac{\partial \mathbf{G}}{\partial z} + \mathbf{S} \tag{3}$$

now allows time accurate simulation while using the preconditioning technique [13, 6]. Using an implicit discretization in artificial time τ and in physical time t

$$\mathbf{\Gamma}\frac{\mathbf{Q_v}^{p+1} - \mathbf{Q_v}^{p}}{\Delta\tau} + \frac{a_1\mathbf{Q_c}^{p+1} + a_2\mathbf{Q_c}^{n} + a_3\mathbf{Q_c}^{n-1}}{\Delta t} = -\frac{\partial \mathbf{E}^{p+1}}{\partial x} - \frac{\partial \mathbf{F}^{p+1}}{\partial y} - \frac{\partial \mathbf{G}^{p+1}}{\partial z} + \mathbf{S}^{p+1} \tag{4}$$

is obtained. The non-linear problem in artificial time is solved by an inner loop for any physical time step (n is the index for a physical time step, p for an inner iteration). A convergent solution of the inner iteration is obtained for $\mathbf{Q_c^{n+1}} = \lim_{p\to\infty} \mathbf{Q_c^{p+1}}$. After spatial discretization and linearization in time the following set of equations has to be solved at any inner iteration step

$$\underbrace{\left[\frac{\mathbf{\Gamma}}{\Delta\tau} + \frac{a_1}{\Delta t}\mathbf{T} + \mathbf{J}_E + \mathbf{J}_F + \mathbf{J}_G + \mathbf{J}_S\right]}_{\mathbf{L+D+U}} \Delta\mathbf{Q_v} = \underbrace{-\left[\frac{a_1\mathbf{Q_c}^{p} + a_2\mathbf{Q_c}^{n} + a_3\mathbf{Q_c}^{n-1}}{\Delta t} + \frac{\partial \mathbf{E}^{p}}{\partial x} + \frac{\partial \mathbf{F}^{p}}{\partial y} + \frac{\partial \mathbf{G}^{p}}{\partial z} - \mathbf{S}^{p}\right]}_{\mathbf{R}} \ . \tag{5}$$

As to allow a computational efficient solution the operator on the left hand side is decomposed into a diagonal operator **D**, a lower triangular operator **L**, and an upper triangular operator **U** [14]. These operators include Jacobians

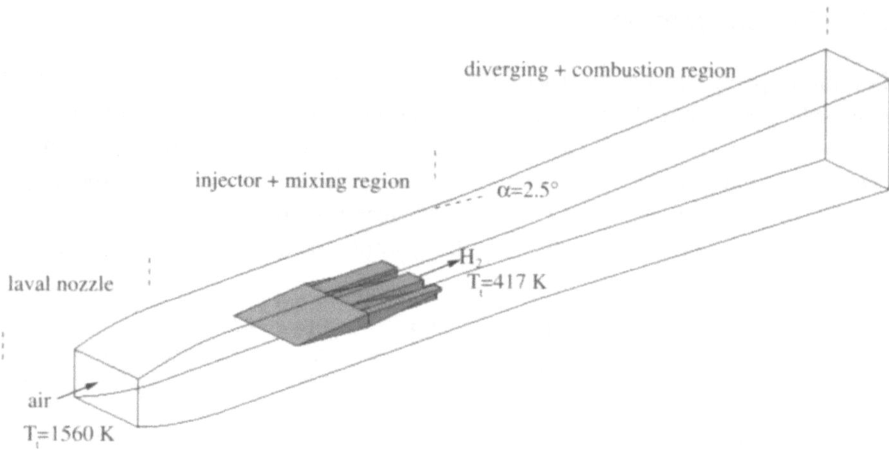

Fig. 1. Schematic representation of the combustion chamber.

of inviscid and viscous fluxes $\mathbf{J}_E, \mathbf{J}_F, \mathbf{J}_G$ and of the source term $\mathbf{J}_S$. The right hand side operator $\mathbf{R}$ contains all explicit terms and represents the residual vector of the implicit scheme. The left hand side operator is approximately factorized by a LU-decomposition and the system is solved by a symmetric Gauss-Seidel iteration [15, 7]. $\Delta\mathbf{Q_v}$ is the update for the solution $\mathbf{Q_c^{n+1}} = \mathbf{Q_c^n} + \mathbf{T}\Delta\mathbf{Q_v}$ during one time step Δt. For convergence acceleration multigrid techniques may be used [8, 9].

3 Supersonic Mixing and Combustion

The objective of this work is the development of a scramjet combustor geometry which allows a stable combustion without artificial flame stabilization. The inflow conditions as well as mass flow rates are defined by the capabilities of the ITLR (Institute of thermodynamics of aerospace engineering) experimental supersonic combustion test facility at the University of Stuttgart. The aim of these investigations is to achieve short combustor lengths while keeping the flow supersonic on its way through the combustor. As to avoid thermal choking expansion is necessary to compansate effects from heat release due to combustion.

In the present case a simple combustor geometry is chosen with a constant angle of expansion downstream of the injector. Figure 1 shows a sketch of the combustor under investigation. Hydrogen is injected through the lobed end of the strut positioned at the channels symmetry axis. Due to the occurance and interaction of shock waves the flowfield is quite complex. The duct height in front of the strut is 36 mm, the expansion angle downstream of the injector 2.5°, and the combustor width is 40 mm, respectively. Expansion starts 48 mm downstream of the injector end. The definition of the expansion angle is

Table 1. Inflow conditions and injector design (sizes in mm)

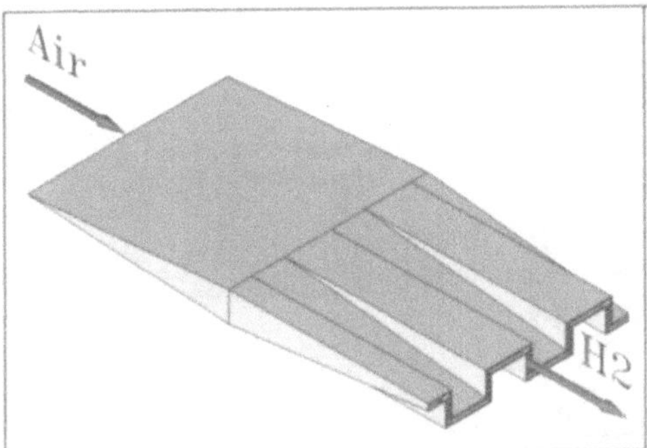

inflow conditions		
	air	hydrogen
static pressure (Pa)	211312	80000
stat. temp. (K)	1300	300
velocity (m/s)	722.72	1850
Mach number	1.0	1.4
H_2 mass fraction	0.0	1.
O_2 mass fraction	0.23	0.0
N_2 mass fraction	0.77	0.0
mass flow (g/s)	327.8	2.87
injection area 0.6 x 8 x 5 (mm^2)		

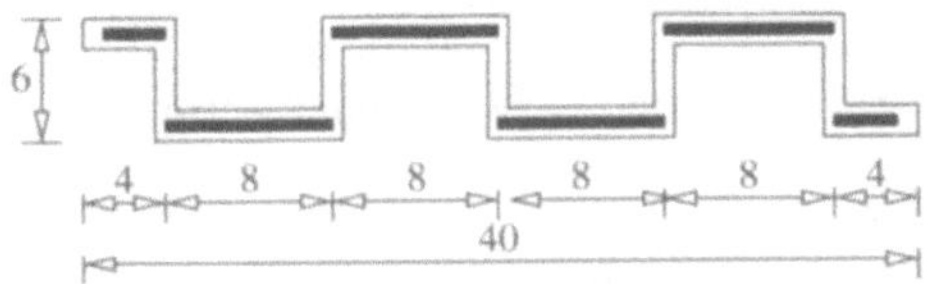

a critical point in achieving a stable combustion. The simulation starts at the nozzle throat for the main air flow. The total temperature of the air is 1560 K and 417 K for the hydrogen, respectively. Because the hydrogen is used as a coolant for the strut a rise total temperature is assumed in comparison to the ambient value in case without combustion. The inflow conditions for both air and hydrogen are summarized in Table 1. Inside the lobed strut injector the hydrogen is accelerated by a nozzle up to its exit Mach number of 1.4.

3.1 Injector Design

As to achieve a rapid mixing of hydrogen and fuel shock waves as well as streamwise vortices may be used. With hydrogen wall injections good mixing may be achieved but usually at the cost of large losses in total pressure and a significant blockage of the flow. Moreover, in case of wall injections it may become difficult to keep the flame away from the wall. Strut injectors offer advantages concerning the last point. In Ref. [16] a number of planar strut geometries is investigated numerically for nonreactive flows. A lobed strut injector which is the basis for the combustions investigations of the present paper has been already investigated numerically for a non reactive flow in Ref. [5, 6]. The cited references also show a comparison with experimental data. Now simulations are performed that include combustion. The shape of double-wedge strut with a lobed end is shown in the figures of Table 1. The geometry is chosen in such a way, that a relatively simple manufacturing and an efficient cooling is possible. One of major disadvantages of planar injectors [16] is the bad mixing in compressible high speed flows. The geometry of the

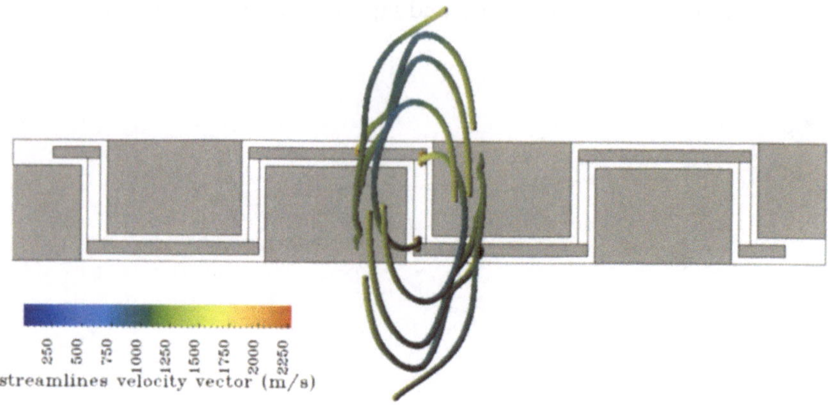

Fig. 2. Streamlines starting from the hydrogen injection zone (view from the back).

lobed strut injector is designed in such a way that strong streamwise vortices are induced which accelerate the hydrogen air mixing.

3.2 Results and Discussion

Figure 2 shows streamlines starting at chosen positions at the hydrogen injection area of the strut. The colours correspond to the magnitude of velocity. The view is from the exit of the channel in upstream direction. From this view it becomes clear that vortices are induced which are maintained up to the channel end. A shorter section of streamlines directly downstream of the strut is given in Fig. 3. It is shown in Ref. [5] that the mixing efficiency is considerably improved by this technique without causing large losses in total pressure. After the advantages of the chosen strut geometry are demonstrated, some results will be given for the channel geometry described at the beginning

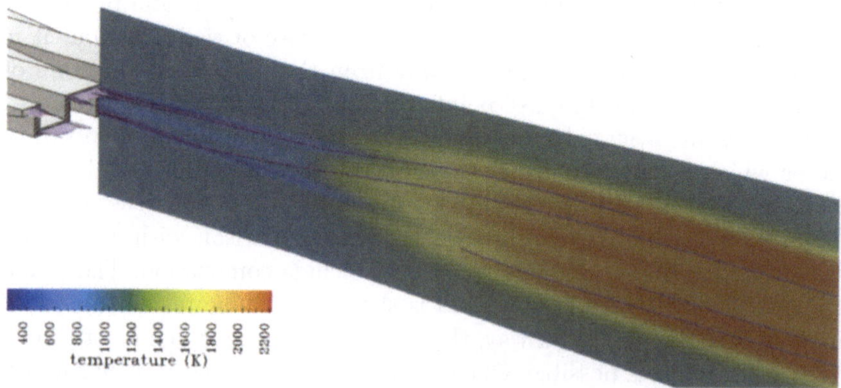

Fig. 3. Streamlines and temperature distribution.

Table 2. CPU time and memory requirements for coarse and fine grid scramjet combustion simulations

grid	processors	multigrid	time (s/iteration)	memory (GByte)
coarse	52	no	13.1	1.04
	104	no	6.88	1.8
	104	yes	39.6	7.36
	208	yes	29.1	9.63
fine	104	no	35.2	8.34

of this section. For the chosen inflow conditions a lifted flame is obtained. The ignition delay is a result of the fast flow velocity in relation to the time needed for mixing and combustion to be initiated. Figure 3 shows the temperature distribution in the central plane of the channel. Ignition takes place about 6 cm downstream of the injector what may be observed by the strong increase in temperature. Due to the three dimensional distribution there is no common point of ignition along the channel depth. To provide a more complete view of the reacting flowfield the distributions of H_2, H_2O and OH-mass fracions are plotted in Fig. 4 (from top to bottom). As may be observed from the hydrogen distribution combustion is not fully completed at the end of the computed channel section. The colour distribution for hydrogen is chosen in such a way that a remaining mass fraction of 0.05 is easy to indicate. Due to the diverging channel geometry the flow is accelerated and temperature is decreasing. If this decrease is not compensated by heat release from combustion the flame may be extinguished. Therefore the choice of an appropiate channel geometry is a crucial factor especially if varying flow conditions are taken into accourt. Numerical simulations are an important tool in achieving optimized scramjet combustor geometries.

4 Numerical and Computational Aspects

The simulation of large problems in CFD is usually performed on parallel systems. In this case a great number of CPUs may be use at the same time. This requires that it is possible to split the discretized problem into a large number of smaller ones to be solved on different nodes. In the present case this is done by domain decomposition. Because any block has the same number of volumes a good and constant load balancing is obtained. A detailed study of the influence of the number of domains on the efficiency of the implicit LU-SGS solver is given in Ref. [17]. The numerical code is written in C++ and is optimized for massively parallel architectures with distributed memory using message-passing (MPI) as communication interface. A semi-coarsening multigrid technique is implemented for convergence acceleration. A detailed overview concerning solver and parallelization can be found in [6] and [17]. The results shown in Sect. 3 are based on numerical simulations using 0.43

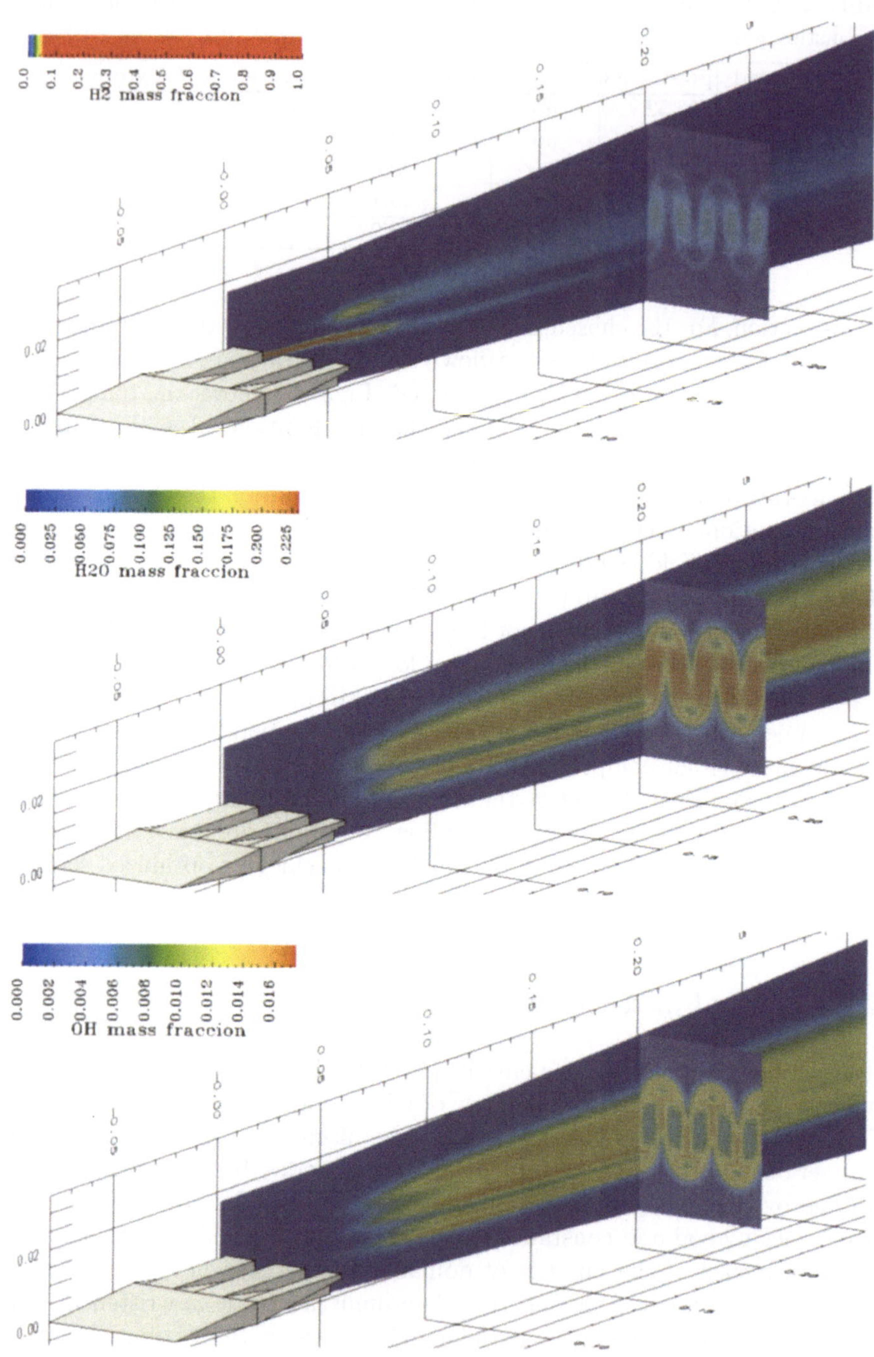

Fig. 4. H_2 (top), H_2O (middle) and OH (bottom) mass fraction distribution.

million grid points. To check grid dependencies simulation using 3.4 million grid points are performed too. Thus the total number of unknows add up to 8 and 48 million, respectively. In both cases the computational grid is divided into 416 blocks. The number of volumes per block is $32 \times 16 \times 16$ for the fine grid and $16 \times 8 \times 8$ for the coarse grid. Depending on the initial conditions 5000 to 6000 iterations were needed for a converged steady solution on the coarse grid, which corresponds to about 900 to 1200 hours in total CPU-time on the CRAY T3E-900 at the HLRS. A multigrid simulation with the finer grid was found to be impossible due to memory limitations. The simulation of finite-rate chemistry requires implicit numerical solvers to deal with stiffness problems. Thus source Jacobians are needed which increase the demand in memory in contrast to explicit solvers. For multigrid simulations with semi-coarsening the memory demand theoretically increases by factor of 8. In the present case a simplified strategy is used which requires for 3D simulations about 4 times more memory than the one grid solution. This was too much in case of the mentioned combustion simulation. The computation times for the scramjet combustion test case as well as memory requirements are given in Table 2.

5 Conclusions

A computational study of a the mixing and combustion process in a model scramjet combustor is given. It is shown that the use of a lobed strut injector achieves a good mixing by inducing strong streamwise vortices. Especially the design of the combustor geometry is a critical point because there is only a limited region of stability in the given high speed flow. Numerical simulations are an important tool in this field even if the CPU and memory requirements for such simulations are very high.

6 Acknowledgments

We wish to thank the Deutsche Forschungsgemeinschaft (DFG) for financial support of this work within the Collaborative Research Center SFB 259 at the University of Stuttgart and the High Performace Computing Center Stuttgart for the generous granting of their computational facilities.

References

1. Sunami, T., Wendt, Michael N., Nishioka, M.: Supersonic Mixing and Combustion Control Using Streamwise Vortices. AIAA paper 98-3271, (1998)
2. Sunami, T. and Scheel, F.: Analysis of Mixing Enhancement Using Streamwise Vortices in a Supersonic Combustor by Application of Laser Diagnostics. AIAA paper 2002-203, (2002)

3. Eklund, D. R., Stouffer, S. D.:A Numerical and Experimental Study of a Supersonic Combustor Employing Swept Ramp Fuel Injectors. AIAA paper 94-2819, (1994)
4. Baurle, R. A., Mathur, T., Gruber, M. R., Jackson, K. R.: A Numerical and Experimental Investigation of a Scramjet Combustor for Hypersonic Missile Applications. AIAA paper 98-3121, (1994)
5. Gerlinger, P., Kasal, P., and Stoll, P.: Experimental and Theoretical Investigation on 2D and 3D Parallel Hydrogen/Air Mixing in a Supersonic Flow. ISABE paper 2001-1019 (2001)
6. Stoll, P.: Entwicklung eines parallelen Mehrgitterverfahrens zur Simulation der Verbrennung in kompressiblen und inkompressiblen Strömungen. VDI Forts.-Ber., Nr. 411, Reihe 7, Strömungstechnik (2001)
7. Jameson, A., Yoon, S.: An LU-SSOR Scheme for the Euler and Navier-Stokes Equations. AIAA paper 87-0600 (1987)
8. Gerlinger, P., Stoll, P., Brüggemann, D.: An Implicit Multigrid Method for the Simulation of Chemically Reacting Flows. J. Comp. Phys. **146** (1998) 322–345
9. Gerlinger, P., Möbus, H., Brüggemann, D.: An Implicit Multigrid Method for Turbulent Combustion. J. Comp. Phys. **167** (2001) 247–276
10. Jachimowski, C. J.: An Analytical Study of the Hydrogen-Air Reaction Mechanism with Application to Scramjet Combustion. NASA-TP-2791, (1988)
11. Coakley, T. J. and Huang, P. G.: Turbulence Modeling for High Speed Flows. AIAA-Paper 98-0436, (1998)
12. Withington, J. P., Shuen, J. S. and, Yang, V.: A Time Accurate Implicit Method for Chemically Reacting Flows at all Mach Number. AIAA-Paper 91-0581, (1991)
13. Hosangadi, A., Merkle, C. L. und Turns, S. R.: Analysis of Forced Combusting Jets. AIAA Journal, **28** (1990) 1473-1480,
14. Shuen, J. S.: Upwind Differencing and LU Factorization for Chemical Non-Equilibrium Navier-Stokes Equations. J. Comp. Phys. **99** (1992) 233–250
15. Jameson, A., Yoon, S.: Lower-Upper Implicit Schemes with Multiple Grids for the Euler Equations. AIAA J. **25**, 929–937 (1987)
16. Gerlinger, P., Brüggemann, D.: Numerical Investigation of Hydrogen Strut Injections into Supersonic Air Flows, J. of Prop. and Power, **16** (2000) 22–28
17. Gerlinger, P., Schneider, F., Aigner, M.: Implicit LU Time Integration Using Domain Decomposition and Overlapping Grids. High Performance Computing in Science and Engineering'02, 311-322 Springer-Verlag, 2002

Numerical Investigation of Semi-Turbulent Pipe Flow

Emad Khalifa[1] and Eckart Laurien[2]

[1] Institute for Nuclear Technology and Energy Systems (IKE), University of Stuttgart, Pfaffenwaldring 31, D-70550 Stuttgart, Germany; khalifa@ike.uni-stuttgart.de

[2] Institute for Nuclear Technology and Energy Systems (IKE), University of Stuttgart, Pfaffenwaldring 31, D-70550 Stuttgart, Germany; laurien@ike.uni-stuttgart.de

Abstract

Surface deformations of a jet ejected from a straight-pipe atomizer may be due to the turbulent fluctuations in the interior of the pipe. At relatively low Reynolds number (Re) near the transition from laminar to turbulent pipe flow (e.g. Re = 3000 based on the mean velocity and the pipe diameter), an unsteady semi-turbulent state exists which differs from the better known fully developed flow at higher Re. In the present work low- and high-Re jets are investigated experimentally by the DLR group. To identify the influence of the inner injector flow condition to the jet surface phenomena a special injector set-up has been designed. With this setup it was possible to eliminate relative velocity effects between jet and ambient fluid. Using shadowgraphy and a novel image processing approach, wavelengths, amplitudes and undisturbed jet length could be determined. The corresponding pipe flow is simulated numerically using Direct Numerical Simulation (DNS) by the University group. A new second order finite difference scheme in space and time for the incompressible Navier-Stokes equations in cylindrical coordinates is applied. The slope of the axial mean velocity profile near the wall is smaller than that for higher Reynolds number. The turbulent intensities are smaller than those at higher Re. The observed jet surface waves agree well with the computed lengths scales of the turbulent structures within the injector. Varicose instability is observed. Atomization is affected when Re is reduced to the semi-turbulent state(Re=3000), due to the thicker laminar envelope.

1 Introduction

Injection is affected by many different phenomena. This study takes a closer look at the effects of turbulence and on the deformation of the jet surface. Using a liquid ethanol and gaseous nitrogen combination makes it possible to see the effects on the surface of the jet from shadowgraph images [1].

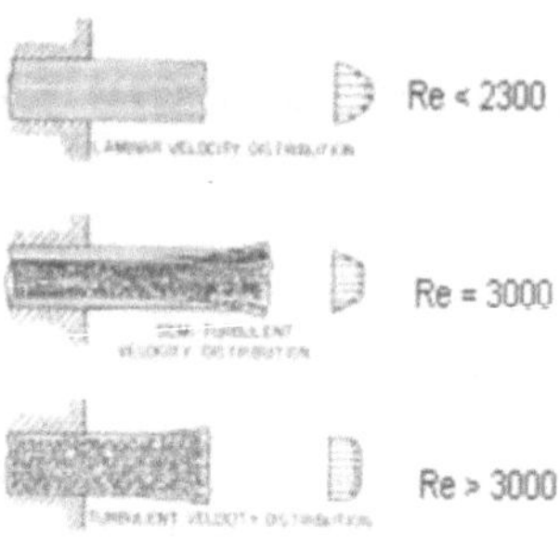

Fig. 1. Various velocity distributions in jets presented in [2]

As sketched in Figure 1, the jet emerges from the nozzle in either a laminar, a semi-turbulent or fully turbulent state [2]. If the flow at the orifice is fully turbulent, the jet will disintegrate under the influence of its own turbulence. A semi-turbulent flow comprises a turbulent core and a laminar envelope. In this case, jet disintegration does not occur close to the orifice exit. However, further downstream, the faster turbulent core outpaces its protective laminar layer and then disintegrates in the normal manner of a turbulent jet.

The flow visualization picture in Figure 2 shows the jet breakup of an ethanol jet at Re $= 3000$, using a straight pipe atomizer. At this low Reynolds number the surface deformations appear only far downstream of the nozzle. This advances the concept that turbulence within the liquid jet leads to the observed instability and breakup, but has to penetrate through the laminar envelope

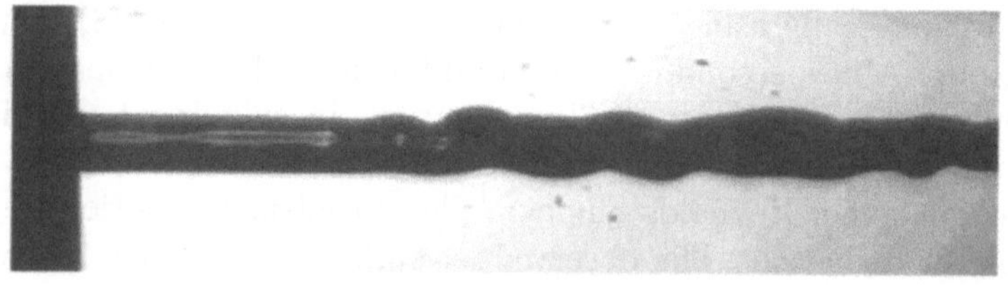

Fig. 2. Visualizations of surface deformation of a free jet at Re=3000 [3]

2 Experimental Setup

Figure 3 shows the pressurized chamber with the injector used in the experiments. The inner diameter of the ethanol injector is 2.2 mm, the outer diameter is 2.5 mm surrounded by the nitrogen coaxial injector with a large gap of 10 mm diameter to achieve constant relative velocities along the liquid jet. The length-to-diameter ratio of the ethanol injector tube is greater than 40 ensuring fully developed pipe flow. Optical access to the chamber is provided by three windows. The figure also shows a representative shadowgraph of the flow and the characteristic wavelength (λ) and amplitude (A). The wavelength is the length of the surface structures along the axial direction (z) and the amplitude is the extension in the radial direction (r) (figure 3).

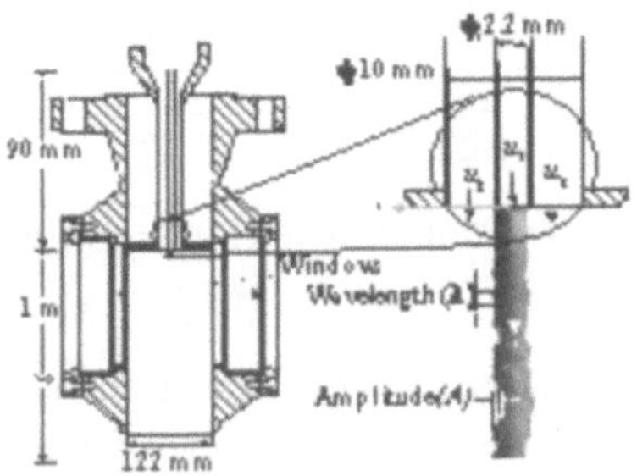

Fig. 3. Test chamber, injector tip design, jet surface characteristics

The tests were performed at a chamber pressure of 1.0 MPa. In order to achieve a wide range of Reynolds numbers, the jet injection velocities were 2.2, 2.4, 2.6, 3.8, 5.0, 5.4 and 9.0 m/s. The Reynolds number is calculated using the dense core flow values for this presentation. The wavelength is calculated as the axial length of a surface structure on the jet and the amplitude is the radial extension of this structure (figure 3). The undisturbed jet length is determined as the smallest axial value, where a wave occurred.

3 Experimental Results

The results of earlier studies at DLR [3] are confirmed by this work. With a constant chamber pressure p = 1 MPa the trend for the wavelength is obvious. The magnitude of wavelength decreases with increasing injection velocity and therefore with the turbulent energy contained in the jet (figure 4 and figure 5). At the higher velocity, the turbulent kinetic energy is creating smaller surface structures and eventually droplets.

In contrast, the behaviour at Re = 3000 (ul=2,6 m/s) is totally different. At the beginning of wave evolution we can recognize wavelengths of the dimension

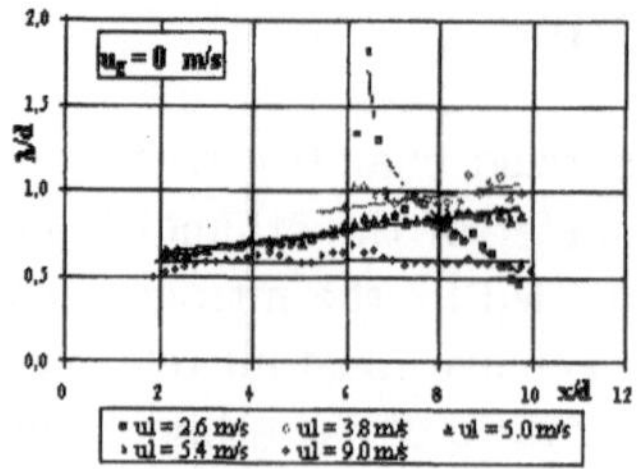

Fig. 4. Wavelength (λ), p = 1MPa

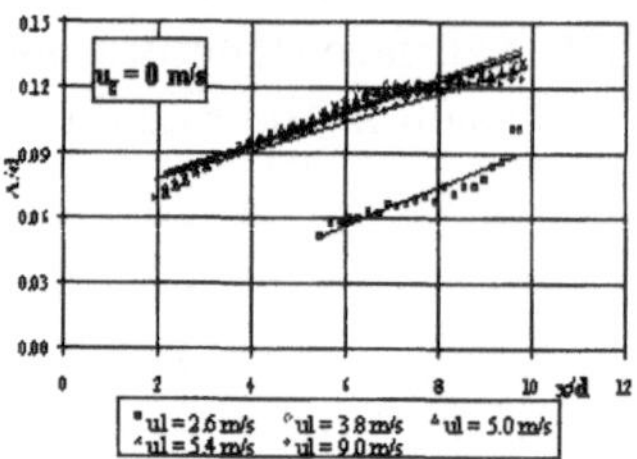

Fig. 5. Amplitude (A), p = 1MPa

nearly twice the injector diameter, also the amplitude is very small (figure 4 and figure 5). Further downstream the wavelength decreases sharply to a dimension similar to the fully turbulent jets at higher Reynolds numbers with an increasing amplitude, which is still smaller than the ones at higher Reynolds numbers (figure 4 and figure 5). The same trend can be seen with coaxial flow and no relative velocity. However the wavelength of a semi-turbulent jet (e.g. Re = 3000) is decreasing more steadily and the amplitude is nearly constant in the axial direction. These characteristics of the surface deformation of the liquid jet can be recognized by comparing the images from figure 6 till figure figure 9.

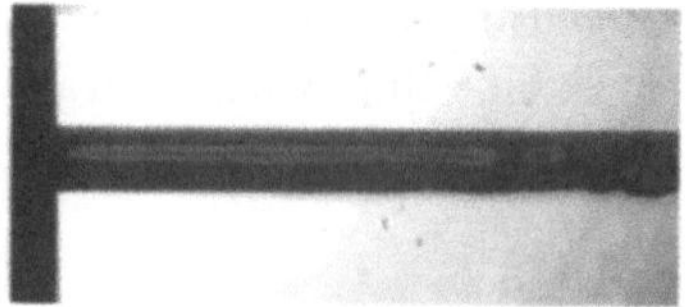

Fig. 6. u_l = 2.6 m/s, u_g = 2.6 m/s (Re=3000)

The undisturbed jet length does not vary significantly with injection velocity or relative velocity (e.g. Weber number) at higher Reynolds number. At lower Re the length varies considerably more. At higher relative velocity (e.g. higher Weber number) the undisturbed jet length increases. This suggests the turbulence level is much lower at low Re and has less influence on the surface

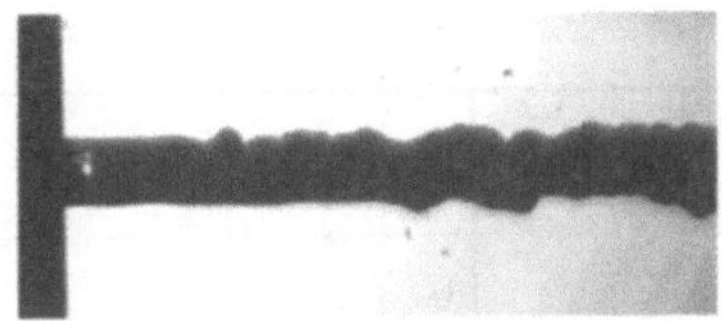

Fig. 7. $u_l = 3.8$ m/s, $u_g = 3.8$ m/s

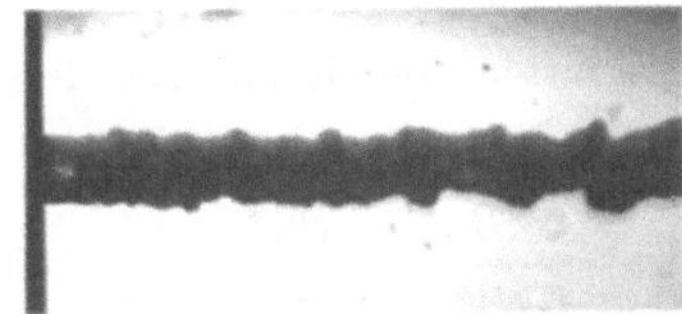

Fig. 8. $u_l = 5.0$ m/s, $u_g = 5.0$ m/s

Fig. 9. $u_l = 9.0m/s, u_g = 9.0m/s$

disturbances, with a stabilising effect of the aerodynamic forces. Additionally the undisturbed jet length is three to four times longer than at higher Re, due to the thicker viscous sublayer at lower Re.

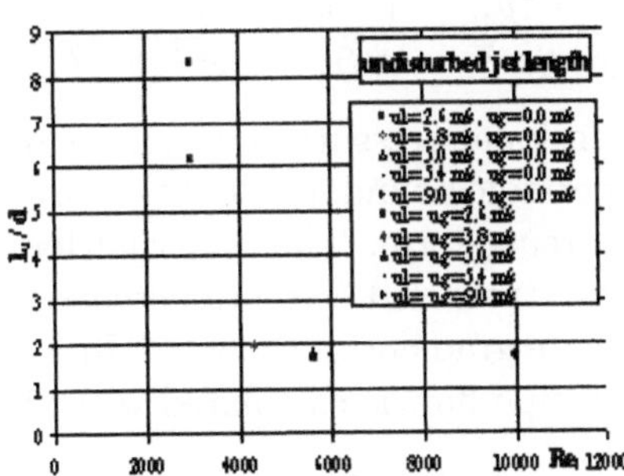

Fig. 10. undisturbed jet length

4 Governing equations

Figure 11 shows the integration domain and the coordinate system. The non-dimensional incompressible Navier-Stokes equations in cylindrical coordinates read [4] :

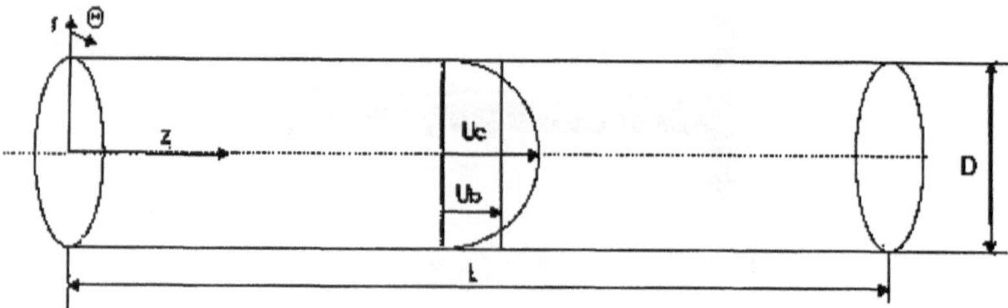

Fig. 11. Integration domain and coordinate system

$$\frac{\partial u_r^*}{\partial r^*} + \frac{1}{r^*}\frac{\partial u_r^*}{\partial \Theta^*} + \frac{\partial u_z^*}{\partial z^*} + \frac{u_r^*}{r^*} = 0$$

$$\frac{\partial u_r^*}{\partial t^*} + u_r^*\frac{\partial u_r^*}{\partial r^*} + \frac{u_\Theta^*}{r^*}\frac{\partial u_r^*}{\partial \Theta^*} + u_z^*\frac{\partial u_r^*}{\partial z^*} - \frac{{u_\Theta^*}^2}{r^*} =$$
$$-\frac{\partial p^*}{\partial r^*} + \frac{1}{Re}\left(\frac{\partial^2 u_r^*}{\partial r^{*2}} + \frac{1}{r^*}\frac{\partial u_r^*}{\partial r^*} - \frac{u_r^*}{r^{*2}} + \frac{1}{r^{*2}}\frac{\partial^2 u_r^*}{\partial \Theta^{*2}} + \frac{\partial^2 u_r^*}{\partial z^{*2}} - \frac{2}{r^{*2}}\frac{\partial u_\Theta^*}{\partial \Theta^*}\right)$$

$$\frac{\partial u_\Theta^*}{\partial t^*} + u_r^*\frac{\partial u_\Theta^*}{\partial r^*} + \frac{u_\Theta^*}{r^*}\frac{\partial u_\Theta^*}{\partial \theta^*} + u_z^*\frac{\partial u_\Theta^*}{\partial z^*} + \frac{u_r^* u_\Theta^*}{r^*} =$$
$$-\frac{1}{r^*}\frac{\partial p^*}{\partial \Theta^*} + \frac{1}{Re}\left(\frac{\partial^2 u_\Theta^*}{\partial r^* 2} + \frac{1}{r^*}\frac{\partial u_\Theta^*}{\partial r^*} - \frac{u_\Theta^*}{r^{*2}} + \frac{1}{r^{*2}}\frac{\partial^2 u_\Theta^*}{\partial \Theta^{*2}} + \frac{\partial^2 u_\Theta^*}{\partial z^{*2}} + \frac{2}{r^{*2}}\frac{\partial u_r^*}{\partial \Theta^*}\right)$$

$$\frac{\partial u_z^*}{\partial t^*} + u_r^*\frac{\partial u_z^*}{\partial r^*} + \frac{u_\Theta^*}{r^*}\frac{\partial u_z^*}{\partial \theta^*} + u_z^*\frac{\partial u_z^*}{\partial z^*} =$$
$$-\frac{\partial p^*}{\partial z^*} + \frac{1}{Re}\left(\frac{\partial^2 u_z^*}{\partial r^{*2}} + \frac{1}{r^*}\frac{\partial u_z^*}{\partial r^*} + \frac{1}{r^{*2}}\frac{\partial^2 u_z^*}{\partial \Theta^{*2}} + \frac{\partial^2 u_z^*}{\partial z^{*2}}\right)$$

where p^* is the non-dimensional pressure and consists of a mean part $\bar{p}$ and a fluctuating part p'. The periodic boundary condition in axial direction enforces the use of an external forcing term. This external forcing term is required to balance the wall shear stress and is represented with the mean pressure gradient. The mean pressure gradient in the axial direction along the pipe for fully developed turbulent pipe flow is calculated with the formula:

$$\frac{\partial \bar{p}^*}{\partial z^*} = \frac{1}{2} \cdot \frac{0.3164}{\sqrt[4]{Re}} \tag{1}$$

t^* is the non-dimensional time. u_z^*, u_r^* and u_Θ^* are the non-dimensional velocity components in the axial, radial and circumferential directions; respectively. z^* , r^* and θ^* are the non-dimensional axial, radial and circumferential directions; respectively.

Reynolds number Re is defined with $Re = \frac{\rho \cdot u_b \cdot D}{\mu}$ where is the density, u_b is the reference velocity (mean(or bulk) velocity) and D is the atomizer diameter.

5 Numerical method

The Navier-Stokes equations in cylindrical coordinates are solved using a new second order finite difference scheme [5]. The temporal differencing is also second order using Crank-Nicolson for the pressure and the linear terms and Adams-Bashforth for the nonlinear terms. The scheme uses a non-staggered, structured grid. The linear equations are solved iteratively using the Preconditioned Biconjugate Gradient Method (PBCG) with left preconditioning[6]. The equation solver stores only non-zero diagonal elements. The solver references the sparse matrix through multiplication of it's transpose and a vector. The three dimensional grid locations are converted into one dimensional storage locations.

At the wall, no slip boundary conditions are imposed. To avoid the singularity at $r^* = 0$ a small "empty" core of radius is imposed, at which an axial "symmetry" boundary condition $\frac{\partial u_z^*}{\partial r^*} = 0$, $\frac{\partial u_\Theta^*}{\partial r^*} = 0$ and $u_r^* = 0$ is imposed. The influence of that artificial boundary is negligible due to the smooth behavior of the flow near the centreline. Periodic boundary conditions for velocity components and pressure are applied in the axial and circumferential directions. The pressure is calculated with extrapolation in the radial direction. The numerical resolution is 40x32x99 in the circumferential, radial and axial directions, respectively, with an equidistant grid. A turbulence field obtained from a high-Reynolds number simulation [7] is imposed at the beginning of the simulation. The final statistical data have been accumulated by spatial averaging in the homogeneous streamwise and circumferential directions and by time averaging over 6 data fields (one data field for every half turn-over period). A nondimensional time step widths of 0.005 is used.

6 Numerical results

Figure 12 shows the streamwise velocity $u_z(N_z, N_\Theta, N_r)$ at $u_z(49, 20, 16)$ recorded as a function of time iterations. In this signal fast' and slow' temporal variations appear which might be associated with small and large eddies respectively.

Figure 13 shows the axial mean-velocity profile normalized by the centreline axial velocity. The numerical result is in very good agreement with detailed turbulence measurements [9]. The slope of the curve near the wall is smaller than that for higher Reynolds number (Re = 5600).

Figure 14 shows the turbulent intensities (root mean square values) of the three fluctuation velocity components normalized with the centreline velocity. The turbulent fluctuations are concentrated further away from the wall, compared with the high Reynolds number (Re = 5600). This is in accordance with the thicker viscous sublayer or envelope of the semi-turbulent case. The fluctuation intensities shown in Figure 14 are nearly of the same order of magnitude in the three directions.

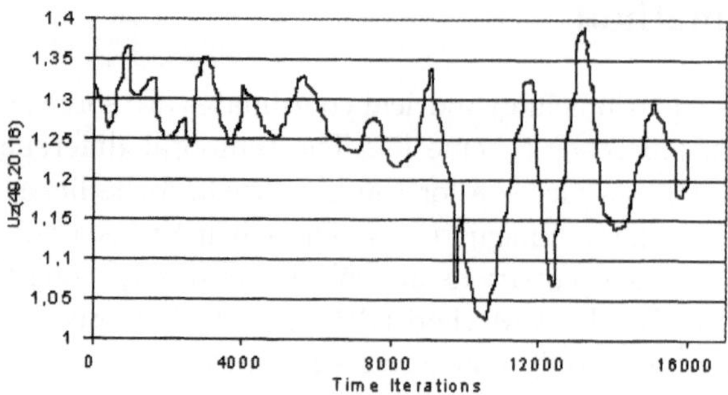

Fig. 12. Evoluation of the streamwise velocity as a function of time iterations

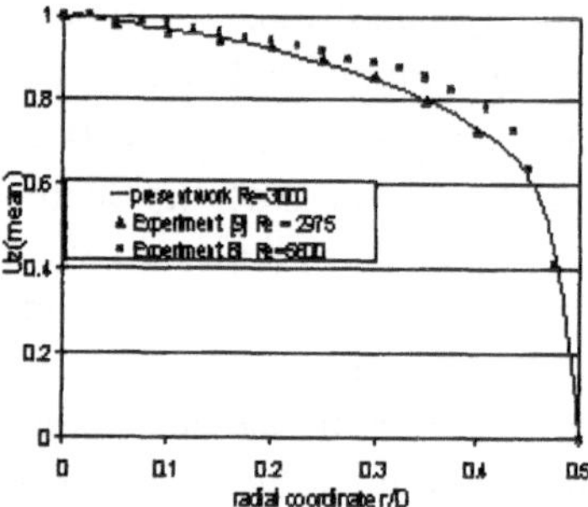

Fig. 13. Axial mean velocity normalized by the centreline velocity

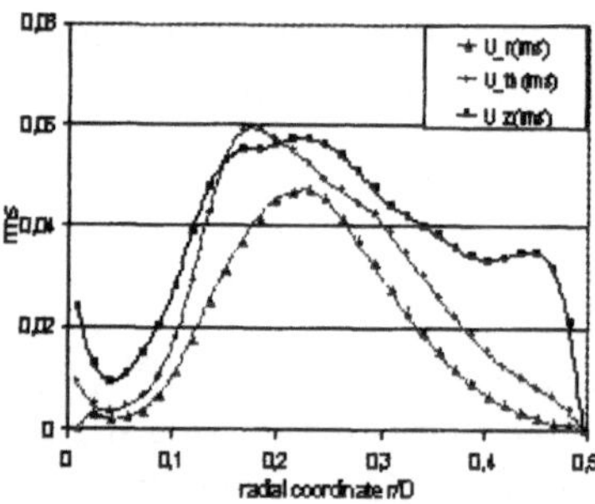

Fig. 14. Root-mean-square (rms) Velocity fluctiations normalized by the centerline velocity

Figure 15,16 show the isolines of the axial velocity fluctuation with respect to the mean velocity in an (z,Θ) plane at r/D=0.45 and 0.3 respectively. In figure 15 one can distinguish several localized regions of very high-speed fluid that form the high-speed streaks. The streaks and any definite organized structure are absent away from the wall as seen in figure 16.

The simulations at Re = 3000 show, that a turbulent core composed of a system of almost axisymmetric eddies exists. This turbulent core is surrounded

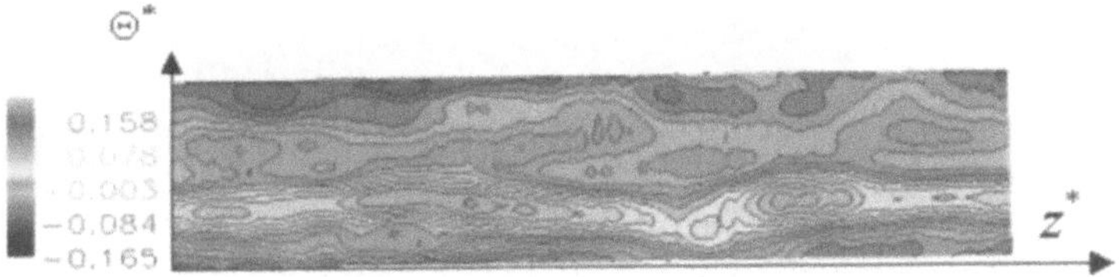

Fig. 15. Isolines of the axial velocity fluctuation ar r/D=0.45

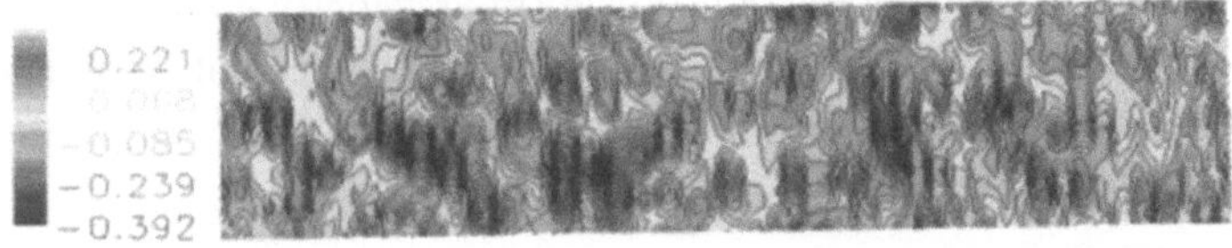

Fig. 16. Isolines of the axial velocity fluctuation ar r/D=0.3

by a laminar envelope as can be seen in Figure 17. The laminar envelope is much thicker than the viscous sublayer existing at higher Reynolds numbers. The sweep and ejection events are clearly illustrated in Figure 18 where the positive and negative radial velocity moves to and from the wall of the atomizer. The Figure 18 is analogous to the experimentally observed "varicose" instability of low-speed jets [10]. The size of the structures is bigger than that achieved at higher Reynolds number[11],[12],[13].

Fig. 17. Simulated turbulent structures of te streamwise axial velocity u_z

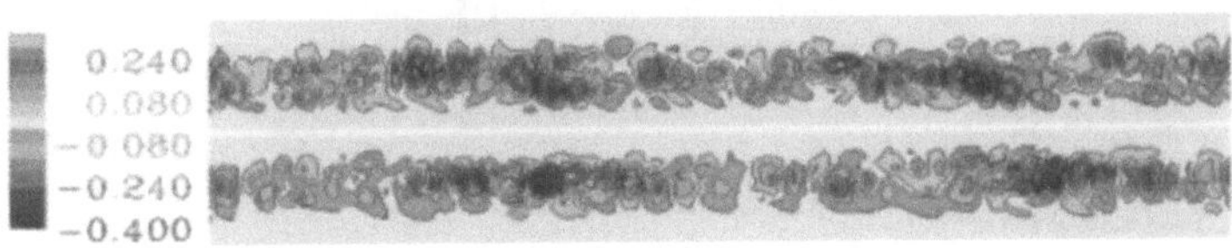

Fig. 18. Simulated turbulent structures of the radial velocity u_r

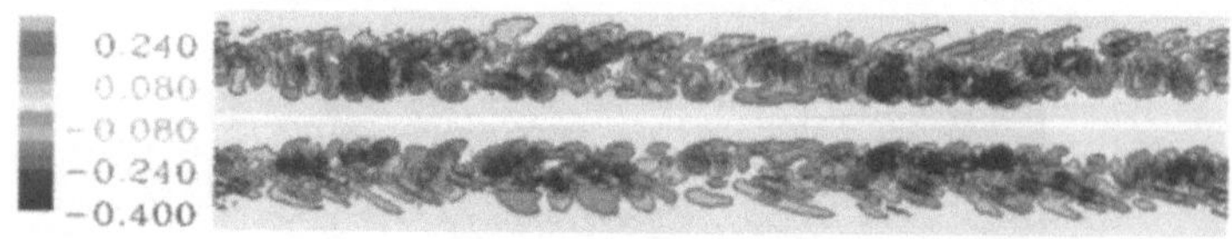

Fig. 19. Simulated turbulent structures of the circumferential velocity u_Θ

Figure 20 shows the vector plot of the three velocity components in a (z,r) plane showing that the turbulent motion are mainly concentrated at r/D=0.25 where the turbulent intensities are nearly maximum.

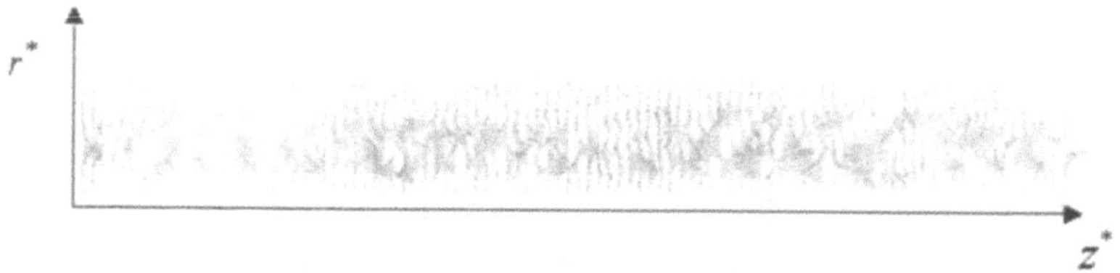

Fig. 20. Vector plot of the three velocity components in a (z,r) plane

Figure 21 shows the energy spectrum showing the turbulent kinetic energy as the y-axis and the number of waves per unit length in a log scale as the x-axis. The energy spectrum is constant in the range from 5 to 20. The range of wavelength is from 1.5 to 0.314. This range includes the range of wavelength observed experimentally in figure 4 for (Re=3000).

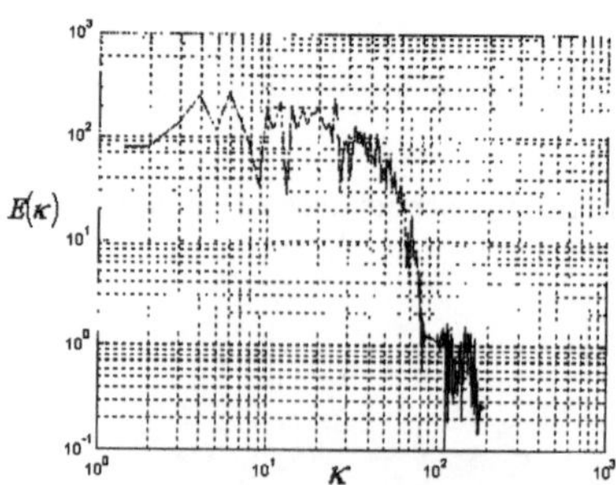

Fig. 21. Energy Spectrum

7 Computational resources

The simulation has been performed on the NEC SX-5 at the High Performance Computing Center Stuttgart. The most time consuming subroutines, for the multiplication of a matrix and it's transpose times a vector, were vectorized with the Jagged Diagonal Format (JAD). A typical simulation for this project takes place on the 40*32*99 with 129720 grid points. The computational time is approximately 10 hours.

8 Conclusion

In coaxial jet flow conditions, the internal turbulence in the flow plays a significant part in the surface deformation. Turbulence atomization performance changes considerably, if Re is reduced from the fully turbulent to the semi-turbulent regime. Experimentally, the undisturbed jet length for the semi-turbulent regime is larger than that at higher Reynolds numbers due to the thicker laminar envelope. From the numerical investigation and at the semi-turbulent state (Re=3000), varicose instability is observed and the turbulent intensities are distributed away from the wall compared with higher Reynolds number(Re=5600). The turbulent eddies, at Re=3000, are widely distributed with less turbulent energy. Thus the occurrence of the surface deformations more downstream compared to high Reynolds numbers can be explained.

Acknowledgements

The present work is supported by the DFG La553/9 and the High Performance Computer Centre in Stuttgart.

References

[1] Hoyt, J.W., Runggaldier, W.J.: Waves on Water Jets. J. Fluid Mech., **83(1)**, 119-127 (1977)

[2] A. Lefebvre.: Atomization and Sprays. Hemisphere Publishing Corp., (1989)

[3] R. Branam, G. Schneider, A. Volpp, W. Mayer: Injection Characteristics on the Surface of a Coaxial Jet. JPC, AIAA Paper, Indianapolis, (2002)

[4] Hoffmann, A.: Computational Fluid Dynamics for Engineers. Engineering Education System USA, (1995)

[5] Strikwerda, J.: Finite difference methods for the stokes and Navier-Stokes Equations. SIAM J. Stat. Comput., **5(1)**, (1984)

[6] E. Khalifa and E. Laurien: Development of a new second and fourth order compact finite-difference scheme for the direct numerical simulation of turbulent pipe flow. GAMM, Augsburg, (2002)

[7] F. O. Albina: Private communications. TUHH. Fluiddynamik und Schifftheorie, (2001)

[8] J.G.M. Egels, F. Unger, M.H. Weiss, J. Westerweel, R.J. Adrian, R. Friedrich and F.T.M. Nieustadt.: Fully developed turbulent pipeflow: a comparison between direct numerical simulation and experiment. J. Fluid Mech., vol. 268, 175-209 (1994)

[9] V.C. Patel and M.R. Head: Some observations on skin friction and velocity profiles in fully developed pipe and channel flow. J. Fluid Mech., **38(1)**, 181-201 (1969)

[10] W. Mayer and G. Kruele: Rocket Engine Coaxial Injector Liquid/Gas Interface Flow Phenomena.AIAA 92-3389., Nashville, (1992)

[11] E. Khalifa and E. Laurien: Numerical Investigation of Turbulence in the Fully Developed Regime in a Straight Pipe Atomizer. SPRAY 2002, Freiberg, (2002)

[12] M. Schueler, W. Mayer, E. Khalifa and E. Laurien: On the surface deformation of a liquid jet ejected from semi-turbulent pipe flow. ICLASS 2003, Sorrento, Italy, (2003)

[13] E. Khalifa and E. Laurien: Numerical Investigation of Semi-Turbulence and turbulence in a non-rotating and a rotating Pipe. SPRAY 2003, Toulouse, (2003)

Numerical Simulation of Forced Breakup of a Liquid Jet

Frank-Olivier Albina[1] and Milovan Perić[2]

[1] AB 3-13 Fluiddynamik und Schiffstheorie, TU Hamburg-Harburg, Germany
albina@tu-harburg.de
[2] CD adapco Group, Dürrenhofstraße 4, 90402 Nürnberg, Germany
milovan@de.cd-adapco.com

1 Introduction

Even nowadays, the physics of jet breakup is still not well understood and this is the reason why a lot of effort has been put recently into the investigation of this kind of flow. Within this framework, special attention has been focused on the use of excitation sources of defined amplitude and frequency in order to understand their effect on the (possible) jet disintegration process [SCO99, CGO+00]. The lack of analytical models for describing the strong non-linear behavior of this type of flow on the one hand, and the difficulty of accessing experimentally flow features like pressure and/or velocities on the other hand, have made it necessary to develop and use numerical methods for this purpose. In the recent past, numerical methods based on the Navier-Stokes equations with the possibility of handling free-surfaces have been developed and proved to fulfill the above requirements; they also have shown to be able to reproduce qualitatively well the jet-breakup behavior observed in the experiments [AMP00]. In the work presented here, after a brief description of the numerical method used, the results of simulations of the forced breakup of a round jet of ethanol into quiescent air will be presented. Three different types of excitation have been used and their effects on the jet breakup have been analyzed. Special attention has been paid to the capture of droplet generation.

2 Numerical method

2.1 Mathematical model

The model used for the fluid flow simulation is based on the Navier-Stokes equations for incompressible media. For an arbitrary control volume V enclosed by the surface ∂V, the integral form of the Navier-Stokes equations is given by equations (1) and (2):

$$\iint_{\partial V} < \mathbf{u}, \mathrm{d}\mathbf{S} >= 0 \tag{1}$$

$$\frac{\partial}{\partial t}\iiint_V \rho\mathbf{u}\,\mathrm{d}V + \iint_{\partial V} \rho\mathbf{u} < \mathbf{u}, \mathrm{d}\mathbf{S} >= -\iint_{\partial V} p\,\mathrm{d}\mathbf{S} + \iint_{\partial V} \overline{\overline{\tau}}\cdot\mathrm{d}\mathbf{S} + \iiint_V \mathbf{b}\,\mathrm{d}V \tag{2}$$

Under the assumption that the fluid is Newtonian, has constant physical properties and is incompressible, the viscous stress tensor $\overline{\overline{\tau}}$ reduces to:

$$\overline{\overline{\tau}} = \mu\left(\overline{\overline{\nabla}}\mathbf{u} + \overline{\overline{\nabla}}\mathbf{u}^{\mathrm{T}}\right) \tag{3}$$

For the modeling of free-surface flows using the *interface-capturing* method [Ubb97, MPSS98], a supplementary conservation equation for the so-called void-fraction c needs to be solved:

$$\frac{\partial}{\partial t}\iiint_V c\,\mathrm{d}V + \iint_{\partial V} c < \mathbf{u}, \mathrm{d}\mathbf{S} >= 0 \tag{4}$$

The fluid is considered to be single-phased but with variable physical properties. The physical properties are then determined by a linear combination of the physical properties of each phase. For a binary system with a gaseous (g) and a liquid phase (ℓ), the physical properties of the resulting effective fluid will be given by:

$$\mu = (1-c)\mu_g + c\mu_\ell \tag{5}$$

$$\rho = (1-c)\rho_g + c\rho_\ell \tag{6}$$

It is also assumed that the different phases do not mix or react with each other. The position of the free-surface is then given by the iso-surface for which $c = 0.5$.

Surface tension is represented by a body force in the momentum equations (2) using the following formulation according to the CSF-model [BKZ92, LNS+94]:

$$\mathbf{b} = \sigma\kappa\boldsymbol{\nabla}c, \tag{7}$$

where κ models the local free-surface curvature and is estimated with:

$$\kappa = - < \boldsymbol{\nabla}, \frac{\boldsymbol{\nabla}c}{||\boldsymbol{\nabla}c||} > \tag{8}$$

The flows studied here are assumed to be isothermal and laminar.

2.2 Numerical procedure

The computational domain is divided into non-overlapping control-volumes of arbitrary shape obtaining thus the so-called numerical grid. For solving eqns.

(2), (1) and (4), a discretization technique of finite-volume type is applied: Volume and surface integrals are estimated on every control volume using a colocated arrangement of the variables. The integral values are obtained with second-order approximations in space: mid-point rule, linear interpolation and central-differences are used [FP01].

Time integration is fully implicit: Volume and surface integral are calculated for the newest time step, whereas the older variable values at previous time-steps are used for the estimation of the time derivative (two-time or three-time level implicit method [FP01]). Since the flow is considered to be incompressible, pressure is obtained by solving a Poisson equation derived from eqns. (2) and (1). Coupling between pressure and velocity fields is realized by using the SIMPLE-algorithm [CGPS72]. The linearised systems of equations are solved with iterative solvers of the conjugate-gradient family [GVL96].

Special attention has to be paid during the solving process to keeping the volume-fraction bounded. The values must lie between the bounds 0 and 1, so that dependent variables like density do not take unphysical values due to over- or undershoots. Moreover, the face-centered values of the void-fraction must remain bounded between the cell-values sharing the common face. These constraints can be ensured by applying special discretization schemes based on the Normalized Variable Diagram [Leo88, GL88]. Here the *High-Resolution Interface-Capturing* (HRIC) scheme is used [MPSS98].

3 Simulation of forced jet-instabilities

3.1 High-frequency forced jet-breakup

At the IMFD (Institut für Mechamik und Fluiddynamik) of the TU (Technical University) Freiberg, experiments on the forced breakup of round liquid jets into quiescent air have been realized [SCO99, CGO+00]. The IMFD has authorized the use of experimental results for comparison purposes with fluid flow simulations. In the following, the focus is on results obtained with the liquid ethanol, with physical properties of table 1.

The jet issues out of a conical nozzle with an outlet diameter of 0.21 mm. The jet-breakup is obtained at different outlet velocities (typically in the range $10\sim40\ \mathrm{m\,s^{-1}}$) by imposing a disturbance of defined frequency and amplitude to the nozzle flow. The disturbance frequency is typically in the range of $\sim$100 kHz. Figure 1 depicts a stroboscopic photography of jet deformations in an experiment with outlet velocity of 21 $\mathrm{m\,s^{-1}}$ and for an excitation frequency of 126 kHz.

3.2 Numerical simulation of forced jet breakup

The flow is considered to be laminar, so that no turbulence model is used. It is also assumed that the numerical grid and the time steps used are fine enough

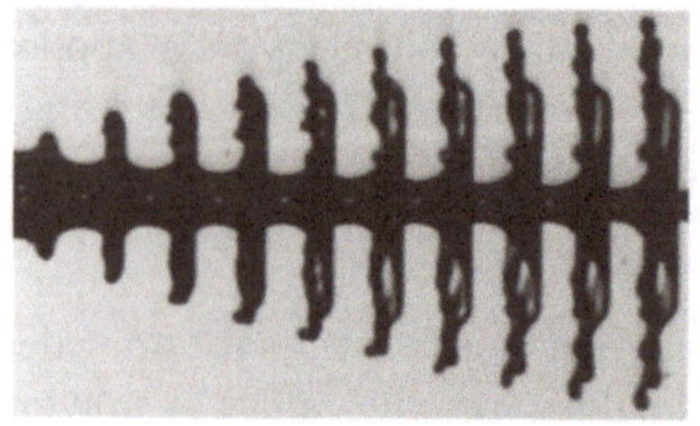

Fig. 1. Experimental, stroboscopic photography of jet deformation at $U_{\mathrm{m}} = 21\,\mathrm{m\,s^{-1}}$ for an excitation frequency of 126 kHz. Courtesy of IMFD, Technical University of Freiberg [SCO99, CGO$^+$00].

Fig. 2. Deformation of the jet free-surface in a two-dimensional simulation for $U_{\mathrm{m}} = 21\,\mathrm{m}\,s^{-1}$, disturbance frequency of 126 kHz, and a disturbance amplitude of 16%. Flow is from left to right.

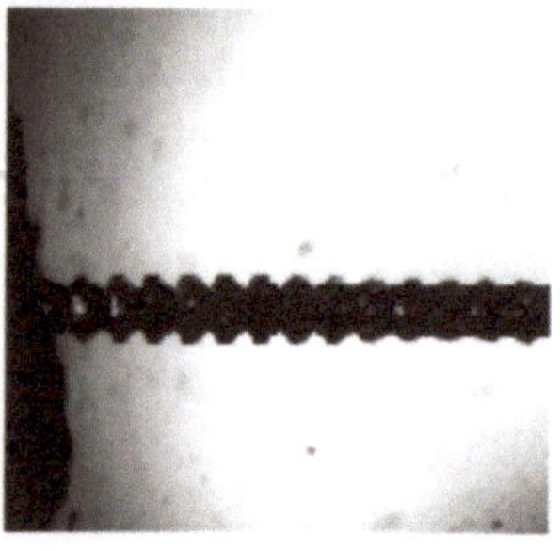

Fig. 3. Stroboscopic view of the jet free-surface deformations for an excitation amplitude of 38mV for the piezoelectric base at 126 kHz and an outlet velocity $U_{\mathrm{m}} = 18.9\,\mathrm{m\,s^{-1}}$; Courtesy of IMFD, TU Freiberg.

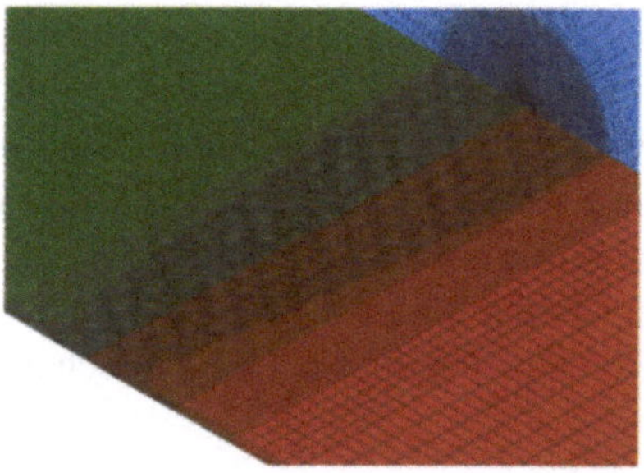

Fig. 4. Detail of the mesh used for the jet-breakup simulations. Cyclic boundary conditions are used on the regions coloured in red and green on the figure.

Table 1. Physical properties of ethanol

$\rho\,[\mathrm{kg\,m^{-3}}]$	$\nu\,[\mathrm{m^2\,s^{-1}}]$	$\sigma\,[\mathrm{kg\,s^{-2}}]$
789	$1.52 \cdot 10^{-6}$	0.0225

to resolve all scales of possible velocity fluctuations. Hence, the simulation approach can be qualified as a direct numerical simulation (DNS).

Simple jet actuation

In the first approach, it has been assumed that the boundary layer effects at the nozzle walls were negligible, so that the velocity at nozzle outlet can be approximated by a block profile and the simulation of the nozzle internal flow

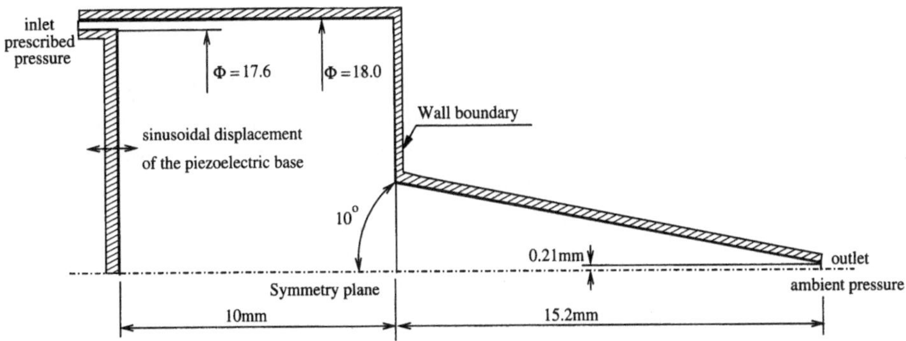

Fig. 5. Nozzle geometry and boundary conditions used for the numerical simulations of forced break-up of round jets.

does not need to be conducted. In order to model the jet disturbance, it has also been assumed that it can be represented by a sinusoidal modulation in time of the block profile according to the following equation:

$$U(r) = U_m\left[1 + \epsilon\ \sin(\omega t)\right] \quad 0 \leq r \leq \frac{D}{2} \tag{9}$$

Here, ϵ is a parameter used for setting the level of actuation for the jet disturbance and is given in percentage of the mean jet velocity U_m; ω represents the excitation frequency corresponding to the one of the experiment of Fig. 1, i.e. 126 kHz. Since the amplitude of excitation ϵ could not be obtained from the experimental setup, two-dimensional numerical simulations of the jet flow with the help of eq. (9) have been undertaken for various excitation amplitudes and have been compared to the experimental photography of Fig. 1. For $\epsilon = 16\%$, good agreement between experiment and simulation could be found, see also Fig. 2.

Nevertheless, three-dimensional effects such as a possible droplet formation cannot be captured in a two-dimensional simulation. Therefore, a three-dimensional numerical mesh has been built with the same mesh density as used for the two-dimensional analysis. In order to save computational effort, only one half of the complete three-dimensional computational domain has been modeled and cyclic boundary conditions have been applied. A close-up of the mesh involving 1,376,862 control volumes is depicted in Fig. 4. Due to a high frequency of disturbance, very small time steps have had to be used (1.e-8 s, i.e. ?? time steps per oscillation period).

Surprisingly enough, the results obtained with the forcing of eq. (9) – with a disturbance amplitude of $\epsilon = 16\%$ – did not lead to the expected formation of droplets; surface tension effects tend to form rings of fluid that detach from the developing wave train, see Fig. 7. The deformations of the jet free-surface are periodic with the same frequency as the one of the excitation source; therefore, results are shown over one period. These rings would break up further downstream if the solution domain were long enough.

Nozzle flow

Because the block velocity-profile does not take into account any radial or azimuthal velocity fluctuations arising from the nozzle internal flow that may be responsible for the generation of droplets out of the resulting deformations of the jet free-surface, it has been decided to simulate the nozzle internal flow for gaining a more realistic velocity profile. This velocity profile will be used later as the source of excitation for the simulation of the jet disintegration. In order to simplify the model, and on the basis of the consideration that the disturbances imposed to the jet do not travel back to their origin but rather in downstream direction (as shown in the experimental work of [CGO+00]), it has been assumed that the nozzle flow simulation could be undertaken independently from the jet flow. The velocity profiles at the nozzle outlet have been then stored for every time-step and reused as an inlet boundary condition for investigating the effects of the nozzle internal-flow on the jet breakup.

The nozzle boundary conditions and geometry are depicted on Fig.5. The inlet pressure was chosen to give a mean outlet velocity of 21 $\mathrm{m\,s^{-1}}$, so that direct comparison of the results with the solution obtained with the block profile can be made. Moreover, the movement of the piezoelectric base as shown on Fig. 5 was imposed to vary according to the following law for a plate under uniform loading [Gou98]:

$$U(r,t) = \Delta U_{\max} \sin\left[\frac{\pi}{2}\left(1 - \frac{r}{\mathrm{R}}\right)\right] \sin(\omega t + \varphi)\,, \tag{10}$$

where $\Delta U_{\max}$ is related to $\Delta X_{\max}$ by the following equation:

$$\Delta U_{\max} = \omega \Delta X_{\max}\,. \tag{11}$$

The displacement level was imposed to give the same amplitude of excitation as in the case of the simple excitation model, so that direct comparison of the resulting jet deformations can be made with the simple excitation model. The numerical mesh used modeled only the half of the complete nozzle and contained 1,324,862 control volumes. The mesh density was chosen to match exactly the mesh density used in the previous two-dimensional calculations [AMP00], so that the same numerical accuracy as with the two-dimensional model is achieved. Moreover, the same mesh density was used at the outlet boundary of the nozzle model as at the inlet boundary of the mesh for the jet breakup simulation, so that outlet velocities from nozzle simulation can be directly used as inlet velocities for the jet flow simulation. Typical velocity profiles in axial and radial direction obtained from the computations are shown in Fig. 6.

Deformations of the jet free-surface obtained from the use of the velocity profiles from the nozzle simulation are shown on Fig. 8. Again, as in the case with the simple excitation model, the flow pattern obtained is periodic with the period of the imposed actuation. Compared to the jet deformations obtained with the block profile, Fig. 7, the free-surface deformation here undergoes a totally different behavior: backward breaking of the growing waves is observed, leading first to a growing wave whose amplitude decreases after the gas-fluid interface has rolled up. This correlates well with experimental results for which the free-surface of the jet, at moderate amplitudes of excitation, is found to develop a similar behavior, Fig. 3. This can be explained by the strong boundary layer effects due to the nozzle walls, for which the radial velocity component can not be neglected, see also Fig. 6.

As in the case of the simple excitation model, no droplet formation is observed.

Azimuthal disturbances

One of the possible reasons for no creation of droplets could be that the velocity profiles used in the previous jet simulations did not have any kind of disturbance in the azimuthal direction. In order to investigate possible effects of azimuthal disturbances on the jet deformation behavior, an azimuthal disturbance has been superimposed onto the block profile for the simple excitation model. The form of the azimuthal disturbance was chosen to have the spatial and temporal divergence-free variation given by:

$$V_r(r,\theta) = \epsilon\, U_m \frac{k_\theta r}{2} \cos(k_\theta \theta) \tag{12}$$

$$V_\theta(r,\theta) = \epsilon\, U_m r \cos(k_\theta \theta) \tag{13}$$

where k_θ is the disturbance wavelength in azimuthal direction and has been chosen arbitrarily to be equal to 8. The maximum amplitude of the disturbance so applied is a fraction of the mean jet velocity, where the fraction is given by the factor ϵ. For the results presented hereafter, ϵ has been set to

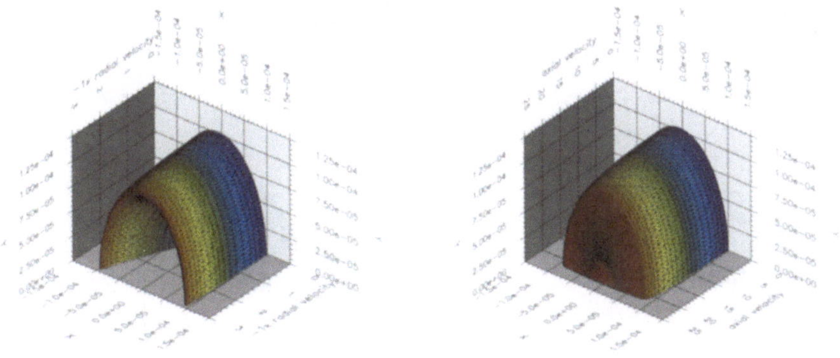

Fig. 6. Typical velocity profiles in radial and axial directions obtained at the nozzle outlet.

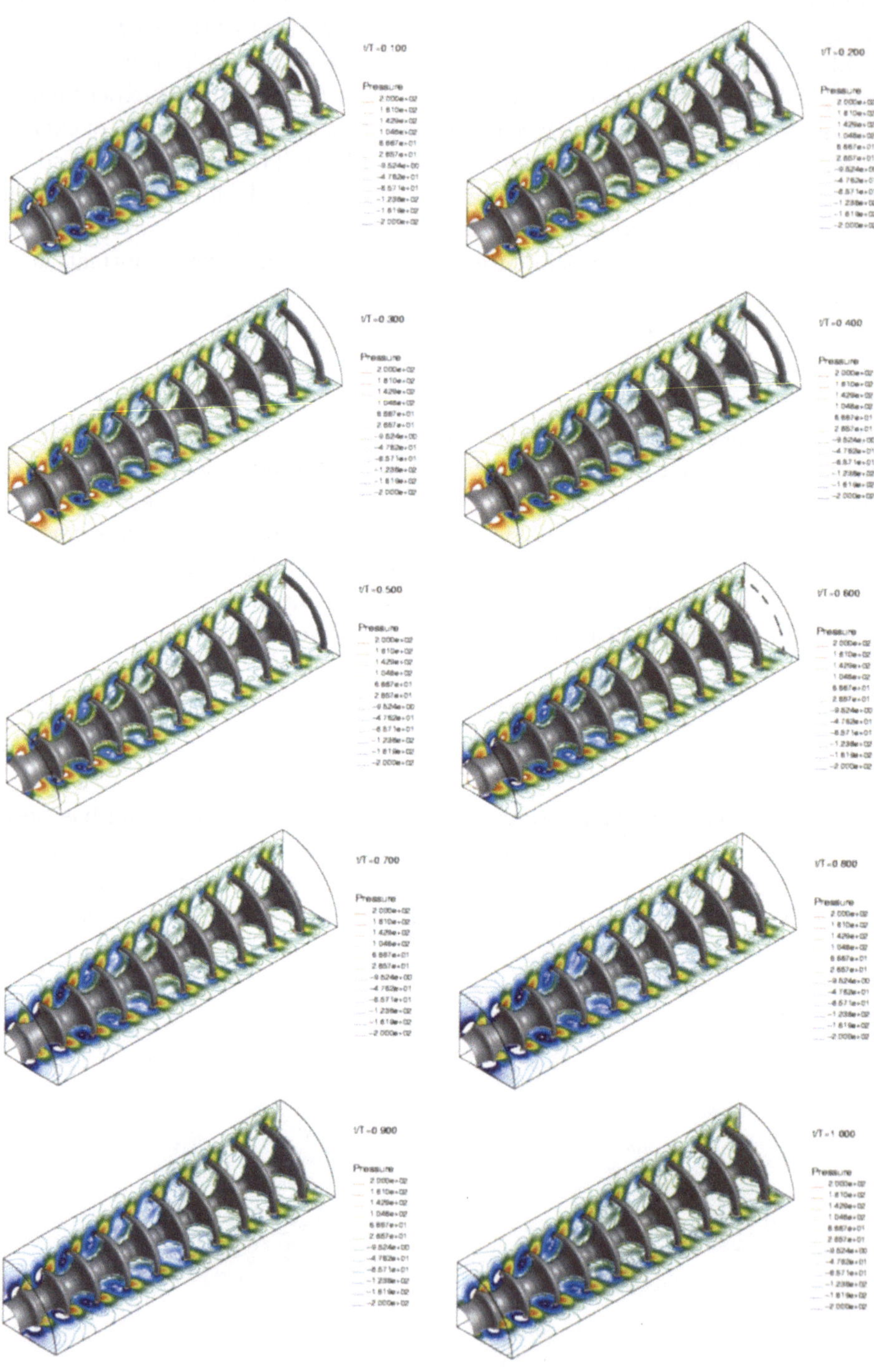

Fig. 7. Jet free-surface deformation with pressure contours at different instants over an excitation period. Inlet velocity is obtained by the modulation of a block-profile. Excitation amplitude is $\epsilon = 16\%$.

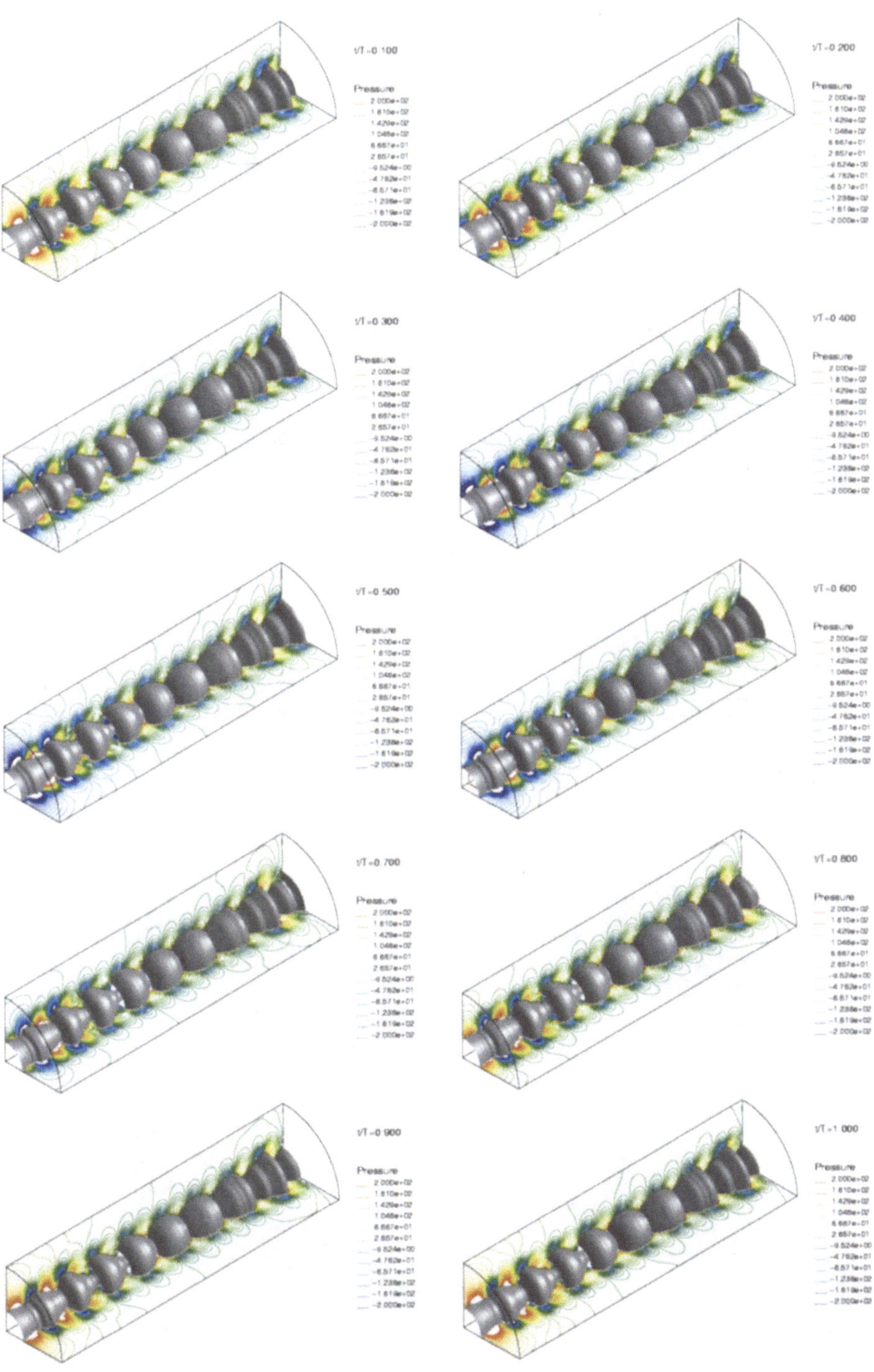

Fig. 8. Jet free-surface deformation with pressure contours at different instants over an excitation period. Inlet velocity is obtained by the simulation of the nozzle internal flow.

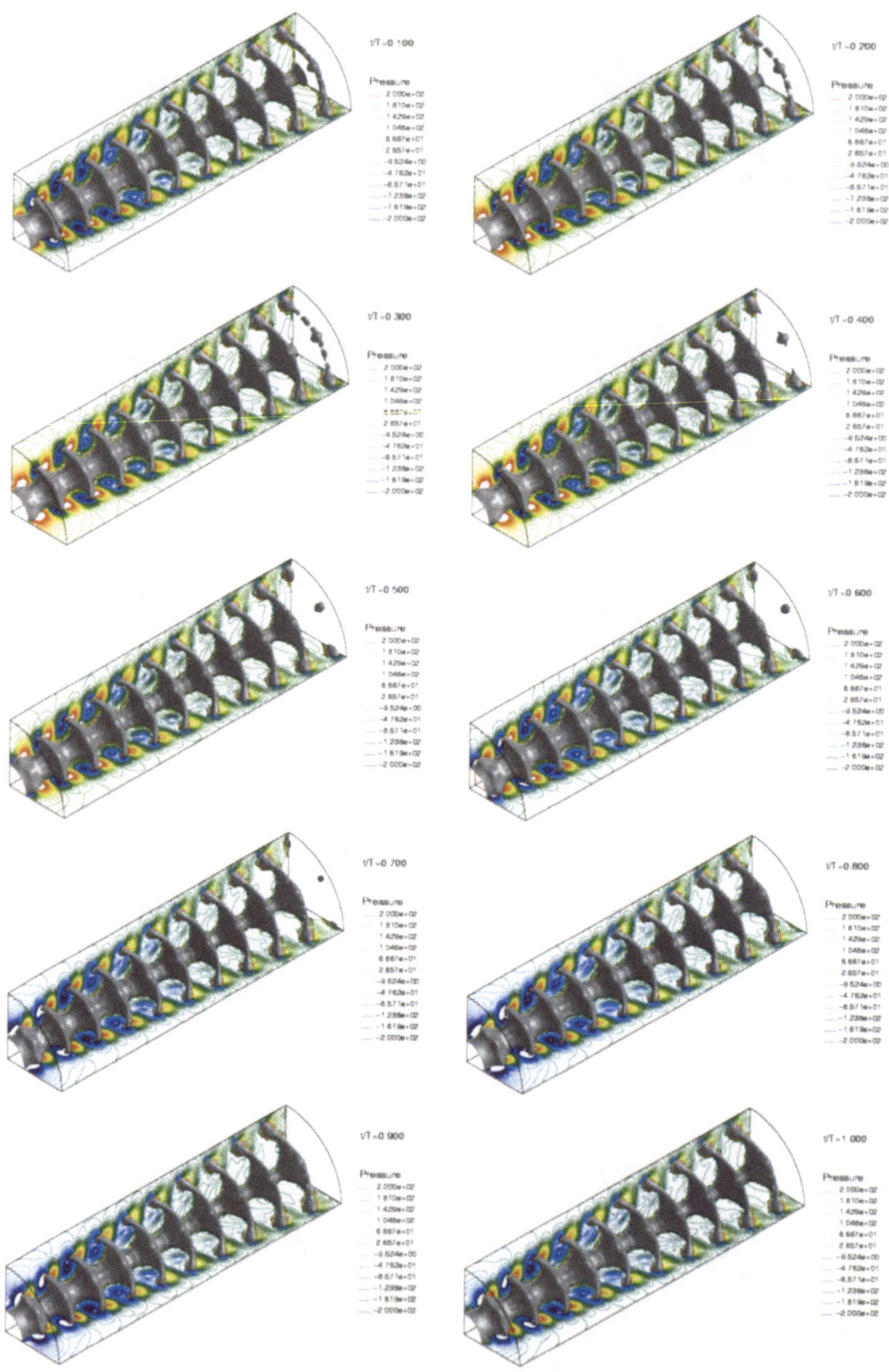

Fig. 9. Jet free-surface deformation with pressure contours at different instants over an excitation period. Inlet velocity is obtained by a modulated block-profile with disturbances in azimuthal direction.

2%. The excitation level in streamwise direction remained fixed to 16%, see eqn. (9). Again, the flow was found to undergo periodic behavior and the free-surface deformations over one period T by steps of $0.1T$ are shown in Fig. 9. As in the case of the simple excitation, waves grow on the jet free-surface and take a similar from to the ones of Fig. 7. However, because of the excitation applied in azimuthal direction, a cross-wave develops on the wave crowns and leads further downstream to the creation of droplets, in the same way as a jet undergoes a breakup of Rayleigh type [LR98]. Thus, it can be concluded that in the investigated cases, droplet formation could be obtained only when a disturbance in azimuthal direction exists.

4 Concluding remarks

The motivation for this study has been to gain detailed understanding in the physical mechanisms of the breakup of a round laminar jet subjected to an excitation of high amplitude and high frequency. The results obtained in the simulations have shown that changes in the boundary conditions have a dramatic influence on the jet disintegration behavior. The boundary layer effects, such as the ones obtained by the wall bounded flow in the used nozzle, cannot be neglected and must be taken into account when analyzing the jet breakup. The numerical method can reproduce qualitatively the jet behavior observed in the experiments. However, quantitative comparisons can only be conducted when the boundary conditions are exactly matched; this was not the case here, because the exact amplitude of the disturbance and its spatial profile were not known. Present simulations have shown that droplet formation during primary jet breakup only takes place when azimuthal disturbances are present.

Acknowledgments

The authors thank the German Research foundation (DFG) for supporting the work presented here under grant Pe-350/13-3. The authors are also very grateful to IMFD of TU Freiberg, especially to Dr. Chaves and Prof. Obermeier, for their helpful hints and for having allowed us to use their experimental results for qualitative comparison with the simulations.

References

[AMP00] F.-O. Albina, S. Muzaferija, and M. Perić. Numerical simulation of jet instabilities. In *Proceedings of the 16th Annual Conference on Liquid Atomization and Spray Systems*, Darmstadt, Germany, September, 11 – 13 2000.

[BKZ92] J.U. Brackbill, D.B. Kothe, and C. Zemach. A continuum method for modeling surface tension. *J. Comput. Phys.*, 100:335–354, 1992.

[CD 01] CD adapco Group. *comet* user manual. Computational Dynamics Germany, Dürrenhofstraße 4, 90402 Nuremberg, Germany, 2001.

[CGO+00] H. Chaves, H. Glate, F. Obermeier, T. Seidel, V. Weise, and G. Wozniak. Disintegration of sinusoidally forced liquid jet. In *ILASS-Europe 2000*, Darmstadt, Germany, September 11–13 2000.

[CGPS72] L.S. Caretto, A.D. Gosman, S.V. Patankar, and D.B. Spalding. Two calculation procedures for steady, three-dimensional flows with recirculation. In *Proc. Third Int. Conf. Numer. Meth. Fluid Dyn.*, Paris, 1972.

[FP01] Joel Ferziger and Milovan Perić. *Computational Methods for Fluid Dynamics.* Springer Verlag, 3rd edition, 2001.

[GL88] P.H. Gaskell and A.K. Lau. Curvature-compensated convective transport: SMART, a new boundedness preserving transport algorithm. *International Journal for Numerical Methods in Fluids*, 8:617–641, 1988.

[Gou98] P. M. Gould. *Analysis of plates and shells.* Prentice Hall, 1998.

[GVL96] G.H. Golub and C.F. Van Loan. *Matrix Computations.* The John Hopkins University Press, third edition, 1996.

[Leo88] B.P. Leonard. Simple high-accuracy resolution program for convective modeling of dicontinuities. *Int. J. Numer. Meth. Eng.*, 8:1291–1318, 1988.

[LNS+94] B. Lafaurie, C. Nardone, R. Scardovelli, S. Zaleski, and G. Zanetti. Modelling merging and fragmentation in multiphase flows with SURFER*. *J. Comput. Phys.*, 113:134–147, 1994.

[LR98] S.P. Lin and D.Z. Reitz. Drop and spray formation from a liquid jet. *Annu. Rev. Fluid Mech.*, 30:85–105, 1998.

[MPSS98] S. Muzaferija, M. Perić, P.C. Sames, and T.E. Schellin. A two-fluid Navier-Stokes solver to simulate water entry. In *Proceedings of the 22nd Symposium on Naval Hydrodynamics*, Washington, D.C., August 9–14 1998.

[SCO99] T. Seidel, H. Chaves, and F. Obermeier. Grundlagenuntersuchungen an einer Einstoffdüse bei hochfrequenter Strahlanregung und stroboskopischer Betrachtung. In *Spray '99*, University of Bremen, Bremen, Germany, October 5–6 1999.

[Ubb97] O. Ubbink. *Numerical prediction of two fluid systems with sharp interfaces.* PhD thesis, University of London, 1997.

List of symbols

Mathematical symbols

$\iint$	Surface integral
$\iiint$	Volume integral
$\frac{\partial}{\partial\phi}$	Partial derivative with respect to ϕ
∇	Nabla operator
$.$	Tensor multiplication
$<,>$	Euclidian scalar product

Greek symbols

Δ	Finite variation
ϵ	Reduced amplitude
φ	phase shift
κ	Curvature
μ	Dynamic viscosity
ρ	Density
σ	Surface tension

$\|\|.\|\|$	Euclidian norm	θ	Azimuthal angle
ϕ	Scalar field	$\overline{\overline{\tau}}$	Viscous-stress tensor
$\boldsymbol{\phi}$	Vector field of ϕ	ω	Excitation frequency
$\overline{\overline{\phi}}$	Tensor/Matrix ϕ		$\omega = 2\pi \mathrm{f}$
$\boldsymbol{\nabla}\phi$	Vector gradient of scalar ϕ		
$\overline{\overline{\nabla}}\boldsymbol{\phi}$	Gradient tensor of $\boldsymbol{\phi}$ $\left(\overline{\overline{\nabla}}\boldsymbol{\phi}\right)_{ij} \stackrel{\text{def.}}{=} \frac{\partial \phi_i}{\partial x_j}$		
∂V	Surface enclosed by the volume V		

Latin symbols

c	Void-fraction field	$\mathrm{d}V$	Volume element
D	Jet diameter	St	Strouhal number $\mathrm{St} = \frac{\pi f D}{U_m}$
f	Frequency	k_θ	Azimuthal wavelength
p	Pressure field	U_m	Jet mean velocity
t	Time	V_r	Radial velocity
r	Radius	$\Delta U_{\max}$	Maximum axial velocity of the piezoelectric base
R	Radius of the piezoelectric base		
$\mathbf{u}$	Velocity vector	$\Delta X_{\max}$	Maximum axial displacement of the piezoelectric base
$\mathbf{b}$	Body-forces field		
$\mathrm{d}\mathbf{S}$	Surface element vector		

The Effect of Impinging Wakes on the Boundary Layer of a Thin-Shaped Turbine Blade

Vittorio Michelassi[1], Jan Wissink, and Wolfgang Rodi

Institute for Hydromechanics, University of Karlsruhe, Kaiserstrasse 12,
76128 Karlsruhe, Germany.
[1] On leave from University Roma Tre, Italy.

A series of Large Eddy Simulations (LES) of flow in a modern, highly loaded low-pressure turbine cascade, with and without oncoming wakes, has been performed. The computations were carried out using 64 processors on the Hitachi-SR8000 F1. The blade shape is that of newly designed turbine rotors which allow controlling the blade weight. Five different operating conditions were investigated by changing strength and frequency of the incoming wakes. The large number of simulations was made possible by a previous careful tuning of the code.

1 Introduction

The accurate prediction of the flow around turbine and compressor blades largely depends on the ability of the computational model to predict transition from laminar to turbulent flow and relaminarisation. As suggested by Emmons [2] and Mayle [7], who reviewed the transition data and models relevant to the turbomachinery field, it is possible to distinguish between three types of transition mechanisms, known as natural, by-pass, and separated flow transition. The natural transition (see [15]), which is driven by Tollmien-Schlichting waves arising in a mildly triggered laminar boundary layer, is very unlikely to occur in a turbomachine due to the large turbulence levels usually encountered in a real multi-stage environment. By-pass transition is characterised by the appearance and subsequent growth of turbulent spots in the laminar boundary layers, and generally occurs at large Reynolds numbers (Re) and turbulence levels (see [2]). This transition mechanism has been successfully studied in steady [14] and unsteady flow environments with incoming wakes over flat plates, both experimentally [6] and numerically by DNS [21]. The by-pass transition scenario, characteristic of attached boundary layers, has also been successfully computed in complex transonic turbine stages (see for example [9]). Most of the computational studies presented so far introduced modifications to existing two-equation turbulence models [14] based on algebraic correlations, sometimes combined with an extra transport equation for intermittency, to predict both the onset and length of transition. In off-design conditions, or in the presence of shocks, the boundary layer on the suction side of a turbine blade often separates while being still laminar. Separation can occur both in the immediate proximity of the leading edge (due to excessive incidence), or further downstream along the suction side where the pressure gradient reverts from favourable to mildly adverse. Walker *et al.* [18] attempted a classification of separated flow transition models used in the literature, and of the related experiments, which revealed the intrinsically unsteady and three-dimensional nature of the bubble under certain operating conditions. Recently, Stadtmüller [17] provided detailed measurements of the flow around the low speed T106 turbine blade with incoming periodic wakes generated by a row of moving bars. The presence of incoming wakes significantly increases the relevance of this test case with respect to previous experimental data sets since it is representative of real off-design conditions and it does provide crucial information on the wake-blade interaction. The flow, measured at $Re = 5.81 \times 10^4$, was successfully computed by DNS, LES, and URANS in previous simulations [19, 10, 12, 13]. These computations illustrated the ability of DNS and LES to highlight relevant phenomena in the wake blade interaction with an acceptable computational effort. These phenomena could be conveniently modelled, with a significant reduction in the computational effort by URANS. Since the first test case under investigation referred to an off-design condition, it was decided to perform further simulations with increased engineering relevance. Compared to the previous simulations, in the simulations presented here the Reynolds number is increased and a detailed study is made of the effects of various strengths and frequencies of incoming wakes, which are representative of various possible design choices. To perform the simulations, a modern low-pressure (LP) turbine rotor blade was selected, which reflects the recent trends in aero-engine turbines [16].

2 The LP rotor turbine blade

The simulations were set up to reproduce some of the experimental conditions described in Schulte and Hodson [16]; in the linear test rig the blade aspect ratio (h/C_{ax}) is 2.37, where $C_{\mathrm{ax}} = 156\,\mathrm{mm}$ is the axial chord length, which allows the mid-span section to be considered nearly two-dimensional. The stagger angle, inlet and outlet blade angles are 35.32°, −30.46°, and 62.86° respectively. The experimental test rig allows the effect of upstream blade rows to be simulated by a moving bar wake generator, located upstream of the leading edge, with a bar diameter of 2 mm or 4 mm. The tangential speed of the bars is related to the axial flow velocity by the flow coefficient, Φ, defined as the ratio of the axial velocity and the bar speed, which varies around the value of 0.7. In the numerical simulations the wakes are not generated by bars, but by introducing an artificial turbulent wake at the inflow plane. The wake data has been kindly made available by X. Wu and P. Durbin of Stanford University [20], who generated the incoming wakes with a preliminary LES that mimics the far-field behaviour of a turbulent wake. We selected a Reynolds number, based on the axial chord length and the mean inflow velocity, of $Re = 10^5$. This value allows real design conditions to be simulated with a reasonable computational effort.

3 Computational details

3.1 Computational grid

The computational domain, displayed in Figure 1, was chosen in accordance with the experiments performed by Schulte and Hodson [16]. The computational domain extends $0.5 \times C_{\mathrm{ax}}$ upstream of the leading edge of the blades and $1.0 \times C_{\mathrm{ax}}$ downstream of the trailing edge to avoid undesirable effects stemming from the inlet and outlet sections being too close to the blade vane. The span-wise width of $h = 0.15 \times C_{\mathrm{ax}}$ was selected to allow three-dimensional flow structures to develop without being effected by the periodic boundary condition in the span-wise direction. The grid was generated using the elliptic method proposed by Hsu and Lee [4], which ensured a nearly orthogonal mesh close to the blade walls. The grid size was selected using experience gained in performing several other LES of similar flows (see for example [11]). The final grid has $614 \times 294 \times 64$ nodes in the stream-wise, pitch-wise, and span-wise directions, respectively. The grid adopted in the present LES places approximately $20-25$ grid nodes inside the boundary layer. This number was deemed suited to provide a reasonably good resolution of the boundary layer and a fair resolution of the incoming wakes vortical structures while controlling the overall computational effort. The cell sizes in wall units in the x and z direction, are below 60 and $12-18$, respectively. The distance in wall units between the wall-nearest grid points and the blade, which is a measure of the effective resolution of the boundary layer, ranges between $2-3$ for $x/C_{\mathrm{ax}} < 0.55$, and reduces to 1 while approaching the trailing edge where the boundary layer thickens due to the effect of the adverse pressure gradient.

3.2 The specification of boundary conditions

A no-slip boundary condition is enforced at the surface of the blade. For $x/C_{\mathrm{ax}} < 0$ and $x/C_{\mathrm{ax}} > 1$ periodic boundary condition are enforced in the pitch-wise direction. In the span-wise direction too, a periodic boundary condition was selected. At the outlet, a convective boundary condition was employed, while at the inlet the wake data of Wu and Durbin [20, 21] was superposed on a uniform flow field. In the current simulations the wake data

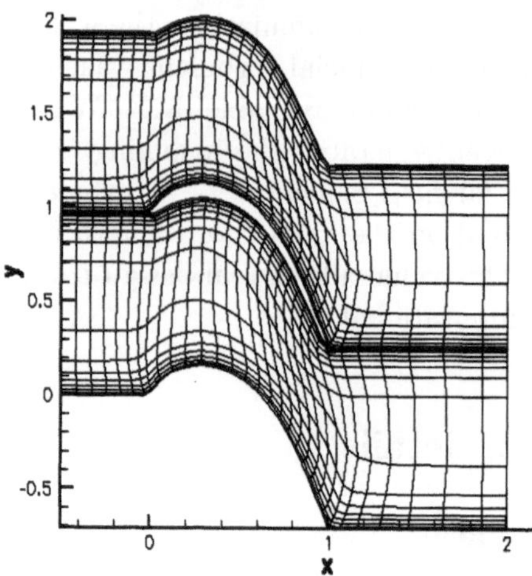

Fig. 1. Elliptic grid used for LES (every 20 nodes are shown in both stream-wise and pitch-wise directions)

closely resemble those of typical wakes generated by upstream stator blades. The calculations refer to five different operating conditions listed in Table 1

With the exception of the base line run A, which has no incoming wakes, all runs attempt to reproduce various conditions taking place inside a multistage axial turbine. In fact the wake-to-blade pitch ratio of 1 refers to a stator-rotor-like interaction, while the ratio of 0.5 to rotor-stator-like interaction. The bar diameter directly controls the strength of the incoming wakes. A bar diameter of $2mm$ is representative of thin-trailing-edge blades at design conditions, while at $4mm$ the strength of the wake is that of off-design conditions, with moderate suction side separation from the preceding blade row, or that of thick inlet guide vanes. This relatively wide scatter of conditions, affordable due to the high performance of the adopted computational platform, allows an interesting screening of the rotor blade performance. In order to properly resolve the temporal fluctuations in the wake and the boundary layer of the blade, a non-dimensional time step of $dt = 7.688 \times 10^{-5}$ was selected. By adopting this value the CFL number is always below unity. The flow was

Table 1. Overview of the simulations performed.

Run	Oncoming Wakes	Bar Diameter (mm)	Wake-to-Blade pitch
A	No	-	-
B	Yes	2	1.0
C	Yes	2	0.5
D	Yes	4	0.5
E	Yes	4	1.0

allowed to develop for 51200 time steps, which corresponds to five periods in Runs B and D and ten periods in Runs C and E. Subsequently, phase-averaging was carried out for 10 further periods. During at least one period, a series of snapshots of instantaneous flow fields was stored to help gain a better understanding of the instantaneous flow structure dynamics. Phase-averaging was performed by dividing each period, T, in 64 equally spaced phases, $\phi = 0/64, \cdots, 63/64$. Instantaneous quantities are referred to using t/T.

3.3 The computer simulation tool and computational effort

The simulations of the flow around the turbine rotor blade has been performed by using the LESOCC code [1]. The incompressible Navier-Stokes equations are discretised by using a cell-centred finite volume approach. The Poisson equation for the pressure correction was solved implicitly by the SIMPLE algorithm, as originally adopted in [1]. A Fourier solver in the span-wise direction [8] allows a substantial reduction of the overall computational effort to enforce mass conservation. The momentum equations were solved by marching in time with a three-stage Runge-Kutta algorithm, where mass conservation is enforced only after the final Runge-Kutta step. In all simulations, the Dynamic Sub-Grid-Scale (SGS) model of Germano et al. [3] was employed. Further details about the code may be found in [1, 8]. The necessity for the flow with incoming wakes to develop a statistically steady periodic behaviour before commencing phase-averaging, substantially increased the computational effort compared to the case in which incoming wakes were absent. During phase-averaging, the flow fields were also averaged in the spanwise direction, such that a total of 10×64 two-dimensional (x, y) fields containing the ten phase-averaged statistics are stored in core memory. Compared to writing phase-averaged data to disk, this choice was found to allow more efficient postprocessing, while still being affordable due to the large core memory available in the $SR8000$ computer. As reported in [11] the first implementation of the code was tested using the grid and operating conditions, described in more detail in [10, 12]. The total number of grid nodes is 1.15×10^7, with 4.26×10^7 unknowns. The problem was split into 64 sub-domains, using 8 processors in both the x-direction and y-direction. The domain was not split in the z-direction to optimise the performance of the Fourier solver in the span-wise direction. The inter-processor communication was performed with standard MPI, and each one of the 64 sub-domains has approximately 1.8×10^5 nodes, which easily fit into the memory of one processor. Of the various possible options to perform the domain decomposition, the one adopted does not necessarily minimise the number of inter-processor communication, but it was found to produce the minimum reduction of the convergence rate. On average, each simulation required a total number of 153.600 time steps. For each run it was often possible to get 64 processors for nearly 24 hours a day for an elapsed time of approximately 650 hours ($\approx$ 28 days). The total CPU time for the five runs on 64 processors was therefore of the order of 40.000 hours. The present version of the code was optimised with the support of the High Performance Computing Centre Stuttgart, as reported in [11], and reaches 14 – 15% of the computer peak performance. Compared to the initial computations a reduction of more than a factor two in the overall computational time was achieved.

4 Results

4.1 The static pressure coefficient around the blade

In Figure 2 and Figure 3, the time-averaged wall static pressure distribution

$$\frac{P_{01} - P}{P_{01} - P_2},$$

where P is the local static pressure, P_{01} is the inlet total pressure, and P_2 is the outlet static pressure, has been plotted as a function of the normalised wall-coordinate, S/S_{max}. The wall coordinate is normalised such that $S/S_{\text{max}} = 0$ corresponds to the leading edge, while $S/S_{\text{max}} = 1$ corresponds to the trailing edge. The time averaged data are obtained by averaging the well developed flow field during 3.15 time-units in Run A and, alternatively, during 10 periods of simulation in Runs $B - E$.

Figure 2 refers to the case without incoming wakes (Run A), while Figure 3 reports the results obtained with the four different combinations of wake strength and frequency, as defined in Table 1 (Runs B to E). The experiments of Schulte and Hodson, which refer to the same blade, although at slightly different operating conditions, are also reported. Figure 2 shows that the agreement between the computations and the measurements for the case without incoming wakes is fairly good. The pressure distribution on the pressure side is quite well reproduced from the leading edge until the trailing edge. This suggests that the development of the boundary layer on this critical

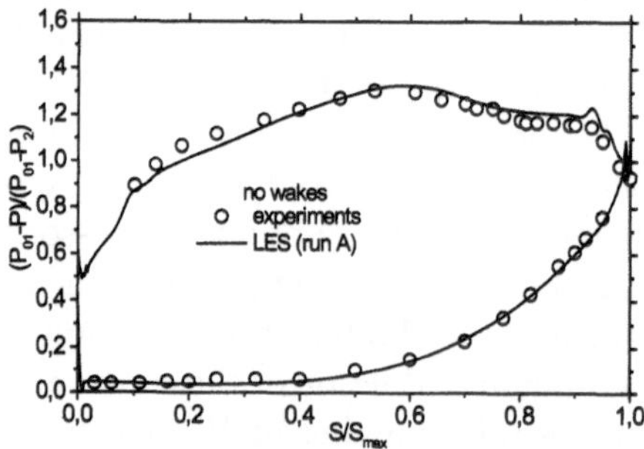

Fig. 2. Static pressure distribution around the blade: no wakes

portion of the thin rotor blade is well captured by the simulations. Observe that the flat static pressure portion on the pressure side, for $S/S_{\max}$ below 0.4, is very well reproduced. The agreement with the experiments does not change significantly on the suction side. The lack of measurements in the immediate proximity of the leading edge prevents a direct evaluation of the quality of the simulations. Nevertheless, for $S/S_{\max} > 0.1$ the predictions agree quite well with the measurements. In the critical region for $S/S_{\max} > 0.6$, where the pressure gradient reverts from favourable to adverse, and where the boundary layer thickness increases considerably, the computations only show very small deviations from experiment. The further runs with variable incoming wakes strength and frequency are compared with the typical experimental results obtained with 2mm diameter bars and with one wake per pitch. Unsurprisingly, the best agreement between the measurements and the computations is found for Run B, which attempts to mimic the same operating conditions adopted in the experiments. All the computed results exhibit a "knee" on the static pressure distribution at approximately $S/S_{\max} = 0.85$. This position on the suction side of the blade corresponds to the transition to turbulence of the boundary layer. Observe that the decrease of the suction side static pressure for $S/S_{\max} > 0.8 - 0.85$ becomes smoother with increasing wake strength. Note that some of the observed differences between the four static pressure distribution can be attributed to the different inlet total pressure. This difference stems from the variable losses induced by the different wakes starting at the inlet section and continuing downstream. In practice, the overall operating conditions of the four runs with variable wakes are similar, but not identical, which partly justifies the overall differences observed in the shape of the suction side static pressure distribution, which is more sensitive to the disturbances convected by the wakes than the pressure side (see also Schulte and Hodson). For the case with incoming wakes the experiments provide several pressure measurements in the immediate proximity of the suction side leading edge. The agreement between computations and experiments in this crucial region is remarkably good, as witnessed by Figure 3 for $S/S_{\max} < 0.15$.

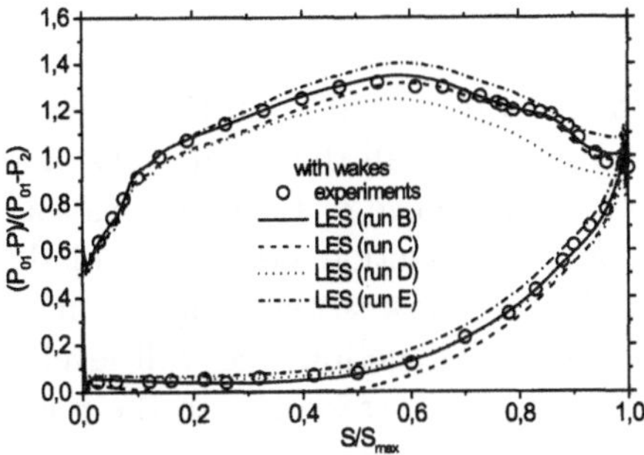

Fig. 3. Static pressure distribution around the blade: with wakes

4.2 The wall shear stress around the blade

The wall shear stress is monitored separately for both the suction side and the pressure side. Figure 4 and Figure 5 report the time averaged value of the computed wall shear stress for the five investigated operating conditions along the axial coordinate x/C_{ax}.

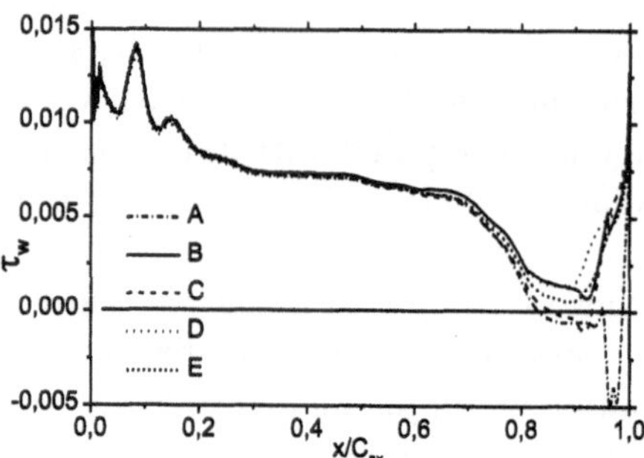

Fig. 4. Time averaged wall shear stress on the suction side of the blade

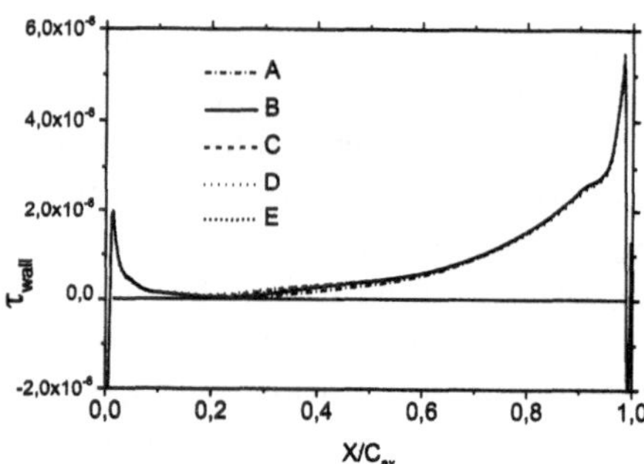

Fig. 5. Time averaged wall shear stress on the pressure side of the blade

Figure 4 shows that for $x/C_{\mathrm{ax}} < 0.1 - 0.15$ the wall shear stress has an up-down shape which is shared by all the simulations. This particular feature is present regardless of the presence of incoming wakes. Hence, it is probably related to the shape of the suction side of the blade rather than to the incoming disturbances. Further downstream, the strong flow acceleration together with the curvature of the suction side wall level the wall shear stress until $x/C_{\mathrm{ax}} = 0.6$ which roughly corresponds to the position where the pressure gradient reverts from favourable to adverse. Here the wall shear stress rapidly decreases, thereby indicating a tendency to develop flow separation. The separation actually develops, although the analysis, not shown here for the sake of brevity, suggests that it is only intermittently present and periodically suppressed by the incoming wakes. Apparently the impinging wakes manage to inhibit boundary layer separation. In the absence of wakes, the wall shear stress is found to become negative for $x/C_{\text{ax}} > 0.82$. Similar behaviour is surprisingly also observed in Run C, where the suction side boundary layer is only partly separated. In the presence of strong wakes, however, separation is completely suppressed. Figure 5 shows that on the pressure side the wall shear stress is far smaller. This is due to the relatively thin shape of the blade, which is designed to promote the growth of the pressure side boundary layer and, consequently, the reduction of the wall shear stress. Regardless of the presence and the intensity of the wakes, the pressure side is nearly stalled for $0.1 < x/C_{\text{ax}} < 0.3$. According to Figure 1, which reports the shape of the blade, the tendency to pressure side stall is very likely induced by the high concave portion of the blade wall.

Note that the flow is not massively stalled, something that would have a detrimental effect on the performance, but rather close to stall with a very thick boundary layer. It is remarkable how this behaviour is common to all the simulations. Only further downstream does the wall shear stress develop larger values due to the strong flow acceleration evidenced by Figure 2 and Figure 3.

4.3 Total pressure losses

A good measure of the effect of the incoming wakes on the performance is provided by the kinetic loss coefficient defined in Figure 6. This coefficient compares the inlet total pressure, P_{01}, with the outlet total pressure, P_{02}. This difference is made non dimensional with respect to the inlet kinetic energy. Figure 6 reports the results of all the runs. The curves refer to the time averaged total pressure losses extracted $0.4 \times C_{\text{ax}}$ downstream of the trailing edge. The peaks refer to the time averaged position of the wakes. Without entering into the details of each simulation, for the sake of brevity, it is worthwhile observing how the largest losses are encountered in Run A, where no

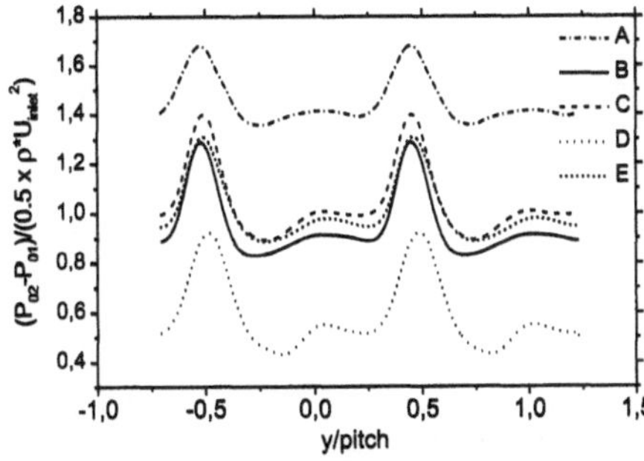

Fig. 6. Kinetic losses $0.4 \times C_{\text{ax}}$ downstream of the trailing edge

wakes are present. This behaviour is not surprising since the incoming wakes are known to promote transition to turbulence of both the suction and pressure side boundary layers, thereby delaying the possible onset of separation. Separation of the boundary layer is the main source of energy losses. Hence, Figure 6 shows that a careful choice of the incoming wake frequency and strength (when possible) may allow a control of the overall profile losses, i.e. the losses produced by the friction between blade and fluid.

4.4 Flow visualisation

In Figure 7, snapshots of the vorticity field of Run B and Run D, respectively, have been plotted. The snapshots clearly show the effect of increased wake-strength and wake-frequency on the transition of the suction side boundary layer flow. Between impinging wakes, in Run B the disturbances previously induced by the wakes are partly convected downstream and partly damped, such that at $t/T = 13.50$ the suction side boundary layer appears to be virtually free of disturbances for a large section somewhat upstream of the trailing edge. Immediately upstream of the trailing edge, a region of transitional boundary layer flow is present, as witnessed by the presence of relatively small scale flow structures. During some of the phases, Λ-vortices can be observed which indicate the presence of a natural transition scenario.

Compared to Run B, the increased wake-strength in Run D leads to a stronger triggering of modes in the suction side boundary layer, while the increased wake-frequency significantly shortens the time for the boundary layer to relax. The corresponding snapshot shows not only a larger transitional region immediately upstream of the trailing edge, but also witnesses a more intense triggering of the boundary layer upstream of the transitional region by the impinging wakes.

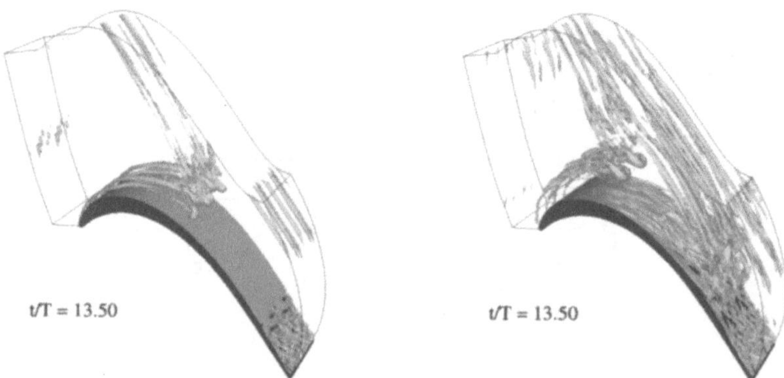

Fig. 7. Instantaneous vorticity field at $t/T = 13.50$ made visible using the λ_2-criterion of Jeong and Hussain [5], left: Run B, right: Run D

Conclusions

The possibility to perform a series of LES on the Hitachi SR8000-F1 supercomputer in Stuttgart, allowed a comprehensive and accurate investigation of the effects of impinging wakes on the boundary layer of a modern low pressure turbine blade.
From the simulations, we reach the following conclusions:

- The wall static pressure distribution of Run A and Run B where found to be in fairly good agreement with the measurements of Schulte and Hodson [16].
- Increasing the strength of the impinging wakes was found to successfully inhibit separation of the suction side boundary layer.
- The total pressure losses were found to be effectively reduced with increasing wake-strength and/or wake-frequency.

Acknowledgements

The authors would like to thank the German Research Foundation (DFG) for funding this project and the steering committee for the supercomputing facilities in Stuttgart for granting computing time on the Hitachi SR8000-F1.

References

1. Breuer, M., Rodi, W.: Large eddy simulation of complex turbulent flows of practical interest, Flow Simulation with High Performance Computers II, Notes on Num. Fluid Mechanics, Vieweg Verlag, (1996).
2. Emmons, H.W.: The laminar-turbulent transition in a boundary layer-Part I, J. of the Aeronautical Sciences, **18**, 490–498 (1951).
3. Germano, M., Piomelli, U., Moin, P., Cabot, W.H.: A dynamic subgrid-scale eddy viscosity model, Physics of Fluids A **3**, number 7, 1760–1765 (1991).
4. Hsu, K., Lee, L., "A numerical technique for two-dimensional grid generation with grid control at all of the boundaries", J. Comput. Phys., **96**, 451-469 (1991).
5. Jeong, J., Hussain, F.: On the identification of a vortex. J. Fluid Mech., **285**, 69–94 (1995).
6. Liu,X., Rodi, W., Experiments on transitional boundary layers with wake-induced unsteadiness, J. of Fluid Mechanics, **231**, 229–256.
7. Mayle, R.E., The role of laminar-turbulent transition in gas turbine engines, ASME J. of Turbomachinery, **113**, pp. 509–537 (1991).
8. Mellen, C.P., Fröhlich, J., Rodi, W.: Computations for the European LESFOIL project. In: E.Krause and W. Jäger (eds.), Scientific Computation in 2000, Springer, (2001).
9. Michelassi, V., Martelli, F., Dénos, T. Arts, C.H. Sieverding: Unsteady Heat Transfer in Stator-Rotor Interaction by Two Equation Turbulence Model, ASME J. of Turbomachinery, **121**, pp. 436-447 (1999).

10. Michelassi, V., Wissink, J.G., Rodi, W.: Analysis of DNS and LES of a Low-Pressure Turbine Blade with Incoming Wakes and Comparison with Experiments, Report. N. 789, Institute for Hydromechanics, University of Karlsruhe, Germany (2002).
11. Michelassi, V., Wissink, J.G., Rodi, W., LES of flow in Low Pressure Turbine with Incoming wakes. In: E. Krause and W. Jäger (eds.), High-Performance Computing in Science and Engineering 2002, Springer, (2002).
12. Michelassi, V., Wissink, J.G., Rodi, W., "Analysis of DNS and LES of Flow in a Low Pressure Turbine Cascade with Incoming Wakes and Comparison with Experiments". Accepted for publication, Flow Turbulence and Combustion, (2003).
13. Michelassi, V., Wissink, J.G., Rodi, W., "DNS, LES, and URANS of periodic unsteady flow in a LP turbine cascade: a comparison". In: Proceedings, 5th European Turbomachinery Conference, (2003).
14. Savill, A.M., A summary report on the COST ERCOFTAC Transition SIG Project evaluating turbulence models for predicting transition, ERCOFTAC Bulletin No. 24, 57-61 (1995).
15. Schlichting, H., Boundary layer theory, McGraw-Hill, New York (1979).
16. Schulte, V., Hodson, H.P., "Unsteady Wake-Induced Boundary Layer Transition in High Lift LP Turbines", ASME Journal of Turbomachinery, **120**, 28-34 (1998).
17. Stadtmüller, P., Investigation of Wake-Induced Transition on the LP turbine Cascade T106A-EIZ, DFG-Verbundproject Fo 136/11, Version 1.0.
18. Walker, G.J., Subroto, P.H., Platzer, M.F., 1988, Transition modelling effects on viscous/inviscid interaction of low Reynolds number airfoil flows involving laminar separation bubbles, ASME Paper No. 88-GT-32.
19. Wissink, J.G., DNS of a separating low Reynolds number flow in a turbine cascade with incoming wakes, Accepted for publication in: Int. J. of Heat and Fluid Flow, (2003).
20. Wu,X., Jacobs, R.G., Hunt, J.R.C., Durbin, P.A., Simulation of boundary layer transition induced by periodically passing wakes, J. of Fluid Mechanics, **398**, pp. 109-153 (1999).
21. Wu, X., P.A., Durbin, "Evidence of longitudinal vortices evolved from distorted wakes in a turbine passage". Journal of Fluid Mechanics, **446**, pp.199-228 (2001).

Numerical High Lift Research II

S. Melber-Wilkending, E. Stumpf, J. Wild, and R. Rudnik

DLR Braunschweig, Institute of Aerodynamics and Flow Technology,
Transport Aircraft, Lilienthalplatz 7, 38108 Braunschweig, Germany

Summary. The project NHLRes is concerned with the simulation of aircraft aerodynamics and thus belongs to the research field of computational fluid dynamics (CFD) for aerospace applications. NHLRes comprises the numerical simulation of the viscous flow around transport aircraft high lift configurations based on the solution of the Reynolds-averaged Navier-Stokes equations. The project NHLRes II, a follow-on activity of the NHLRes project [1], consists of three parts representing a analysis of complex 3D-flow features, wake vortex simulations and an optimization task for selected three-dimensional high lift flow problems.

1 Introduction

1.1 Overview of the project NHLRes

The project NHLRes II is a follow-on activity of the HLRS project "Numerical High Lift Research – NHLRes" [1]. The activity belongs to the research field of computational fluid dynamics for aerospace applications. The main project objective is to push forward three-dimensional numerical investigations of transport aircraft high lift configurations based on methods for the solution of the Reynolds-averaged Navier-Stokes equations.

The follow-on project NHLRes II continues the preceding investigations in the areas of analysis of wing root aerodynamics of high lift aircraft configurations and numerical optimisation. For this purposes a finite volume parallel solution algorithm with an unstructured data concept (DLR TAU code) has been utilized.

Further on the project is intended to extend the scope of research to the field of wake vortex studies. This topic focuses on possible means to minimize the wake vortex hazard behind transport aircraft at take-off and landing to alleviate the airport capacity problem by shorter aircraft separation distances. One mean to achieve this objective is the modification of the high lift system, and thus such type of research is closely related to the other activities carried out in NHLRes II. For this application the EuLag-code[1] is used to simulate the development of vortices in the in the farfield.

All three research fields are characterized by high requirements of computational resources: the analysis task due to the high total grid point number of roughly 12 million nodes, the optimisation task due to the fact that the analysis is embedded in an optimisation procedure and thus is repeated several times within an optimisation run, and finally the wake vortex investigations due to the time-accurate viscous computations. The three sub-activities of NHLRes II and the obtained results are described in the following sections.

2 Wing-Root Aerodynamics at High Lift

2.1 Overview

The preceding activities within NHLRes in the year 2001-2002 concern a systematic numerical study and variation of geometric features at the wing/fuselage intersection, especially the slat end plate devices and the wing/root fairing. The objective has been to determine the impact of such type of devices on the maximum attainable lift and the validation of numerical results against wind-tunnel data.

After a detailed analysis of simulations with geometry variations, differences between measurement and simulation occurred, that could not be explained on the basis of the available results. In the annual report some possible reasons are investigated: differences between semi-span and full-span model tests, vortex motion of the slat-horn- and body-vortex and improvement of the solution resolution with preconditioning.

As the outcome of these investigations is not considered to be fully satisfying, the clarification of the differences between measurement and simulation are to be continued in the first part of the project NHLRes II. First, a measurement of slat/flap position of the wind-tunnel model was completed to update the CAD-description of the numerical grids. This new CAD-model is the basis for subsequent simulations and should improve the simulation validity. Another important item was the investigation of the turbulence-modeling with respect to free vortices and use of improved turbulence models. Both tasks are discussed in this paper. Another task presented here is an direct application of simulating high lift configurations: the aerodynamic investigation of engine integration in high lift configurations.

Currently in progress is an investigation about hysteresis-effects in 3D-high lift simulations: In the experiment a high lift investigation usually starts at moderate angles of attack. The polar measurement is performed then with a subsequent step by step increase to reach a fully developed flow field at high angles of attack. Currently, in numerical simulations the "history" of the flow

[1] National Center for Atmospheric Research, Boulder, Colorado

is neglected, when searching for global steady state solutions. The question to answer is if this simulation procedure is valid or whether the experimental procedure has to be taken into account.

The study of the above mentioned topics shall yield an improved understanding of the simulation capabilities of wing-root aerodynamics at high lift condition. Based on this thorough CFD verification activity, the study of configurative variations, according to existing detailed wind tunnel data appears to be feasible to be computed within the remaining project time.

2.2 Aircraft Configuration and Computational Grids

The same DLR ALVAST wing/fuselage wind-tunnel model [2] as in NHLRes [1] with deployed slat and flaps in take-off configuration is selected as the baseline high lift configuration, Figure 1. The geometry specifications are similar to an AIRBUS A320. For the present investigation the take-off configuration is considered, characterized by a continuous slat with a deflection angle of $\delta_s = 20.0°$ and a single slotted flap with a deflection angle of $\delta_s = 19.5°$. The flap is departed in span-wise direction by a thrust gate.

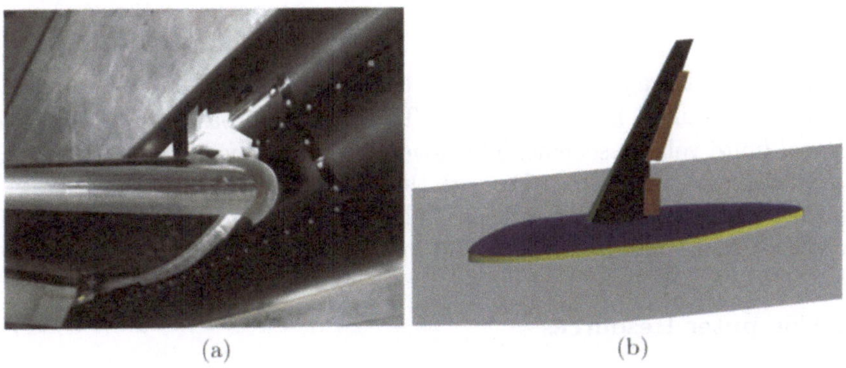

(a) (b)

Fig. 1. (a) Wind-tunnel model of ALVAST high lift configuration with slat-horn at wing/body junction. (b) CAD-geometry of ALVAST high lift configuration with peniche (yellow) and tunnel-wall (grey).

The hybrid unstructured grids are generated with the code system CENTAUR [3]. The grid consists of two parts: a quasi-structured prismatic cell layer is utilized in order to achieve an appropriate resolution of the viscous effects inside the boundary layer. In contrast to this, tetrahedral cells are used to fill the outer domain of the flow-field. To resolve all features of the wind-tunnel geometry, a typical grid consists of about $12 \cdot 10^6$ points.

2.3 Improved CAD Geometry

For the planned configuration variations on the wing-root in NHLRes II an improved CAD geometry of the ALVAST configuration was built. Thereby the positions of the main-wing, the 5 slat- and 3 flap-parts of the wind-tunnel model were measured on a 3D-coordinate measuring device. The resulting nearly 1500 points were used to reposition the corresponding parts in the CAD model by using a numerical optimization tool of DLR. The CPU-time of this task was about one week on a Linux-PC. Now the CAD-description captures most details of the wind-tunnel model, e.g. slat-end-plate, slat-stump and wing-root fairing and the position of the parts of the high lift system are equal to the wind-tunnel model, Figure 1. In the CAD-model the flap- and slat-track fairings are not included.

Because of the differences between full- and half model measurements as shown in NHLRes [1], the planned configuration variations will include the peniche between body and wind-tunnel floor, Figure 1b. The wind-tunnel was likewise included to have a direct comparison to the half-model measurements in DLR-NWB, e.g. [4], without relying on wind-tunnel corrections for high lift configurations [5].

2.4 Numerical Methods

The Reynolds-averaged Navier-Stokes equations (RANS) are solved by the hybrid unstructured flow solver DLR TAU, which is based on a three-dimensional finite volume scheme. The governing equations are solved on a dual background grid, which, together with the edge-based data structure, allows to run the code on any type of grid cells. The solver is part of the MEGAFLOW-project [6], more details can be found in NHLRes [1].

2.5 Computer Resources

In the time-frame from 01.08.2002 to 01.04.2003 the part "Wing-Root Aerodynamics at High Lift" has consumed 800 CPU-hours on NEC-SX5 of HLRS. The main application "Configuration Variations at the Wing-Root" will consume the remaining 3200 CPU-hours on NEC-SX5 of this subtask in the remaining time-frame from 01.04.2003 to 01.08.2003. The results achieved up to now are discussed in detail in the following sections.

2.6 Numerical Results

Vortex Correction:

By simulating complex configurations using Reynolds-averaged Navier-Stokes equations problems appear both in the grid generation and in the solution of the flow equations themself, which among other aspects are related to the resolution and discretisation of local flow phenomena. As an example, the simulation of a high lift configuration at high angles of attack and the numerical discretisation of possibly existing flow separation areas may be mentioned. Further on vortex-dominated phenomena, e.g. the nacelle vortex at an engine [1] or the vortices above a delta-wing at high angles of attack are examples. Vortex dominated phenomena can play an important role in simulating high lift configurations, since the flow condition of such configurations are possibly dominated by vortices. As an example the slat horn vortex, as shown in [1], has an important influence on the attainable maximum lift of such a configuration.

From the literature it is a well known fact that standard turbulence models produce an unphysical high degree of turbulence in vortex cores because of their formulation with vorticity or shear rate, e.g. [7]. On the other hand the insight from measurements and theory is that real vortices have a laminar vortex core. Thus the usage of such turbulence models can lead to an early vortex dissipation, a smear out of the vortex caused by an intense turbulent mixture of the flow or a suppression of the vortex breakdown because of too high dissipation in the vortex core [8].

A possible way out is the usage of a so called "Reynolds-stress model" with a implicit treatment of rotation and curvature effects instead of a simpler "eddy viscosity model". Indeed a "Reynolds-stress model" is unsuitable for complex configurations at the moment because of inadequate robustness and a significantly increased numerical effort.

Another way is the usage of empirical modifications in robust eddy viscosity models concerning the deactivation of the production term in areas of vortex dominated flow. In this section the effect of such a vortex correction is demonstrated for the example of a three dimensional fuselage configuration at high angles of attack with the Spalart-Allmaras turbulence model including the Edwards modification (SAE) [9],[10]. This configuration was already investigated in NHLRes [1] with respect to the peniche influence. Now only the flow separation and the resulting vortex is of interest.

Details regarding this configuration and the construction of the vortex correction can be found in [11], where different modifications concerning the improved prediction of vortex dominated flow with one- and two-equation models on the basis of literature are shown and a resulting modification for the Spalart-Allmaras turbulence model was chosen and demonstrated on the above mentioned example.

All simulations are carried out for a Mach-number of $M_\infty = 0.22$ and a Reynolds-number of $Re_\infty = 2 \cdot 10^6$. The angle of attack was in all cases $\alpha = 21°$. For the SAE turbulence model the variable $\tilde{\nu}$ (a measure of the turbulent eddy viscosity) is shown in Figure 2a in a cut perpendicular to the fuselage axis at 80 percent of fuselage length. Clearly visible is an increased production of $\tilde{\nu}$ in the area of the vortex. This increased eddy viscosity appears over the complete extension of the vortex and leads to an early dissipation of the vortex. In contrast the effect of the SAE turbulence model with the vortex

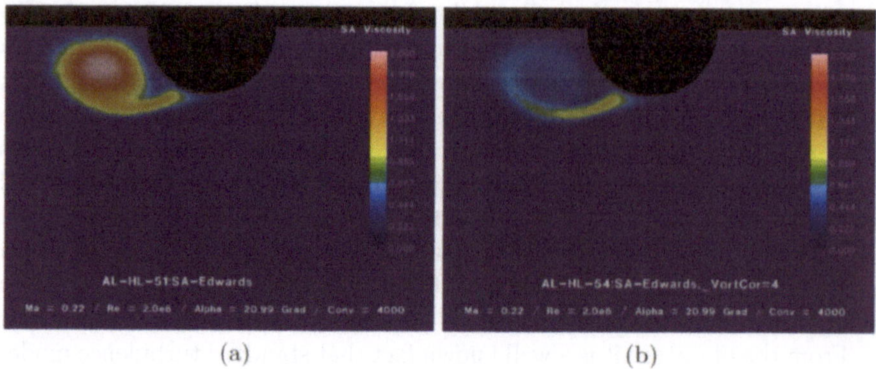

(a) (b)

Fig. 2. Cut perpendicular to the fuselage axis at 80 percent of fuselage length, SA-eddy viscosity $\tilde{\nu}$, simulation with SA-turbulence model with Edwards modification (a) without vortex correction, (b) with vortex correction.

correction is displayed in Figure 2b: A clear reduction of the eddy viscosity in the vortex core can be found, while the flow in the boundary layers remains unchanged.

The presented vortex correction requires only a minor additional effort for the numerical simulation and is easy to implement in existing codes because of its simple formulation. As shown in the literature (e.g. [7]) the Spalart-Allmaras turbulence model with vortex correction (SARC) leads to a comparable solution quality in the simulation of free vortices compared to more complex turbulence models e.g. DES, LES or EARSM, whereas the SARC turbulence model is preferred because of its small additional effort and its stability. The presented vortex correction is an important module to yield an improved understanding of the flow physics of complex high lift flows.

Engine Integration in High Lift Configurations:

As a continuation of the first applications of the DLR TAU-code to the problem of engine/airframe integration for high lift configurations in NHLRes [1], in the current chapter results of engine/airframe integration on the ALVAST high lift configuration with two types of engines are presented. Special attention is paid to possible reductions of the lift due to strong three-dimensional effects of the engine installation [12]. A more detailed analysis of this topic can be found in [13].

The two engine concepts mounted on the ALVAST [2] are first the so called VHBR[2]-engine, Figure 3a), which agrees to a CFM56-5A1 engine and has a bypass ratio of 10. The second engine concept is a UHBR[3]-engine simulator, Figure 3b), which is based on the CRISP concept of MTU Munich. It has a

[2] Very High Bypass Ratio
[3] Ultra High Bypass Ratio

ultra high bypass ratio of 16 in order to improve the propulsion efficiency of the engine. As an adverse effect, the increased engine diameter, which is more than 50 percent larger than that of the VHBR-engine, causes aerodynamic losses due to the negative influence on the wing aerodynamics.

The presented flow computations were performed for a free stream Mach-number of $M_\infty = 0.22$ and a Reynolds-number of $Re_\infty = 2 \cdot 10^6$ corresponding to a mean chord length of $l_\mu = 0.40992\ m$. All calculations were done for an angle of attack of $\alpha = 12°$. To enable a realistic experimental simulation, the engine is powered by compressed drive air in the wind tunnel. This jet effect is modeled numerically by special inflow and outflow conditions, neglecting the inner parts of the engine. The jets are characterized by total pressure and temperature boundary conditions. Details can be found in [13].

In Figure 3 the pressure coefficient c_p is shown on the surface of both configurations. A slight difference in the pressure coefficient can be found on the wing in the area of the engine. A closer look into the differences between VHBR and UHBR-configuration permits the lift distribution in span-wise direction on the components in Figure 4: Because of the bigger diameter the UHBR-nacelle produces more than double the lift of the VHBR-nacelle, see Table 1. On the other hand the bigger diameter of the UHBR-nacelle reduces the lift on the wing in the engine area by shadowing this area against the incoming flow. The same shielding effect reduces the lift on the slat for the UHBR-configuration, except a small peak on the inboard side of the engine. This peak results from a so called nacelle-vortex, which is described in more detail in [1] and [13].

An increased lift can be found for the inboard-flap in the UHBR-case, the lift on the outboard flap is reduced. The reason of the reduction is the shielding effect of the engine due to the wing, slat and flap. The reason of the additional lift inboard can be found in Figure 5: The jet of the VHBR-case is

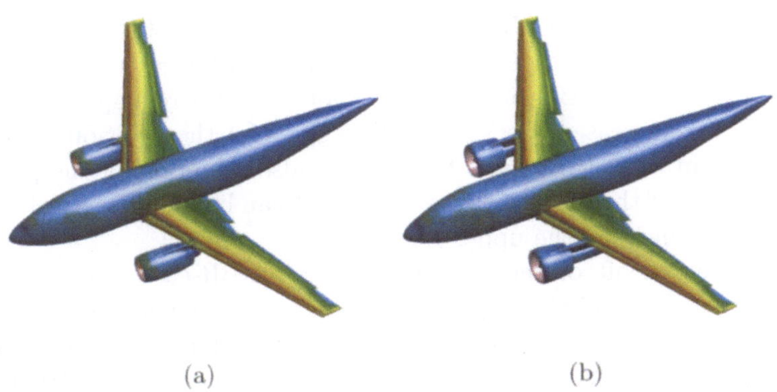

Fig. 3. Pressure coefficient c_p on surface for ALVAST high lift configuration with (a) VHBR-engine and (b) UHBR-engine.

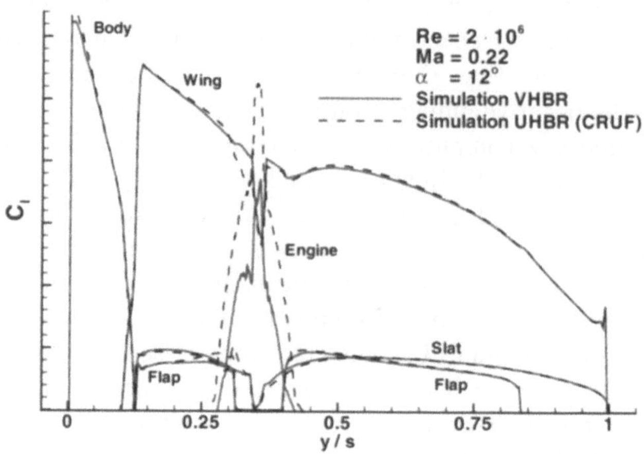

Fig. 4. Lift distribution of VHBR/UHBR-configuration from simulation.

Table 1. Lift distribution over components of ALVAST-high lift with VHBR/UHBR-engine at the same onflow-conditions.

Component	VHBR C_l	C_l/C_L [%]	UHBR C_l	C_l/C_L [%]
Wing	1.2300	59.86	1.2291	56.64
Slat	0.2929	14.25	0.2877	13.26
Flaps	0.1663	8.09	0.1735	7.99
Body	0.2595	12.63	0.2609	12.02
Nacelle	0.1062	5.17	0.2188	10.08
Total	2.0549		2.1700	

smaller than the jet of the UHBR-configuration because of the smaller engine diameter. According to that geometrical difference the jet runs much closer to the flaps in the UHBR-case especially for the inboard flap (due to wing-dihedral). This increased flow speed is the reason for the additional lift on the inboard flap in the UHBR-case. In Figure 5 also the effect of the pylon and the down-wash of the wing and flaps on the jet can be found as a deformation of the jet boundary on the upper side.

The increased lift of the fuselage for the UHBR-case is caused by the displacement of the bigger engine-diameter and the channel-effect between engine and fuselage. But if the integral lift of the component body (Table 1) is considered the portion of lift produced by the body in the UHBR-case is lower than in the VHBR-case because of the bigger portion of engine-nacelle lift in the UHBR-case.

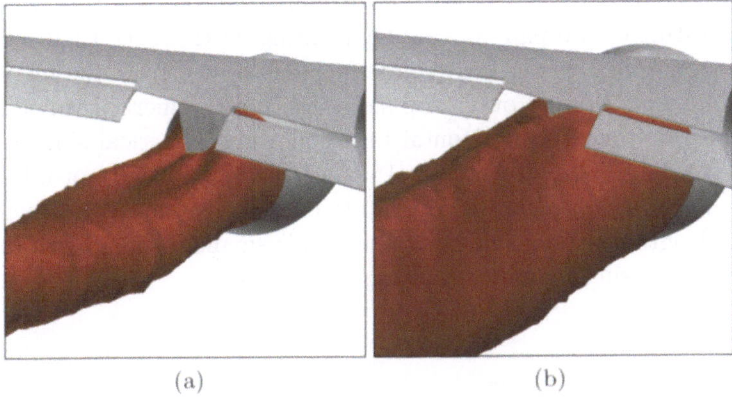

Fig. 5. Engine jet behind the (a) UHBR-configuration and (b) VHBR-configuration.

Overall, the differences between ALVAST high lift VHBR- and UHBR-configuration are constricted to vicinity of the engines. On the outer wing no influence could be found for high lift cases – in contrast to cruise flight, where the engine influence is propagated due to the supersonic flow character over the whole upper wing area [12].

This investigation shows the capabilities of the hybrid unstructured DLR TAU-code with respect to engine/airframe integration, especially for complex high lift configurations with different engine types. Several sources of lift reduction due to engine integration are discussed (see also [13]). It is found numerically that the UHBR-configuration produces more lift than the VHBR-configuration despite of the bigger engine diameter at the same onflow conditions.

3 Wake Vortex Simulation

3.1 Introduction

World airports face a long-term capacity problem due to the separation standards presently in use. These air traffic control (ATC) separation regulations determine the minimum longitudinal distance between two aircraft on the same flight path dependent on the maximum take-off weights of both aircraft in order to prevent the following aircraft from encountering potentially hazardous wake turbulence.

For an alleviation of wake vortices of a subsonic transport aircraft it seems most promising to make use of inherent vortex system instabilities: Either by triggering actively the most unstable wavelength or by designing a highly unstable vortex system for which natural perturbations from the aircraft boundary layer or weak atmospheric turbulence are sufficient for an onset of instability. For the investigation of the latter approach the complex topology of a high lift configuration vortex wake is generally reduced to a four-vortex system, consisting of two pairs of counter-rotating vortices, see Figure 6. These represent on either side of the symmetry plane respectively the merged vortices of the wing tip and outer flap side-edge and the merged vortices of the inner flap side-edge and horizontal tail. Following numerical simulations by Crouch [14] and Fabre & Jacquin [15] the amplification rate of inherent instabilities on such vortex systems is up to one order of magnitude higher and the wave length nearly one order of magnitude smaller compared to the Crow instability on a single counter-rotating vortex pair.

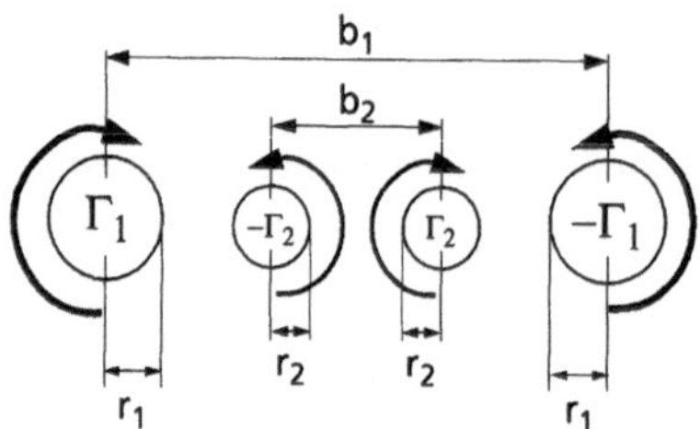

Fig. 6. Sketch of 4-vortex set-up

3.2 Configuration and Computational Grid

Two 4-vortex configurations are chosen to be investigated. As shown in Figure 6 a 4-vortex system is defined by four parameters: the vortex spacing ratio $B = b_2/b_1$, with b_1 resp. b_2 being the spacing of the counter-rotating vortices of the two pairs, $G = \Gamma_2/\Gamma_1$ the ratio of their circulations and the normalized vortex core radii r_1/b_1 and r_2/b_1. Identical in both calculations are the normalized vortex core radii $r_1/b_1 = 5\%$ and $r_2/b_1 = 2.5\%$, the main vortex spacing $b_1 = 2.8\,m$, the spacing ratio $B = 0.35$ and the total circulation $\Gamma = \Gamma_1 + \Gamma_2 = 30.0\,m^2/s$. These values correspond roughly to a scaled down Airbus A340 in landing configuration ($b = 3.56\,m$, $u_\infty = 143\,kts$, $M = 0.21$, $c_L \approx 1.8$, ambient pressure and temperature).

The loading of the horizontal tail plane (HTP) of a subsonic transport aircraft varies typically in the range $-0.5 < c_{L-HTP} < 0$. This translates into a HTP-vortex with a maximum circulation of $\Gamma_{HTP} \approx -0.175\,\Gamma_{wing}$. Accordingly, the parameters for configuration 1 are given by $B = 0.35$; $G = -0.175$. In order to study the potential benefit of a strong inner vortex pair in the wake evolution for configuration 2 the circulation ratio has been doubled ($B = 0.35$; $G = -0.35$).

Taking an analysis by Steijl [16] the Crow wave length corresponding to the radius of the outer vortex pair $r_1/b_1 = 0.05$ is $\lambda = 9.4285\,b$. The simulation domain comprises one Crow wave length in axial direction resolved by 151 points. In vertical direction 201 points (domain height 2.5b) and in lateral direction 401 points (domain width 5b) are distributed, slightly clustered in the center region.

3.3 Numerical Method

The flow solver used is NCAR-EuLag. EuLag was developed at NCAR for meteorological applications and is based on the nonoscillatory advection scheme MPDATA by Smolarkiewicz [17]. EuLag is a pure advection scheme, no diffusive fluxes are explicitly considered. The advection is calculated in a first pass by a first order accurate upstream differencing predictor step. A second pass, as a corrector step, increases the accuracy of the calculation to second order by estimating and compensating the truncation error of the first pass. For the results presented in the following sections the Adams-Bashforth [18] time integration scheme is used, the boundary conditions specified in all calculations are periodic in axial direction and free slip walls in vertical and horizontal direction.

3.4 Computer Resources and Performance

For the presented temporal simulations of the 4-vortex configurations, first presented at [19], a mesh with $151 \times 401 \times 201 \approx 12 \times 10^6$ points is used. This is equivalent to a memory requirement of $8\,GB$ on NEC-SX5. The simulated time for each configuration is $2\,sec$ which corresponds to a normalized time of $t^* = 1.76$. The normalized time t^* is defined as $t^* = (\Gamma_1/(2\pi b_1^2))t$. In total the time needed for the simulation of one configuration is $200\,h$ in single processor mode on a NEC-SX5. This is equivalent to 50 iterations per hour. The peak performance of EuLag on the NEC-SX5 is 1900 MFLOPS.

3.5 Results

Figure 7 show vorticity iso-surfaces (displayed vorticity level is $|\omega| = 0.05 \times max(|\omega_{ini}|)$) of the two generic wake vortex models evolving in time. The rotation rate of the inner around the outer vortices is identical in both configurations. The upcoming instability develops at the same time ($t^* = 0.7$) and shows in the linear phase a comparable topology. On the inner vortices Ω-loops develop which wrap around the outer vortices. The main difference is the vortex interaction. While in configuration 1 the Ω-loops are small and confined to each side of the symmetry plane in configuration 2 the vorticity stretching produces large loops that collide in the symmetry plane. After the dissipation of the scattered vorticity ($t^* = 1.41$) it becomes visible that, just as in configuration 1, the main vortices survive the chaotic process.

An analysis of the temporal rolling moment evolution is given for configuration 1 in Figure 8a) and respectively for configuration 2 in Figure 8b) up to $t^* = 1.76$ ($\widehat{=}$ 2.89 NM for a real size Airbus A340 at 143 kts). For vortex encounter the considered scenario is a medium-class airplane (with half the span

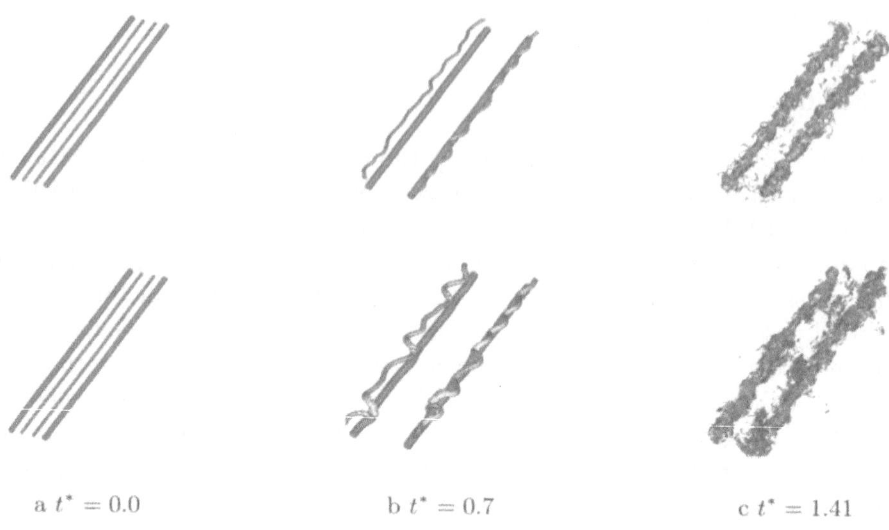

Fig. 7. Temporal development of iso-surfaces of vorticity of configuration 1 (top) and 2 (bottom).

of the wake generator) flying into the vortex center of one of the two main vortices (which represent a heavy-class airplane wake). The ATC separation in this case is 5 NM. In order to account for numerical dissipation effects the calculated rolling moment is analyzed relative to the simulation result for a 2-vortex reference system which was obtained on a mesh with the same resolution as used for the 4-vortex configurations. Total circulation and vortex spacing b_1 of the reference 2-vortex system and the 4-vortex configurations are identical. The rolling moment c_l of configuration 1 and 2 are normalized with the value of the rolling moment $|[c_l]_2|$ of the reference 2-vortex system at a downstream distance equal to the separation distance (i.e. 5 NM). The magnitude of normalized maximum rolling moment coefficient changes periodically, i.e. it is maximal whenever the orbiting inner vortices are in the horizontal plane with the outer vortices.

In Figure 8 the upper curves correspond to the maximal rolling moment found in the left main vortex, the lower curves respectively to the right main vortex. The lower abscissa shows the normalized time t^*, the upper one gives the traveled distance for a real size Airbus A340. In the simulation result for configuration 1 no axial gradient is found up to $t^* \approx 0.8$, Figure 8a) . This changes with the onset of 3D instabilities where a slightly accelerated decrease, different in magnitude in each analyzed plane, is observable. Until $t^* = 1.76$ the maximum rolling moment of configuration 1 has decreased to $\approx 88\%$ of the maximum rolling moment of the reference 2-vortex system. As known from the vorticity iso-surfaces the instability develops more rigorous in configuration 2. Here, the sudden decrease of rolling moment occurs at $t^* = 0.6$ and leads to a drop down to 45%-55% of the maximum rolling moment of the reference 2-vortex system until $t^* = 1.76$, see Figure 8b). This leads to an considerable increase of the airport capacity.

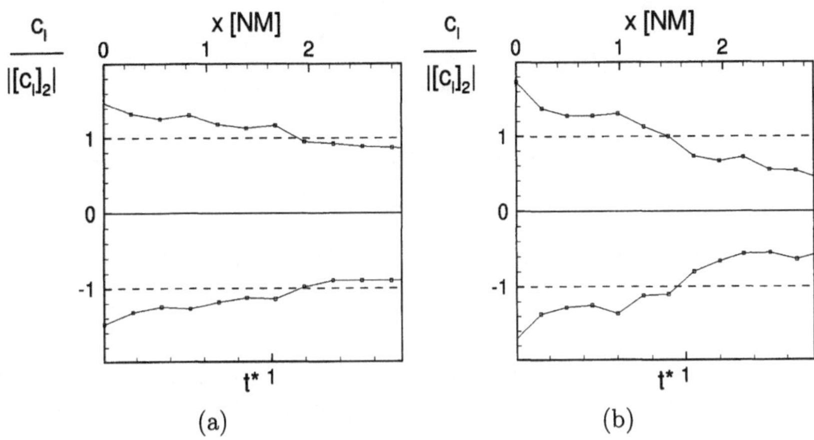

Fig. 8. Rolling moment for configuration (a) 1 and (b) 2.

4 Numerical Optimisation of High Lift Configurations

The objective of the activity "3D High lift optimization" within NHLRes has been an extension of the DLR optimization system for 2D multi-element airfoils to full 3D wing-body high lift configurations. The goals of the project could not be totally reached due to work load as explained in the last annual project report [1]. On the other hand, first sensitivity studies carried out in NHLRes, showed, that the grid generation approach is unpracticable for optimization. The sensitivities of the targeted AST-wing configuration were not captured as expected.

The project NHLRes II is targeted to solve the problems encountered in the preceding project runtime. Shortcomings of the grid generation system have been fixed in the workstation-version of the DLR grid generator MegaCads [20]. Now, instead of using hybrid grids with prisms near the walls and triangles on the surface a semi-structured grid approach is followed, which consists of structured grid-like hexaedral elements in the boundary layers leading to quadrilateral elements on the surface.

As an example a Navier-Stokes grid for the M6-wing is shown in Figure 9 with a boundary layer resolution of 64 cells normal to the surface and a farfield distance of 10 wing-spans. Figure 9a) shows a hybrid grid generated with the code system CENTAUR [3], which has $5.32 \cdot 10^6$ points and $10 \cdot 10^6$ elements.

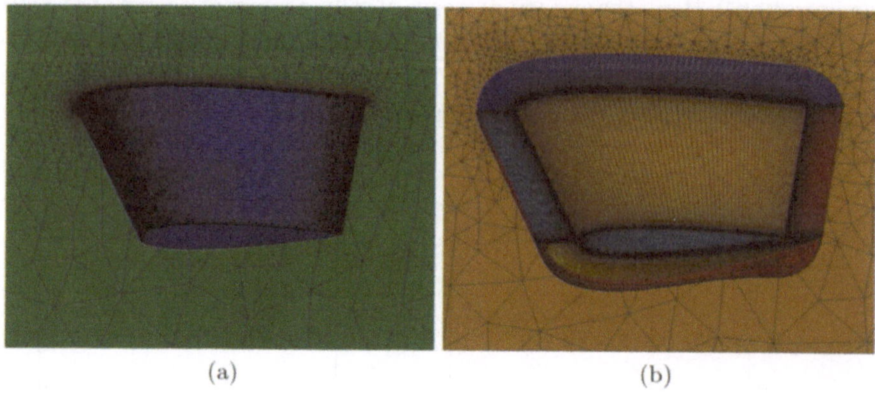

(a) (b)

Fig. 9. Navier-Stokes grid for M6-wing: (a) hybrid unstructured (b) semi-structured.

In Figure 9b) for the same configuration a semi-structured grid generated with MegaCads is displayed with only $1.15 \cdot 10^6$ points and $1.32 \cdot 10^6$ elements.

As shown on this example the technique of semi-structured grids saves computational points due to the high stretching ratios allowed for this type of cells. Additionally a better resolution of the boundary layer is reached giving probably the right sensitivities on geometric changes. With this technique 3D high lift optimization seems to be practicable and will be continued in the remaining project run-time.

5 Summary

The first part of the project NHLRes II "Wing-Root Aerodynamics at High Lift" continues the work from the first project NHLRes to clarify differences between measurement and simulation. The improved CAD geometry and improvements concerning vortex correction for the Spalart-Allmaras turbulence model (SARC) are presented. Further more, the current abilities of 3D high lift simulation are demonstrated on the example of engine integration on high lift configurations. Based on this thorough CFD verification activity, the study of configurational variations, according to existing detailed wind tunnel in the remaining project run-time appears now feasible.

The second part of the project deals with wake vortex studies. The effect of a stronger inner vortex pair of a 4-vortex configuration was demonstrated on the evolution of vorticity in time and the resulting rolling moment of an aircraft flying in the vortex center.

Concerning the optimization activities preparing studies on a grid generation technique have been carried out. As shown semi-structured grids save computational points due to the high stretching ratios allowed for this type of cells and allows a cost efficient optimization loop for 3D high lift configurations.

Acknowledgment

The numerical simulations reported in this paper were performed at the High-Performance Computing-Center Stuttgart. The authors would thank the HLRS for the generous granting of their computational facilities for the project NHLRes.

References

1. Melber, S.; Wild, J.; Rudnik, R.: Numerical High Lift Research - NHLRes. Annual Review 2001. High Performance Computing in Science and Engineering '02, Springer-Verlag Berlin, Heidelberg, New York, 2002, pp. 406–421.
2. Kiock, R.: The ALVAST Model of DLR. DLR IB 129-96/22, 1996.
3. Kallinderis, Y.: Hybrid Grids and Their Applications. Handbook of Grid Generation, CRC Press, Boca Raton / London / New York / Washington, D.C., pp. 25-1 - 25-18, 1999.
4. Puffert-Meissner, W.: ALVAST Half-Model wind-tunnel Investigations and Comparison with Full-Span Model Results. DLR IB 129-96/20, 1996.
5. Rogers, S.E.; Roth, K.: CFD Validation of High-Lift Flows with significant Wind-Tunnel Effects. AIAA paper 2000-4218, 2000.
6. Kroll, N.; Rossow, C.-C.; Becker, K.; Thiele, F.: MEGAFLOW - A Numerical Flow Simulation System. 21st ICAS congress, 1998, Melbourne, 13.9.–18.9.1998, ICAS-98-2.7.4, 1998.
7. Dacles-Mariani, J.; Zilliac, G.G.: Numerical/Experimental Study of a Wingtip Vortex in the Near Field. AIAA Journal, No. 4, April 1996.
8. Morton, S.A.; Forsythe, J.R.; Mitchell, A.M.; Hajek, D.: DES and RANS Simulations of Delta Wing vortical Flows. AIAA-paper 2002-0587, 2002.
9. Spalart, P. R.; Allmaras, S.R.: A One-Equation Turbulence Model for aerodynamic Flows. La Recherche Aerospatiale, Nr. 1, 5–21, 1994.
10. Edwards, J.R.; Chandra, S.: Comparison of Eddy Viscosity-Transport Turbulence Models for Three-Dimensional, Shock-Separated Flowfields. AIAA Journal, No. 4, April 1996.
11. Melber, S.: Wirbelkorrektur fuer Ein- und Zweigleichungs-Turbulenzmodelle und Implementation fuer das Spalart-Allmaras Turbulenzmodell in den Stroemungsloeser DLR-TAU. DLR IB 124-2002/17, 2002.
12. Hoheisel, H.: Aerodynamische Effekte bei der Zelle-Triebwerksintegration von Verkehrsflugzeugen. DLR IB 129-99/29, 1999.
13. Melber, S.: 3D RANS-Simulations for High-Lift Transport Aircraft Configurations with Engines. DLR IB 124-2002/27, 2002.
14. J.D. Crouch: Instability and transient growth of two trailing-vortex pairs. J. Fluid Mech., Vol. 350, pp. 311–330, 1997.
15. D. Fabre, L. Jacquin: Stability of a four-vortex aircraft wake model. Phys.Fluids, Vol. 12, No. 10, pp. 2438–2443, 2000.

16. R. Steijl: Computational Study of Vortex Pair Dynamics. Ph.D. thesis, University of Twente, Netherlands, 2001.
17. P.K. Smolarkiewicz, L.G. Margolin: MPDATA: A finite-difference solver for geophysical flows. J.Computational Physics, No. 140, pp. 459–479, 1998.
18. L.F. Sampine, M.K. Gordon: Computer Solution of Ordinary Differential Equations: the Initial Value Problem. Freeman, 1975.
19. E. Stumpf: Numerical Study of Four-Vortex Aircraft Wakes. Accepted for publication in Notes Num. Fluid Mech., 2003.
20. Brodersen, O., Hepperle, M., Ronzheimer, A., Rossow, C.-C., Schoning, B.: The Parametric Grid Generation System MEGACADS. Proc. of the 5th Intern. Conference on Numerical Grid Generation in Computational field Simulations 1996, Mississippi, Ed.: Soni, B.K., Thompson, J.F., Hauser, J., Eisemann, P., pp. 353–362, 1996.

Prediction of the Model Deformation of a High Speed Transport Aircraft Type Wing by Direct Aeroelastic Simulation

C. Braun, A. Boucke, M. Hanke, A. Karavas, and J. Ballmann

Lehr- und Forschungsgebiet für Mechanik, RWTH Aachen,
D-52062 Aachen, Germany
carsten@lufmech.rwth-aachen.de

1 Introduction

The aerodynamic performance, maneuverability and flight stability of aircrafts are highly dependent on the deformation of their wings under aerodynamic loads. The accurate prediction of aeroelastic properties, such as aeroelastic equilibrium configurations under cruise conditions, is therefore crucial in early design stages. Due to increasing computer power and further development of numerical methods, direct numerical aeroelastic simulation, in which the governing equations for the fluid and the structure are solved consistently in time, has become feasible [1, 2]. In the collaborative research center SFB 401 "Flow Modulation and Fluid-Structure Interaction at Airplane Wings" at Aachen University the numerical aeroelastic method SOFIA (SOlid Fluid InterAction) for direct numerical aeroelastic simulation is being progressively developed [4, 5]. According to the coupled field approach, three modules can be identified in SOFIA: flow solver, structural solver and a grid deformation tool, which is necessary since the boundaries of the computational grid for the flow solver always have to coincide with the deforming aerodynamic surface of the structure.

The aeroelastic code SOFIA has been applied within the HiReTT project (High Reynolds Number Tools and Techniques) which is partly funded by the European Union [6]. The project's main objectives are to understand the Reynolds number and Mach number effects as well as the influence of model elasticity on the performance of a clean wing-body configuration and furthermore, to assess the ability of current CFD analysis methods to predict these effects. For this purpose an extensive wind tunnel data base has been available, which has been collected within the frame of the HiReTT project in the European Transonic Wind Tunnel (ETW) in Cologne for a modern high speed transport aircraft type configuration.

In order to achieve a good agreement between experimental data and results from pure CFD methods, the deformation of the wind tunnel model under aerodynamic loads has to be determined, e.g. by using a FEM method applying loads according to measured pressure distributions. The computed wing deformation has then to be considered during the generation of the CFD grids.

The second approach within the HiReTT programme is to combine the numerical prediction of model deformation and flow solution by application of aeroelastic methods such as SOFIA. The goal is to predict experimental data correctly starting from the undeformed configuration without any initial information about the model deformation.

In the present work, computational results for the model deformation obtained with SOFIA are presented. Furthermore, computed pressure distributions are compared with experimental data for $Re{=}8.1{\cdot}10^6$ and $Re{=}32{\cdot}10^6$ at three angles of attack and a Mach number of 0.85, and the influence of model deformation is demonstrated.

2 Physical Models and Numerical Methods

2.1 Fluid Flow

The governing equations for the flow field are the three-dimensional Reynolds-averaged Navier-Stokes equations which are derived from the Navier-Stokes equations by introducing a time averaging procedure. The integral form of these equations, which form the basis for the finite-volume technique, read

$$\frac{\partial}{\partial t}\int\limits_{V(t)} \mathbf{U} dV + \int\limits_{\partial V(t)} \mathbf{F}\mathbf{n} dS = 0\,, \tag{1}$$

where $\mathbf{U} = (\rho, \rho u, \rho v, \rho w, \rho e)^T$ is the algebraic vector of conserved quantities: density, Cartesian components of momentum and specific total energy. $\mathbf{F} = \mathbf{F}^c + \mathbf{F}^d$ represents the flux function with its parts $\mathbf{F}^d$ denoting the diffusive part and $\mathbf{F}^c$ designating the convective terms including pressure.

In the present method, the flow solver FLOWer, developed by DLR [7], is employed to solve the governing flow equations on block-structured grids. Because a finite volume technique with control volumina dependent on time is used, FLOWer is applicable for aeroelastic problems where a temporal deformation of the computational mesh is required. Central differences are used for the spatial discretization. For steady state calculations, time integration is performed by a 5-step Runge-Kutta method. To achieve a good convergence rate acceleration techniques such as local time stepping, implicit residual smoothing, and multigrid algorithms are used. In case of unsteady flow computations, the implicit dual-time stepping scheme is employed [8], [9].

Several turbulent models are available in FLOWer. For the computation of the results shown in the present work only the Wilcox $k - \omega$ model has been applied.

2.2 Beam Model

In SOFIA the elastic wing is modeled by a Timoshenko-like beam structure with six degrees of freedom for a material cross-section and non-coinciding centerlines of gravity, bending, and torsion. In contrast to the often used Euler-Bernoulli beam theory, the Timoshenko approximation with its two more degrees of freedom concerning the shear deformation exhibits no effects of anomalous dispersion and thus describes unsteady deformation in a physically reasonable way.

The governing equations can be found via Hamilton's principle. Based on the resulting variational formulation, a system of second order in time ordinary differential equations (ODEs) is derived by applying the finite element method in the sense of Ritz/Kantorowitsch [10, 11]. Discretization is done by iso-parametric, two–noded elements. A reduced integration scheme avoids shear locking. The set of ODEs is integrated by Newmark's method, where the resulting linear system of equations is solved directly using LU-decomposition. The external forces are assumed to vary linearly during every time-step. Alternatively, the system of ODEs is diagonalised by solving the generalised eigenvalue problem (EVP), then the time integration is done by evaluation of Duhamel's integral [10].

2.3 Grid Deformation Tool

In every time step of an aeroelastic computation the computational mesh for the flow solver has to be updated accordingly to the change of shape of the wing surface. Therefore an algorithm has been developed, in which the block boundaries and a certain number of grid lines, which depends on the grid topology, are modeled as a fictitious framework of elastic beams. These beams are considered rigidly fixed together in points of intersection and to surfaces as well, such that angles are preserved where beams (=grid lines) intersect or emerge from a solid surface. The deformation of the framework, which is due to displacements of the surface grid nodes, is calculated by a finite-element solver. The new positions of grid points in the interior of the domain which are not included in the beam-framework are found by interpolation.

3 Solution Strategy and Coupling

The solution schemes for the flow field and the structure are combined in an iterative process. The central time loop works as follows:

1. Step: Starting with the initial geometry, the flow field is computed for the first time step using the FLOWer code.

2. Step: The aerodynamic forces resulting from pressure and friction (in case of viscous flows) on each grid node on the wing surface are calculated and projected via the principle of virtual work onto the nearest finite element of the beam under consideration of rigid cross-section behavior in the beam. Thus, an equivalent load distribution for forces and moments how they act on the beam is determined and the structural solution for the current time step can be computed. In order to calculate the deformation of the wing surface consistently, every node of the fluid mesh on the surface is assumed being rigidly connected to the nearest finite element of the beam. By that assumption the instantaneous dynamic state of deformation concerning position and velocity of each surface grid node can be updated corresponding to the instantaneous dynamic solution for the beam structure.

3. Step: The third step is performed to update the computational grid for the fluid solver accordingly to the change of shape of the wing surface.

In static aeroelastic analysis the equilibrium configuration can be calculated in a sequence of instantaneous equilibrium configurations applying in each iteration step the aerodynamic loads according to the converged stationary solution for the flow field about the actual configuration. A necessary prerequisite for this method to converge to the correct solution is the absence of negative aerodynamic damping. This quasistatic approach, if admissible, is particularly suitable for costly problems with a high number of grid points and was therefore applied for the computation of the results which are presented in this paper.

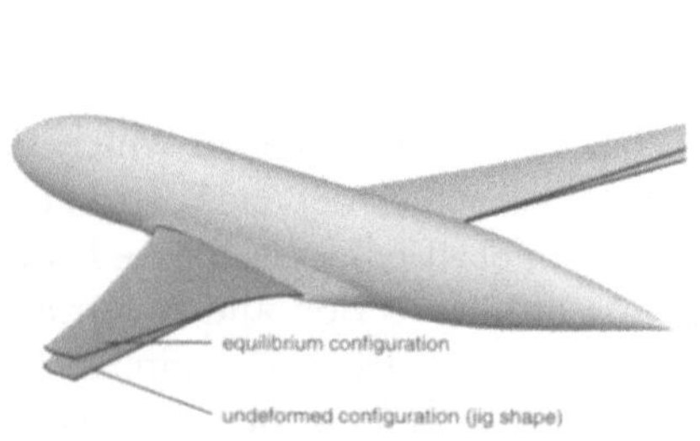

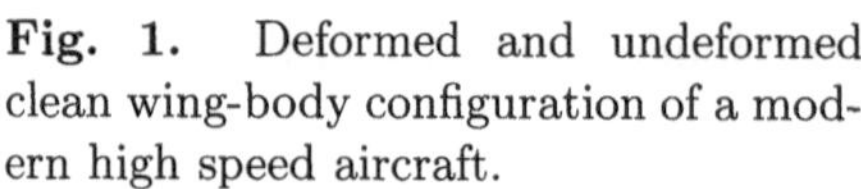

Fig. 1. Deformed and undeformed clean wing-body configuration of a modern high speed aircraft.

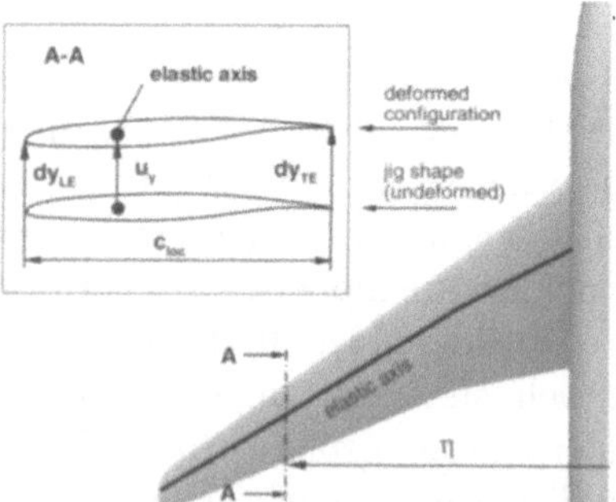

Fig. 2. Definition of the aerodynamic twist.

4 Results

During wind tunnel tests, the wings of wind tunnel models bend upward under the aerodynamic loads, Fig. 1. For swept back wings this leads to a reduction of the local angle of incidence, Fig. 2. Besides bending, also torsional deformation may be relevant. The degree of deformation depends on the flexibility of the model and the total pressure of the incoming flow. Especially for high Reynolds-number test conditions in cryogenic wind tunnels the twist distribution of the wing can be changed considerably. This leads to pressure distributions and an aerodynamic performance different from those for the undeformed configuration (=*jig shape*).

Since experimental data in stationary flow is always measured for the configuration which a model assumes under wind tunnel conditions (static aeroelastic equilibrium configuration), the model deformations have to be considered when trying to predict the experimental data numerically. One approach within the HiReTT programme is the direct aeroelastic simulation using SOFIA. Starting from the undeformed configuration, the static aeroelastic configuration and the corresponding flow solution are computed in an iterative process as described in chapter 3. The structural data for the Timoshenko-like beam structure of the wind tunnel model, which is a two-piece full material assembly with cavities only for the measurement equipment like probe tubes and wiring, have been identified applying a three-dimensional finite element code for long elastic solids of full material having constant cross-sections according to the actual cross-sections of the wing model at different span-wise stations.

In this work, the predicted model deformations are presented and computed pressure distributions for the deformed model are compared with data measured in experiments in the European Transonic Wind Tunnel (ETW). Table 1 gives an overview of the test cases considered in this paper. Test

Table 1. Flow parameters and global coefficients, Re=$8.1 \cdot 10^6$.

Re=$8.1\cdot10^6$, Ma=0.85, q/E=$0.44\cdot10^{-6}$, Trans.:15%/5%			
	case 1 (Fig. 4)	case 2 (Fig. 5)	case 3 (Fig. 6)
α	$0.2086\cdot\alpha_{ref}$	$1.0576\cdot\alpha_{ref}$	$2.0066\cdot\alpha_{ref}$
Δc_l	4.78%	1.52%	1.89%
Δc_d	10 dc	9 dc	25 dc
$\varphi_{z,max}$	$0.87\cdot\alpha_{ref}$	$1.16\cdot\alpha_{ref}$	$1.58\cdot\alpha_{ref}$

Table 2. Flow parameters and global coefficients, Re=32.5·10^6.

Re=32.5·10^6, Ma=0.85, q/E=0.44·10^{-6}, Trans.: free			
	case 4 (Fig. 7)	case 5 (Fig. 8)	case 6 (Fig. 9)
α	0.1456·α_{ref}	1.0000·α_{ref}	1.7948·α_{ref}
Δc_l	1.55%	0.20%	2.16%
Δc_d	5 dc	4 dc	18 dc
$\varphi_{z,max}$	0.89·α_{ref}	1.19·α_{ref}	1.51·α_{ref}

cases 1–3 are for a Reynolds number of 8.1·10^6, test cases 4–6 for a Reynolds number of 32.5·10^6 at a Mach number of 0.85.

It should be noted that angles of attack as well as twist angles are normalised with a reference angle α_{ref}, which is the angle of attack of case 5 and represents a typical incidence for a large transport aircraft at cruise conditions. Furthermore, no absolute values for lift or drag coefficients are presented, only differences to the experimental results in percent (lift) and drag counts (1 dc = 0.0001).

4.1 Lift and Drag Coefficients

Tables 1 and 2 show the differences between computed and measured lift coefficients (Δc_l) and drag coefficients (Δc_d), respectively. The computed results reproduce the experimental values for all test cases well. However, the agreement for the *Re*=32.5·10^6-test cases concerning lift and drag is better. The most complicated test cases are case 3 and case 6. In these cases flow separation occurs which cannot be perfectly captured by the used turbulence model. This deficiency may result in slightly shifted shock positions.

4.2 Model Deformation

The deformation of the wing at each span-wise position is predominantly characterised by two values: vertical displacement of the elastic axis u_y (see Fig. 2) and aerodynamic twist φ_z. The aerodynamic twist is not the torsional deformation of the wing (which would be the twist around the elastic axis of the wing), but represents the twist of the aerodynamic cross-sections of the wing, i.e. the change of the local angle of incidence of a wing cross-section in the vertical plane spanned by the undisturbed flow direction and the vertical axis. Thus, because of the sweep angle of the wing, it includes the corresponding parts of torsional twist and bending. For a positive sweep angle and upward bending, φ_z appears as a reduction of the local angle of incidence compared to the undeformed geometry. The aerodynamic twist φ_z is defined by

$$\varphi_z(\eta) = \arctan \frac{dy_{TE}(\eta) - dy_{LE}(\eta)}{c_{loc}(\eta)} \quad ,$$

with η as a non-dimensional span-wise coordinate. $dy_{LE}(\eta)$ and $dy_{TE}(\eta)$ denote the displacements of the leading and trailing edge, respectively (see Fig. 2). $c_{loc}(\eta)$ represents the local chord length of the wing. Hence, if φ_z is positive, the local angle of attack is reduced.

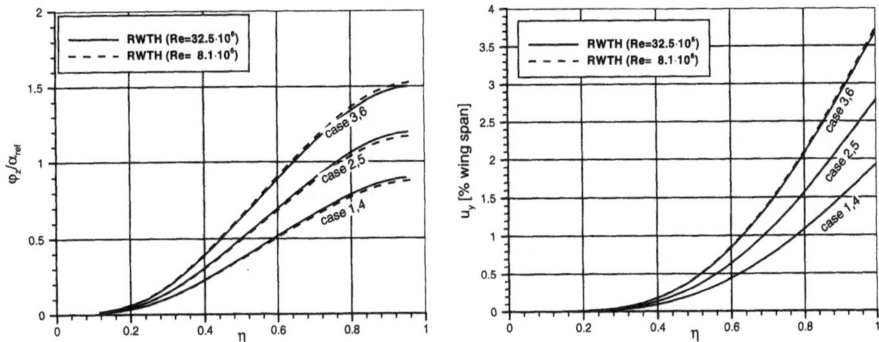

Fig. 3. Deformation of the wing. Left: aerodynamic twist φ_z. Right: Displacement of the elastic axis u_y.

Figure 3 shows the computational results for φ_z (normalised with α_{ref}) and u_y (in % wing span) for cases 1-6 plotted over η. The lines in Fig. 3, top, are not continued up to η=1 (=wing tip). This is due to the fact that the evaluation of φ_z implies straight grid lines of the CFD mesh on the wing surface, which is not the case at the wing tip.

As can be seen, the results show a considerable twist deformation of the wing of up to 1.6 times α_{ref}, which means that the change of the local angle of attack at the wing tip due to wing deformation is of the magnitude of the global angle of attack. The maximum twist $\varphi_{z,max}$, which occurs at the wing tip, is also listed in Tables 1 and 2. The bending displacement u_y reaches values of 3.7 % of the wing span for cases 3 and 6. The computed results agree well with results from FEM analysis performed by Airbus UK [12]

The aeroelastic results indicate only a very small effect of the Reynolds number on the wing deformation. The differences in Fig. 3 are probably not only due to variation in Reynolds number but also due to slightly different angles of attack, see Tables 1 and 2.

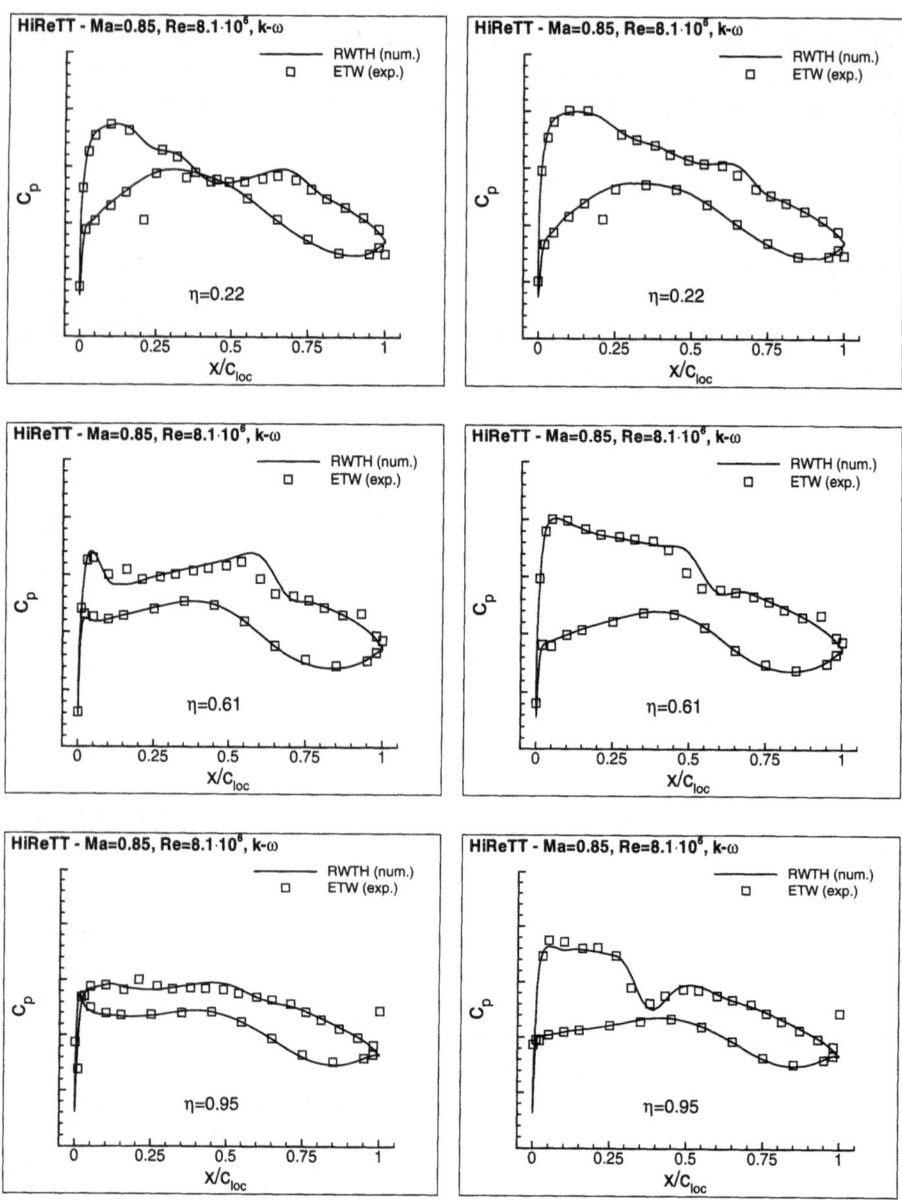

Fig. 4. Pressure distribution (case 1: Re=8.1·10^6).

Fig. 5. Pressure distribution (case 2: Re=8.1·10^6).

4.3 Pressure Distributions

Pressure distributions for all test cases are presented for different span-wise locations in Figs. 4–9. Symbols denote the experimental data whereas the solid lines represent the computed distributions for the static aeroelastic equilibrium configurations. The dashed lines, shown in Figs. 7–9, indicating the solu-

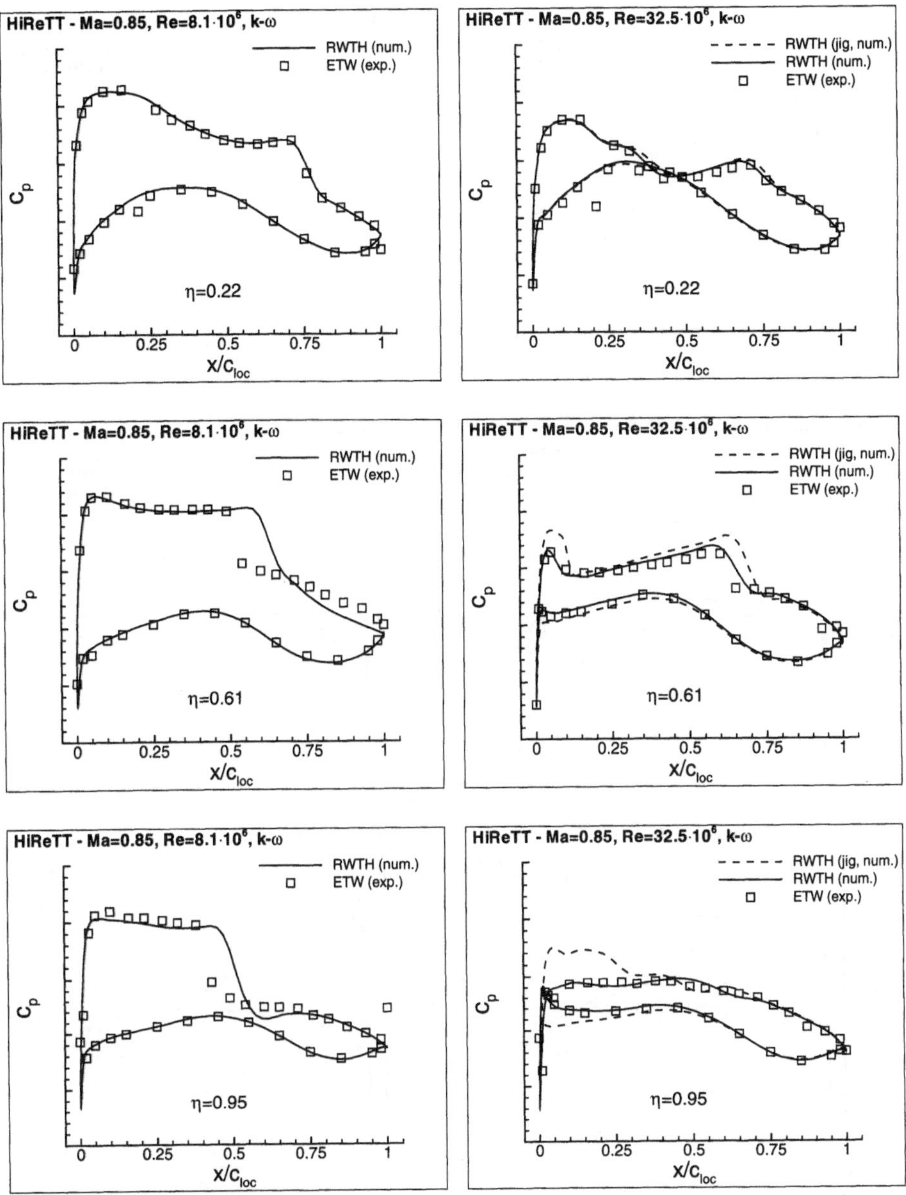

Fig. 6. Pressure distribution (case 3: Re=$8.1 \cdot 10^6$).

Fig. 7. Pressure distribution (case 4: Re=$32.5 \cdot 10^6$).

tion for the undeformed geometry (jig shape), are included to demonstrate the overall effect of the wing deformation. According to the large twist deformations, the pressure distributions for undeformed and deformed configuration differ considerably.

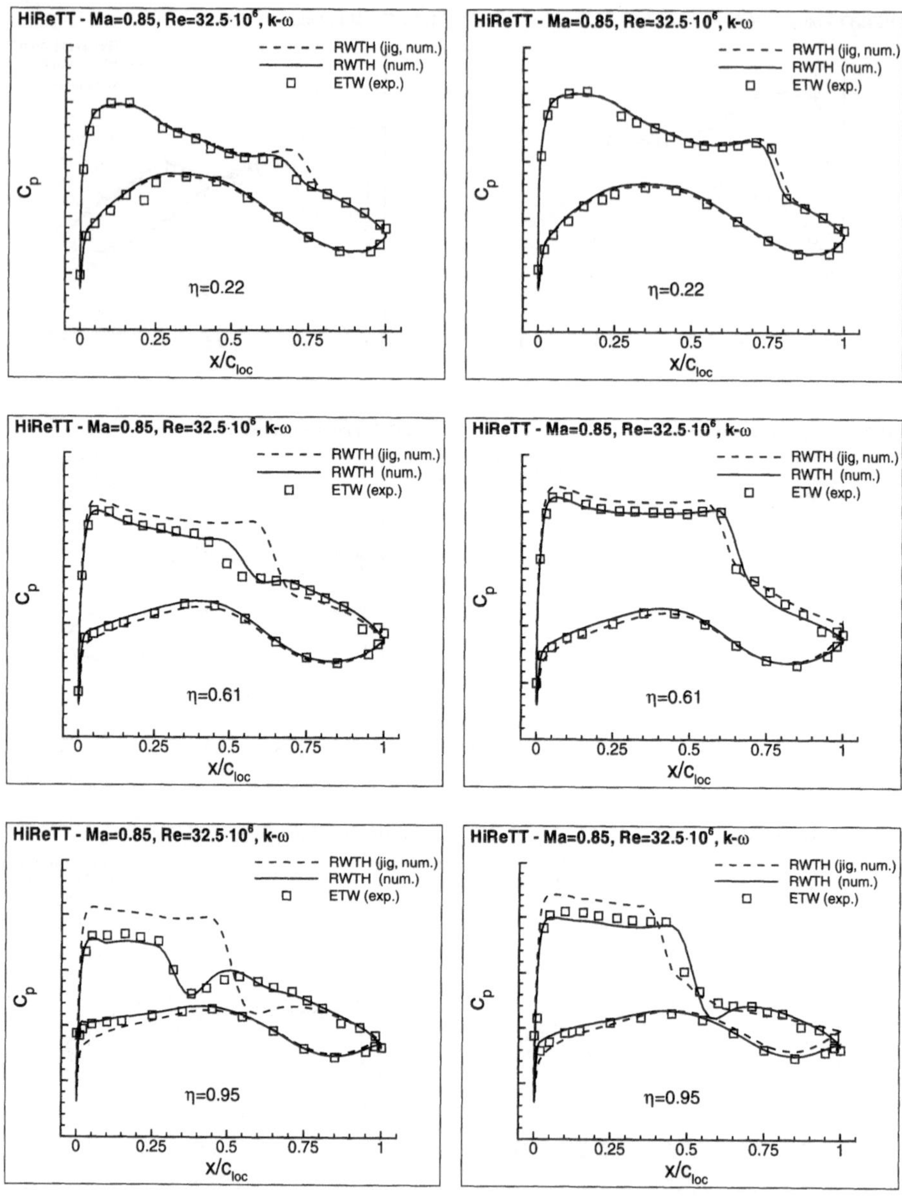

Fig. 8. Pressure distribution (case 5: Re=32.5·10^6).

Fig. 9. Pressure distribution (case 6: Re=32.5·10^6).

For all cases (except case 3), the agreement is well. The pressure plateaus on the upper surface as well as the distributions on the lower surface are captured accurately. The main difference can be found regarding the shock position on the upper side, particularly for cases 3 and 6 where separation was found.

5 Conclusions

Numerical results are presented for a wind tunnel model of a wing body configuration, how it was investigated in the European Transonic Wind Tunnel (ETW) within the frame of the HiReTT programme at high Mach numbers and Reynolds numbers. A comparison of computational results and measured data for the static aeroelastic problem of the deforming wind tunnel model under aerodynamic loads shows a good agreement. Comparison with computed pressure distributions for the undeformed wing show that the neglect of the aeroelastic deformation leads to completely wrong results.

6 Code performance

During aeroelastic computations using a beam model to represent the aircraft structure, only approximately 1% of the cpu time is consumed for calculating the structural deformation and 99% for the computation of the flow field. Since the FLOWer code has been highly optimised for vector architectures, a code performance of about 800 MFLOPS is achieved on a single NEC SX4 processor. Steady aeroelastic analysis of a wing body configuration using Navier-Stokes equations to model the fluid flow, as presented in this paper, takes 15 hours and 2 GB memory on a single NEC SX5 processor for a computational mesh of 4 mio. grid points. A more frequent utilisation of the FLOWer code with its parallel option using the clic3d library will be required as soon as dynamic aerodynamic problems under consideration of viscous effects are investigated. In order to obtain acceptable turn-around times hardware architectures like NEC SX4 and NEC SX5 with powerful processors are of a great importance for this project.

7 Acknowledgements

This work has been partly supported by the Deutsche Forschungsgemeinschaft (DFG) in the Collaborative Research Center SFB 401 "Flow Modulation and Fluid-Structure Interaction at Airplane Wings" at Aachen University. Computations were performed using the NEC-facilities of the High-Performance Computing-Center Stuttgart (HLRS). We would like to express our gratitude to Dr. R. Heinrich from the German Aerospace Research Establishment (DLR/Braunschweig) and all partners within the project MEGAFLOW developing the FLOWer code as well as all partners within the HiReTT project, especially Airbus UK, EADS Airbus, Aerospatiale Matra-Airbus, ETW, who provided the opportunity for testing SOFIA against transonic high Reynolds number experiments. The HiReTT project, coordinated by Airbus UK, is partly founded by the European Commission within the 5th Frame work programme.

References

1. Beckert, A.: Coupling Fluid (CFD) and Structural (FE) Models using Finite Interpolation Elements, Aerosp. Sci. Technol. 4 (2000), pp. 13–22
2. Henke, H.H.: The Viscous-Coupled 3D Euler Method EUVISC and its Aeroelastic Application, CEAS International Forum on Aeroelasticity and Structural Dynamics, IFASD2001, Madrid, Vol. 2, pp. 95–107
3. Ballmann J.: Flow Modulation and Fluid-Structure Interaction at Airplane Wings – Survey and Results of the Collaborative Research Center SFB 401, DGLR-2002-009, 2002
4. Ballmann, J., Britten, G., Hurka, J.: Fluid-Structure Interaction at Panels and Large Span Wings with Computational Aeroelasticity. CEAS International Forum on Aeroelasticity and Structural Dynamics, IFASD 2001, Madrid, Vol. 2, pp. 51-62
5. Kämpchen, M., Dafnis, A., Reimerdes, H.-G., Britten, G., Ballmann, J.: Dynamic Aero-Structural Response of an Elastic Wing Model, accepted for publication in Journal of Fluids and Structures, 2002
6. Rolston, S., Elsholz, E.: Initial Achievements of the European High Reynolds Number Aerodynamic Research Project 'HiReTT', AIAA paper 2002-0421, 2002
7. Kroll, N., Rossow C.-C., Becker, K., Thiele, F.: The MEGAFLOW project, Aerosp. Sci. Technol. 4 (2000), pp. 223–237
8. Heinrich, R., Pahlke, K., Bleecke, A.: Three-Dimensional Dual-Time Stepping for the Solution of the Unsteady Navier-Stokes Equations, Proceedings RAS "Unsteady Aerodynamics" Conference, London 1996, pp.5.1–5.12
9. Jameson, A. J.: Time dependent calculation using multigrid, with applications to unsteady flows past airfoils and wings, AIAA paper 91-1596, 1991
10. Nellessen, D.: Schallnahe Strömungen um elastische Tragflügel, VDI Fortschrittsberichte Reihe 7: Strömungstechnik, Nr. 302, 1996
11. Britten, G.: Numerische Aerostrukturdynamik von Tragflügeln großer Spannweite, Dissertation RWTH Aachen, Shaker Verlag, Aachen, 2002
12. Gibson, T. M.: Investigation of Wind Tunnel Model Deformation under High Reynolds Number Aerodynamic Loading, AIAA paper 2002-0424, 2002

Rayleigh-Bénard Convection at Large Aspect Ratios

T. Hartlep and A. Tilgner

Institute of Geophysics, University of Göttingen, Herzberger Landstr. 180, 37075 Göttingen

Summary. Rayleigh-Bénard convection is simulated in a plane layer with periodic boundary conditions in the horizontal directions. A spectral method allows us to reach Rayleigh numbers up to 10^7 even for an aspect ratio of 10.

1 Introduction

Convection is a ubiquitous phenomenon. It occurs in many engineering applications, in the atmosphere, the oceans, and on a long time scale even in the mantle of the Earth. Most applications deal with turbulent flows. Research on turbulent flows focuses on at least two fairly well separated sets of questions, i.e. the study of the "small scales" of the flow and its "coherent structures".

The interest in the small scales stems from the hope that these are "universal", i.e. identical from one flow to the other. If the behavior at small scales is indeed independent of large scales, it should be possible to parameterize the effect of the small scales on the large scales so that it would not be necessary to spatially resolve the smallest scale of the flow in numerical simulations. "Large eddy simulations" (LES) are one class of methods exploiting this idea.

Coherent structures on the other hand are organized features of the flow which are distinguished from the turbulent background. These structures can be of either large or small spatial scale. Prominent examples include streamwise vortices in turbulent boundary layers and thermal plumes.

This paper makes a contribution to the study of the small scales in turbulent convection in section 4 by extracting statistical quantities useful for the design and validation of LES methods. Section 3 on the other hand deals with large scale patterns which persist in spite of turbulent fluctuations and which can be classified as coherent structures even though this term is not commonly used for large scale convective patterns.

We study here turbulent convection by numerical simulation in the simplest possible geometry: A plane layer heated from below and cooled from above. Periodic lateral boundary conditions are used. All variables are discretized in space with a spectral method. This is possible because the geometry is simple, and there does not seem to be any better method. Grid methods become necessary if one wants to use adaptive grids. However, we are interested in turbulent flows which require high resolution everywhere, so that there is no point in using adaptive grids. Only the horizontal boundaries of the plane layer are special because boundary layers form there. A finer resolution at the boundaries is automatically achieved by employing Chebychev polynomials to discretize the vertical coordinate. Using a Fourier decomposition for the horizontal coordinates leads to a method with the same operational cost per time step as a grid method except for a logarithmic factor. On the other hand, spectral methods require fewer grid points at equal accuracy than standard grid methods because of the higher order of spectral methods. The spectral method is therefore more effective. The high order of spectral methods becomes an obstacle in flows with discontinuities or shock fronts. In this paper however, the flow is resolved down to the viscous scales so that the solution is smooth at the resolution employed and the high order of the spectral method becomes an advantage. The numerical method is described in the next section.

2 Numerical Method

The problem of Rayleigh-Bénard convection considers convection in a fluid confined to a plane layer of thickness d bounded in the z-direction (figure 1). Gravitational acceleration g is acting towards the negative z-direction. The temperatures of the top and bottom boundaries are held fixed and differ by ΔT.

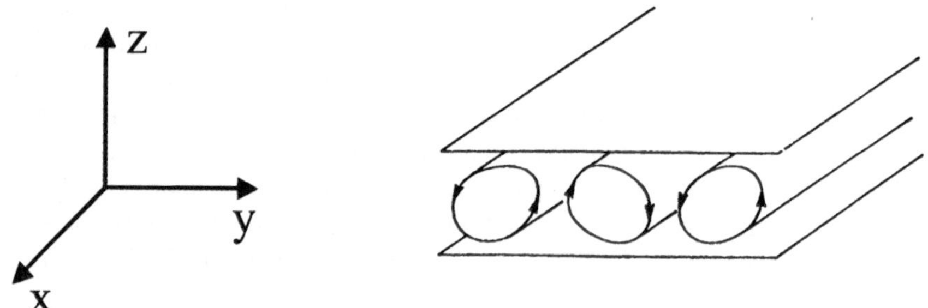

Fig. 1. Sketch of Rayleigh-Bénard convection

2.1 Basic Equations

A commonly accepted model, adequate for describing most laboratory experiments, is the "Boussinesq approximation" which assumes that all material properties of the convecting fluid except its density are independent of temperature and pressure. The variations of density are only retained in the buoyancy term. The fluid is therefore specified by its viscosity ν, thermal diffusivity κ, and expansion coefficient α. As usual in fluid mechanics, the equations of motion are best formulated in dimensionless variables. With d, d^2/κ, ΔT and $\rho\kappa^2/d^2$ as units of length, time, temperature, and pressure, the equations for the dimensionless velocity $\mathbf{v}(x,y,z,t)$ and temperature $T(x,y,z,t)$ read:

$$\frac{\partial}{\partial t}\mathbf{v} + (\mathbf{v}\cdot\nabla)\mathbf{v} = -\nabla p + Pr\ \nabla^2\mathbf{v} + Ra\ Pr\ T\hat{\mathbf{z}}, \tag{1}$$

$$\frac{\partial}{\partial t}T + (\mathbf{v}\cdot\nabla)T = \nabla^2 T, \tag{2}$$

$$\nabla\cdot\mathbf{v} = 0. \tag{3}$$

Two control parameters occur in these equations, the Rayleigh number Ra and the Prandtl number Pr:

$$Ra = \frac{g\alpha\Delta T d^3}{\kappa\nu}, \quad Pr = \frac{\nu}{\kappa}. \tag{4}$$

Ra quantifies the forcing of the fluid, whereas Pr is a material property of the fluid. Gases for instance have $Pr = 0.7$, water at 20^oC is characterized by $Pr = 7$ and high Pr are realized in oils, for example.

Periodic boundary conditions are imposed in x- and y-directions, which introduces another control parameter: The “aspect ratio” is the ratio of the periodicity length and the layer thickness. The remaining boundary conditions are:

$$\begin{aligned} T(z=0) = 1\ , \quad T(z=1) = 0, \\ \mathbf{v}(z=0) = \mathbf{v}(z=1) = 0. \end{aligned} \tag{5}$$

The continuity equation (3) suggests a decomposition of the velocity into toroidal and poloidal fields ψ and ϕ. If we demand these fields to be periodic in x-and y-directions we need an additional third component: the “mean flow” $\mathbf{U}$. A unique decomposition of the velocity is [6]:

$$\mathbf{v}(x,y,z,t) = \nabla\times[\psi(x,y,z,t)\hat{z}] + \nabla\times\nabla\times[\phi(x,y,z,t)\hat{z}] + \mathbf{U}(z,t), \tag{6}$$

where $\mathbf{U}$ has only non-vanishing x-and y-components and depends only on z. Equations of motion for ϕ, ψ and $\mathbf{U}$ are obtained from the z-component of the curl of the curl of (1), the z-component of the curl of (1), and the average over horizontal planes of (1), respectively.

Another decomposition is used for the temperature field. T is expressed in terms of the static temperature profile $T_s(z) = 1 - z$ and the deviation θ from the static profile, i.e.

$$T(x,y,z,t) = \theta(x,y,z,t) + T_s(z). \tag{7}$$

2.2 Fourier Decomposition

As already mentioned a Fourier representation is used in the horizontal directions x and y. The resulting equations for the unknown variables are:

$$[\partial_z^2 - k^2]\Big[\partial_t - Pr[\partial_z^2 - k^2]\Big]\hat{\phi}_{\mathbf{k}}(z,t) = \mathcal{R}_{\hat{\phi}_{\mathbf{k}}}(z,t), \tag{8}$$

$$\Big[\partial_t - Pr[\partial_z^2 - k^2]\Big]\hat{\psi}_{\mathbf{k}}(z,t) = \mathcal{R}_{\hat{\psi}_{\mathbf{k}}}(z,t), \tag{9}$$

$$\Big[\partial_t - [\partial_z^2 - k^2]\Big]\hat{\theta}_{\mathbf{k}}(z,t) = \mathcal{R}_{\hat{\theta}_{\mathbf{k}}}(z,t), \tag{10}$$

$$[\partial_t - Pr\partial_z^2]\mathbf{U}(z,t) = \boldsymbol{\mathcal{R}}_{\mathbf{U}}(z,t), \tag{11}$$

where $\hat{\phi}_{\mathbf{k}}$, $\hat{\psi}_{\mathbf{k}}$ and $\hat{\theta}_{\mathbf{k}}$ denote the Fourier components of ϕ, ψ and θ, respectively. $\mathbf{k} = k_x\hat{x} + k_y\hat{y}$ is the wave vector in the horizontal plane. The corresponding right hand sides $\mathcal{R}$ read:

$$\begin{aligned}
\mathcal{R}_{\hat{\phi}_{\mathbf{k}}} &= \frac{1}{k^2}\Big\{\hat{z}\cdot\Big[\nabla\times\nabla\times[(\nabla\times\mathbf{v})\times\mathbf{v}]\Big]\Big\}_{\mathbf{k}} - RaPr\hat{\theta}_{\mathbf{k}},\\
\mathcal{R}_{\hat{\psi}_{\mathbf{k}}} &= -\frac{1}{k^2}\Big\{\hat{z}\cdot\Big[\nabla\times[(\nabla\times\mathbf{v})\times\mathbf{v}]\Big]\Big\}_{\mathbf{k}},\\
\mathcal{R}_{\hat{\theta}_{\mathbf{k}}} &= k^2\hat{\phi}_{\mathbf{k}} - \Big\{\nabla\cdot(\mathbf{v}\theta)\Big\}_{\mathbf{k}},\\
\boldsymbol{\mathcal{R}}_{\mathbf{U}} &= -\langle(\nabla\times\mathbf{v})\times\mathbf{v}\rangle_{x,y}.
\end{aligned} \tag{12}$$

Here $\{...\}_{\mathbf{k}}$ and $\langle...\rangle_{x,y}$ denote the Fourier component and the average over x and y, respectively.

We first discuss the treatment of the equations for $\hat{\psi}_{\mathbf{k}}$, $\hat{\theta}_{\mathbf{k}}$ and $\mathbf{U}$ (9–11), which are of same form and can therefore be treated together. They can be written in a more general form as:

$$[\partial_t + C_0 - C_2\partial_z^2]f(z,t) = \mathcal{R}_f(z,t), \qquad f = \hat{\psi}_{\mathbf{k}}, \hat{\theta}_{\mathbf{k}}, \mathbf{U}, \tag{13}$$

with some constants C_0 and C_2, and some right hand side $\mathcal{R}$. Since the boundary conditions:

$$f(z=0) = 0, \quad f(z=1) = 0 \tag{14}$$

are also the same for those three fields, exactly the same method can be used.

2.3 Chebyshev Method

We use a Chebyshev representation for the z-dependence which allows simple and fast computation of derivatives and a fast transformation from spectral coefficients to physical space through a cosine transformation. The decomposition

$$f(z,t) = \sum_{j=0}^{N_z-1} c_j(t) T_j(\tilde{z}), \quad \tilde{z} \equiv 2z - 1 \tag{15}$$

is used throughout most of the program. T_j denotes the Chebyshev polynomial of order j. An expansion like (15) still allows for arbitrary boundary conditions. A more adapted decomposition [5, 3]

$$f(z,t) = \sum_{i=0}^{N_z-3} a_i(t) F_i(\tilde{z}) \tag{16}$$

is used during the time step. The functions:

$$F_i(\tilde{z}) = (1 - \tilde{z}^2) T_i(\tilde{z}) \tag{17}$$

automatically satisfy the boundary conditions (14), and can be written as a linear combination of Chebyshev polynomial using the recursion formula $T_{n+1}(x) = 2xT_n(x) - T_{n-1}(x)$ $(n \geq 1)$:

$$\begin{aligned} F_0 &= \frac{1}{2}T_0 - \frac{1}{2}T_2, \\ F_1 &= \frac{1}{4}T_1 - \frac{1}{4}T_3, \\ F_i &= -\frac{1}{4}T_{i-2} + \frac{1}{2}T_i - \frac{1}{4}T_{i+2} \qquad i \geq 2. \end{aligned} \tag{18}$$

Both representations (15) and (16) are equivalent for all $f(z,t)$ that satisfy the boundary conditions (14). If we express the right hand sides of (13) in terms of Chebyshev polynomials:

$$\mathcal{R}_f(z,t) = \sum_{j=0}^{N_z-1} b_j(t) T_j(\tilde{z}), \tag{19}$$

then the differential equation (13) reads:

$$\sum_{i=0}^{N_z-3} [\partial_t + C_0 - C_2\partial_z^2] a_i(t) F_i(\tilde{z}) = \sum_{j=0}^{N_z-1} b_j(t) T_j(\tilde{z}). \tag{20}$$

2.4 Time Discretization

The next step is to discretize time using some appropriate time step h, i.e.

$$t_n = t_{n-1} + h. \tag{21}$$

h is dynamically adjusted during the simulation using a Courant-Friedrich-Levy (CFL) criterion [2, 3]. An implicit Crank-Nicolson method [7] is used for the left hand side of equation (20):

$$[C_0 - C_2\partial_z^2]a_i(t) \to [C_0 - C_2\partial_z^2]\frac{a_{i,t_n} + a_{i,t_{n-1}}}{2}, \tag{22}$$

$$\partial_t a_i(t) \to \frac{a_{i,t_n} - a_{i,t_{n-1}}}{h}, \tag{23}$$

where a_{i,t_n} and $a_{i,t_{n-1}}$ denote the coefficients at new time t_n and the previous time t_{n-1}, respectively. The right hand side, however, contains the nonlinear terms. This allows only for an explicit method like the second order Adams-Bashforth [7] method we are using. This yields the replacement:

$$b_j(t) \to \frac{1}{2}(3b_{j,t_{n-1}} - b_{j,t_{n-2}}). \tag{24}$$

The general diffential equation for $\hat{\psi}_{\mathbf{k}}$, $\hat{\theta}_{\mathbf{k}}$ and $\mathbf{U}$ (13) therefore reads in discretized form as:

$$\begin{aligned}
&\sum_{i=0}^{N_z-3}\left[(1+\frac{h}{2}C_0)F_j(\tilde{z}) - \frac{h}{2}C_2\partial_z^2F_j(\tilde{z})\right]a_{i,t_n} = \\
&\sum_{i=0}^{N_z-3}\left[(1-\frac{h}{2}C_0)F_j(\tilde{z}) + \frac{h}{2}C_2\partial_z^2F_j(\tilde{z})\right]a_{i,t_{n-1}} + \\
&\sum_{j=0}^{N_z-1}\frac{h}{2}(3b_{j,t_{n-1}} - b_{j,t_{n-2}})T_j(\tilde{z}).
\end{aligned} \tag{25}$$

Lastly, an appropriate projection is necessary to yield a set of linear equations for the unknown coefficients a_{i,t_n}. Following [5] we use the projection functions:

$$P_i(\tilde{z}) = \frac{T_{i-1}(\tilde{z}) - T_{i+1}(\tilde{z})}{2i\sqrt{1-\tilde{z}^2}} \qquad i = 1, \ldots, N_z - 2. \tag{26}$$

Multiplying (25) with these projections functions and integrating from $\tilde{z} = -1$ to 1 yields the following linear system of equations:

$$\begin{aligned}
&\left[(1+\frac{h}{2}C_0)\mathcal{M}_1 - \frac{h}{2}C_2\mathcal{M}_2\right]\mathbf{a}_{t_n} = \\
&\left[(1-\frac{h}{2}C_0)\mathcal{M}_1 + \frac{h}{2}C_2\mathcal{M}_2\right]\mathbf{a}_{t_{n-1}} + \\
&\frac{h}{2}\mathcal{M}_3(3\mathbf{b}_{t_{n-1}} - \mathbf{b}_{t_{n-2}}).
\end{aligned} \tag{27}$$

The coefficients a_{i,t_n} and b_{j,t_n} are written in vector form $\mathbf{a}_{t_n} = (a_{0,t_n}, \ldots, a_{N_z-3,t_n})^\top$ and $\mathbf{b}_{t_n} = (b_{0,t_n}, \ldots, b_{N_z-1,t_n})^\top$ for better readability. The matrices

$$\left(\mathcal{M}_1\right)_{i=0,\ldots,N_z-3}^{j=0,\ldots,N_z-3} = \int_{-1}^{1} d\tilde{z}\, P_{i+1}(\tilde{z})F_j(\tilde{z}),$$

$$
\begin{aligned}
\Big(\mathcal{M}_2\Big)_{i=0,\dots,N_z-3}^{j=0,\dots,N_z-3} &= \int_{-1}^{1} d\tilde{z}\, P_{i+1}(\tilde{z})\partial_z^2 F_j(\tilde{z}), \\
\Big(\mathcal{M}_3\Big)_{i=0,\dots,N_z-3}^{j=0,\dots,N_z-1} &= \int_{-1}^{1} d\tilde{z}\, P_{i+1}(\tilde{z})T_j(\tilde{z})
\end{aligned}
\tag{28}
$$

that occur in (27) can be analytically computed using the definition and the orthogonality relation of the Chebyshev polynomials. It turns out, that the matrices are of band diagonal form. $\mathcal{M}_1$ is shown exemplarily:

$$
\pi \begin{pmatrix}
\frac{3}{8} & 0 & -\frac{1}{4} & 0 & \frac{1}{16} & 0 & 0 & 0 & \dots \\
0 & \frac{1}{16} & 0 & -\frac{3}{32} & 0 & \frac{1}{32} & 0 & 0 & \\
-\frac{1}{24} & 0 & \frac{1}{16} & 0 & -\frac{3}{48} & 0 & \frac{1}{48} & 0 & \\
0 & -\frac{1}{64} & 0 & \frac{3}{64} & 0 & -\frac{3}{64} & 0 & \frac{1}{64} & \\
0 & 0 & -\frac{1}{80} & 0 & \frac{3}{80} & 0 & -\frac{3}{80} & 0 & \\
0 & 0 & 0 & -\frac{1}{96} & 0 & \frac{3}{96} & 0 & -\frac{3}{96} & \\
0 & 0 & 0 & 0 & -\frac{1}{112} & 0 & \frac{3}{112} & 0 & \\
0 & 0 & 0 & 0 & 0 & -\frac{1}{128} & 0 & \frac{3}{128} & \\
0 & 0 & 0 & 0 & 0 & 0 & \frac{1}{144} & 0 & \\
0 & 0 & 0 & 0 & 0 & 0 & 0 & -\frac{1}{160} & \\
0 & 0 & 0 & 0 & 0 & 0 & 0 & 0 & \\
\vdots & & & & & & & & \ddots
\end{pmatrix} .
$$

$\mathcal{M}_2$ is of similar form, but is even more sparsely occupied. Only the diagonal and the second superdiagonal contain non-zero elements. Inverting the combined matrix

$$
\mathcal{M} = (1 + \frac{h}{2}C_0)\mathcal{M}_1 - \frac{h}{2}C_2\mathcal{M}_2 \tag{29}
$$

is the last step to solve equation (27). C_0 and C_2 are different for $\hat{\psi}_{\mathbf{k}}$, $\hat{\theta}_{\mathbf{k}}$ and $\mathbf{U}$ and in general depend on the absolute value of the wave vector for that Fourier component (see equations (9),(10) and (11)). Since typically 256 or 512 Fourier components are used in each of the horizontal directions, several thousand different matrices have to be inverted. Due to memory limitations we cannot compute all inverses at the beginning of the simulation and store them in main memory. Fortunately, $\mathcal{M}$ can be very efficiently inverted due to the band diagonal structure of the matrices $\mathcal{M}_1$ and $\mathcal{M}_2$. The numerical expense is much less then for fully occupied matrices. This makes on the fly composition and inversion of the matrix $\mathcal{M}$ a viable option.

Finally, we make a small change to equation (27) in the actual implementation of the presented method to reduce computational expense. As previously mentioned we want to represent all fields in terms of Chebyshev polynomials throughout the program. The decomposition into F_i (16) is only used during the time step to account for the boundary conditions (14). The fields at the previous time step t_{n-1} are therefore stored as Chebyshev coefficients. There is no need to convert these into coefficients of an expansion in F_i's. We can just rewrite the corresponding terms of the equation (27):

$$\left[(1-\frac{h}{2}C_0)\mathcal{M}_1+\frac{h}{2}C_2\mathcal{M}_2\right]\mathbf{a}_{t_{n-1}}=\mathcal{M}_3\left[(1-\frac{h}{2}C_0)\boldsymbol{\alpha}_{t_{n-1}}+\frac{h}{2}C_2\boldsymbol{\alpha}''_{t_{n-1}}\right], \quad (30)$$

were $\boldsymbol{\alpha}_{t_{n-1}}$ and $\boldsymbol{\alpha}''_{t_{n-1}}$ denote the Chebyshev coefficients at $t=t_{n-1}$ of the function f and its second derivative $\partial_z^2 f$, respectively. This substitution saves us one matrix multiplication as can be seen in the final equation:

$$\mathcal{M}\mathbf{a}_{t_n}=\mathcal{M}_3\left[(1-\frac{h}{2}C_0)\boldsymbol{\alpha}_{t_{n-1}}+\frac{h}{2}C_2\boldsymbol{\alpha}''_{t_{n-1}}+\frac{h}{2}(3\mathbf{b}_{t_{n-1}}-\mathbf{b}_{t_{n-2}})\right]. \quad (31)$$

2.5 Poloidal Field

The method for the equation of motion of the poloidal field is outlined below. It is analogous to the previous method except for a different set of expansion and projection functions. The boundary conditions for $\hat{\phi}_{\mathbf{k}}$ are:

$$\begin{aligned}\hat{\phi}_{\mathbf{k}}(z=0)=0\,, \quad \hat{\phi}_{\mathbf{k}}(z=1)=0,\\ \partial_z\hat{\phi}_{\mathbf{k}}(z=0)=0\,, \quad \partial_z\hat{\phi}_{\mathbf{k}}(z=1)=0,\end{aligned} \quad (32)$$

which makes the previously used F_i functions not a good choice for an expansion. Again, following [5] the functions

$$G_i(\tilde{z})=(1-\tilde{z}^2)^2T_i(\tilde{z}) \quad (33)$$

are chosen. They obviously obey the boundary conditions (32). The corresponding expansion of $\hat{\phi}_{\mathbf{k}}$ is

$$\hat{\phi}_{\mathbf{k}}(z,t)=\sum_{i=0}^{N_z-5}c_i(t)G_i(\tilde{z}). \quad (34)$$

The resulting matrices have again a sparse band diagonal structure if the functions

$$Q_i(\tilde{z})=\frac{1}{4\sqrt{1-\tilde{z}^2}}\left[\frac{T_{i-2}(\tilde{z})}{i(i-1)}-\frac{2T_i(\tilde{z})}{(i+1)(i-1)}+\frac{T_{i+1}(\tilde{z})}{i(i+2)}\right] \quad (35)$$

are used for the projection. The final equation is:

$$\begin{aligned}\mathcal{N}\mathbf{c}_{t_n}=\mathcal{N}_4\Big[&(-k^2+\frac{h}{2}Prk^4)\gamma_{t_{n-1}}+\\ &(1-hPrk^2)\gamma''_{t_{n-1}}+\frac{h}{2}Pr\gamma^{(4)}_{t_{n-1}}+\frac{h}{2}(3\mathbf{d}_{t_{n-1}}-\mathbf{d}_{t_{n-2}})\Big],\end{aligned} \quad (36)$$

with

$$\mathcal{N}=\left[(-k^2-\frac{h}{2}Prk^4)\mathcal{N}_1+(1+hPrk^2)\mathcal{N}_2-\frac{h}{2}Pr\mathcal{N}_3\right] \quad (37)$$

and

$$\left(\mathcal{N}_1\right)_{i=0,\ldots,N_z-5}^{j=0,\ldots,N_z-5} = \int_{-1}^{1} d\tilde{z}\, Q_{i+2}(\tilde{z}) G_j(\tilde{z}),$$

$$\left(\mathcal{N}_2\right)_{i=0,\ldots,N_z-5}^{j=0,\ldots,N_z-5} = \int_{-1}^{1} d\tilde{z}\, Q_{i+2}(\tilde{z}) \partial_z^2 G_j(\tilde{z}),$$

$$\left(\mathcal{N}_3\right)_{i=0,\ldots,N_z-5}^{j=0,\ldots,N_z-5} = \int_{-1}^{1} d\tilde{z}\, Q_{i+2}(\tilde{z}) \partial_z^4 G_j(\tilde{z}),$$

$$\left(\mathcal{N}_4\right)_{i=0,\ldots,N_z-5}^{j=0,\ldots,N_z-1} = \int_{-1}^{1} d\tilde{z}\, Q_{i+2}(\tilde{z}) T_j(\tilde{z}). \tag{38}$$

Even though these matrices have more non-zero entries than $\mathcal{M}_1$, $\mathcal{M}_2$ and $\mathcal{M}_3$, they are still band diagonal. 13, 9, 5 and 5 main and secondary diagonals need to be considered for $\mathcal{N}_1$, $\mathcal{N}_2$, $\mathcal{N}_3$ and $\mathcal{N}_4$, respectively.

2.6 Parallelization

A fully parallel version of the presented method has been implemented. The code allows the use of either the message passing interface (MPI) routines available on most parallel architectures, or the logically shared, distributed memory access (SHMEM) routines available on Cray supercomputers. SHMEM routines are used on HLRS's Cray T3E, since they provide better communication performance on that architecture.

The scaling behaviour of the simulation program on the Cray T3E has been investigated using test simulations with 128×128 Fourier modes in the horizontal plane and 129 Chebyshev modes in the vertical direction. Computation time for a 16 processor run was 6.9 times longer than for a job with 128 processors in parallel, both without counting time for disk I/O at the beginning and the end of the jobs. An increase of the number of processors by a factor of 8 therefore yielded a satisfying speed up factor of 6.9.

3 Large Scale Patterns

Motion in the fluid occurs only if *Ra* is larger than some threshold. The movement is organized in rolls as depicted in figure 1 at the onset of convection. There is an ongoing debate as to what happens to these structures once the flow is turbulent. Fitzjarrald [1] reports on a length scale derived from the spatial distribution of heat transport which is thought to represent the roll size and which increases with increasing *Ra*. Krishnamurti and Howard [4] on the other hand claim that organized large scale structures disappear at $Ra \approx 10^6$

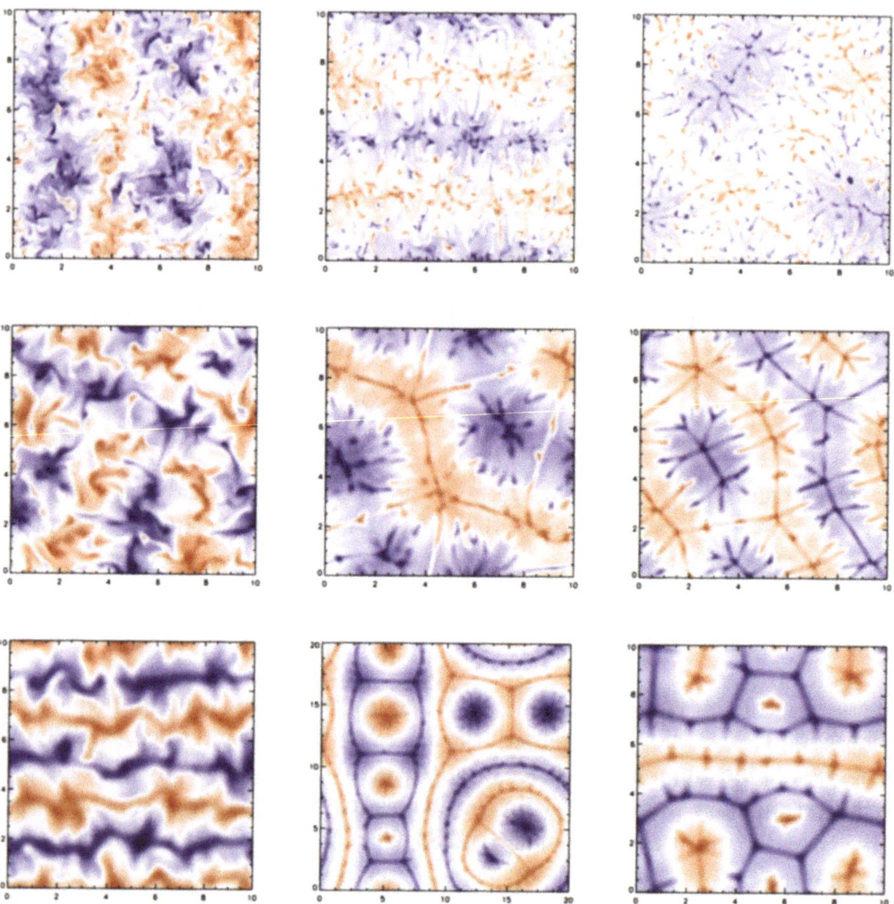

Fig. 2. Plots of the temperature field in the plane $z = 0.5$. Red and blue colors indicate warm and cold fluid, respectively. Pr is constant within columns. Pr equals 0.7, 7, and 30 for the left, middle, and right column. Each row corresponds to a different Ra. Ra is 3.2×10^4, 10^5, and 10^6 in the bottom, middle, and top row, respectively. All simulations have used an aspect ratio of 10 except the one at $Pr = 7$, $Ra = 3.2 \times 10^4$, which is for an aspect ratio of 20.

and reappear at higher Ra. Since these Ra can be reached numerically, it seems worthwhile to investigate the issue by direct numerical simulation.

The study of large scale patterns requires large aspect ratios. We have up to now little knowledge of Rayleigh-Bénard convection at both large Ra and large aspect ratios. The reason is that ever thinner boundary layers develop with increasing Ra which need to be resolved in the z-direction, so that fewer grid points can be invested in the horizontal directions, thus limiting the accessible aspect ratio. The spectral method described in the previous section is powerful enough to allow simulations at Ra up to $10^6 - 10^7$ in layers of aspect ratio $10-20$ which is enough to address the issue of large scale pattern formation in turbulent convection.

Figure 2 gives an impression of convective flows at different Ra and Pr. The figure shows snapshots of temperature in the midplane of the layer. In the case of regular rolls as sketched in figure 1, one expects warm fluid in the region of rising motion and cold fluid in downwellings. This corresponds to parallel bands colored in red and blue in figure 2. It is seen that at $Pr = 0.7$, a recognizable roll structure, distorted by turbulent fluctuations, survives up to the highest Ra simulated. The roll structure is even more apparent in animations of the temperature field in which the eye can subjectively average the pattern over time. Rolls, however, are not the only possible form of organization. Cellular structures dominate at higher Pr. The border between rolls and cells lies around $Pr \approx 7$. At all Pr, the average separation between rolls or the average cell size increases with increasing Ra. The large scale patterns therefore exist despite turbulent fluctuations and increase in size with increasing Ra.

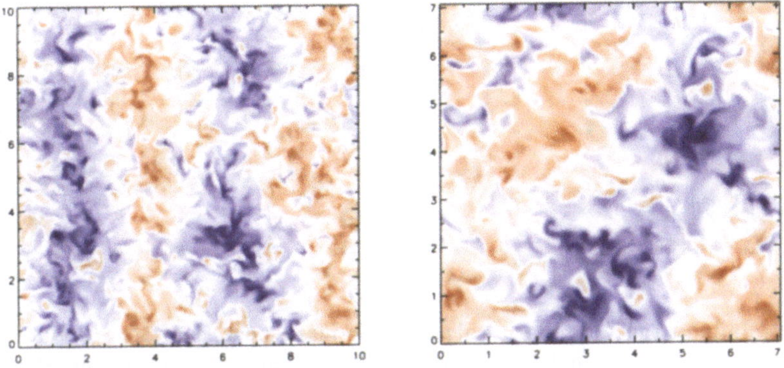

Fig. 3. Plots of the temperature field in the plane $z = 0.5$ for $Pr = 0.7$ and $Ra = 10^6$. The aspect ratio is 10 in the left panel and $10/\sqrt{2}$ in the right panel.

Figure 3 demonstrates that the selected size of the large scale pattern is intrinsic to the flow and is little influenced by the boundary conditions. The left panel shows two pairs of rolls aligned with the z-axis in a box of aspect ratio 10. In the right panel, the same parameters are used except that the box now has an aspect ratio of $10/\sqrt{2}$. The rolls are now aligned with a diagonal, thus keeping their separation constant. This comparison shows that the shape and size of the computational volume do not impose a particular large scale pattern.

Finally, figure 4 shows the temperature field at the edge of the bottom thermal boundary layer for the same parameters for which the temperature

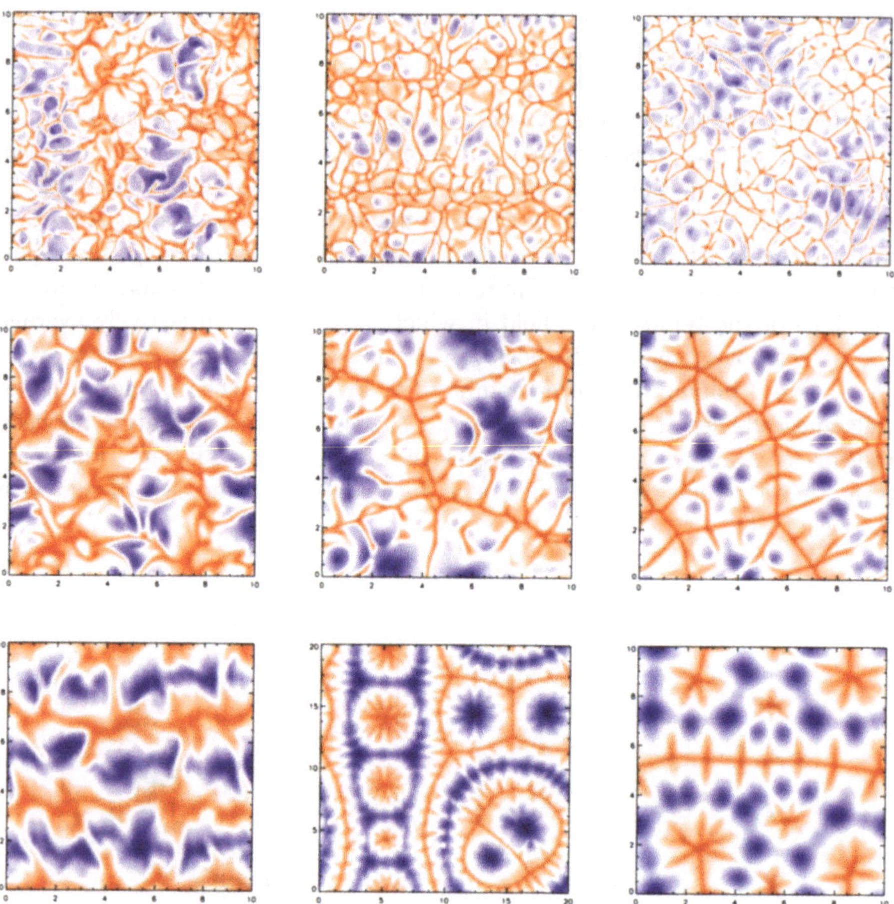

Fig. 4. Plots of the temperature field in the plane near the bottom boundary in which the temperature fluctuations are maximum. This plane marks the edge of the thermal boundary layer (see figures 5 and 6). The other parameters are exactly the same as in figure 2.

in the midplane is shown in figure 2. One recognizes the same large scale patterns in both figures, but more detailed features appear in the boundary layer which are smeared out in the middle of the fluid layer.

4 Statistics

Several statistical quantities are of interest in connection with the design of LES methods. These quantities are averages over horizontal planes and time. Denote this average by $< ... >$, and define:

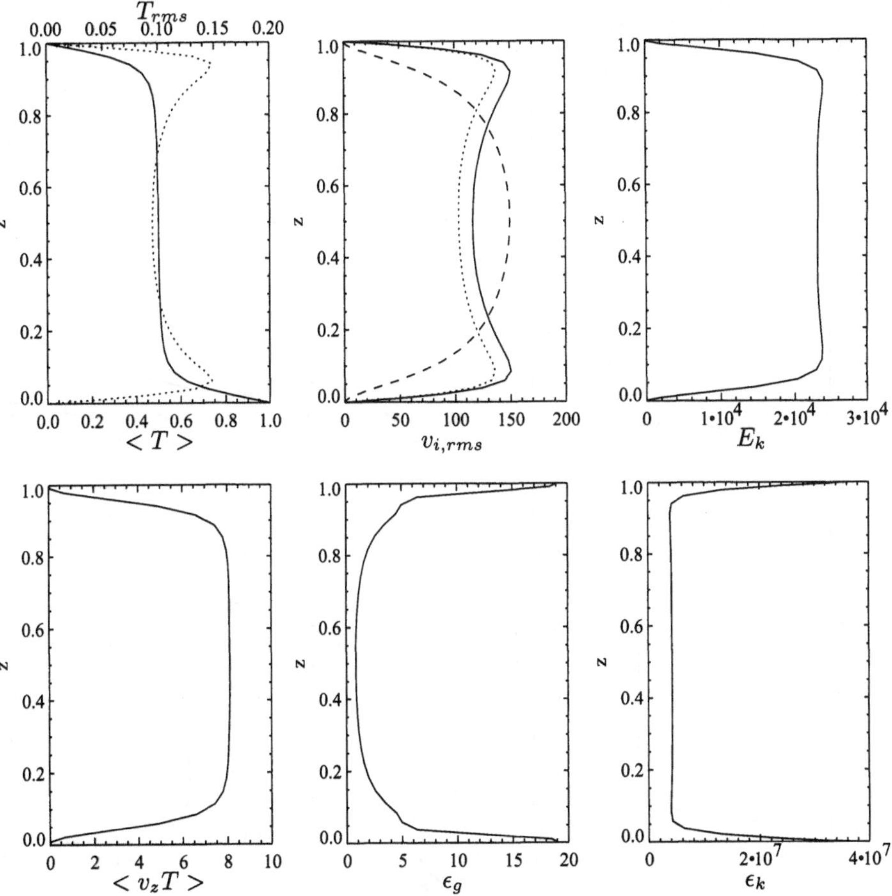

Fig. 5. Various statistical quantities defined in the text as a function of z for $Pr = 0.7$ and $Ra = 10^6$. In the upper left panel, the continuous and dashed lines indicate $< T >$ and T_{rms}, respectively. The x, y and z components of the velocity fluctuations are given by the continuous, dashed, and long dashed lines, respectively.

$$T' = T - < T >, \quad v_i' = v_i - < v_i >, \tag{39}$$

$$T_{rms} = \sqrt{< (T')^2 >}, \tag{40}$$

$$v_{i,rms} = \sqrt{< (v_i')^2 >}, \tag{41}$$

$$E_k = \frac{1}{2} \sum_i < (v_i')^2 >, \tag{42}$$

$$\epsilon_k = Pr \sum_{i,j} < (\partial_i v_j')^2 >, \tag{43}$$

$$\epsilon_g = \sum_i < (\partial_i T')^2 > . \tag{44}$$

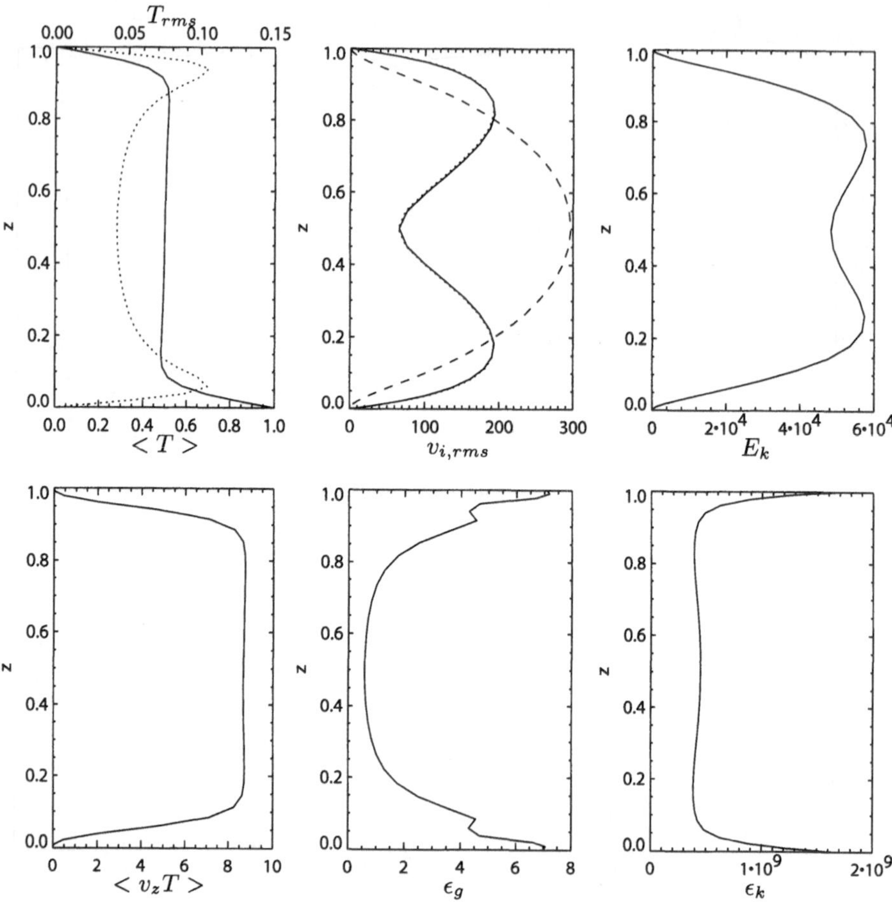

Fig. 6. Various statistical quantities defined in the text as a function of z for $Pr = 60$ and $Ra = 10^6$. The line styles are chosen as in the previous figure.

The averages depend only on z. The z-dependence is shown in figures 5 and 6 for two different sets of parameters. One immediately recognizes boundary layers in the plots of $< T >$, T_{rms}, $< v_{i,rms} >$, and $< v_z T >$. As expected, thermal and viscous boundary layers are about the same size for $Pr = 0.7$, but the viscous layer is much thicker than the thermal one for $Pr = 60$. $< v_z T >$ gives the advective heat transport. It is zero at the boundaries, where the heat flux is entirely diffusive. The mean temperature gradient is very small in the bulk of the fluid so that the heat transport is almost completely advective there. $< v_z T >$ must thus be nearly constant in the bulk in a stationary state so that heat does not accumulate at any height z.

ϵ_g and ϵ_k are included in the figures for later reference. The relevance of the dissipation rates ϵ_g and ϵ_k for turbulence modeling is explained in reference [8].

All plots must be symmetric about the plane $z = 0.5$, except for the average temperature $< T >$, which has to be antisymmetric. Deviations from this symmetry indicate insufficient statistical accumulation. The runs in figures 5 and 6 covered roughly 30 turnover times in the statistically steady state, which is apparently a long enough accumulation time to obtain the required symmetry.

5 Summary

A spectral method has proven to be efficient in simulating Rayleigh-Bénard convection. Despite frequently expressed opinions to the contrary, the spectral method has a speed up factor on parallel machines which is comparable to those achieved with grid methods. The spectral method allows us to investigate turbulent Rayleigh-Bénard convection even in containers of large aspect ratio.

References

1. D. E. Fitzjarrald. An experimental study of turbulent convection in air. *J. Fluid Mech.*, 73: 693–719, 1976.
2. D. Gottlieb and S. A. Orszag. Numerical analysis of spectral methods: Theory and applications. *Society for Industrial and Applied Mathematics*, 1977.
3. R. M. Kerr. Rayleigh number scaling in numerical convection. *J. Fluid Mech.*, 310: 139–179, 1996.
4. R. Krishnamurti and L. N. Howard. Large scale flow generation in turbulent convection. *Proc. Natl. Acad. Sci. USA*, 78: 1981–1985, 1981.
5. R. D. Moser, P. Moin, and A. Leonard. A spectral numerical method for the navier stokes equations with application to taylor couette flow. *J. Comp. Phys.*, 54: 524–544, 1983.
6. B. J. Schmitt and W. von Wahl. Decomposition into poloidal fields, toroidal fields, and mean flow. *Differential and Integral Equations*, 5: 1275–1306, 1992.
7. A. Tilgner. Spectral methods for the simulation of incompressible flows in spherical shells. *International Journal for Numerical Methods in Fluids*, 30: 713–724, 1999.
8. Q. Y. Ye, M. Wörner, and G. Grötzbach. Modelling turbulent dissipation correlations for rayleigh-bénard convection based on two-point correlation technique and invariant therory. *Forschungszentrum Karlsruhe*, FZKA 6103, 1998.

Chemistry

Christoph van Wüllen

Institut für Chemie, Sekr. C3, Technische Universität Berlin,
Straße des 17. Juni 135, D-10623 Berlin,
Germany

Advances both in methodology and computer hardware are constantly enlarging the domain of the problems that can be addressed by quantum chemical calculations. This is demonstrated by the vast area of chemistry spanned by the reports from quantum chemical research projects which can be found in this volume.

Calculations of the geometric and electronic structure of individual molecules have a long history. It is however still a challenge to perform reliable calculations on transition metal complexes because of the diversity of bonding situations and reactivity patterns which these compounds exhibit. It is important that quantum chemistry provides data beyond what can be measured, at least in a simple experimental setup. Fruitful chemical concepts, like the classification of bonds being covalent or ionic, can hardly be cast as a quantum mechanical observable. To obtain not only numbers, but also insight, the bonding situation can be clarified by wave function analysis or numerical experiments like the energy decomposition analysis (EDA) applied by Erhardt and Frenking. A prototypical question is how pnictogen dimers X_2 (X= N, P, As, Sb, Bi) bind to transition metals, and what is the reason behind the preference for different coordination modes observed in the calculation.

The treatment of boron clusters embedded in a solid silicon matrix requires computational methods which have their roots in solid-state physics and are somewhat different from methods traditionally used in *molecular* quantum chemistry. For modern semiconductor technology, it is important to know what boron clusters in a silicon environment do. Signatures from vibrational spectroscopy can experimentally be obtained with reasonable effort, but the analysis of the bands found in the spectrum is greatly facilitated by quantum chemical calculations. The calculations performed by Deak, Gali and Pichler involve high computational effort since the boron clusters in the computational model have to be quite dilute in the silicon matrix to avoid unrealistic interactions between such clusters. In these periodic calculations this translates to large unit cells, which is the main reason why supercomputers are needed to perform such calculations. A large set of calculations, each calculating the vibrational spectrum of different boron clusters, then provide a database which can then be used to interpret or assign experimental vibrational spectra. Such joint theoretical and experimental investigations, where each side contributes what it can do better than the other side, are more efficient than theory or experiment alone.

While application of quantum mechanical methods in solid-state and molecular physics have a long history, biochemical or even medicinal applications of quantum chemistry are somewhat younger. The reason for this is not only that the systems under investigation are much larger, but that optimising the geometry of a single conformer becomes meaningless. The combination of density functional theory and molecular dynamics, as implemented in the Car-Parinello approach, is therefore better adapted to biochemical simulations. Even today, these calculations would not be feasible without combining quantum mechanical and force field models. The reaction centre, where bond breaking and making takes place, is described by quantum chemical methods, while the protein environment and surrounding water molecules are treated using molecular mechanics methods. The simulation by Klein is targeted at finding stable structures (in contrast to shallow local minima) for fumagillin bound to methionine aminopeptidases, an important class of enzymes which are involved in the final step of protein synthesis in living organisms. Molecules such as fumagillin which inhibit the enzyme have a high potential as an anticancer drug, therefore understanding the mechanism at a molecular level is quite important.

The simulations just described mark the interface of chemistry with biology and medicine. Nano-chemistry on the other hand, which deals with molecular devices and machines, enters the field of engineering sciences. In the contribution of Stampfuß et al., the conducting properties of single molecules are investigated. To become faster and faster, electronic devices must also become smaller and smaller because of the finite speed of light. This eventually leads to a situation where single molecules have to used as "wires". On this scale, quantum effects play a role, and the transport of electrons has to be modeled by quantum chemical calculations. Large metal clusters are used to describe the contacts which are connected by the molecular wire, and this is mainly the reason for the high computational effort.

The four following articles show how supercomputer facilities are used to deepen our understanding of chemistry and to extend the realm of computational quantum chemistry in the fields of solid state chemistry, life sciences and nano-scale technology.

Quantum Chemical Calculations of Transition Metal Complexes

Stefan Erhardt and Gernot Frenking

Fachbereich Chemie, Phillipps-Universität Marburg, Hans-Meerwein-Straße, 35032 Marburg

1 Introduction

Transition metal complexes show a wide variety of chemical reactions. To gain insight into the bonding situation of these complexes and the transition states involved in these reactions is not only crucial for understanding the underlying principles, but even more for finding new reaction pathways or optimizing reaction conditions in chemical industry. Where experiments fail to obtain the needed results, modern quantum chemical approaches can be utilized to investigate chemical systems and predict their properties. This is a challenging task for computational chemists and demanding in computational resources. Such calculations have been carried out in order to predict geometries, bond energies, and Lewis basicity of various transition metal complexes. A number of projects are still in progress which are including CCSD(T) and MP2 calculations that have been carried out on the HLR. The following chapters give an overview about the research of our group using computational resources of the HLR Stuttgart.

2 Why is BCl_3 a stronger Lewis Acid with Respect to Strong Bases than BF_3?

2.1 Abstract

In this work, calculations at the density functional theory or DFT (PW91) and at *ab initio* levels (MP2 [1] and CCSD(T) [2, 3]) using large basis sets were performed. Several levels of theory with various basis sets were used to ensure that our conclusions were not dependent on the theoretical methods.

2.2 Method

The program package ADF [4, 5] was used for the DFT calculations and Gaussian 98 [6] for the MP2 and CCSD(T) methods. At the DFT PW91/QZ4P level, we calculated the geometries and the energies of the complexes. PW91 is a gradient-corrected density functional using the exchange-correlation functional by Perdew and Wang [7]. QZ4P basis sets have a quadruple-ζ quality with four polarization functions and employ uncontracted Slater-type orbitals (STOs) [8]. All geometries were also optimized at the MP2 level with the Pople basis sets 6-311G(2d) [9, 10] and with the Dunning's correlation consistent basis sets cc-pVTZ [11]. At the MP2/6-311G(2d) level, we calculated the vibrational frequencies and at the MP2/Aug-cc-pVTZ and CCSD(T)/cc-pVTZ we obtained improved energies.

2.3 Discussion

Table 1 gives calculated energies of Cl_3B-NH_3 and F_3B-NH_3 which are relevant for the discussion. The MP2 values are given with three different basis sets in order to show the changes which are given when the basis set becomes augmented with addition ploarization functions and diffuse functions. We will discuss the MP2/Aug-cc-pVTZ results because they have been obtainted with the largest basis set that has been employed in the ab initio calculations.

The calculated bond dissociation energy (BDE) of Cl_3B-NH_3 is predicted at all levels of theory to be higher than for F_3B-NH_3. The difference between the D_e values is slightly larger at the ab initio levels (6.2 kcal/mol at MP2/Aug-cc-pVTZ, 4.2 kcal/mol at CCSD(T)/cc-pVTZ) than at DFT (2.5 kcal/mol at PW91/QZ4P) but the order is the same. Table 1 gives also the preparation energies ΔE_{prep} of BX_3 and NH_3, i.e. the energy difference be-

Table 1. Calculated interaction energies ΔE_{int} between the frozen fragments, preparation energies ΔE_{prep} of the fragments and bond dissociation energies D_e and ZPE corrected values D_0. All values are given in kcal/mol.[a] Using MP2/cc-pVTZ optimized geometries. [b] ZPE corrections are taken from unscaled vibrational frequencies at MP2/6-311G(2d).

Method	ΔE_{int}	ΔE_{prep} X_3B	ΔE_{prep} NH_3	D_e	D_0[b]
		Cl_3B-NH_3			
MP2/6-311G(2d)	58.0	24.4	0.7	32.9	29.0
MP2/cc-pVTZ	53.0	23.5	0.4	29.1	25.1
MP2/Aug-cc-pVTZ[a]	52.8	23.1	0.2	29.5	25.5
CCSD(T)/cc-pVTZ[a]	49.9	22.4	0.3	23.4	19.4
PW91/QZ4P	45.6	21.9	0.3	23.4	19.4
		F_3B-NH_3			
MP2/6-311G(2d)	58.0	24.4	0.7	32.9	29.0
MP2/cc-pVTZ	53.0	23.5	0.4	29.1	25.1
MP2/Aug-cc-pVTZ[a]	52.8	23.1	0.2	29.5	25.5
CCSD(T)/cc-pVTZ[a]	49.9	22.4	0.3	23.4	19.4
PW91/QZ4P	45.6	21.9	0.3	23.4	19.4

tween the acceptor and donor moeties in the equilibrium geometries of the free species and in the complexes. The preparation energies of BX_3 are quite large while the ΔE_{prep} values for NH_3 are negligible. The differences between the calculated data for BCl_3 and BF_3 are crucial for answering the title question. The theoretical preparation energies for BCl_3 at MP2/Aug-cc-pVTZ (23.1 kcal/moland PW91/QZ4P (21.9 kcal/mol) are a bit larger than for BF_3 (22.3 kcal/mol at MP2/Aug-cc-pVTZ, 21.8 at PW91/QZ4P). The calculations at CCSD(T)/cc-pVTZ give a slightly smaller value for BCl_3 (22.4 kcal/mol) than for BF_3 (23.4 kcal/mol) but the energy difference of 1.0 kcal/mol is not enough to compensate for the large BDE of Cl_3B-NH_3. The calculated interaction energy ΔE_{int} is always higher for Cl_3B-NH_3 than for F_3B-NH_3. Thus, BCl_3 is also an intrinsically stronger Lewis acid with respect to NH_3 than BF_3. This is predicted at all three levels of theory.

The bond dissociation energy of the complex Cl_3B-NH_3 is greater than the one of F_3B-NH_3. Moreover, the deformation energy from the equilibrium geometry to the pyramidal form in the former complexes requires nearly the same energy. Thus, the higher acid strength of BCl_3 compared to BF_3 into the donor-acceptor complexes X_3B-NH_3 is an intrisic property of the molecule. The energy decomposition analysis (EDA) [12, 13] enables to see that Cl_3B-NH_3 exhibits a stronger bond than F_3B-NH_3 due to stronger covalent interactions between the Lewis acid and the Lewis base, which can be explained thanks to the energetically lower lying LUMO of BCl_3 (see Table 2).

Table 2. Energy levels of the lowest unoccupied molecular orbitals (LUMO) of BCl_3 and BF_3 [eV]. [a]This value was calculated using the frozen geometry in the complex.

Method	BCl_3	$BCl_3{}^a$	BF_3	$BF_3{}^a$
MP2/Aug-cc-pVTZ	1.091	0.076	1.234	0.405
PW91/QZ4P	-2.817	-4.302	-0.799	-3.625

3 Quantum Chemical Investigations of Transition Metal Nitrido Complexes with a TM-N-E Linkage (E = Main Group Element)

3.1 Abstract

Five different projects around the transition metal nitrido complexes with TM-N-E linkage were under investigation:

1. Comparison of the N-donor ability in the nitrido complexes Cl_3W-N-ECl_n where ECl_n = NaCl, $MgCl_2$, $AlCl_3$.
2. Comparative analysis of the bonding situation in Cl_4W-N-X where X = Na, MgCl, $AlCl_2$, $SiCl_3$, PCl_2, SCl, Cl.

3. Comparison of the structure and bonding in $Cl_5W\text{-}NPH_3$, $Cl_5W\text{-}OPH_3^+$, $Cl_4W\text{-}(NPH_3)(OPH_3)^+$.
4. Comparative analysis of the bonding situation in $Cl_5Ta\text{-}OPH_3$, $Cl_5W\text{-}NPH_3$,$Cl_5Re\text{-}CPH_3$.
5. Energy decomposition analysis of the bonding of the isolobal ligands NPH_3 and Cp with WCl_5.

Here, due to the space limitation, we only report the results concerning the fifth part of the work.

3.2 Method

Using the program package Gaussian 98 [6], projects 1. to 4. have been performed at the B3LYP level, which is a gradient-corrected DFT method using Becke-Lee-Yang-Parr three-parameter fit B3LYP of the exchange and correlation functionals [14, 15, 16]. The atomic basis sets include a quasi-relativistic small-core ECP [17] for the transition metal with a (441/2111/21) valence basis set and 6-31G(d,p) all electron basis sets for the other atoms [18, 19]. This combination is our standard basis set II [20]. The character of the stationary points on the potential energy surface was examined by vibrational frequency analysis.

For the fifth project, the calculations were performed with the program package ADF at the non-local DFT level of theory using the exchange functional of Becke [21] and the correlation functional of Perdew [22] (BP86). With the zero-order regular approximation (ZORA) [23, 24] we took into account the scalar relativistic effects. BP86 was associated with basis sets of triple-ζ quality augmented by two sets of polarization functions.

3.3 Discussion

The concept of isolobal analogy was introduced by Hoffmann in 1981 [25, 26, 27]. The negatively charged ligand NPR_3^- is isolobal with the cyclopentadienyl anion ligand Cp^-[1]. Figure 1 shows the most important orbitals of these two ligands that may interact with a metal in complex. Figure 2 shows the equilibrium geometries and the most important bond lengths and bond angles at BP86/TZ2P. Both molecules have C_s symmetry.

In order to analyze the bonding interactions between the metal fragment WCl_5 and the ligands NPH_3 and Cp one has to define the electronic configuration and the electronic state of the interacting species. We decided to investigate the dependency of the results of the EDA from the choice of the electronic states. Therefore, we carried out the analysis using four different electronic states or configurations of the fragmants. First, we analyzed the interactions between the closed-shell species WCl_5^+ and L^-. Next, we used the fragments with a single occupied π orbital. We solved the problem that the fragments WCl_5^+ and L^- have degenerate π orbitals by performing the

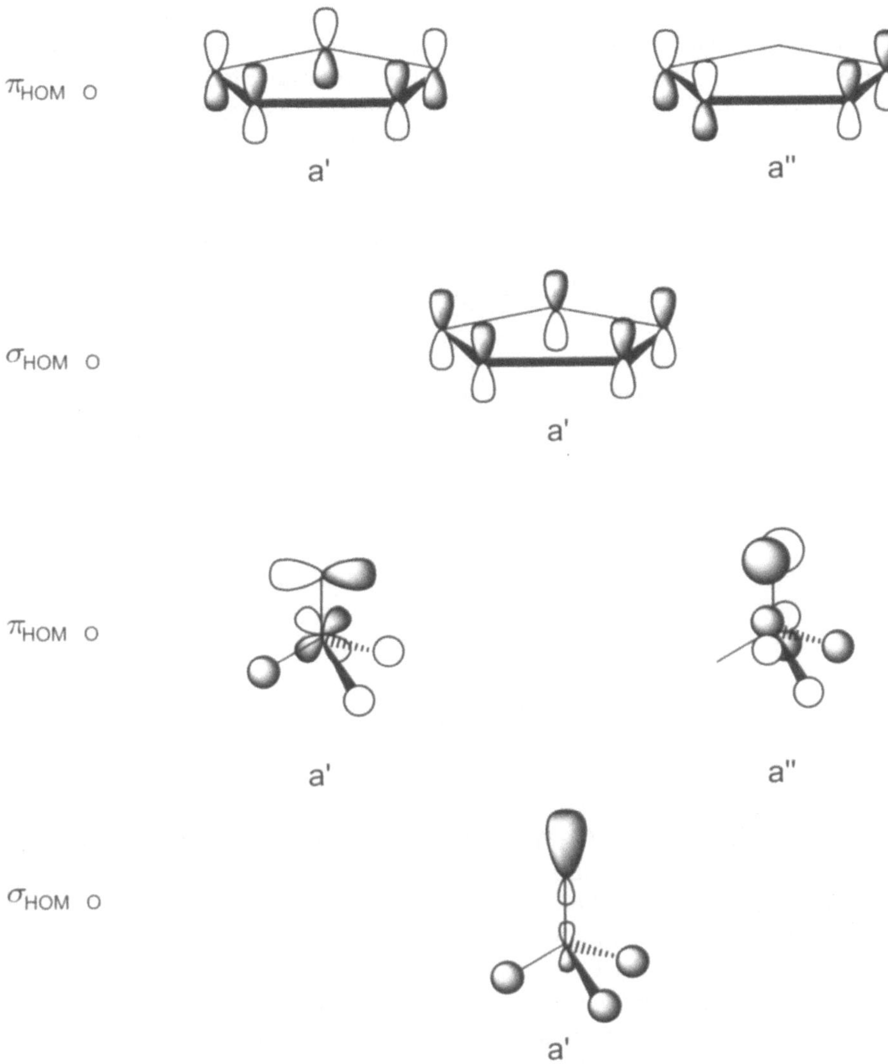

Fig. 1. Schematic representation of the most important occupied orbitals of the ligands Cp^- and NPH_3^-. In molecule with C_s symmetry, the degenerate π orbitals split into a' and a".

EDA calculations twice. Since the molecules have C_s symmetry, the degenerate π orbital splits into a' (in-plane) and a" (out-of-plane) components. The a' orbital is actually not a guenuine π orbital but it may be considered as pseudo π-type orbital. In one analysis we took the fragments with a singly occupied a" (π) orbital and in the other analysis we took the fragments with a singly occupied a' (pseudo π) orbital. The fourth energy analysis was done in the same way as chosen by Diefenbach and Bickelhaupt [28], i.e. the frag-

Table 3. Energy decomposition analysis of the equilibrium structure of H_3PN-WCl_5 using the fragments H_3PN and WCl_5. The energy values are given in kcal/mol. [a] The value in parentheses gives the percentage contribution to the total attractive interactions. [b] The value in parentheses gives the percentage contribution to the total orbital interactions. [c] The a' orbital in the π system was singly occupied in both fragments. [d] The a" orbital in the π system was singly occupied in both fragments. [e] The a' orbital in the σ system was singly occupied in both fragments.

Term	A H_3PN^-+ WCl_5^+ $(\sigma^2\pi^4)+(\sigma^0\pi^0)$	B[c] H_3PN+ WCl_5 $(\sigma^2\pi^3)+(\sigma^0\pi^1)$	C[d] H_3PN+ WCl_5 $(\sigma^2\pi^3)+(\sigma^0\pi^1)$	D[e] H_3PN+ WCl_5 $(\sigma^1\pi^4)+(\sigma^1\pi^0)$
ΔE_{int}	-286.2	-116.8	-120.6	-202.7
ΔE_{Pauli}	365.0	276.8	296.5	345.7
ΔE_{elstat}[a]	-382.5	-180.1	-208.0	-232.6
	(58.7%)	(45.7%)	(49.8%)	(42.4%)
ΔE_{orb}[a]	-268.7	-213.3	-209.1	-315.8
	(41.3%)	(54.3%)	(50.2%)	(57.5%)
a'[b]	-190.5	-171.5	-127.0	-272.1
	(70.9%)	(80.4%)	(60.6%)	(86.2%)
a"[b]	-78.1	-41.9	-82.1	-43.7
	(29.1%)	(19.6%)	(39.4%)	(13.8%)
σ	-112.4	-87.6	-85.9	-228.3
	(41.8%)	(41.1%)	(41.1%)	(72.3%)
π	-156.3	-125.7	-123.2	-87.5
	(58.2%)	(58.9%)	(58.9%)	(27.7%)

ments have singly occupied σ(a') orbitals while the π orbitals are filled. Table 3 gives the EDA results of Cl_5W-NPH_3 using the four different fragmentation patterns **A** - **D** in the equilibrium geometry. Table 4 show the EDA results of Cl_5W-NPH_3 (**A** - **D**) where the ligand is attached to the metal in a linear fashion, i.e. the bonding angle W-N-P was constrained to 180 degree. Table 5 gives the EDA results for Cl_5W-Cp for **A** - **D**.

By comparison between the results of the two complexes, Cl_5W-NPH_3 and Cl_5W-Cp, we will have information about the contributions of the σ and π orbitals to ΔE_{orb} and about the validity of the isolobal model. The EDA results of the latter complex using charged fragments (model **A**) indicate that the bonding is 52.8% electrostatic and 47.2% covalent. Thus, the metal-ligand interactions between WCl_5^+ + Cp^- are slightly more electrostatic than covalent which is similar to the analysis of the WCl_5^+ + $NPH_3{}^-$ interactions. Table 5 shows that the covalent contributions in model **A** are clearly dominated by the π bonding. The covalent interactions have 80.8% character and only 19.2% σ character. A comparison with the data for WCl_5^+ + NPH_3^- shows that the latter interactions have also much higher π character than σ character, although the values for the relative π contributions are somewhat less. This shows that the qualitative model of isolobal analogy between Cp^- and NPH_3^- is supported and qualitatively enhanced by the EDA calculations. The

Table 4. Energy decomposition analysis of linear H_3PN-WCl_5 using the fragments H_3PN and WCl_5. The energy values are given in kcal/mol. [a] The value in parentheses gives the percentage contribution to the total attractive interactions. [b] The value in parentheses gives the percentage contribution to the total orbital interactions. [c] The a' orbital in the π system was singly occupied in both fragments. [d] The a" orbital in the π system was singly occupied in both fragments. [e] The a' orbital in the σ system was singly occupied in both fragments.

Term	**A** H_3PN^-+ WCl_5^+ $(\sigma^2\pi^4)$+$(\sigma^0\pi^0)$	**B**[c] H_3PN+ WCl_5 $(\sigma^2\pi^3)$+$(\sigma^0\pi^1)$	**C**[d] H_3PN+ WCl_5 $(\sigma^2\pi^3)$+$(\sigma^0\pi^1)$	**D**[e] H_3PN+ WCl_5 $(\sigma^1\pi^4)$+$(\sigma^1\pi^0)$
ΔE_{int}	-273.7	-111.3	-111.3	-190.7
ΔE_{Pauli}	298.4	239.8	239.8	262.2
ΔE_{elstat}[a]	-332.9 (58.1%)	-162.9 (46.3%)	-162.9 (46.3%)	-166.2 (36.6%)
ΔE_{orb}[a]	-239.2 (41.9%)	-188.2 (53.7%)	-188.2 (53.7%)	-286.8 (63.4%)
a'[b]	-168.7 (70.5%)	-152.9 (81.2%)	-106.2 (56.4%)	-248.6 (86.7%)
a"[b]	-70.4 (29.5%)	-35.4 (18.8%)	-82.0 (43.6%)	-38.2 (13.3%)
σ	-97.3 (40.7%)	-70.8 (37.6%)		-172.3 (60.1%)
π	-140.8 (59.3%)	-117.4 (62.4%)		-76.3 (39.9%)

same conclusion can be made from the EDA results of the neutral fragments (see Table 5).

Elucidating the nature of the chemical bond in transition metal nitrido complexes with TM-N-E linkage was one of the aim of the five research projects. It became clear that we could gain insight into electronic structure and binding interactions of the molecules and also that it is possible to link numerically exact calculations and heuristic qualitative bonding models. That was particularly obvious in the fifth project we decided to specifically summarize here.

4 Energy Decomposition Analysis of the Chemical Bond in the Main Group

4.1 Abstract

The energy decomposition analysis (EDA) has been used to investigate the nature of the chemical bond in the main group compounds diborane(4) compounds, X_2B-BX_2 (X=H, F, Cl, Br, I). The results give deep insight into the nature of the chemical interactions. The EDA gives quantitative results

Table 5. Energy decomposition analysis of the equilibrium structure of Cp-WCl_5 using the fragments Cp and WCl_5. The energy values are given in kcal/mol. [a] The value in parentheses gives the percentage contribution to the total attractive interactions. [b] The value in parentheses gives the percentage contribution to the total orbital interactions. [c] The a' orbital in the π system was singly occupied in both fragments. [d] The a" orbital in the π system was singly occupied in both fragments. [e] The a' orbital in the σ system was singly occupied in both fragments.

Term	**A** $Cp^- + WCl_5^+$ $(\sigma^2\pi^4)+(\sigma^0\pi^0)$	**B**[c] Cp+ WCl_5 $(\sigma^2\pi^3)+(\sigma^0\pi^1)$	**C**[d] Cp+ WCl_5 $(\sigma^2\pi^3)+(\sigma^0\pi^1)$	**D**[e] Cp+ WCl_5 $(\sigma^1\pi^4)+(\sigma^1\pi^0)$
ΔE_{int}	-242.7	-85.9	-85.9	-197.9
ΔE_{Pauli}	259.5	227.7	227.9	282.1
ΔE_{elstat} [a]	-265.4 (52.8%)	-126.2 (40.2%)	-126.6 (40.3%)	-176.5 (36.6%)
ΔE_{orb} [a]	-236.8 (47.2%)	-187.4 (64.4%)	-187.2 (59.7%)	-305.5 (63.4%)
a' [b]	-141.1 (59.6%)	-120.5 (64.3%)	-98.6 (52.7%)	-241.1 (78.9%)
a" [b]	-95.7 (40.4%)	-66.9 (35.7%)	-88.7 (47.3%)	-64.4 (21.1%)
σ	-45.4 (19.2%)	-31.8 (17.0%)		-176.7 (57.8%)
π	-191.4 (80.8%)	-155.5 (83.0%)		-128.8 (42.2%)

about the strength of the covalent and electrostatic interactions and about the contributions of σ and π electrons to the covalent bond.

4.2 Methods

The geometries of the molecules have been optimized at the non-local DFT level of theory using the exchange functional of Becke [21] in conjunction with the correlation functional of Perdew [22] (BP86). Uncontracted Slater-type orbitals (STOs) were employed as basis functions for the SCF calculations [29]. The basis sets have triple-ζ quality augmented by two sets of polarization functions, i.e. p and d functions on hydrogen and d and f on the other atoms. This level of theory is denoted BP86/TZ2P. All structures have been verified as minima on the potential energy surface by calculating the Hessian matrices. The atomic partial charges have been calculated with the Hirshfeld partitioning scheme [30]. All calculations were carried out with the program package ADF [4, 5]. The EDA method is based on the ideas of Morokuma [12, 31] and Ziegler [13].

4.3 Discussion

This project aims at a quantitative analysis of the chemical bond in terms of electrostatic versus covalent interactions and at a reliable estimate of the relative degree of multiple bonding of the covalent term. The method of choice was the EDA, which has been used in our previous studies to investigate the nature of the chemical bond in transition metal complexes with various ligands [32, 33, 34] and more recently in main-group elements [35, 36]. The project is a two part project: The first part deals with the analysis of the unpolar bond in the diborane(4) compounds, X_2B-BX_2 (X=H, F, Cl, Br, I). We investigate the bonding interactions between the open shell fragments BX_2 in the planar (D_{2h}) and perpendicular (D_{2d}) equilibrium form of the diboranes.

We were able to optimize the geometries of the diboranes(4) with planar (D_{2h}) and perpendicular (D_{2d}) orientation of the BX_2 groups. The orthogonal equilibrium structures (number of imaginary frequencies i = 0) are always lower in energy than the planar species which are transition states (i = 1) for rotation about the B-B bond except for B_2F_4, which has a very flat potential energy surface and the Hessian matrix gave positive eigenvalues for the D_{2h} and D_{2d} forms. Hyperconjugation interaction and less steric repulsion of the atom X are usually believed to be the cause of the energy lowering and the shorter B-B bonds of the orthogonal structure.

Table 6. Calculated bond length B-B and B-X [Å] and bond angles B-B-X [degree] of the planar (D_{2h}) and perpendicular (D_{2d}) structures of X_2B-BX_2 (X = H, F, Cl, Br, I). Relative energies ΔE of the planar and perpendicular forms [kcal/mol].

Symmetry	Variable	H	F	Cl	Br	I
D_{2h}	r(B-B)	1.752	1.734	1.719	1.716	1.705
	r(B-X)	1.203	1.329	1.754	1.922	2.141
	ϑ(B-B-X)	116.1	117.2	118.6	118.0	117.3
	ΔE	17.3	0.2	2.2	3.6	6.3
D_{2d}	r(B-B)	1.623	1.725	1.688	1.678	1.654
	r(B-X)	1.206	1.330	1.756	1.922	2.137
	ϑ(B-B-X)	115.0	116.8	119.6	120.0	122.0
	ΔE	0.0	0.0	0.0	0.0	0.0

The results of the EDA calculations are shown in Table 7. Unpolar bonds are believed to be purely covalent by orthodox bonding models. Surprisingly, we have found that the boron-boron bond is ˜half electrostatic and ˜half covalent and that the covalent bond comes mainly from the σ orbitals. The differences between the planar and orthogonal structures are due to the changing of the strength of the hyperconjugative stabilization. The π contribution increases more in the D_{2d} forms of the heavier halodiboranes than in planar forms. This correlates nicely with the energy differences between the planar

Table 7. Energy decomposition analysis [kcal/mol] of the B-B bond and atomic partial charges q(E) of the D_{2h} and D_{2d} structures of B_2X_4

Symm.	Term	X = H	X = F	X = Cl	X = Br	X = I
D_{2h}	$\Delta E_{int.}$	-96.8	-102.3	-96.7	-96.0	-96.4
	ΔE_{Pauli}	125.0	119.6	159.8	168.4	192.4
	$\Delta E_{elstat.}$ [a]	-110.7 (49.9%)	-114.2 (51.5%)	-131.9 (51.4%)	-127.7 (48.3%)	-141.9 (49.5%)
	$\Delta E_{orb.}$ [a]	-111.1 (50.1%)	-107.7 (48.5%)	-124.6 (48.6%)	-136.7 (51.7%)	-144.9 (50.5%)
	A_1	-108.3	-138.6	-117.2	-126.8	-131.6
	A_2	0.0	-0.1	-0.2	-0.2	-0.2
	$B_1(\pi_\perp)$	0.0	-1.9	-4.2	-5.5	-8.0
	$B_2(\pi_\parallel)$	-2.8	-2.1	-3.0	-4.2	-5.1
	ΔE_σ [b]	-108.3 (97.5%)	-103.7 (96.3%)	-117.4 (94.2%)	-127.0 (92.9%)	-131.8 (91.0%)
	ΔE_π [b]	-2.8 (2.5%)	-4.0 (3.7%)	-7.2 (5.8%)	-9.7 (7.1%)	-13.1 (9.0%)
	q(E)	0.08	0.25	0.08	0.04	-0.02
D_{2d}	$\Delta E_{int.}$	-114.5	-102.6	-98.5	-98.7	-98.4
	ΔE_{Pauli}	161.7	123.8	175.8	187.2	223.6
	$\Delta E_{elstat.}$ [a]	-140.5 (50.9%)	-116.5 (51.5%)	-141.6 (51.6%)	-139.5 (48.7%)	-162.6 (50.5%)
	$\Delta E_{orb.}$ [a]	-135.7 (49.1%)	-109.9 (48.5%)	-132.7 (48.4%)	-146.6 (51.3%)	-159.4 (49.5%)
	A_1	-119.7	-105.1	-123.1	-133.9	-141.2
	A_2	0.0	-0.2	-0.2	-0.3	-0.2
	B_1	-8.0	-2.3	-4.7	-6.2	-9.0
	B_2	-8.0	-2.3	-4.7	-6.2	-9.0
	ΔE_σ [b]	-119.7 (88.2%)	-105.3 (95.8%)	-123.3 (92.9%)	-134.2 (91.5%)	-141.4 (88.7%)
	ΔE_π [b]	-16.0 (11.8%)	-4.6 (4.2%)	-9.4 (7.1%)	-12.4 (8.5%)	-18.0 (11.3%)
	q(E)	0.08	0.25	0.08	0.04	-0.02
	$\Delta E_{prep.}$	6.1	4.1	5.2	5.3	5.4
	$-D_e$	108.4	98.5	93.3	93.4	93.0

[a] The value in parentheses gives the percentage contribution to the total attractive interaction.
[b] The value in parentheses gives the percentage contribution to the total orbital interaction.

and orthogonal forms. Thus, the lower energy of the D_{2d} structure of the B_2X_4 than that of the form for X = H, Cl, Br, I and the nearly degenerate energies of the two structures of B_2F_4 can be explained with the difference between the hyperconjugation in the perpendicular structures and the conjugation in the planar structures.

5 Comparison of Side-on and End-on Coordination of E_2 Ligands in Complexes $W(CO)_5E_2$

5.1 Abstract

Density functional theory at the BP86 level of theory with valence sets of TZP quality were used to optimise complexes of W(CO)5 with the neutral diatomic pnictogen ligands N_2, P_2, As_2, Sb_2 and Bi_2 and the anionic group ligands Si_2^{-2}, Ge_2^{-2}, Sn_2^{-2}, Pb_2^{-2} coordinated in both side-on and end-on ways. Bond energies have been calculated to compare the preferential binding modes of each respective ligand. Additionally an energy decompostition analysis was performed to gain further quantitative information about the strength of the covalent and the electrostatic interactions between the metal and the ligand. The results show, that all the studied ligands bind in a side-on coordination mode, with the exception of N_2, which prefers to coordinate in an end-on mode. The negatively charged complexes with the group-14 ligands neither prefer the side-on nor the end-on coordination mode.

5.2 Method

The calculations were performed with the programs Gaussian98 [6] and ADF [4, 5] at the nonlocal density functional theory level of theory using the exchange functional of Becke [21] and the correlation functional of Perdew [22] (BP86). For the Gaussian98 calculations the 6-31G(d) basis-set was used for C and O for the neutral complexes and the 6-31G+(d) basis-set for the anionic complexes. For W the LANL2DZ [37, 38, 17] basis set was used. The Stuttgart basis sets with effective core potentials and one polarisation function were used for N, P, As, Sb, Bi; Si, G, Sn and Pb. The exponents of the polarisation functions were optimized by using numerical interpolation of the calculated atomic energies at CISD level, according to the method of Höollwarth [39]. One set of diffuse s and p functions were added to Si, Ge, Sn and Pb; the exponents were determined following the method suggested by Lee & Schaefer [40].

5.3 Discussion

The heavier element homologues to dinitrogen, i.e. P_2, As_2, Sb_2 and Bi_2 are only found as free molecules at high temperatures in the gas phase. A stabilisation effect is observed by coordination of two or more organometallic species, typically metal carbonyls [41, 42, 43, 44, 45, 46, 47, 48, 49]. The bonding mode of the heavier diatomics is substantially different to N_2. The heavier homologues prefer side-on coordination to the metal and act as four, six or eight-electron donor species, where as N_2 binds in general in an end-on manner [50] and only a few complexes containing N_2 coordinated side-on to two organometallic fragments have been reported [51, 52].

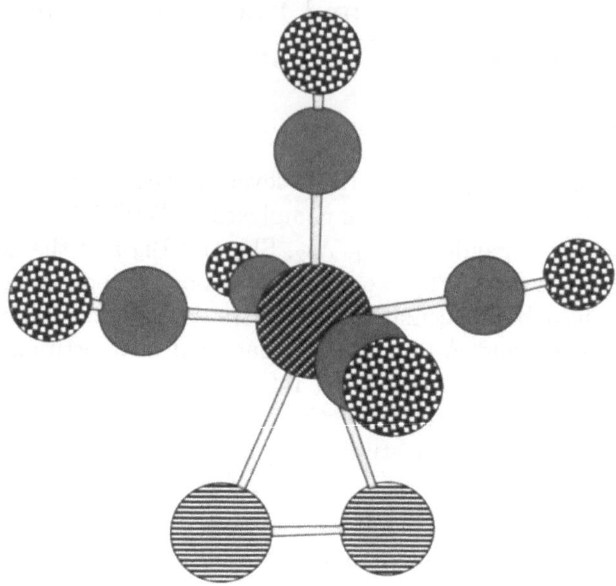

Fig. 2. Side-on coordination mode

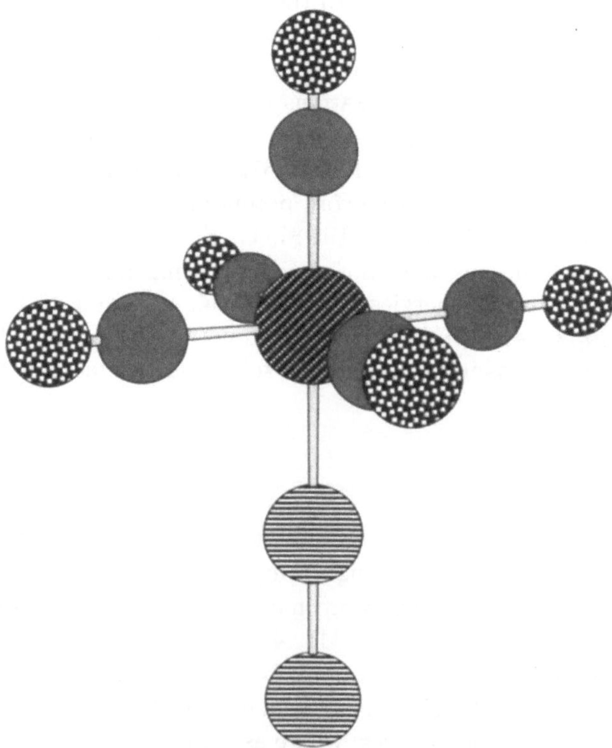

Fig. 3. End-on coordination mode

Table 8. Selected bond lengths (Å) and angles (°) of $W(CO)_5(E_2)$ complexes and free ligands calculated at BP86/TZP level of theory. The BP86/LANL2DZ values are given in parentheses.

		E = N (1)	E = P (2)	E = As (3)	E = Sb (4)	E = Bi (5)
Free ligand $_{exp.}$	E - E	1.098 [54]	1.893 [54]	2.103 [54]	2.48 [55]	2.660 [54]
Free ligand	E - E	1.104 (1.107)	1.935 (1.937)	2.161 (2.124)	2.579 (2.506)	2.728 (2.665)
Side-on	E - E	1.129 (1.129)	2.001 (2.007)	2.250 (2.204)	2.674 (2.592)	2.824 (2.752)
coordination	E - W	2.474 (2.519)	2.683 (2.695)	2.815 (2.791)	3.020 (2.995)	3.129 (3.064)
	W - C_{trans}	1.979 (1.981)	2.016 (2.021)	2.010 (2.019)	2.005 (2.014)	1.999 (2.010)
	W - C_{cis}	2.059 (2.051)	2.065 (2.067)	2.063 (2.066)	2.061 (2.064)	2.059 (2.063)
	∠W - E - E	76.2 (77.1)	68.1 (6.1)	66.4 (66.8)	63.7 (64.4)	63.2 (63.3)
End-on	E - E	1.120 (1.121)	1.934 (1.938)	2.161 (2.121)	2.564 (2.496)	2.714 (2.651)
coordination	E - W	2.117 (2.124)	2.434 (2.449)	2.585 (2.556)	2.779 (2.757)	2.913 (2.840)
	W - C_{trans}	2.019 (2.022)	2.027 (2.029)	2.010 (2.019)	2.002 (2.012)	1.990 (2.003)
	W - C_{cis}	2.060 (2.062)	2.062 (2.064)	2.061 (2.064)	2.059 (2.062)	2.058 (2.061)

The behaviour of the group-15 diatomic molecules E_2 (E = P - Bi) as ligands in transition metal complexes is also found for the isoelectronic group-14 diatomic dianions E_2^{-2} (E = Si - Pb). The synthesis of a complex containing the Pb_2^{-2} fragment coordinated to four $W(CO)_5$ fragments leads to a structure whose Pb-Pb bond length is very short; despite the Pb_2^{-2} ligand acting as an eight electron donor [53]. These short bond lengths are also found for the group-15 diatomic molecules if they are coordinated to a transition metal.

Although the complexes of the diatomic ligands of the group-14 and group-15 elements have been studied experimentally, no attempt has been made to study the bonding situation by modern quantum chemical methods, even

Table 9. Absolute energies (in a.u.), bond dissociation energies (D_e in kcal/mol) and zero-point corrected energies (D_0 in kcal/mol) of the $W(CO)_5(E_2)$ molecules calculated at BP/TZP. The BP86/LANL2DZ values are given in parentheses. The energery difference between the end-on and side-on coordination modes ($\Delta E_{coord.}$ in kcal/mol) is also given.

		E = N (1)	E = P (2)	E = As (3)	E = Sb (4)	E = Bi (5)
Free ligand	energy	-0.60644 (-19.95899)	-0.31418 (-13.13543)	-0.26074 (-12.50531)	-0.20031 (-10.96672)	-0.18382 (-10.90759)
Side-on coordination	energy	-3.83795 (654.64015)	-3.59457 (-647.86455)	-3.53949 (-647.23646)	-3.48060 (-645.70029)	-3.46245 (-645.64391)
	D_e	-8.3 (-9.2)	-39.0 (-39.3)	-38.0 (-40.6)	-38.9 (-42.1)	-37.9 (-43.8)
	D_0	-7.4 (-8.2)	-38.0 (-38.4)	-37.3 (-39.9)	-38.3 (-41.5)	-37.4 (-43.3)
End-on coordination	energy	-3.86440 (-654.66809)	-3.58314 (-647.85349)	-3.51952 (-647.21758)	-3.45804 (-645.67679)	-3.43349 (645.61269)
	D_e	-24.9 (-26.7)	-31.8 (-32.6)	-25.4 (-28.7)	-24.8 (-27.4)	-19.7 (-24.2)
	D_0	-23.3 (-25.2)	-30.9 (-31.4)	-24.8 (-28.1)	-24.3 (-26.9)	-19.2 (-23.7)
	$\Delta E_{coord.}$	-16.6 (-17.5)	7.2 (6.9)	12.5 (11.8)	14.2 (14.8)	18.2 (19.6)

Table 10. Wiberg bond indices (WBI) and atomic partial charges on E and W (q(E) and q(W) of $W(CO_5)(E_2)$ molecules from NBO analysis calculated at BP86/LANL2DZ.

		E = N (1)	E = P (2)	E = As (3)	E = Sb (4)	E = Bi (5)
Free ligand	WBI	3.02	3.00	3.00	3.00	3.00
	Charge	0	0	0	0	0
Side-on coordination	WBI E-E	2.76	2.25	2.21	2.15	2.13
	WBI E-W	0.23	0.42	0.41	0.40	0.38
	q(E)	0.01	0.09	0.12	0.17	0.20
	q(W)	-0.59	-0.79	-0.80	-0.82	-0.80
	No. electrons donated from E-E $\pi_{\parallel}$ orbital	0.08	0.43	0.48	0.57	0.61
	No. electrons donated from E-E $\pi_{\perp}$ orbital	0.01	0.05	0.05	0.06	0.07
	No. electrons accepted from E-E π^* orbital	0.17	0.36	0.35	0.33	0.31
End-on coordination	WBI E-E	2.76	2.68	2.66	2.66	2.61
	WBI E-W	0.50, 0.16	0.62, 0.15	0.57, 0.15	0.53, 0.14	0.48, 0.16
	q(E)	-0.05, 0.08	0.12, 0.06	0.13, 0.08	0.18, 0.09	0.17, 0.11
	q(W)	-0.57	-0.83	-0.83	-0.83	-0.81
	No. electrons donated from E lone pair orbital	0.26	0.41	0.34	0.59	0.25
	No. electrons accepted into E-E π^* orbital	0.12	0.13	0.12	0.11	0.09

though the experiments raise very interesting questions. In the presented project we tried to get insights into the metal-E_2 interaction in terms of covalent and electrostatic bonding and which orbitals are involved. We were also interested in the reasons for the different preference of the end-on and side-on coordination mode by N_2 and the heavier diatomic homologues, respectively. Furthermore we wanted to know the differences between the group-15 ligands and the isoelectronic group-14 dianions in their complexes with transition metals.

To achieve this goal we studied first how one $W(CO)_5$ interacts with the different diatomic ligands, in both ways side-on and end-on coordination

Table 11. Energy decomposition analysis of the $W(CO_5)(E_2)$ molecules calculated at BP86/TZP. The symmetry point group for the side-on coordination mode is C_{2v} and C_{4v} for the end-on coordination mode. All values are in kcal/mol.

		E = N (1)	E = P (2)	E = As (3)	E = Sb (4)	E = Bi (5)
Side-on	$\Delta E_{int.}$	-10.1	-43.7	-43.1	-44.0	-42.4
coordination	ΔE_{Pauli}	43.5	128.2	117.4	116.3	110.0
	$\Delta E_{elstat.}$	-18.2 (33.9%)	-88.6 (51.5%)	-83.7 (52.1%)	-81.9 (51.0%)	-81.3 (53.3%)
	$\Delta E_{orb.}$	-35.4 (66.0%)	-83.4 (48.4%)	-76.8 (47.8%)	-78.4 (48.9%)	-71.1 (46.6%)
	a_1	-14.4 (40.5%)	-35.3 (42.3%)	-35.8 (46.5%)	-42.4 (54.2%)	-42.4 (59.7%)
	a_2	-1.1 (3.1%)	-3.6 (4.4%)	-3.0 (3.9%)	-2.4 (3.1%)	-1.8 (2.5%)
	$b_1(\pi_\perp)$	-1.54 (4.3%)	-4.9 (5.9%)	-3.7 (4.8%)	-3.6 (4.6%)	-3.1 (4.4%)
	$b_2(\pi_\parallel)$	-18.43 (52.0%)	-39.5 (47.4%)	-34.4 (44.8%)	-29.9 (38.1%)	-23.8 (33.5%)
End-on	$\Delta E_{int.}$	-27.1	-33.8	-26.8	-25.9	-20.4
coordination	ΔE_{Pauli}	68.2	79.5	61.1	58.6	45.9
	$\Delta E_{elstat.}{}^{a}$	-43.8 (45.9%)	-47.5 (41.8%)	-37.6 (42.7%)	-40.5 (47.9%)	-30.2 (45.5%)
	$\Delta E_{orb.}{}^{a}$	-51.4 (54.0%)	-65.9 (58.1%)	-50.3 (57.2%)	-44.0 (52.0%)	-36.1 (54.4%)
	$a_1(\sigma)$	-23.8 (46.3%)	-34.7 (52.8%)	-30.2 (60.1%)	-29.8 (67.7%)	-27.0 (74.7%)
	a_2	0	0	0	0	0
	$b_1(\pi_\perp)$	-0.1 (0.1%)	-0.2 (0.2%)	-0.1 (0.1%)	0.0 (0.0%)	0.0 (0.1%)
	$b_2(\pi_\parallel)$	0.0 (0.1%)	-0.2 (0.2%)	-0.1 (0.2%)	0.0 (0.1%)	0.0 (0.1%)
	e_1	-27.5 (53.5%)	-30.8 (46.8%)	-19.9 (39.6%)	-14.2 (-32.3%)	-9.1 (25.2%)

mode. We also calculated the free, uncoordinated E_2 (E = P - Bi) molecules to determine the change in the bonding situation after coordination has taken place. However, in this report, due to the space limitations we will only comment on the group-15 results.

The minimum energy conformation for side-on coordination of the $W(CO)_5$ complexes with the neutral E_2 ligands has an eclipsed C_2 symmetry, whereas complexes containing E_2 fragments in the end-on coordination mode have C_{4v} symmetry, as shown in Figure 2. Some selected geometrical parameters for the optimized structures are listed in Table 8.

Comparison of the calculated values in Table 8 shows that upon side-on coordination the E-E bond becomes increasingly lengthened from that in the uncoordinated molecule along the series **1s-5s** (0.022 for 1s to 0.096). In the case of the end-on structures **1e-5e**, however; only the N-N bond lengthens by 0.013 upon metal coordination. The P-P and As-As bonds remain unaffected by coordination, while the Sb-Sb and Bi-Bi bonds are even shortened, by 0.015 and 0.014 respectively.

In Table 8 the absolute energies, bond dissociation energies (D_e) and zero-point corrected energies (D_0) are given for the various neutral complexes. Since the respective BP86/TZP and BP86/LANL2DZ results agree very well, the discussion will compare results of the BP86/TZP obtained values only. The predicted D_e indicates that the N_2 ligand is much more strongly bound in the end-on than in the side-on complex. In contrast to that for all higher homologues the side-on is highly preferred. However, the D_e values for the end-on complexes for P_2 and As_2 are even higher than for N_2, whereas Sb_2 and Bi_2 are more weakly bonded than N_2 . The explanation for the strong increase in the D_e value from **1s** to **2s-5s** while the change in D_e from **1e** to **2e-5e** is much less is found in the metal-ligand orbital interaction. σ donation from the in-plane π-orbital of the ligand (which has σ symmetry in the complex) takes places to the empty orbitals of the metal and in-plane π-backdonation ($\pi_{\|}$)from the filled d(π) atom orbital of the metal into the empty π^* orbital of the E_2 . The ligands with a formal triple bond $E \equiv E$ may also donate from the out-of-plane $\pi_{\perp}$orbital into the empty d(π) atomic orbital of the metal.

The calculated Wiberg bond indices (WBI) values (Table 10) for the E-E bonds in the complexes (**1s-5s** and **1e-5e**) are lower (between 2.76 for both **1s** and **1e** and 2.13 and 2.61 for **5s** and **5e**, respectively) than the value of 3.0 found in **1** - **5**. This decrease in the bond order agrees with previous proposals[4,5]. The WBI values show a significant difference between the N_2 complexes and the complexes containing the heavier homologues. For both N_2 complexes, side-on and end-on, the identical WBI value of 2.76 was found, whereas for the heavier homologues the WBI values are clearly indicating a smaller bond order for the side-on complexes in comparision to the end-on complexes. By a natural bond orbital (NBO) analysis (Table 10) it was found, that there is very little charge donation from the N_2 $\pi_{\|}$ orbital into the $W(CO)_5$ acceptor orbital in the side-on complex, but significant charge donation from the $\pi_{\|}$ orbital of the heavier homologues, 0.08e and 0.43 - 0.61e respectively. At the same time, the charge acceptance of the π^* orbital in the side-on complexes is clearly less for N_2 than for the heavier E_2 ligands, however, the π^* acceptance of the end-on coordinated E_2 ligands remains small for the heavier ligands.

The WBI values and the charge distribution suggest that the metal-ligand interactions in the side-on coordinated species **2s** - **5s** clearly become stronger than in **1s** , while the differences between **1e** and **2e** - **5e** appear to not be very large. From the data it is not obvious, however, why N_2 clearly prefers the end-on over the side-on coordination.

To solve this question and to gain more insight into the bonding situation in the studied complexes an EDA (Table 11) was performed. The EDA shows clearly that the end-on bonded N_2 complex **1e** has a higher degree of electrostatic bonding than the side-on isomer **1s**, whereas in the heavier homologues **2e** - **5e** the bonding exhibits less electrostatic character than in **2s** - **5s**.

By EDA we could show, that the main reason for the differences in the coordination of N_2 and the heavier homologues is due to the electrostatic interaction between the ligand and the metal fragment. We predicted the prefered coordination mode for N_2 to be end-on coordination, whereas P_2, As_2, $Sb_2$and Bi_2 prefer the side-on coordination mode, as indicated in the literature.

This work thus shows that it should, in principle, be possible to obtain complexes of P_2, As_2, $Sb_2$and Bi_2 coordinated to a single metal fragment in a side-on coordination mode since the theoretically predicted binding energies are rather high.

References

1. C. Moller and M. S. Plesset. *Phys. Rev.*, 46:618, 1934.
2. K. Raghavachari, G. W. Trucks, J. A. Pople, and M. Head-Gordon. *Chem. Phys. Lett.*, 157:479, 1989.
3. J. D. Watts, J. Gauss, and R. J. Bartlett. *J. Chem. Phys.*, 98:8718, 1993.
4. F. M. Bickelhaupt and E. J. Baerends. *Rev. Comput. Chem.*, volume 15, page 1. Wiley-VCH, New-York, 2000.
5. G. te Velde, F. M. Bickelhaupt, E. J. Baerends, S. J. A. van Gisbergen, C. Fonseca Guerra, and T. Ziegler. *J. Comput.Chem.*, 22:931, 2001.
6. M. J. Frisch, G. W. Trucks, H. B. Schlegel, P. M. W. Gill, B. G. Johnson, M. A. Robb, J. R. Cheeseman, T. A. Keith, G. A. Petersson, J. A. Montgomery, K. Raghavachari, M. A. Al-Laham, V. G. Zakrzewski, J. V. Ortiz, J. Cioslowski, B. B. Stefanov, A. Nanayakkara, M. Challacombe, C. Y. Peng, P. Y. Ayala, W. Chen, M. W. Wong, J. L. Andres, E. S. Replogle, R. Gomperts, R. L. Martin, D. J. Fox, J. S. Binkley, D. J. Defrees, J. Baker, J. J. P. Stewart, M. Head-Gordon, C. Gonzalez, J. A. Pople, R. E. Stratmann, J. C. Burant, S. Dapprich, J. M. Millam, A. D. Daniels, K. N. Kudin, M. C. Strain, O. Farkas, J. Tomasi, V. Barone, M. Cossi, R. Cammi, B. Mennucci, C. Pomelli, C. Adamo, Q. Cui, S. Clifford, J. Ochterski, K. Morokuma, D. K. Malick, A. D. Rabuck, J. B. Foresman, , G. Liu, A. Liashenko, P. Piskorz, I. Komaromi, and G. E. Scuseria. *Gaussian 98 (Revision A.1)*, 1998.
7. J. W. Perdew and Y. Wang. *Electronic Structure of Solids '91.* Number p.11. Akademie-Verlag, 1991.
8. J. G. Snijders, E. J. Baerends, and P. Vernooijs. *At. Nucl. Data Tables*, 26:483, 1982.
9. A. D. MacLean and G. S. Chandler. *J. Chem. Phys.*, 72:5639, 1980.
10. K. Raghavachari, J. S. Binkley, R. Seeger, and J. A. Pople. *J. Chem. Phys.*, 72:650, 1980.
11. R. A. Kendal, T. H. Jr. Dunning, and R. J. Harrison. *J. Chem. Phys.*, 96:6796, 1992.
12. K. Morokuma. *J. Chem. Phys.*, 55:1236, 1971.
13. T. Ziegler and A. Rauk. *Theor. Chim. Acta*, 46:1, 1977.

14. A. D. Becke. *J. Chem. Phys.*, 98:5648, 1993.
15. C. Lee, W. Yang, and R. G. Parr. *Phys. Rev. B*, 37:785, 1988.
16. P. J. Stevens, F. J. Devlin, C. F. Chablowski, and M. J. Frisch. *J. Phys. Chem.*, 98:11623, 1994.
17. P. J. Hay and W. R. Wadt. *J. Chem. Phys.*, 82:299, 1985.
18. R. Ditchfield, W. J. Hehre, and J. A. Pople. *J. Chem. Phys.*, 54:724, 1971.
19. W. J. Hehre, R. Ditchfield, and J. A. Pople. *J. Chem. Phys.*, 56:2257, 1972.
20. G. Frenking, I. Antes, M. Boehme, S. Dapprich, A. W. Ehlers, V. Jonas, A. Neuhaus, M. Otto, R. Stegmann, A. Veldkamp, and S. F. Vyboishchikov. *Reviews in Computational Chemistry*, volume 8, pages 63–144. VCH, New-York, 1996.
21. A. D. Becke. *Phys. Rev. A*, 38:3098, 1988.
22. J. P. Perdew. *Phys. Rev. B*, 33:8822, 1986.
23. J. G. Snijders. *Mol. Phys.*, 36:1789, 1978.
24. J. G. Snijders and P. Ros. *Mol. Phys.*, 38:1909, 1979.
25. R. Hoffmann. *Science*, 211:995, 1981.
26. R. Hoffmann. *Angew. Chem.*, 94:725, 1982.
27. R. Hoffmann. *Angew. Chem. Int. Ed. Engl.*, 21:711, 1982.
28. A. Diefenbach and F. M. Bickelhaupt. *Z. Anorg. Allg. Chem.*, 625:892, 1999.
29. J. G. Snijders, E. J. Baerends, and P. Vernooijs. *At. Nucl. Data Tables*, 26:483, 1982.
30. E. L. Hirshfield. *Theor. Chim. Acta*, 44:129, 1977.
31. K. Kitaura and K. Morokuma. *Int. J. Quantum Chem.*, 10:325, 1976.
32. A. Diefenbach, F. M. Bickelhaupt, and G. Frenking. *J. Am. Chem. Soc.*, 122:6449, 2000.
33. Y. Chen and G. Frenking. *J. Chem. Soc., Dalton Trans*, page 434, 2001.
34. M. Doerr and G. Frenking. *Z. allg. anorg. Chem.*, 628:843, 2002.
35. V.M. Rayon and G.Frenking. *Chem. Eur. J.*, 8:4693, 2002.
36. C. Loschen, K. Voigt, J. Frunzke, A. Diefenbach, M. Diedenhofen, and G. Frenking. *Z. allg. anorg. Chem.*, 628:1294, 2002.
37. P. J. Hay and W. R. Wadt. *J. Chem. Phys.*, 82:270, 1985.
38. P. J. Hay and W. R. Wadt. *J. Chem. Phys.*, 82:284, 1985.
39. A. Hoellwarth, M. Boehme, S. Dapprich, A. W. Ehlers, A. Gobbi, V. Jonas, K. F. Koehler, R. Stegmann, A. Veldkamp, and G. Frenking. *Chem. Phys. Lett.*, 208:237, 1993.
40. T. J. Lee and H. F. SchaeferIII. *J. Chem. Phys.*, 83:1784, 1985.
41. A. S. Foust, M. S. Foster, and L. F. Dahl. *J. Am. Chem. Soc.*, 91:5633, 1969.
42. G. Huttner, U. Weber, B. Sigwarth, and O. Scheidsteger. *Angew. Chem. Suppl.*, 411, 1982.
43. G. Huttner, U. Weber, and L. Zsolnai. *Z. Naturforsch.*, 37b:707, 1982.
44. G. Huttner, B. Sigwarth, O. Scheidsteger, L. Zsolnai, and O. Orama. *Organometallics*, 4:326, 1985.
45. B. Sigwarth, L. Zsolnai, H. Berke, and G. Huttner. *J. Organomet. Chem.*, 226:C5, 1982.
46. P. J. Sullivan and A. L. Rheingold. *Organometallics*, 1:1547, 1982.
47. L. Yoong Goh, R. C. Wong, W.-H. Yip, and T. C. W. Mak. *Organometallics*, 10:875, 1991.
48. W. J. Evans, S. L. Gonzales, and J. W. Ziller. *J. Am. Chem. Soc.*, 113:9880, 1991.

49. L. Yoong Goh, W. Chen., and R. C. Wong. *J. Organomet. Chem.*, page 47, 1995.
50. A. D. Allen and F. Bottomley. *Accounts Chem. Res.*, 1:360, 1968.
51. C. Krueger and Y.-H. Tsay. *Angew. Chem. Int. Ed. Engl.*, 12:998, 1973.
52. M. D. Fryzuk, T. S. Haddad, M. Mylvaganam, D. H. McConville, and S. J. Rettig. *J. Am. Chem. Soc.*, 115:2782, 1993.
53. P. Rutsch and G. Huttner. *Angew. Chem. Int. Ed. Engl.*, 39:3697, 2000.
54. K. P. Huber and G. Herzberg. *Constants of Diatomic Molecules (data prepared by J. W. Gallagher and R. D. JohnsonIII).* Number 69. NIST Chemistry WebBook, National Institute of Standards and Technology, Gaithersburg MD, 20899, July 2001. http://webbook.nist.gov.
55. H. Sontag and R. Weber. *J. Mol. Spectrosc.*, 91:72, 1982.

Quantum Mechanical Studies of Boron Clustering in Silicon

Péter Deák[1], Ádám Gali[1], Peter Pichler[2], and Heiner Ryssel[3,2]

[1] Department of Atomic Physics, Budapest University of Technology and Economics, Budapest, Budafoki út 8., H-1111, Hungary
[2] Fraunhofer-Institut für Integrierte Systeme und Bauelementetechnologie, Schottkystrasse 10, 91058 Erlangen, Germany
[3] Lehrstuhl für Elektronische Bauelemente, Universität Erlangen-Nürnberg, Cauerstrasse 6, 91058 Erlangen, Germany

1 Introduction

Boron-interstitial clusters (BICs) are known to be a key problem to controlling diffusion and activation of ultra-shallow boron implants in ULSI silicon device technology. During post-implantation annealing the self-interstitials, which had been created by the radiation damage, mediate fast transient diffusion of boron, during which stable and metastable BICs are formed. The BICs are either electrically inactive or the number of holes they can provide per number of boron atoms is significantly less than one. This causes a significant decrease in the activation rate. Therefore, sophisticated annealing strategies have to be developed to regain isolated boron substitutionals from BICs.

The standard approach for studying the deactivation of boron due to clustering is to try to elucidate defect properties from comparison of SIMS diffusion profiles and spreading resistance profiles. While the formation and dissolution energies of the possible clusters have been obtained earlier as fitting parameters of kinetic models (see, e.g., ref. [1]), in recent years attempts have been made to determine these data a priori from first principles theoretical calculations [2, 3, 4, 5, 6, 7, 8]. For that purpose "wholesale" calculations have been carried out on a large number of different BICs consisting of up to 8 or more boron and self-interstitial atoms. The deviation of the calculated values from the fitted ones are generally about 25% [7, 8] but differ substantially (sometimes by 100%) from one calculation to another on the same cluster even if the same method (actually the same computer code) was used [2, 3, 4, 5, 7, 8]. The reason is the inherent uncertainty of the applied corrections to density functional theory (DFT) and the supercell approach which are used to predict the energetics of defects [9]. There are even differences in the predicted transformations of the simplest cluster, BI, during diffusion

[10, 11, 12]. Calculations also predict the dominance of "pure" B clusters after annealing at about 800°C, while parametric models [13, 14] emphasize clusters with an equal number of B-s and I-s. DFT calculations are regarded superior to empirical tight binding ones even though (unlike the latter [15]) they predict complexes of substitutional B atoms energetically unfavorable despite of the fact that the calculated vibration modes of B_{Si}-B_{Si} could have been identified with bands of the boron related infrared spectrum in silicon [16]. Apparently, energy calculations alone are not sufficient to establish the key players in the clustering process of boron. An alternative strategy would be to identify the most important BICs based on their spectroscopic properties. For that purpose, comparison of the calculated properties of BICs with the infrared and Raman spectra as well as with the DLTS spectra of B-implanted samples at various stages of annealing is necessary. Apart from early studies on the substitutional B, and the B+I and B+B pairs (see refs. [17, 18, 19] and references therein), no report of such investigations can be found in the literature. A single (combined) study was reported in 1993 [20] leading to the establishment of a 12 atom boron cluster [21] in as-implanted samples.

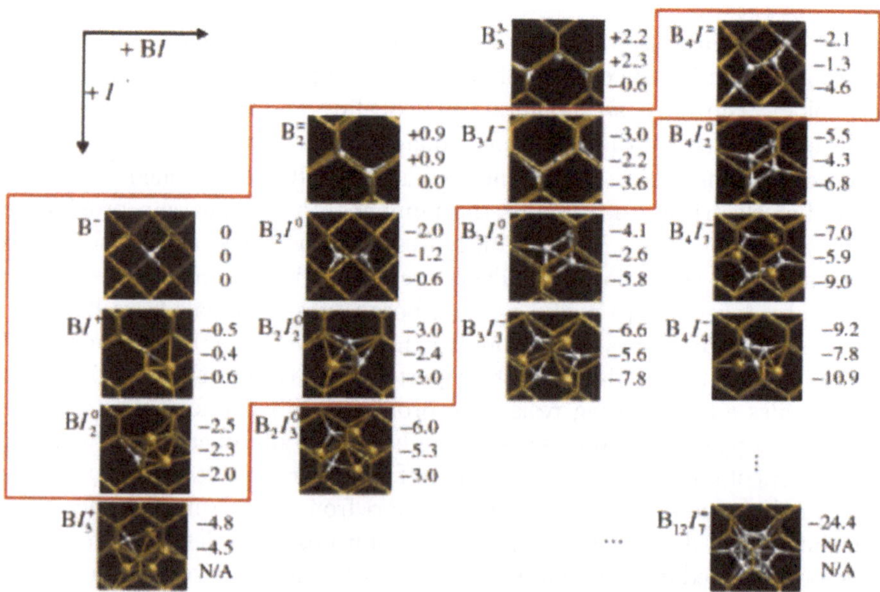

Fig. 1. Structure and energetics of the BICs considered by Windl et al. [8]. The small white balls are B atoms, the large balls are Si atoms involved in the cluster. All other Si atom are shown as a stick-only network. The energy values (eV) next to each picture are, top to down, corrected formation energies from GGA, LDA, and the fitted values from Ref. [1]. The BICs marked by the red polygon have been considered in this study.

The present study describes the first attempt of a systematic first principles quantum mechanical calculation of the characteristic vibration frequencies of all the BICs which have been suggested so far to be important in the de/reactivation process of boron. Figure 1 shows the BICs selected to be the subjects of the first series of calculations from the list published in ref [8]. In addition, some isomers of these BICs have also been studied in the first year of the investigation. In the second series of calculations the remaining BICs and all the possible isomers will be studied, while the third series of calculations will be directed to obtain other spectroscopic properties – like electrical levels and hyperfine tensors – for the most important BICs. The usual nomenclature of BICs, B_nI_m is based on the number n of boron atoms and the number m of interstitials involved in the cluster, irrespective of the fact that the interstitial is a boron or a silicon atom. In this notation, BI may as well mean an interstitial boron or a silicon self-interstitial next to a substitutional boron. Such systems are configurational isomers with the same number of atoms. In order to be able to differentiate among the possible isomers the present study uses the notations Si_i, B_i, and B_s for interstitial silicon and boron, and for substitutional boron, respectively.

2 Methods

First principles supercell calculations have been carried out using the SIESTA code [22]. SIESTA implements Density Functional Theory (DFT) combined with the pseudopotential approximation, and uses numerical atomic orbitals as a basis set. It is aimed at large-scale calculations with linear-scaling simulations but is also capable of employing conventional diagonalization methods. In this work, only the latter capability was used. Norm-conserving pseudopotentials have been generated according to the Troullier-Martins [23] scheme, in the Kleinman-Bylander [24] separable form. Core radii of 1.78 and 1.89 bohr were used for B and Si, respectively. The program requires the use of a grid to compute some of the contributions to the matrix elements and total energy and also for performing the Fourier transforms needed to evaluate the Hartree potential and energy by solving Poisson's equation in reciprocal space. A grid fine enough to represent plane waves with kinetic energy up to 90 Ry have been used. Calculations have been carried out using the Generalized Gradient Approximation (GGA) with the Perdew-Burke-Ernzerhof functional [25, 26]. A high quality basis set was used, consisting of double-ζ plus polarization functions for the valence electrons of all atom types. The maximum extent of these functions was 5.965 Å. Structural relaxations have been performed by means of the conjugated gradient algorithm until the forces were smaller than 0.04 eV/Å. To obtain the vibrational normal modes in the harmonic approximation, the second derivative of the total energy has been calculated by way of finite differences. Since these are the only values of interest, the total energy has not been corrected either for the DFT gap error, or for the supercell dispersion error. Due to the same reason, no charge correction has been applied either. The vibration calculations are computationally very demanding.

In order to keep the usage of resources within the available framework without compromising the predictive power of the results, careful convergence tests had to be carried out. First, the effect of the size of the supercell and of the quality of the Brillouin zone (BZ) summation had to be checked. BZ summations in supercell calculations are usually replaced by a weighted sum over a representative Monkhorst-Pack (MP) k-point set [27]. The formation energies in Figure 1 have been obtained by Windl et al. [8] on a 64 atom supercell in a plane wave supercell calculation using ultrasoft pseudopotentials and a 4^3 MP set (these values contain also a charge correction). Table 1 shows SIESTA results in comparison. As can be seen, formation energies, obtained

Table 1. Comparison of BIC formation energies obtained with different supercells and MP sets within the GGA approximation. The values contain no corrections (those of Windl et al. have been obtained from Figure 1 by subtracting the charge correction). The formation energies of Si_i (I) and B_s^- (B^-) have been calculated with respect to the chemical potential of the bulk materials, while those of all other complexes with respect to I and B^-. Note the positive formation energy of B_2^-.[5]

BIC	Present		Windl et al. [4]	Present	
	64 atom cell 1^3 MP set	64 atom cell 2^3 MP set	**64 atom cell 4^3 MP set**	216 atom cell 2^3 MP set	216 atom cell 1^3 MP set
I	2.53	3.65		3.61	3.35
B^-	$-\mu_B$-80.33	$-\mu_B$-80.46		$-\mu_B$-80.53	$-\mu_B$-80.50
BI^+	-0.24	-0.69	**-0.5**		-0.42
BI_2	-2.14	-2.30	**-2.4**		
B_2^{2-}	+0.87	+0.76	**+0.9**		+0.88
B_2I	-1.07	-1.71	**-1.7**		-1.44
B_2I_2	-1.98	-2.67	**-2.7**	-2.53	-2.35
B_3I^-	-1.98	-2.67	**-2.7**		-2.30
B_4I^{2-}	-1.31	-1.92	**-2.1**	-1.66	-1.53

with the 2^3 MP set in the 64 atom unit cell, are within 0.2 eV to those of the 4^3 set. This justifies restriction to the 2^3 set, leading to a factor of 2 to 16 (depending on the symmetry of the BIC) saving in computational resources. Tests with the 216 atom cell show that the effect of the appropriate BZ summation is much more significant than that of the size of the unit cell, justifying the use of the 64 atom cell. Note, that the divacancy in silicon has been regarded as a worst case test where the correct configuration of the defect could only be obtained with more than 200 atoms in a molecular cluster calculation [28]. Our 216 atom, single k-point supercell calculation reproduces the large pairing distortion of the neighbors of the divacancy, and the geometry does not change essentially in a 512 or a 1000 atom supercell. The fact that the geometry itself does not change more than 0.01 Å between the 64 and 216 atom calculations for these BICs, ensures that the calculated vibration frequencies be close to convergence.

In fact the geometry is much less sensitive to the BZ summation than the formation energy, so the question arises whether the 1^3 set (Γ point approximation) would not suffice for vibration calculations. An additional parameter for the latter is the number of atoms which are allowed to vibrate in the frozen framework of the rest. For localized vibration modes with frequencies well apart from the phonon continuum it is usually sufficient to consider the vibrations of those atoms only which belong directly to the defect cluster. Since many of the normal modes of the BICs are expected to be close to the Raman frequency of Si (520 cm^{-1}), a stronger mixture of the localized vibration modes (LVM) with the continuum modes might occur. Table 2 shows results obtained with allowing different number of shells around the complex to vibrate. As can be seen, the restriction to 2nd neighbor shell (2NN) provides

Table 2. Comparison ^{11}B vibration frequencies (cm^{-1}) calculated by allowing a different number of next neighbor (NN) shells around the complex to vibrate.

BIC	64, 1^3 MP	64, 1^3 MP	64, 1^3 MP	**64, 2^3 MP**
	All atoms vibrating	3NNs vibrating	2NNs vibrating	**2NNs vibrating**
B_s^-	610	609	610	**607**
B_2^{2-}	618		618	**627**
	590		589	**577**
B_2I_2		880	880	**926**
		669	670	**684**
		575	576	**565**
		556	554	**539**
B_4I^{2-} (C_{2v})		961	961	**1019**
		771	771	**744**
		727	729	**730**
		655	657	**652**
		643	642	**641**
		606	607	**609**

sufficiently convergent results even for the lowest LVMs. On the other hand, the use of the 2^3 MP set for the vibration calculation seems unavoidable. As a result of the convergence tests, the following calculations have been carried out on the 64 atom cell with the 2^3 MP set and 2NN shells vibrating. This is more or less what we can afford to do at most, considering the large number of complexes we have to investigate. The tests above provides a basis to judge the convergence of the results. According to earlier experience, the typical accuracy of such calculated vibration frequencies is about ± 30 cm^{-1}. The substitutional boron acceptor has an experimentally well established LVM at 623 cm^{-1} (see in Section 4), to be compared with our prediction of 607 cm^{-1}.

3 Usage of the High Perfomance Computing Center Stuttgart

The calculation of local vibrational modes is extremely time demanding process. In addition, we have investigated more than 10 complexes in their different charge states and configurations. The usage of supercomputers is necessary to accomplish this task. We have used the CRAY-T3E machine with MPI library in the High Perfomance Computing Center Stuttgart to execute the program SIESTA parallel.

First, the geometry was optimized in 64 and 216 atom supercells with different K-point sets (single Γ-point (1^3 MP) and four K-points (2^3 MP)). The execution time scales linearly with the number of K-points. Typically, 64 processors have been used for 64 atom – 1^3 MP calculations for geometry optimization by requiring 10 SUs per processors, which means about **640 SUs** as total usage. After obtaining the optimized geometry in 64 atom supercell, we have used this as an input in 216 atom supercell. For 216 atom – 1^3 MP calculations 128 processors needed with typically 7.5 SUs per processors, which means about **960 SUs** as total for a single geometry optimization run.

The calculation time of second derivatives scales linearly with the number of vibrating atoms. For $(B_s)_2$ defect, the number of the atoms up to the second neighborhood (2NN) is 17. The calculation of the (2NN) second derivatives within 1^3 MP K-point in a 64 atom supercell required about **224 SUs** (using 64 processors). This means that a typical run with 2^3 MP K-points is in the order of **1000 SUs**.

It is apparent that the supercomputer resources of HLRS indeed helps to carry out the necessary calculations needed to investigate the behavior of BICs.

4 Results in the first year of the investigation

In Table 3 the calculated vibration frequencies of the BIC-related LVMs are given assuming the presence of either only ^{11}B or only ^{10}B (the latter values in parentheses). From the two possible isomers of BI, B_i does not give rise to LVMs above the phonon continuum. Even though B_2, alias $(B_s)_2$, appears to be metastable with respect to two isolated B_s atoms, it might be created in the dissolution process of larger BICs. It has a double acceptor level above that of the single boron acceptor, therefore, both the neutral and the double negative charge state has been calculated, for the former should occur in p-type (where the level of boron activation is high) and the latter in compensated samples (or where the activation level of boron is low). Among the possible isomers of B2I the following were considered: the boron pair splitting a substitutional site in a [001] dumbbell configuration and two B_s on neighbor sites with a Si_i between them (D_{3d} in the neutral, and a puckered C_{1h} symmetry in the doubly negative charge state) Further $2B_s$+Si_i complexes are possible in other charge

Table 3. LVMs of BICs calculated in the 64 atom supercell with the 2^3 MP set, 2NN shells around the BIC vibrating. The numbers in italics- and bold-type correspond to IR and Raman active modes, respectively, while bolded-italics numbers denote LVMs which are both IR and Raman active. Values above the phonon continuum are given for isotopically pure complexes of ^{11}B (^{10}B) in (cm^{-1}).

BIC		LVM	Frequency	LVM	Frequency
B_s^-	B^- T_d	T_2 (IR/R)	***607(632)***		
B_i^+	BI^+ T_d		<520		
$(B_s+Si_i)^+$	BI^+ C_{1h}	A'(IR/R)	***693(723)***		
$[(B+Si)_s+Si_i]^-$	BI_2^- C_{1h}	A'(IR/R)	***671(698)***	A"(IR/R)	***658(684)***
$(B_s)_2^0$	B_2^0 D_{3d}	A_{1g} (R)	**644(674)**	E_u (IR)	*527(547)*
$(B_s)_2^{2-}$	B_2^{2-} D_{3d}	A_{1g} (R)	**627(657)**	E_u (IR)	*577(602)*
$(B_2)_s^0$	B_2I^0 D_{2d}	A_1 (R)	**1099(1152)**	E (IR/R)	***577(602)***
$(B_s$-Si_i-$B_s)^0$	B_2I^0 D_{3d}	A_{1g} (R)	**672(706)**	A_{2u} (IR)	*882(908)*
		E_g (R)	**586(608)**	E_u (IR)	*604(628)*
$(B_s$-Si_i-$B_s)^{2-}$	B_2I^{2-} C_{1h}	A' (IR/R)	***821(852)***	A" (IR/R)	***719(748)***
			666(692)		***582(604)***
			573(594)		***569(589)***
$[(B_2)_s+Si_i]^0$	$B_2I_2^0$ C_{1h}	A' (IR/R)	***926(969)***	A" (IR/R)	***684(710)***
			565(585)		***539(560)***
$(B_s+B_i+B_s)^-$	B_3I^- D_{3d}	A_{1g} (R)	**842(882)**	A_{2u} (IR)	*1264(1326)*
		E_g (R)	**612(635)**	E_u (IR)	*652(679)*
$[(B_s(B_2)_sB_s]^0$	B_4I^{2-} C_2	A' (IR/R)	***1095(1148)***	A" (IR/R)	***814(853)***
			729(763)		***637(663)***
			639(665)		
			566(589)		
			564(587)		
$[(B_s(B_2)_sB_s]^0$	B_4I^{2-} C_{2v}			A_1 (IR/R)	***1019(1068)***
				B_1 (IR/R)	***744(780)***
				B_2 (IR/R)	***730(759)***
		A_2 (R)	**641(667)**	B_2	***652(679)***
				A_1 (IR/R)	***609(637)***

states. $(B_2)_s^0$ is more stable than $(B_s\text{-}Si_i\text{-}B_s)^0$ by 0.64 eV, but the activation energy for rearrangement appears to be significant. Among the higher BICs isomerism has only been considered for B_4I, for the arrangement of the two B_s atoms around the central $(B_2)_s$ dumbbell in a C_{2v} and in a C_2 geometry differs only by 0.35 eV in energy (in favor of the latter). The LVMs of complexes with mixed isotopes have also been calculated and are given in Table 4.

As mentioned before, the expected accuracy of the calculated vibration frequencies are about ± 30 cm^{-1}. Therefore, identification of experimentally observed vibrational centers with the models, based on the calculated values, can only be trusted if more than one mode of the center is known and possibly the effect of isotope substitution has also been measured. Unfortunately, the available experimental information about the vibrations of BICs is scarce. The LVM of the isolated boron acceptor has been measured by both infrared (IR)

Table 4. LVMs of BICs calculated in the 64 atom supercell with the 2^3 MP set, 2NN shells around the BIC vibrating. The numbers in italics- and bold-type correspond to IR and Raman active modes, respectively, while bolded-italics numbers denote LVMs which are both IR and Raman active. Values above the phonon continuum are given for isotopically mixed complexes in (cm^{-1}).

BIC		LVM	Frequency	LVM	Frequency
$(B_s)_2^0$	B_2^0 C_{3v}	A_1 (IR/R)	***660***	E(IR/R)	***540***
$(B_s)_2^{2-}$	B_2^{2-} C_{3v}	A_1 (IR/R)	***642***	E(IR/R)	***592***
$(B_2)_s^0$	B_2I^0 C_{2v}	A_1 (IR/R)	***1125***	B_1 (IR/R)	***738***
				B_1 (IR/R)	***738***
$(B_s\text{-}Si_i\text{-}B_s)^0$	B_2I^0 C_{3v}	A_1 (IR/R)	***688***	A_1 (IR/R)	***897***
		E (IR/R)	***592***	E (IR/R)	***622***
$(B_s\text{-}Si_i\text{-}B_s)^-$	B_2I^- C_{1h}	A' (IR/R)	***851***	A" (IR/R)	***721***
			666		***593***
			588		***582***
$[(^{11}B\text{-}^{10}B)_s+Si_i]^0$	$B_2I_2^0$ C_{1h}	A' (IR/R)	***941***	A" (IR/R)	***696***
			585		***553***
$[(^{10}B\text{-}^{11}B)_s+Si_i]^0$	$B_2I_2^0$ C_{1h}	A' (IR/R)	***954***	A" (IR/R)	***699***
			565		***545***
$(^{11}B_s+^{10}B_i+^{11}B_s)^-$	B_3I^- D_{3d}	A_{1g} (R)	**841**	A_{2u} (IR)	*1304*
		E_g (R)	**635**	E_u (IR)	*678*
$(^{11}B_s+^{11}B_i+^{10}B_s)^-$	B_3I^- C_{3v}	A_1 (IR/R)	***861***	A_1 (IR/R)	***1277***
		E (IR/R)	***621***	E (IR/R)	***668***
$(^{10}B_s+^{10}B_i+^{11}B_s)^-$	B_3I^- C_{3v}	A_1 (IR/R)	***862***	A_1 (IR/R)	***1315***
		E (IR/R)	***620***	E (IR/R)	***669***

and Raman [29, 30] spectroscopy but further boron related vibrations are only known from IR spectra of electron irradiated compensated samples [31, 32, 33]. (Compensation is necessary to get rid of the free carrier absorption which would mask the LVMs. In case of neutron irradiation the created damage provides for compensation [34].) Matters are complicated by the presence of oxygen in Cz-Si samples which gives rise to a huge band between 1000 and 1100 cm^{-1} in the IR spectrum making the observation of other features in this region extremely difficult. Based on circumstantial evidence (number of modes, splitting on isotope substitution, stability range) the boron related centers in the IR spectrum (called P-, Q-, R- and S-lines) have been tentatively assigned to various models [31].

In Table 5 we have compared the calculated LVMs of these models (in a charge state appropriate for compensated samples) with the observed data. The calculated frequency of the single acceptor, 607 cm^{-1} with an isotope shift of 25 cm^{-1}, against the observed 623 cm^{-1} with an isotope shift of 23 cm^{-1} marks the accuracy of the calculations. The R line was assumed to originate from a complex of substitutional boron with one ore more self-interstitials. At present we only have the LVMs of the $(B_s+Si_i)^+$, alias BI^+, and the $[(B+Si)_s+Si_i]^-$, alias BI_2^- complex, at 693(723) cm^{-1} and 671(698) cm^{-1}

Table 5. Comparison of early experiments on *e*-irradiated compensated or *n*-irradiated material with calculated frequencies of those BICs which have been suggested as the origin of the observed vibrational centers. Only pure ^{11}B BICs are considered here. The numbers in italics- and bold-type correspond to IR and Raman active modes, respectively, while bolded-italics numbers denote LVMs which are both IR and Raman active. The I2 center has been observed in photoluminescence.

Center	acceptor	R-lines	S-lines	Q-lines	P-lines	I2-center
Stability range (°C)		< -40	-40 – -10	-10 – 200	> 300	> 200
Expt. LVM	***623***	*730*	*903* *599*	*733*	*553*	843 242
Assigment	B_s^-	B_s+nSi_i	2B trig. sym.	B_i	$(B_s)_2$	$2B_s+Si_i$ low. sym.
Model	B_s^-	$(B_s+Si_i)^+$ $(B_s+2Si_i)^-$	$B_s+Si_i+B_s$	$(B_2)_s$	$(B_s)_2$ or $(B_s)_2^{2-}$	$(B_s+Si_i+B_s)^{2-}$
Yamauchi et al. (theory) [16]	T_2 ***612***			A_1 **1026** E ***760***	A_{1g} **585** E_g **603** E_u *530*	
PRESENT (theory)	T_2 ***607***	A_1 ***693*** A_1 ***671***	A_{2u} *882* A_{1g} **672** E_u *604* E_g **586**	A_1 **1099** E ***738***	(0) / (2-) A_{1g} **644/627** E_u *527/577*	A ***821*** ***719*** ***666*** ***582*** ***573*** ***569***

for the isotope ^{11}B (^{10}B), respectively. The deviation from the observed 730(757) cm^{-1} is not small enough in either case, to make the identification convincing based on just a single mode. The Q-line was assigned to interstitial boron [31] but Yamauchi et al. suggested the $(B_2)_s$ complex (alias B_2I) instead, based on their calculations. Our result for the IR mode, 738(767) cm^{-1} is even closer to the experimental 733(760) cm^{-1} but Raman measurements on Fz-Si would be needed to confirm the assignment by finding the high-frequency Raman-active mode.

In case of the S-lines, with two known IR modes, the situation is more favorable. The suggestion of ref. [31] was a complex of two boron atoms with trigonal symmetry. Since B_2 (i.e., $(B_s)_2$) has no high frequency modes the B_2I isomer with trigonal symmetry, $B_s+Si_i+B_s$ can be considered. The calculated IR modes, at 882(908) and 604(628) cm^{-1} are reasonably close to the observed values 903(928) and 599(-) cm^{-1} for the isotopically pure complexes. Also the calculated modes for the mixed complexes, 897 and 622 cm^{-1} fit the observed 917 and 603 cm^{-1} agreeably. It should be noted, though, that the mixed complex has lower symmetry (no inversion) and the other two modes which are Raman-active in the isotopically pure complex should become visible in IR around 688 and 592 cm^{-1}. From the boron related centers in the IR spectrum (called P-, Q-, R- and S-lines), only the Q-line is stable at room temperature and only the P-line survives annealing up to 300°C. The P-lines have been assigned to a pair of substitutional boron atoms [31]. Despite of finding the $(B_s)_2$ complex metastable, Yamauchi et al. [16] confirmed that the calculated IR-active mode of $(B_s)_2^0$, 530 cm^{-1}, is reasonably close to the experimentally observed 553 cm^{-1}. Our result for the IR mode, 527 cm^{-1}, is close to that of ref. [16] (even though we only get one Raman active mode, E_g, above the continuum with a frequency much higher than that of Yamauchi et al.) We note, however, that in compensated samples the stable charge state should be $(B_s)_2^{2-}$, for which we obtain 577(602) cm^{-1} for the isotopically pure complexes and 592 cm^{-1} for the mixed one. (Note that in the mixed complex an additional mode at around 642 cm^{-1} should become IR-visible.) The observed values, 553(570) cm^{-1} for the pure and 560 cm^{-1} for the mixed complex are still within the accuracy of the calculation. Additional support for the assignment stems from the report of a feature at 615 cm^{-1} correlating with the P-lines. This could be the mode which is only IR-active in the mixed complex. Still, a definite identification required the knowledge of the corresponding Raman spectra. In addition to IR centers, LVM assisted sidebands of the boron related photoluminescence (PL) center, called I2, has also been observed [35, 36]. Originally, this center has been attributed to two boron and a self-interstitial atom with trigonal symmetry [35] but later a reduced symmetry was reported [36]. The calculated frequencies of the neutral B_2I (D_{3d}) isomer $(B_s\text{-}Si_i\text{-}B_s)^0$ obviously do not fit the modes of the I2 center. The highest frequency of the doubly negative (C_{1h}) isomer is, however, quite close to the observed I2 mode. Since we cannot, at present, calculate intensities, it is not clear why the other modes were not observed, if this complex is the origin of the I2 center.

The results mentioned above are summarized in a paper entitled "Studies of boron – interstitial clusters (BIC) in Si" by P. Deák, A. Gali, A. Sólyom, P. Ordejón, K. Kamarás, and G. Battistig which is published in Journal of Physics: Condensed Matter [37].

5 Plan for the next year of investigation

In the second series of calculations the remaining BICs and all the possible isomers will be studied. As can be seen in Fig. 1 those defects involves at least three boron and/or interstitial Si atoms. For those defects a bigger supercell than 64 atom cell may be necessary to provide accurate values of the corresponding local vibrational modes. This task could be accomplished by using the resource of supercomputers. We would like to use the resource of HLRS further in order to continue our investigation on BICs.

References

1. L. Pelaz, G. H. Gilmer, J.-J. Gossmann, and C. S. Raferty. *Appl. Phys. Lett.*, 74:3657, 1999.
2. J. Zhu, T. Diaz de la Rubia, L. H. Yang, C. Malhoit, and G. H. Gilmer. *Phys. Rev. B*, 64:4741, 1996.
3. M. J. Caturla, M. D. Johnson, and T. Diaz de la Rubia. *Appl. Phys. Lett.*, 72:2736, 1998.
4. J. Zhu. *Comput. Mater. Sci.*, 12:309, 1998.
5. T. J. Lenosky, B. Sadigh, M.-J. Caturla S. K. Theiss, and T. Diaz de la Rubia. *Appl. Phys. Lett.*, 77:1834, 2000.
6. S. Chakravarthi and S. T. Dunham. *J. Appl. Phys.*, 89:3650, 2001.
7. X.-Y. Liu, W. Windl, and M. P. Masquelier. *Appl. Phys. Lett.*, 77:2018, 2000.
8. W. Windl, X.-Y. Liu, and M. P. Masquelier. *Phys. Stat. Sol. (b)*, 226:37, 2001.
9. P. Deák. In *Computational Materials Science*, volume in print of *Proc. of the NATO ASI*, Dordrecht, Sept. 2001. Kluwer Acad. Publ.
10. B. Sadigh, Th. J. Lenosky, S. K. Theiss, M.-J. Caturla, T. Diaz de la Rubia, and M. A. Foad. *Phys. Rev. Lett.*, 83:4341, 1999.
11. W. Windl, M. M. Bunea, R. Stumpf, S. T. Dunham, and M. P. Masquelie. *Phys. Rev. Lett.*, 83:4345, 1999.
12. M. Hakala, M. J. Puska, and R. M. Nieminen. *Phys. Rev. B*, 61:8155, 2000.
13. L. Pelaz, V. C. Venezia, J.-J. Gossmann, G. H. Gilmer, A. T. Fiory, M. Jaraiz, and J. Barbolia. *Appl. Phys. Lett.*, 75:662, 1999.
14. M. Uematsu. *J. Appl. Phys.*, 84:4871, 1998.
15. W-W. Luo and P. Clancy. *J. Appl. Phys.*, 89:1596, 2001.
16. J. Yamauchi, N. Aoki, and I. Mizushima. *Phys. Rev. B*, 63:073202–1, 2001.
17. P. J. M. Smulders, D. O. Boerma, B. Bech Nielsen, and M. L. Swanson. *Nucl. Instr. & Methods*, 45:438, 1990.
18. J. F. Angress, A. R. Goodwin, and S. D. Smith. *Proc. Roy, Soc. London Ser. A*, 287:64, 1965.
19. R. D. Harris, J. L. Newton, and G. D. Watkins. *Phys. Rev. B*, 36:1094, 1987.
20. I. Mizushima, M. Watanabe, A. Murakoshi, M. Hotta, M. Kashiwagui, and M. Yoshiki. *Appl. Phys. Lett.*, 63:373, 1993.
21. M. Okamoto, K. Hashimoto, and K. Takaynagi. *Appl. Phys. Lett.*, 70:978, 1997.
22. E. Artacho, D. Sánchez-Portal, P. Ordejón, A. García, and J. M. Soler. *Phys. Stat. Sol. (b)*, 215:809, 1999.
23. N. Troullier and J. L. Martins. *Phys. Rev. B*, 43:1993, 1991.
24. L. Kleinman and D. M. Bylander. *Phys. Rev. Lett.*, 48:1425, 1982.
25. J. P. Perdew, K. Burke, and M. Ernzerhof. *Phys. Rev. B*, 77:3865, 1996.
26. J. P. Perdew, K. Burke, and M. Ernzerhof. *Phys. Rev. B*, 78:1396, 1997.
27. H. J. Monkhorst and J. K. Pack. *Phys. Rev. B*, 13:5188, 1976.
28. S.Ögut and J. R. Chelikowsky. *Phys. Rev. Lett.*, 83:3512, 1999.
29. S. D. Smith and J. F. Angress. *Phys. Letters*, 6:131, 1963.
30. M. Chandrasekhar, H. R. Chandrasekhar, M. Grimsditch, and M. Cardona. *Phys. Rev. B*, 22:4825, 1980.
31. A. K. Tipping and R. C. Newman. *Semicond. Sci. & Technol.*, 2:389, 1987.
32. R. C. Newman and R. S. Smith. *Phys. Letters*, 24A:671, 1967.
33. R. S. Bean, S. R. Morrison, R. C. Newman, and R. S. Smith. *J. Phys. C: Sol. State Phys.*, 5:379, 1972.

34. K. Laithwaite, R. C. Newman, and D. H. J. Totterdell. *J. Phys. C: Sol. State Phys.*, 8:236, 1975.
35. K. Thonke, J. Weber, J. Wagner, and R. Sauer. *Physica B*, 116:252, 1983.
36. K. Thonke, N. Bürger, G. D. Watkins, and R. Sauer. *Proc. 13th Int. Conf. On Defects in Semicond.*, page p. 823, 1984.
37. P. Deák, A. Gali, A. Sólyom, P. Ordejón, K. Kamarás, and G. Battistig. *J. Phys: Cond. Matter*, 15:4967, 2003.

Protonation States of Methionine Aminopeptidase Studied by QM/MM Car-Parrinello Molecular Dynamics Simulations

Christian D.P. Klein

Pharmazeutische und Medizinische Chemie
Universität des Saarlandes
Postfach 151150
D-66041 Saarbrücken
E-mail: cklein@pharma.ethz.ch

1 Introduction

Methionine aminopeptidases (MetAPs) play a central role for in vivo protein synthesis as they remove the starter methionine from newly synthetized proteins. MetAPs are metaldependent enzymes. It is not clear which metal activates the MetAPs *in vivo*. For *in vitro* experiments, cobalt is commonly used because it activates all known MetAPs and the cobaltsubstituted enzymes are usually the most active. [1] Zinc and iron(II) have also been shown to activate some MetAPs. [2] The metalchelating residues in all known MetAPs are two glutamates, two aspartates and one histidine. The geometric arrangement of these residues is practically identical in all MetAP x-ray structures. [3]

The natural product fumagillin covalently modifies one of the activesite histidines in the eukaryotic methionine aminopeptidase II (MetAPII) [4] and other MetAPs (cf. Figs. 1 and 2). Fumagillin inhibits the growth of vessels in tumors and a derivative of the compound has been evaluated in clinical trials as an anticancer drug. Besides from being anticancer drug targets, MetAPs have the potential to become the target proteins of antibacterial substances, because the MetAP functionality is essential for cell growth and bacteria possess only one of the two known MetAP subtypes. [5]

Several threedimensional structures of MetAPs have been determined by xray diffraction methods, including the structure of human MetAPII with a covalently bound fumagillin molecule. [6]

Density functional theory (DFT) has long been used to study the reactivity of organic molecules and recently began to find its way into biochemistry. [7] We present here an example in which the application of DFT to a biochemical problem led to a hypothesis which was subsequently verified by experiments.

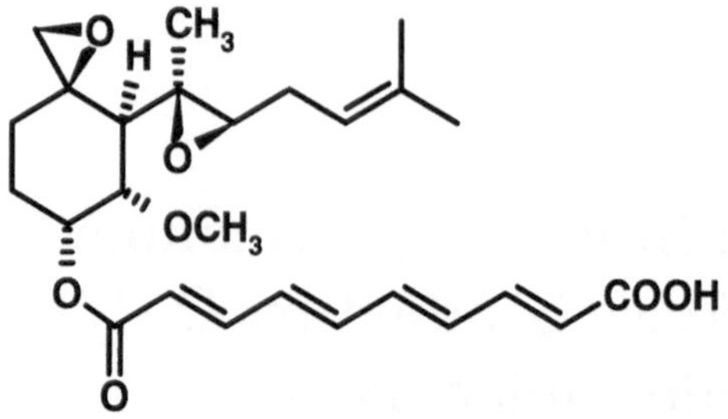

Fig. 1. Fumagillin.

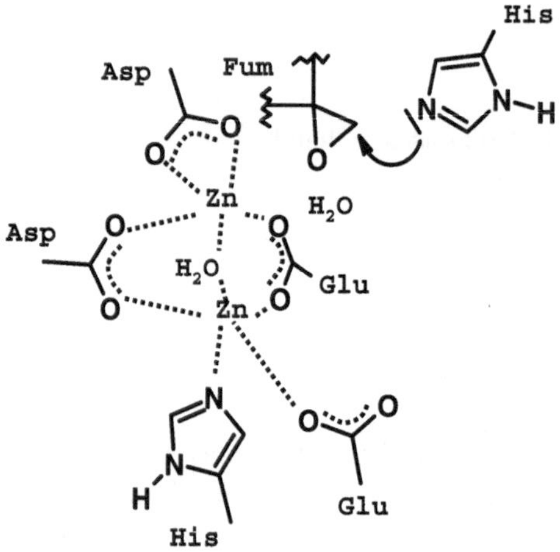

Fig. 2. Binding of fumagillin (Fum) in the MetAP active site.

Our results do not only explain a couple of biochemical phenomena related to MetAPs, but also demonstrate the impact that modern quantum-chemical methods can have on the study of biological systems, where they can help fill the gap between mere conjectures and physical reality.

Our motivation for examining the MetAPs from a theoretical point of view stems from the highly selective, irreversible inhibition mechanism of fumagillin and related epoxides. Many enzyme inhibitors that form a covalent adduct with their target proteins (e.g., the betalactam antibiotics and acetyl salicylic acid) are important drugs. We believe that the fumagillin/MetAP example is a good test case for theoretical methods that aim at rationalizing the development of covalent enzyme inhibitors, because a large amount of high-quality x-ray structural data, also with bound inhibitors, has been published over the last few years.

2 Methods

As a first step towards an understanding of the catalytic and inhibitor-binding mechanisms of MetAPs, we studied different protonation states of the active site water molecules by means of molecular dynamics simulations in a combined quantum-mechanical/molecular-mechanical (QM/MM) framework. In the 1.9Å resolution x-ray structure of E. coli MetAP (pdb code: 2mat), one water molecule (or hydroxide ion) is bridging the two cobalt ions and another water molecule is bound to the cobalt ion that is not coordinated to the histidine. Three different protomeric states were examined: one with two bound water molecules, one with a bridging hydroxide ion and a water molecule, and one with two hydroxide ions. For the quantum-mechanical part of the system, the ab-initio molecular dynamics method described by Car and Parrinello (CPMD) was used. [8] Because an accurate treatment of the open-shell cobalt(II) ion is theoretically and computationally quite demanding we decided to use zinc(II) ions as the active-site metals in the CPMD simulations. Given the fact that at least some MetAPs have been shown to be active with zinc, and considering the results of a thorough theoretical analysis of the zinc- and cobalt-substituted truncated MetAP active sites, [9] we assume that the replacement of cobalt by zinc does not affect the validity of our results. Furthermore, preliminary CPMD simulations on a di-cobalt MetAP active site did not show geometric or dynamic differences to a di-zinc system (data not shown).

3 Results and Discussion

In the CPMD simulations, the system with the two water molecules coordinated to the zinc ions showed a pronounced movement of the coordinating water molecules away from the x-ray structure of E. coli MetAP that has one water molecule located between the metals. The system evolved to a structure in which there was no more "bridge" between the metals and one water molecule started to form a strong hydrogen bond to a carboxylic acid. The simulation was stopped after about 5000 steps of simulation (0.85 ps), because the geometry remained essentially stable. In contrast, the structure with one bridging hydroxide was stable during the whole simulation time (about 10000 steps), showing only the expected thermal motions. A picture of the final state of the simulation is shown in Fig. 3. This figure also shows the HOMO–1 orbital, which is localized at the bridging hydroxide, the species that very likely acts as the nucleophilic agent in the substrate hydrolysis reaction. The HOMO orbital, which is about 0.5 eV higher in energy, is localized at one of the carboxylic acid groups.

The third protonation state – two coordinating hydroxide ions – turned out to be very unstable. After only about 1000 steps of simulation, large deviations from the crystal structure were observed. After 3000 steps, the

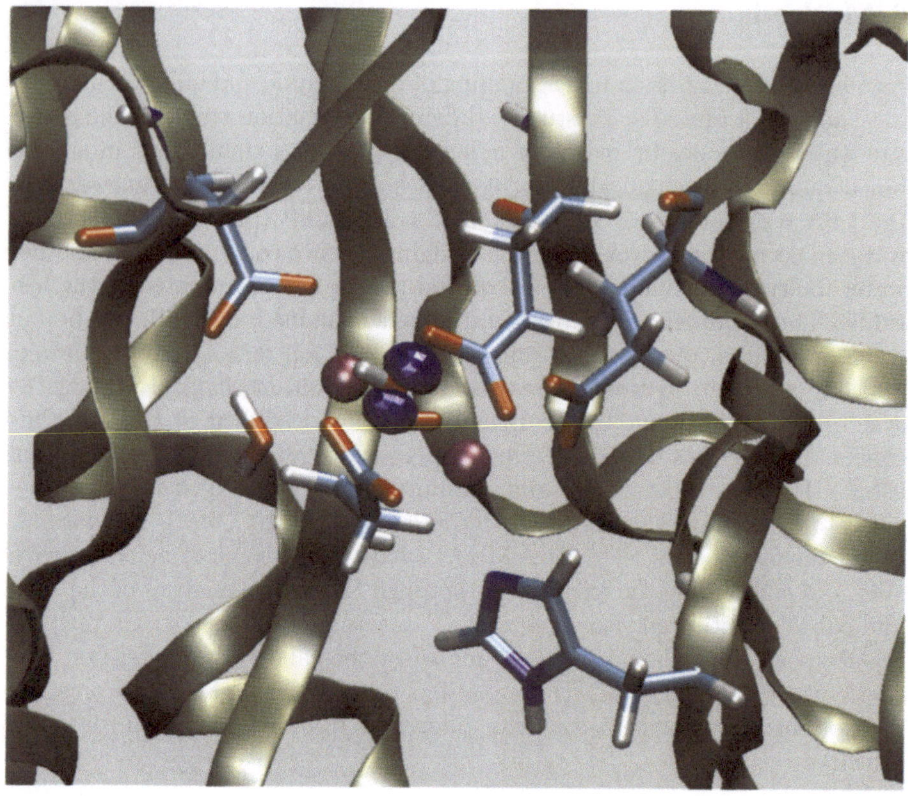

Fig. 3. Final structure of the active site with one hydroxide ion and one water molecule coordinated to the metals (purple spheres). The HOMO–1 orbital, located at the bridging hydroxide ion, is also shown (blue ellipsoids). The cutoff for the visualization of the HOMO–1 electron density was 1.3 e/au^3.

metal-coordinating residues were considerably displaced and the simulation was stopped.

These results allow the following conclusions: In the x-ray structure of E. coli MetAP (2mat), the briding water molecule is most likely deprotonated. However, an active site with two (fully protonated) water molecules coordinated to the metal ions is also stable – albeit with a different coordination geometry. The x-ray structure of the E. coli MetAP (2mat) was determined at a pH larger than 7, whereas several other MetAP structures were determined at a more acidic pH. [10] Examining these, we found that no bridging water molecule is present between the two metals.

Taken together, the x-ray structures of MetAP and the results of the CPMD simulations indicate that two protonation states are possible without disrupting the active site geometry. At more basic pH one water molecule is deprotonated and bridging the two metal ions. From the electronic structure calculations this protonation state is expected to be relevant for the catalytic process. At more acidic pH two water molecules are present in the active site each one coordinating a different metal ion.

The covalent binding of fumagillin to MetAP requires protonation of the epoxide oxygen. Proton donation from an active site group, in particular from one of the active site water molecules, is more likely in the protonation state with two water molecules (i.e. at more acidic pH). Indeed the x-ray structure of the MetAP–fumagillin adduct (resolved at acidic pH) reveals that a water molecule is bridging the two metal ion. Our calculations suggest that this water molecule is most likely a hydroxide ion, and that it was generated during the reaction between MetAP and fumagillin.

In other words: our calculations and analysis of the crystal structures indicate that the active site of MetAP is characterized by the presence of a water molecule with a pKa of $\sim$ 7. [11] Binding of fumagillin is expected to be favored at more acidic pH while a more basic pH would be required for catalysis.

We then decided to validate these results by experiment. Our first idea was to use spectroscopic methods to determine the cobalt coordination geometry at different pH values. However, none of the readily available spectroscopic methods (UV, NMR) can easily be applied to this problem. Cobalt, though NMR-active with a spin of $I=7/2$, displays very broad lines due to its quadrupolar nature, which dramatically reduces sensitivity and resolution of the spectra. Another difficulty with both spectroscopic methods is the presence of unbound cobalt in the medium – in order to get a fully active enzyme, excess cobalt has to be present in the incubation buffer.

We have therefore determined the pH–profile of the fumagillin inhibition reaction and the enzymatic activity of MetAP (see Fig. 4). The results give substantial support to our reasoning. It is obvious that the binding of the inhibitor is favored under more acidic conditions as compared to those that are optimal for catalysis. This observation provides strong substantiation to our interpretation of the QM/MM CPMD simulation results.

Fumagillin and its congeners inhibit the growth of vessels in tumors. Considering the fact that the extracellular pH in tumors is more acidic than in normal tissues, [12] and assuming that the endothelial cells of vessels in tumors are under the influence of the acidic extracellular tumor pH, one may reason that the pH profile of the fumagillin–MetAP binding reaction leads to selectivity (or targeting) of the fumagillin effect to tumor vessels.

In conclusion, we have clarified the protonation behavior of the MetAP active site and made an experimental proof of a hypothesis that was generated on the basis of QM calculations.

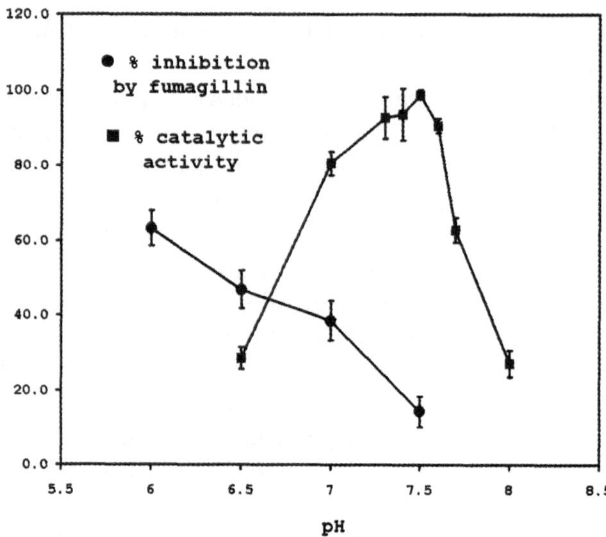

Fig. 4. pH-profile for the substrate hydrolysis and fumagillin binding reactions of E. coli MetAP (mean values with standard deviation error bars).

4 Experimental Section

4.1 Theoretical Methods

The DFT-based Car-Parrinello [8] molecular dynamics program CPMD v. 3.5 was used for the QM/MM simulations of MetAP. [13] AMBER 94 [14] force field parameters were used for the MM part. All systems were fully hydrated with TIP3P waters in a periodic box and equilibrated by extensive classical-mechanical molecular dynamics (MD) simulations prior to the ab-initio MD simulations. [15] The quantum parts included the five metal-binding amino acids (truncated to acetate ions and imidazole, respectively; free valences were capped by adding hydrogen atoms), the two zinc ions and the two water/hydroxide molecules coordinating to them. The QM parts were minimized (annealed to a temperature below 0.1 K) before the start of the production MD phase. CPMD parameters were: Isolated system calculations; gradient-corrected exchange-correlation functionals due to Becke and Lee, Yang, and Parr (BLYP); plane waves basis set; kinetic energy cutoff: 70 Ry; soft norm-conserving Troullier-Martins pseudopotentials for the core electrons; timestep: 6 a.u.; Nosé-Hoover thermostat at 300 K, coupling frequency of 500 cm^{-1}; fictitious electron mass 800 a.u.. [16] The size of the orthorhombic QM box was $13.2 \times 15.9 \times 17.2$ Å, corresponding to a minimum image distance of 6.4 Å. Simulations with cobalt were performed with a kinetic energy cutoff of 90 Ry and Becke-Perdew (BP) functionals. [17]

4.2 Experimental Methods

E. coli MetAP was purified from an overproducer strain that was kindly provided by Drs. Lowther and Matthews. Assays were performed in 96-well microtiter plates using the method of Yang et al. with MGMM as substrate and an enzyme concentration of 12 nM. [18] The fumagillin concentration in the inhibition experiments was 66 μM, which is slightly below the IC_{50} of fumagillin at pH 7 under otherwise identical conditions. Tris/maleate buffer was used for the incubation of MetAP with and without fumagillin at pH 5–8.5 (volume: 20 μL; 15 min.; 37 °C). For the ensuing MGMM hydrolysis and detection reactions, the pH was adjusted to pH 7.5 by adding 170 μL of four-fold concentrated tris/maleate buffer. The fluorescent reaction product (resorufin) was measured using a Wallac microtiter plate reader. All experiments were performed in triplicate.

4.3 Use of HLRS Computer Resources for the CPMD Simulations

We have used a version of the CPMD code that allows the sub-division of the simulation system in one (larger) part that is treated by a computationally cheap classical "Newtonian" molecular dynamics methods (the MM-part), and a smaller "core" part that is treated by density-functional methods as implemented in CPMD (the QM-part). The program code, therefore, consists of a "classical" MD section and a CPMD ab-inito MD section. The CPMD part is parallelized effectively and shows reasonable parallel scaling on 32 or 64 processors of a Cray supercomputer. The MM part of the program, however, is not parallelized. This is not a problem as far as CPU cycle consumption is concerned (because classical MD is much cheaper than CPMD), but can cause serious difficulties if the size of the simulation system exceeds a certain limit which does no longer fit into the RAM of a single Cray PE (128 MB minus the RAM that is consumed by the operating system etc.). This problem actually occured several times and much time was spent in trying to fit the classical MD system into the RAM of a single PE. Another RAM-related problem arose with simulation systems in which the RAM demand for the CPMD simulation box (depends on the size and dynamics of the QM part) could not be satisfied by the combined physical memory of 32 or even 64 Cray PEs. In summary, the usage of the HLRS Cray T3E was sometimes hampered by RAM problems. A typical job (with usually 500 steps of molecular dynamics) took about 6 hours on 64 PEs of the Cray T3E. About 20 such jobs were needed to study one simulation system.

As mentioned above, some CPU time was also supplied by the swiss research computer center CSCS in Manno. Until the summer of 2002, CSCS only offered CPU time on a NEC-SX5, which was of no use for this project. In summer 2002, an IBM SP4 was installed at CSCS. This machine became fully functional at the end of 2002. A part of the simulations described in the manuscript have been performed on that machine. The main advantage of the SP4 as compared to the HLRS Cray T3E is that there are no more memory limitation problems, at least for the simulation systems studied by us. Another advantage of the SP4 is the support from the CPMD user and developer community in compiling and optimizing the code on this platform (very few people seem to be using the T3E for CPMD recently).

Some attempts were also made with the Hitachi SR8000 at HLRS. Unfortunately, the current CPMD code with the QM/MM extensions appears to have problems with the architecture of that machine. It seems as though the MM code (which is based on a somewhat old-fashioned MM program named Gromos) has an extremely bad performance on the SR8000. We hope that future versions of the CPMD QM/MM program will incorporate a more efficient and portable MM code.

References

[1] Ben-Bassat, A., Bauer, K., Chang, S. Y., Myambo, K., Boosman, A., Chang, S. (1987): Processing of the Initiation Methionine from Proteins: Properties of the Escherichia coli Methionine Aminopeptidase and Its Gene Structure. *J. Bacteriol.*, **169,** 751-757

[2] D'Souza, V. M., Holz, R. C. (1999): The Methionyl Aminopeptidase from Escherichia coli Can Function as an Iron(II) Enzyme. *Biochemistry*, **38,** 11079-11085

Walker, K. W., Bradshaw, R. A. (1998): Yeast methionine aminopeptidase I can utilize either Zn2+ or Co2+ as a cofactor: A case of mistaken identity? *Protein Science*, **7,** 2684-2687

[3] The interested reader is referred to the review by Lowther and Matthews for comprehensive informations on MetAPs: Lowther, W. T., Matthews, B. W. (2000): Structure and function of the methionine aminopeptidases. Biochim. Biophys. Acta, 1477, 157-167

[4] Ingber, D., Fujita, T., Kishimoto, S., Sudo, K., Kanamaru, T., Brem, H., Folkman, J. (1990): Synthetic analogues of fumagillin that inhibit angiogenesis and suppress tumour growth. Nature, 348, 555-557

Sin, N., Meng, L., Wang, M. Q. W., Wen, J. J., Bornmann, W. G., Crews, C. M. (1997): The anti-angiogenic agent fumagillin covalently binds and inhibits the methionine aminopeptidase, MetAP-2. *Proc. Natl. Acad. Sci. U.S.A.*, **94,** 6099-6103

Griffith, E. C., Su, Z., Niwayama, S., Ramsay, C. A., Chang, Y. H., Liu, J. O. (1998): Molecular recognition of angiogenesis inhibitors fumagillin and ovalicin by methionine aminopeptidase 2. *Proc. Natl. Acad. Sci. U.S.A.*, **95,** 15183-15188

Griffith, E. C., Su, Z., Turk, B. E., Chen, S., Chang, Y. H., Wu, Z., Biemann, K., Liu, J. O. (1997): Methionine aminopeptidase (type 2) is the common target for angiogenesis inhibitors AGM-1470 and ovalicin. *Chem. Biol.*, **4,** 461-471

The antiangiogenic effect of fumagillin and other inhibitors of MetAPII has been attributed to the inhibition of the Ets1 transcription factor expression and the activation of the p53 pathway:

Wernert, N., Stanjek, A., Kiriakidis, S., Hügel, A., Jha, H. C., Mazitschek,

R., Giannis, A. (1999): Inhibition of Angiogenesis In Vivo by ets-1 Antisense Oligonucleotides - Inhibition of Ets-1 Transcription Factor Expression by the Antibiotic Fumagillin. Angew. Chem. Int. Ed. Engl., 38, 3228-3231
Zhang, Y., Griffith, E. C., Sage, J., Jacks, T., Liu, J. O. (2000): Cell cycle inhibition by the anti-angiogenic agent TNP-470 is mediated by p53 and p21WAF1/CIP1. *Proc. Natl. Acad. Sci. U.S.A.*, **97,** 6427-32

[5] Human MetAP-II: Li, X., Chang, Y. H. (1995): Amino-terminal protein processing in Saccharomyces cerevisiae is an essential function that requires two distinct methionine aminopeptidases. *Proc. Natl. Acad. Sci. U.S.A.*, **92,** 12357-12361
E.coli MetAP: Roderick, S. L., Matthews, B. W. (1993): Structure of the Cobalt-Dependent Methionine Aminopeptidase from Escherichia coli: A New Type of Proteolytic Enzyme. *Biochemistry*, **32,** 3907-3912
Lowther, W. T., Orville, A. M., Madden, D. T., Lim, S., Rich, D. H., Matthews, B. W. (1999): Escherichia coli Methionine Aminopeptidase: Implications of Crystallographic Analyses of the Native, Mutant, and Inhibited Enzymes for the Mechanism of Catalysis. *Biochemistry*, **38,** 7678-7688

[6] Liu, S., Widom, J., Kemp, C. W., Crews, C. M., Clardy, J. (1998): Structure of Human Methionine Aminopeptidase-2 Complexed with Fumagillin. *Science*, **282,** 1324-1327

[7] Carloni, P., Rothlisberger, U., Parrinello, M. (2002): The role and perspective of a initio molecular dynamics in the study of biological systems. *Accounts of Chemical Research*, **35,** 455-464

[8] Car, R., Parrinello, M. (1985): Unified Approach for Molecular-Dynamics and Density-Functional Theory. *Phys. Rev. Lett.*, **55,** 2471-2474; for a review of CPMD applications in biochemistry, see citn. [7]

[9] Jorgensen, A. T., Norrby, P.-O., Liljefors, T. (2002): Investigation of the metal binding site in methionine aminopeptidase by density functional theory. *J. Comput.-Aided Mol. Design*, **16,** 167-179
This article presents a detailed in vacuo DFT study of different electronic configurations and protonation states for the di-cobalt active site, which are compared to a di-zinc system. In short, the authors conclude that there are no major geometrical differences between a di-zinc and a di-cobalt system. The authors also conclude that the bridging water in the simulation system with cobalt is most likely deprotonated.

[10] Human MetAP-II: Liu, S., Widom, J., Kemp, C. W., Crews, C. M., Clardy, J. (1998): Structure of Human Methionine Aminopeptidase-2 Complexed with Fumagillin. *Science*, **282,** 1324-1327
The pH in the crystallization medium can not be determined exactly, but the experimental (buffer) conditions described by Liu et al. indicate that the pH was below 7.

[11] This is at the lower border of the pKa values reported for waters coordinating to zinc ions in organic complexes: I. Bertini, C. Luchinat, in Bioinorganic Chemistry (Eds.: I. Bertini, H. B. Gray, S. J. Lippard, J. S. Valentine), University Science Books, Sausalito, California, U.S.A., 1994, pp. 37.

[12] Gillies, R. J., Raghunand, N., Karczmar, G. S., Bhujwalla, Z. M. (2002): MRI of the tumor microenvironment. *J Magn Reson Imaging*, **16,** 430-50

[13] CPMD. J. Hutter, A. Alavi, T. Deutsch, M. Bernasconi, S. Goedecker, D. Marx, M. Tuckerman, M. Parrinello. MPI fr Festkrperforschung and IBM Zrich

Research Laboratory, 1995-2001.
Implementation of the QM/MM interface: Colombo, M. C., Guidoni, L., Laio, A., Magistrato, A., Maurer, P., Piana, S., Rohrig, U., Spiegel, K., Sulpizi, M., VandeVondele, J., Zumstein, M., Rothlisberger, U. (2002): Hybrid QM/MM Car-Parrinello simulations of catalytic and enzymatic reactions. *Chimia*, **56,** 11-17

[14] Cornell, W. D., Cieplak, P., Bayly, C. I., Gould, I. R., Merz, K. M., Ferguson, D. M., Spellmeyer, D. C., Fox, T., Caldwell, J. W., Kollman, P. A. (1995): A 2nd Generation Force-Field for the Simulation of Proteins, Nucleic-Acids, and Organic-Molecules. *J. Am. Chem. Soc.*, **117,** 5179-5197
D. A. Case, D. A. Pearlman, J. W. Caldwell, T. E. Cheatham, W. S. Ross, C. L. Simmerling, T. A. Darden, K. M. Merz, R. V. Stanton, A. L. Cheng, J. J. Vincent, M. Crowley, V. Tsui, R. J. Radmer, Y. Duan, J. Pitera, I. Massova, G. L. Seibel, U. C. Singh, P. K. Weiner, P. A. Kollman, AMBER 6, University of California, San Francisco, USA, 1999.

[15] AMBER was used for the molecular mechanical MD simulations: see [14]. Structures were visualized with VMD: Humphrey, W., Dalke, A., Schulten, K. (1996): VMD - Visual Molecular Dynamics. *J. Molec. Graphics*, **14,** 33-38

[16] BLYP: Becke, A. D. (1988): Density-functional exchange-energy approximation with the correct asymptotic behaviour. *Phys.Rev.A*, **38,** 3098-3100; Lee, C. L., Yang, W., Parr, R. G. (1988): Development of the Colle-Salvetti correlation-energy formula into a functional of the electron density. *Phys.Rev.B*, **37,** 785-789
Trouiller-Martins PP: Troullier, N., Martins, J. L. (1991): Efficient pseudopotentials for plane-wave calculations. *Phys.Rev.B*, **43,** 1993-2006
Nose-Hoover thermostat: Nose, S. (1984): A molecular dynamics method for simulations in the canonical ensemble. *Mol.Phys.*, **52,** 255-268
Isolated system plane waves calculations: Barnett, R. N., Landman, U. (1993): Born-Oppenheimer molecular dynamics simulations of finite systems: structure and dynamics of (H2O)2. *Phys.Rev.B*, **48,** 2081-2097

[17] Perdew, J. P. (1986): Density-functional approximation for the correlation energy of the inhomogeneous electron gas. *Phys. Rev. B.*, **33,** 8822-8824
The BP pseudopotential for cobalt was kindly supplied by Dr. Hutter: Rovira, C., Kunc, K., Hutter, J., Parrinello, M. (2001): Structural and Electronic Properties of Co-corrole, Co-corrin, and Co-porphyrin. *Inorg. Chem.*, **40,** 11-17

[18] Yang, G., Kirkpatrick, R. B., Ho, T., Zhang, G. F., Liang, P. H., Johanson, K. O., Casper, D. J., Doyle, M. L., Marino, J. P., Thompson, S. K., Chen, W. F., Tew, D. G., Meek, T. D. (2001): Steady-state kinetic characterization of substrates and metal- ion specificities of the full-length and N-terminally truncated recombinant human methionine aminopeptidases (type 2). *Biochemistry,* **40,** 10645-10654

Molecular Transport Through Single Molecules

P. Stampfuß[1], J. Heurich[2], M. Wegewijs[3], M.Hettler[1], J. C. Cuevas[2], H. Schoeller[3], W. Wenzel[3], and G. Schön

[1] Forschungszentrum Karlsruhe, Institut für Nanotechnologie; Postfach 3640, 76021 Karlsruhe wenzel@int.fzk.de
[2] Institut für Theoretische Festkörperphysik, Universität Karlsruhe, 76128 Karlsruhe
[3] Theoretische Physik A, RWTH Aachen, 52074 Aachen

1 Introduction

Present trends in the miniaturization of electronic devices suggest that ultimately single molecules may be used as electronically active elements in a variety of applications [1, 2]. Recent advances in the manipulation of single molecules now permit to contact an individual molecule between two electrodes (see Fig. 1) and measure its electronic transport properties [3, 4, 5, 6, 7, 8]. Interesting and novel effects, such as negative differential conductance [9], were observed in some of these experiments, which still, by-and-large, beg theoretical explanation. In addition to generic principles of nanoscale physics, e.g. Coulomb blockade [6, 10, 11], the chemistry and geometry of the molecular junction emerge as the fundamental tunable characteristics of molecular junctions [3, 4, 12, 13, 14, 15, 16].

When the molecule is coupled *weakly* to the electrodes, i.e. via electron tunneling, charging effects, semi-classically determined by the small capacitance of the molecule, become important. The interplay of charging effects with the specific structure of the molecular orbitals leads to nontrivial current voltage (I-V) characteristics [9, 17, 18]. Very recent experiments demonstrated both Coulomb blockade and the Kondo effect in three terminal transport through a single molecular level [19, 20].

When the molecule is coupled *strongly* to the electrodes, the electron transport is ballistic. The hybridization between molecular and electrode orbitals leads to a broading of the former, which can lead to interesting effects in the current-voltage characteristic . In the following we summarize our investigations of these two theoretically accessible limits. We present an atomistic theory that bridges traditional concepts of mesoscopic and molecular physics to describe transport through single organic molecules in qualitative agreement with recent break-junction experiments [21]. We show how the specific properties of individual MOs are reflected in their contribution to the cur-

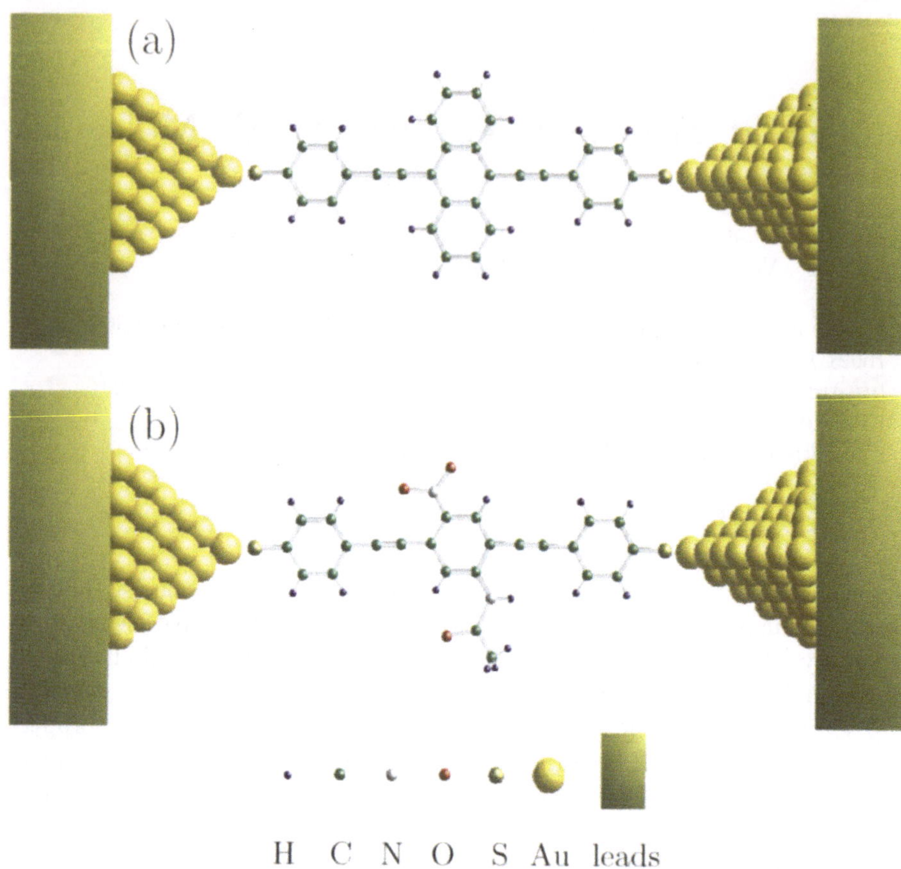

Fig. 1. Scheme of the single-molecule contacts analyzed in this work. The two organic molecules attached to gold electrodes, which were experimentally investigated in Ref. [8], are referred to as: (a) "symmetric molecule", and (b) "asymmetric molecule".

rent. Secondly we investigate conduction through the hydrogen molecule [22] in order to explain a recent break-junction experiment [23].

In the *weak-coupling* limit we present a theoretical model that predicts strong negative differential conductance in tunneling transport through benzene [18] and elucidate the physical mechanism responsible for this effect.

2 Ballistic Transport

2.1 Model

We calculate the current through a single molecule attached to metallic electrodes using Landauer transport theory in analogy to the analysis of transport in atomic-size contacts [24]. Since the conductance is mainly determined by the narrowest part of the junction, only the electronic structure of this "central cluster" must be resolved in detail. It is therefore sensible to decompose the overall Hamiltonian of the molecular junction as

$$\hat{H} = \hat{H}_L + \hat{H}_R + \hat{H}_C + \hat{V}, \tag{1}$$

where $\hat{H}_C$ describes the "central cluster" of the system, $\hat{H}_{L,R}$ describe the left and right electrode respectively, and $\hat{V}$ gives the coupling between the electrodes and the central cluster (see Figure (1)).

The electronic structure of the "central cluster" is calculated with density functional (DFT) methods, while the left and right reservoirs are modeled as two perfect semi-infinite crystals of the corresponding metal using a tight-binding parameterization [25]. $\hat{V}$ is the coupling matrix element that describes the coupling between the leads and the central cluster. The "central cluster" is not necessarily confined to the molecule, but may, in principle, contain arbitrary parts of the metallic electrode. The inclusion of part of the leads in the *ab initio* calculation was shown to improve the description of the molecule-leads coupling [14], in particular regarding charge transfer between the molecule and the electrodes. The Fermi energy of the overall system is determined by the charge neutrality condition of the central cluster.

In order to obtain the current for a constant bias voltage, V, between the leads, we make use of non-equilibrium Green function techniques. Since the Hamiltonian of Eq. (1) does not contain inelastic interactions, the current follows from the Landauer formula [26]:

$$I = \frac{2e}{h} \int_{-\infty}^{\infty} d\epsilon \, \mathrm{Tr}\left\{\hat{t}\hat{t}^{\dagger}\right\} \left[f(\epsilon - eV/2) - f(\epsilon + eV/2)\right], \tag{2}$$

where f is the Fermi function and $\hat{t}$ is the energy and voltage dependent transmission matrix. The details of the formalism and of the calculation are reported in [21].

The understanding of the mechanism of electronic transport is aided the definition of *conduction channels* as eigenfunctions of $\hat{t}\hat{t}^{\dagger}$. Such an analysis allows to quantify the contribution to the transport of *every individual molecular level.* The channels arise as a linear combination of the molecular orbitals $|\phi_j\rangle$ of the central cluster, i.e. $|c\rangle = \sum_j \alpha_{cj}|\phi_j\rangle$, and the corresponding eigenvalues determine their contribution to the conductance.

2.2 Results

Transport through organic molecules

We now use the method to analyze the current through the two organic molecules shown in Figure (1) [8, 21] The gold electrodes are described in a basis of atomic-like $5d, 6s, 6p$ orbitals; for the central cluster we use the

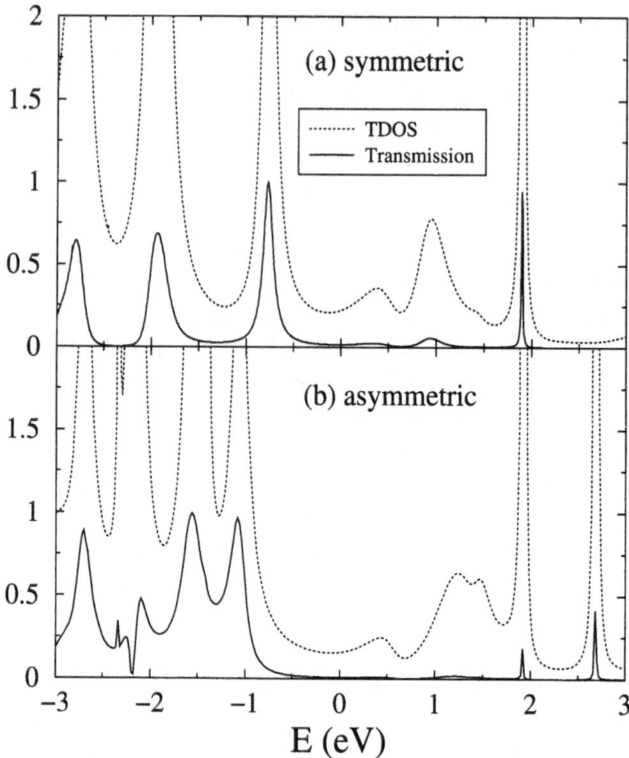

Fig. 2. Total density of states (TDOS) of the molecule and zero-bias total transmission as a function of the energy for both molecules. The Fermi energy is set to zero.

LANL2DZ basis [27] for all atoms. The DFT calculations were performed using the B3LYP functional [28]. In the calculations reported here one additional gold atom was included on either side of the molecule.

Experimentally, both molecules were contacted several times and the nature of the I-V characteristics was found to vary with the quality of the contact. For this reason, theory can presently aim to elucidate important reproducible features of the experiment under the assumption that the contact to the electrodes is well defined. Since there is no direct experimental information regarding the geometry of the molecule and its attachment to the leads, the overall geometry of the central cluster was relaxed without additional constraints in our calculations, resulting in the Au atom being out of the molecular plane.

We start by analyzing the linear response regime: Figure (2) shows the total density of states (TDOS) of the molecule and the zero-bias total transmission as a function of energy. As can be seen in the TDOS, in the covalent bond between Au and S results in a strong hybridization between the molecu-

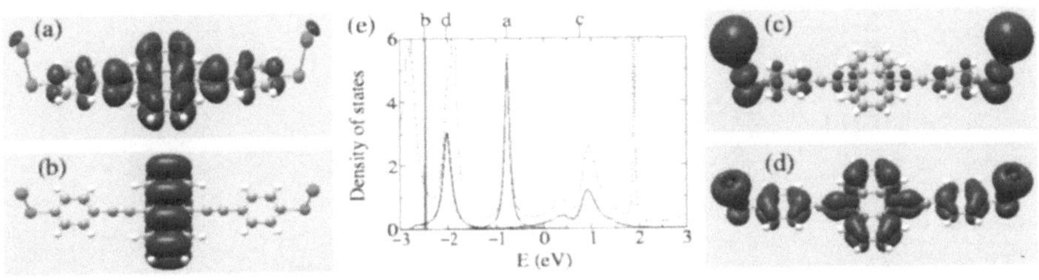

Fig. 3. (a–d) Charge-density plots of four molecular orbitals of the central cluster for the symmetric molecule. Panel (a) displays the HOMO and (c) the LUMO, which is twofold degenerate. (b) shows a confined orbital that contributes little to the current, while the MO in (d) is almost as important as the LUMO despite its difference in energy. Panel (e) shows the total density of states of the central cluster (dotted line) and the individual contributions of the four molecular orbitals (color lines). The level positions are indicated on top of this panel. The contributions of the different MOs to the conduction channel at the Fermi energy (set to zero) are: $|\alpha_a|^2 = 0.007$, $|\alpha_b|^2 = 10^{-11}$, $|\alpha_c|^2 = 0.06$, $|\alpha_d|^2 = 0.02$.

lar orbitals and the extended states of the metallic electrodes. The formation of wide energy bands suggests the absence of Coulomb blockade in this type of molecular junctions.

The zero-bias total transmission as a function of energy follows closely the TDOS. The transmission is dominated overwhelmingly by a single channel in the energy window shown in Fig. 2, and the corresponding eigenvalues of $\hat{t}\hat{t}^\dagger$ at the Fermi energy are $T_{sym} = 0.014$ and $T_{asym} = 0.006$. The decomposition of this channel into molecular orbitals provides us information on the relevance of the different molecular levels. Figure (3) (a)–(d) show charge-density plots for some representative MOs and (e) shows their individual contribution to the TDOS. Figure (3)(a) indicates that the highest occupied molecular orbital (HOMO) is confined to the interior of the molecule and its weight at the gold atoms is rather small. Consequently, in spite of its privileged energy position, the HOMO does not give a significant contribution to the current. The lowest unoccupied molecular orbital (LUMO), see Figure (3)(c), exhibits the opposite behavior, i.e. it is very well coupled to the leads through the $6s$ atomic orbital of the gold atoms (notice that it has a width of about $4\ eV$ in the density of states), but the charge is mainly localized on the Au and S atoms. The interplay of these two factors yields a contribution of $\approx 6\%$ of the total current. Figure (3)(b,d) shows two further MOs with similar energy but very different contribution to the channel. While the localized MO (b) carries almost no current, the extended and well coupled MO (d) has significant weight. Consequently there are three ingredients which determine the contribution of a MO to the current: (i) its energy position (distance to the Fermi energy), (ii) its bridging extent (whether it is extended or localized), and (iii) its coupling to the leads. Our analysis provides a counterexample

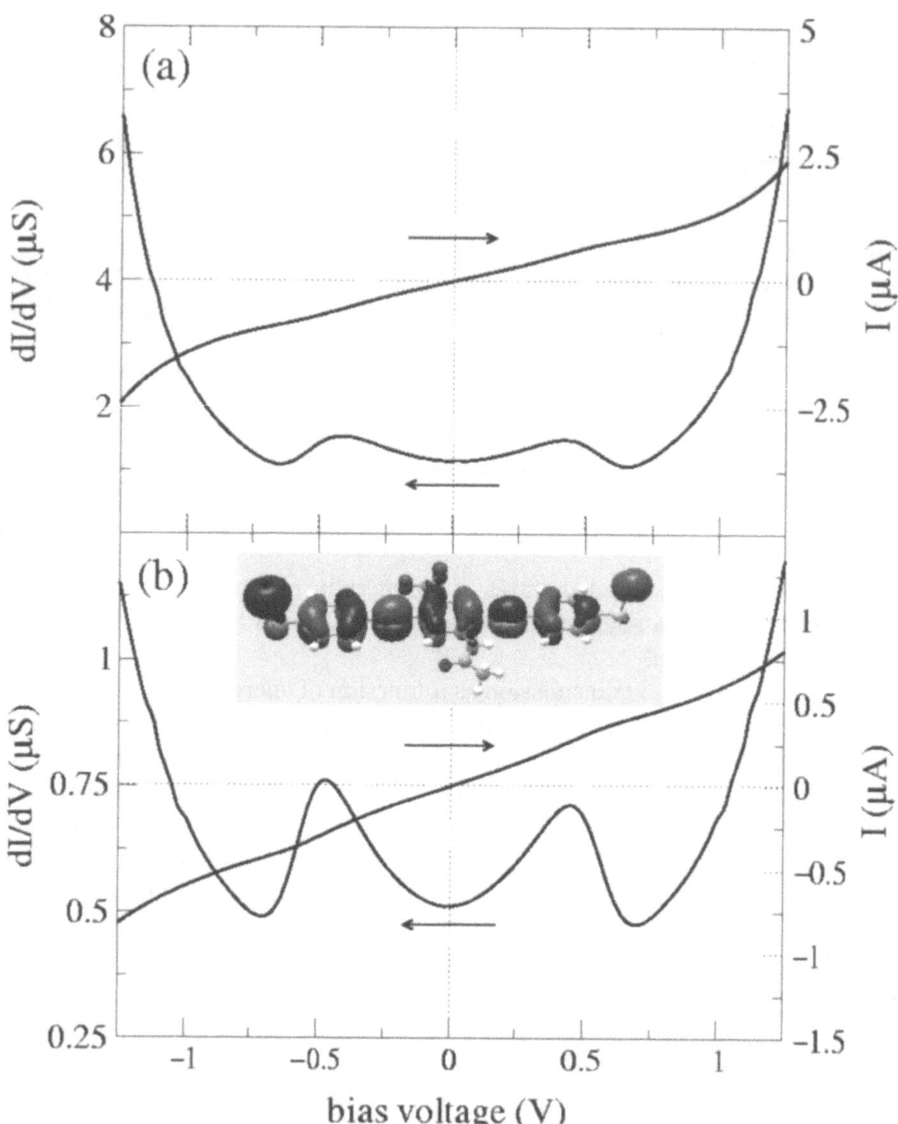

Fig. 4. I-V characteristics and differential conductance for the symmetric (a) and asymmetric (b) molecules. The inset in (b) shows the charge-density plot of the HOMO for the asymmetric molecule. Notice the intrinsic asymmetry of the charge distribution in the gold atoms.

to the conventional wisdom that the HOMO and the LUMO dominate the transport properties.

Figure (4) shows the I-V curves for both molecules in the experimental voltage range. Both the order of magnitude and shape of the current and conductance agree qualitatively with the experimental results. There is no pronounced voltage dependence of the transmission due to the smooth density of states of the gold electrodes in the energy region explored here. The non-linearities in these I-V curves can be then understood by a simple inspection of the energy dependence of the zero-bias transmission. For instance, the pronounced increase in the conductance around 1 V is due to the fact that we approach the resonant condition for the HOMO and LUMO. The current-voltage characteristic in the figure were computes using the zero-field molecular spectra. At a finite bias one should in principle determine how the voltage modifies the molecular spectra, which in turn control the current. We performed DFT calculations at a fixed electric field and found no significant differences in the transmission in the voltage range explored in the experiments.

In agreement with the experiment, the I-V of the symmetric molecule is symmetric with respect to voltage inversion, while the one of the asymmetric molecule is asymmetric. The asymmetry of the I-V curve arises from the left/right asymmetry of the scattering rates. This asymmetry can be due to an asymmetry of either the leads or of the coupling of the leads to the molecule. The latter is influenced by intrinsic properties of the molecule, such as the charge distribution of the contributing MOs is asymmetric (see inset Figure (4)(b)). In the case of the symmetric molecule, we were able to induce asymmetries into the I-V characteristic by distorting the geometry of one of the lead fragments in the central cluster. This fact was nicely demonstrated in the experiment (see Fig. 5 in Ref. [8]). We investigated several other scenarios regarding the number of gold atoms, their geometry and the coupling and found predictable variations of the I-V's with theses changes.

Transport through hydrogen

The measurement of the conductance of an individual hydrogen molecule [23] provides a valuable opportunity to analyze the emerging concepts on the electrical conduction in single-molecule devices in the perhaps simplest possible system. In contrast to results for organic molecules described above, molecular hydrogen has a conductance close to one quantum unit, carried by a single channel. This result belies the conventional wisdom because the closed-shell configuration of H_2 results in a huge gap between its bonding and antibonding states, making it a perfect candidate for an insulating molecule.

In order to investigate the transport through the hydrogen bridge we have performed DFT calculations of the linear conductance using the method described above [21].The most probable configuration is shown in the inset of Figure (5), where the H_2 is coupled to a single Pt atom on either side (top

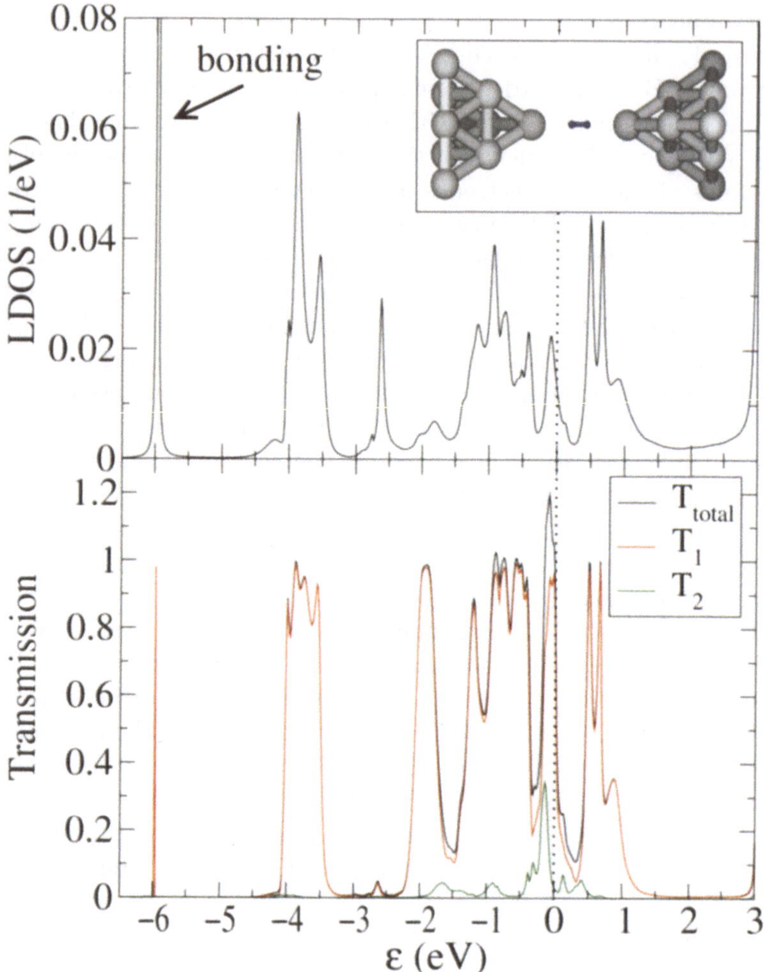

Fig. 5. Transmission and LDOS projected into one of the H atoms as a function of energy for the Pt-H_2-Pt structure, the central cluster of which is shown in the inset. At E_F $T_1 = 0.83$ and $T_2 = 0.03$. The H-H and Pt-H distances are 0.8 Å and 2.1 Å respectively. We use the cc-pVDZ basis set for H.

position). In this geometry the vibrational mode of the center of mass motion of H_2, has an energy of 55.6 meV comparable with the experimental values. In Figure (5) we show the transmission and the LDOS projected into the orbitals of one of the H atoms. The total transmission at the Fermi energy is $T_{tot} = 0.86$ and it is largely dominated by a single channel, in agreement with experimental results.

In principle there are other geometries compatible with the vibrational modes analysis. However, based on the channel analysis performed in the

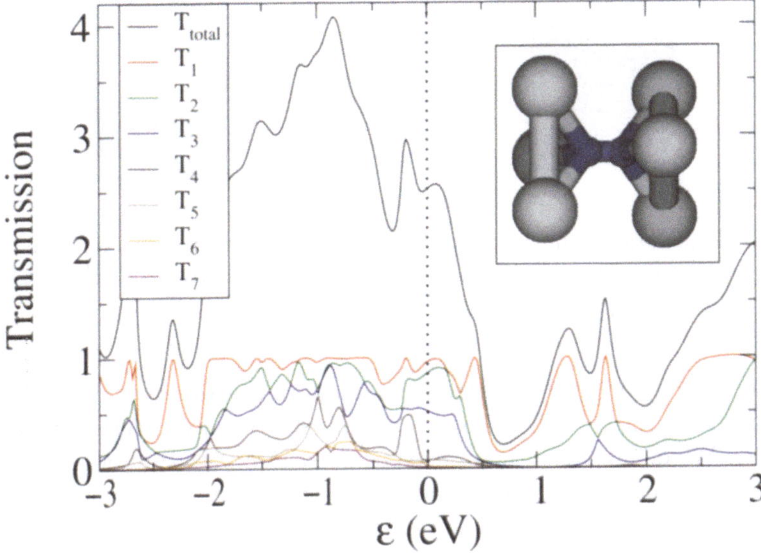

Fig. 6. Transmission as a function of energy for the structure shown in the inset. The H-H and Pt-H distances are 0.8 Å and 1.86 Å respectively. The energy of the vibrational mode of the center of mass motion of H_2 is 65 meV.

experiment many geometries can be ruled out. As an example we consider the situation sketched in the inset of Fig. 6 where each H is bound to three Pt atoms (hollow position). Indeed this configuration is suggested in studies of the chemisorption of H on Pt surfaces [29]. The conductance is carried by up to 7 individual channels (see Fig. 6). Due to the short distance between the Pt leads most of the current is bypassing the H_2 going directly from Pt to Pt. This analysis allows us to conclude that this type of geometries is not realized in the experiment and illustrates the importance of the channel analysis.

3 Weak Coupling

Using benzene as a prototypical example we investigate novel effects that arise when the molecule is weakly coupled to the electrodes (see Figure (7)) and transport through several competing electronic configurations becomes possible [18]. We derive a semi-quantitative model for the conducting many-body states of the system from electronic structure calculations. We compute the transport properties within the golden rule approximation (sequential tunneling) and include screening of the applied electric field as well as radiative transitions between the electronic states of the molecule. We predict a current collapse in the current-voltage characteristic and strong negative differential conductance (NDC) due to the occurrence of a "blocking" state when the

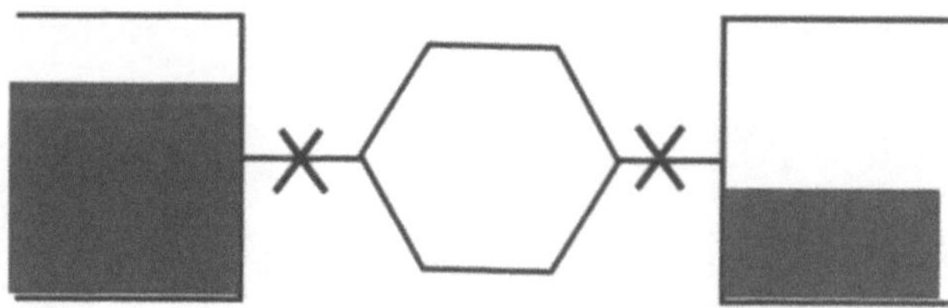

Fig. 7. Schematic of the setup considered. A single benzene molecule is copuled at the para position via two tunnel couplings (X) to electrodes at the left and the right.

molecule is coupled to the electrode at the para-position. For coupling at the meta-position, the current-voltage characteristic displays a series of steps, but no NDC. We demonstrate how the specific *spatial* structure of the molecular orbitals qualitatively determines electronic transport. Finally, we discuss the limits of the model and the impact of disorder and symmetry breaking effects likely encountered in an experimental realization.

3.1 Model

To perform transport calculations in the weakly coupled regime, we extract an effective model from electronic structure calculations of the molecule. For benzene transport is assumed to be dominated by the π-electron system, but generalizations are straightforward. We first perform Hartree-Fock calculations in a suitable basis and then transform the Hamiltonian to the molecular orbital basis. We then integrate out the σ electrons of the system arriving at an effective *interacting* model Hamiltonian for the π electrons of the system in the presence of the atomic cores and the "frozen" density of the σ electrons:

$$\mathcal{H}_\pi = \sum_{ij\sigma} \epsilon_{ij} c^\dagger_{\sigma,i} c_{\sigma,j} + \sum_{ijkl\sigma\sigma'} U_{ijkl} c^\dagger_{\sigma,i} c^\dagger_{\sigma',j} c_{\sigma,k} c_{\sigma,l} \tag{3}$$

The second quantized operators $c^\dagger_{\sigma,i}, c_{\sigma,i}$ create/destroy electrons of spin σ in orthogonalized Wannier-like orbitals centered at the carbon atoms. While this model neglects σ-π mixing for certain excited states of the molecule [30, 31], its parameters account for the detailed electronic structure of the molecule. For the current work, we compute the model parameters only once, using an augmented double-zeta quality atomic natural orbital (ANO) basis set that was truncated to contain only one 2p-shell for each carbon. When applying a bias over the molecule this neglects the higher order effect of field screening by σ electrons and its impact on the π electrons.

We find that the low-energy spectrum of benzene obtained from the diagonalization of the effective Hamilton operator eq. 3 compares favorably with the spectrum directly obtained from accurate multi-reference configuration interaction calculations [31, 32]. The remaining differences can be understood by the lack of σ-π mixing and do not qualitatively affect the transport properties. The restriction to a pure π-electron system is more severe for the charged states, as discussed in some detail below.

To account for the effect of an external bias potential V^{ext} on the electrons we include a term

$$\mathcal{H}_{bias} = e \sum_{ij\sigma\sigma'} V_{ij}^{ext} c_{\sigma,i}^{\dagger} c_{\sigma,j} \tag{4}$$

with $V_{ij}^{ext} = \int dr \Phi_i(r) V^{ext}(\hat{r}) \Phi_j(r)$ and $V^{ext}(\hat{r}) = (V_L + V_R)/2 - V_{bias}(\hat{r})$ in the Hamiltonian. $V_{L,R}$ is the chemical potential in the left/right electrode, distances are measured from the center of the molecule. The matrix elements V_{ij}^{ext} also result from the electronic structure calculation described above. We consider a bias $V_{bias} = V_L - V_R$ aligned with the transport direction (x-direction), unless otherwise specified.

To compute the transport properties of the system in the weak coupling limit we diagonalize the Hamiltonian eq. 3 in the appropriate charge, spin and symmetry sectors using a recently developed massively parallel implementation of the MRD-CI method [32, 33]. Care is needed to handle degeneracies properly. After diagonalization of the Hamiltonian we have the many-body eigenstates $|s\rangle$ with the corresponding energies E_s and their total spin S_s. We use a Master equation approach [17] for the occupation probabilities P_s in a stationary state. The transition rate $\Sigma_{ss'}$ from state s' to s is computed in perturbation theory using the golden rule. The "perturbation" is the coupling of the molecule to the leads

$$H_{mol-leads} = (\frac{\Gamma}{2\pi\rho_e})^{1/2} \sum_{\mathbf{k}\sigma\alpha l} \left(c_{l\sigma}^{\dagger} a_{\mathbf{k}\sigma\alpha} + h.c. \right) , \tag{5}$$

and (optionally) the coupling to electromagnetic fields (photons). Γ is the coupling strength (in units of energy) of leads to the benzene and ρ_e is the density of states of the electrons in the electrode (assumed constant). The operators $a_{\mathbf{k}\sigma\alpha}$ and their hermitian conjugates destroy/create electrons with momentum $\mathbf{k}$ and spin σ in electrode α = left/right For simplicity, we assume that tunneling is only possible through two "contact" carbon atoms which we choose to be at the 1 and 4 (para) positions unless noted otherwise.

As we do not consider the leads microscopically, the coupling of molecule states $|s\rangle$ is determined by the overall coupling strength Γ and the relative wave function amplitude of the state $|s\rangle$ at the coupling carbon site l. For the transition rates we have $\Sigma_{ss'} = (\sum_{\alpha,p=\pm} \Sigma_{ss'}^{\alpha p}) + \Sigma_{s,s'}^{d}$ where $\Sigma_{ss'}^{\alpha p}$ is the tunneling rate to/from electrode α for creation ($p = +$) or destruction ($p = -$) of an electron on the molecule. We have

$$\Sigma_{ss'}^{\alpha+} = \Gamma f_\alpha(E_s - E_{s'}) \sum_{\sigma} | \sum_{l} < s | c_{l\sigma}^{\dagger} | s' > |^2 , \tag{6}$$

and a corresponding equation for $\Sigma_{s's}^{\alpha-}$ by replacing $f_\alpha \rightarrow 1 - f_\alpha$, where f_α is the Fermi function. When including relaxation by radiative transitions, we use the dipole approximation with dipole transition moments

$d_{i,j} = \int d\mathbf{r}\Phi_i(\mathbf{r})\hat{\mathbf{r}}\Phi_j(\mathbf{r})$ obtained from the electronic structure calculations. The corresponding transition rates are

$$\Sigma^d_{ss'} = \frac{4e^2}{3\hbar^3 c^3}(E_s - E_{s'})^3 N_b(E_s - E_{s'}) \,|< s|d|s' >|^2 \qquad (7)$$

where $N_b(E)$ denotes the equilibrium Bose function. Note that for emission $E_s - E_{s'}$ is negative, and $N_b(-|E|) = -(1 + N_b(|E|))$.

The total transition matrix $\Sigma_{ss'}$ consists of blocks connecting N and $N \pm 1$ electron states (tunneling processes) and blocks from the radiative transitions that do not change the electron number on the molecule. Taking only one member of the subspace of spin and energy degenerate states into account, the rank r of the transition matrix is 1716. The stationarity condition $\dot{P}_s = 0$ can be written as $\sum_{s'} A_{ss'} P_{s'} = 0$ with the matrix $A_{ss'} = \Sigma_{ss'} - \sum_{s''} \Sigma_{s''s}\delta_{ss'}$. This implies that $A_{ss'}$ has an eigenvector with zero eigenvalue, which is the wanted solution for P_s. Rather than computing this eigenvector by brute force, to speed up the calculation we make use of $\sum_s P_s = 1$ to eliminate one row/column, thus reformulating the eigenvector problem into one of solving an inhomogeneous linear system of rank $r - 1$. Eventually, the current in the left and right electrode is calculated via

$$I_\alpha = e \sum_{s,s'} (\Sigma^{\alpha+}_{ss'} P_{s'} - \Sigma^{\alpha-}_{s's} P_s) \; . \qquad (8)$$

3.2 Results

To elucidate the impact of various effects on the current we have performed transport calculations with and without radiative transitions (relaxation) and with and without effect of the applied bias. On the left panel of Fig. 8 we present the current-voltage characteristic obtained without the effect of the applied field (zero bias electronic structure). Without radiative relaxation (solid line) the current-voltage characteristic consists of a series of steps of which only a few are well resolved on this scale. The first step is associated with the population of the first π^* orbital of molecule (molecular charge $= -e$), an electron hops onto the lowest available level and then hops off again. At slightly larger bias, a transition of the anion to the first excited state of the neutral molecule becomes possible, resulting in a slight increase of the current. If the bias is sufficiently large this excited state may now accept another electron to populate higher excited states of the anion or low-lying states of the di-anion, resulting in a rapidly growing cascade of transitions between literally hundreds of states of the system. In the model considered here the growth of this cascade leads to quasi-ohmic behavior above 3.6 eV. In our calculation the first states of the di-anion become occupied at about 4.5 eV.

The inclusion of radiative transitions has a dramatic effect on the current-voltage characteristic (dashed line). We observe a collapse of the current over

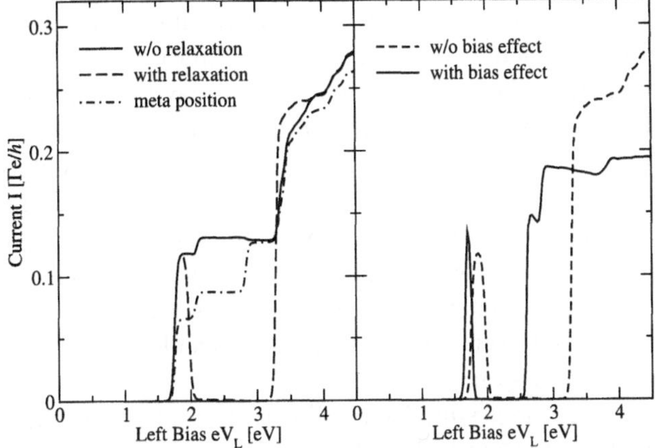

Fig. 8. $I - V$ characteristics for symmetric bias. Left Panel: current-voltage characteristic without (solid line) and with radiative relaxation (dashed line) without inclusion of bias effect. The current collapses at a bias when the antisymmetric anion state can become occupied by radiative relaxation from an excited anion state. For coupling of the right electrode at the meta position (dash-dotted line) the radiative relaxation leads to no cascade into a blocking state and no current collapse is observed. Right Panel: With inclusion of the bias effect (solid line) the onset of current is generally shifted to lower bias, but the current collapse remains and additional weak NDC occurs at larger bias.

a substantial range of the applied bias (2.1-3.4 eV). The reason for this collapse is the population of a "blocking" state in the cascade of transitions that becomes possible when exited states of the neutral molecule and anion become accessible. Above approximately 2.1 eV bias an excited state of the anion at about 5.6 eV (see Fig. 9), becomes partially populated in the transport cascade. This state can decay by photon emission to either a symmetric or an

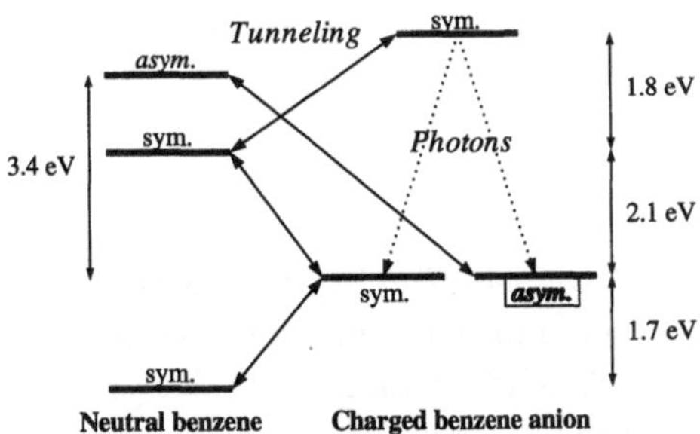

Fig. 9. Sketch of the energetics and symmetry of the relevant neutral and anion states. At a left bias of 2.1 eV the antisymmetric blocking state becomes occupied via radiative relaxation. At 3.4 eV electrons can escape the blocking state via tunneling to the antisymmetric state of neutral benzene.

antisymmetric many body state (with respect to the plane through the transport axis and perpendicular to the molecular plane) of the anion. In the bias range of the current collapse this state cannot decay by coupling to the leads, because the lowest neutral states are symmetric (see Fig. 9) and the tunneling preserves the symmetry. Since there are no further radiative transitions possible on the molecule, the rate equations contain no draining term from this state. As a result, in the stationary state, the probability of occupying the "blocking" state is unity and the current ceases to flow. At a larger bias (3.4 eV), the first escape channel opens, and the system can decay to the first antisymmetric excited state of the neutral molecule, which can then decay further by photon emission.

The above calculations simplify the illustration of the mechanism of NDC, because the energies of the participating states are independent of the applied bias. In the physical system the various states couple differently to the applied field. The solid line in the right panel of Fig. 8 shows the current-voltage characteristic of the benzene with radiative relaxation and with the effect of the external potential applied parallel to the transport axis. We note a shift of the first step in the current-voltage characteristic , which is due to differential screening effects between the ground state of the neutral molecule and the anion. At larger bias, the differential effects on all of the states in the cascade result in a significant renormalization of the current-voltage characteristic . However, the current collapse still occurs, although the voltage window is reduced. The onset of quasi-ohmic behaviors occurs at higher voltage and a series of weak NDC effects occurs due to redistribution of occupation probabilities in favor of states with slightly smaller "transmission".

If we couple the right electrode to the meta position of benzene (left panel, dash-dotted curve) we find no current collapse with or without radiative relaxation or bias effect. This can be readily understood by the fact that at the meta position the wave function of the formerly blocking state is non-vanishing and electrons can tunnel out to the right electrode. Consequently, we observe a series of current steps similar (but not identical) to the case of coupling to the para position without relaxation.

4 Conclusions

We have presented an atomistic semi-quantitative description of non-linear transport through a single molecule junction. We were able to attribute distinctive features of the I-V of the symmetric and asymmetric molecule to their individual molecular levels obtained from *ab initio* calculations. The resolution of conductance into conduction channels permits a quantitative analysis of the contributions of individual orbitals to the overall transport. We have also presented a theory for the conductance of a hydrogen bridge between Pt contacts explaining the experimental observations. Furthermore we we able to describe transport in an idealized, though semi-quantitative, π-electron model of weakly coupled benzene and found a dramatic suppression of the current in a finite voltage window. In weakly coupled molecules the concept of the "blocking" state suggests new opportunities for the design of single molecule electronic devices.

These results were made possible through the use of large scale computational resources, such as those provided by the IBM-SP2 of the HLRZ Karlsruhe. In the strongly coupled limit it is advantageous to use as large electrode fragments as possible in the central cluster for which the DFT calculation is performed. For non-hybrid functionals the TURBOMOLE electronic structure program serves as a efficient, stable, parallel platform to carry out such calculations. In the weakly coupled limit, complex many-particle problems must be solved for all energetically relevant electronic levels of the molecule. These calculations are performed with our massively parallel configuration selecting MRD-CI package. In both cases, these calculations should ideally be repeated for every applied voltage to account for charging effects. The required computational performance for such calculations presently requires the dedicated use of supercomputer facilities.

Regarding transport through the organic molecules we are grateful for many stimulating discussions with D. Beckmann, M. Mayor, H. Weber and F. Weigend. Regarding the transport through hydrogen we are grateful for many stimulating discussions with R.H.M. Smit, C. Untied and J.M. van Ruitenbeek. JCC acknowledges funding by the EU TMR Network on Dynamics of Nanostructures, and WW by the German National Science Foundation (We 1863/10-1), the BMBF and the HLRZ Karlsruhe.

References

1. M. Petty and M. Bryce, eds., *An Introduction to Molecular Electronics* (Oxford University Press, New York, 1995).
2. C. Joachim, J. Gimzewski, and A. Aviram, Nature **408**, 541 (2000).
3. C. J. et. al., PRL **74**, 2102 (1995).
4. S. Datta, W. Tian, S. Hong, R. Reifenberger, J. I. Henderson, and C. P. Kubiak, Phys. Rev. Lett. **79**, 2530 (1997).
5. M. A. Reed, C. Zhou, C. J. Muller, T. P. Burgin, and J. M. Tour, Science **278**, 252 (1997).
6. C. Kergueris, J.-P. Bourgoin, S. Palacin, D. Esteve, C. Urbina, M. Magoga, and C. Joachim, PRB **59**, 12505 (1999).
7. D. Porath, A. Bezryadin, S. de Vries, and C. Dekker, Nature **403**, 635 (2000).
8. J. Reichert, R. Ochs, D. Beckmann, H. B. Weber, M. Mayor, and H. v. Lhneysen **88**, 176804 (2002).
9. J. Chen, M. A. Reed, A. M. Rawlett, and J. M. Tour, Science **286**, 1550 (1999).
10. U. Banin, Y. Cao, D. Katz, and O. Millo, Nature (1999).
11. H. Park, J. Park, A. K. L. Lim, E. H. Anderson, A. P. Alivisatos, and P. L. McEuen, Nature **407**, 52 (2000).
12. E. Emberly and G. Kirczenow, Phys. Rev. B. **58**, 10911 (1998).

13. S. N. Yaliraki, A. E. Roitberg, C. Gonzalez, V. Mujica, and M. Ratner, J. Chem. Phys. **111**, 6997 (1998).
14. M. Di Ventra, S. T. Pantelides, and N. D. Lang, Phys. Rev. Lett. **84**, 979 (2000).
15. J. Taylor, H. Guo, and J. Wang **63**, 121104R (2001).
16. J. J. Palacios **64**, 115411 (2001).
17. M. Hettler, H. Schoeller, and W. Wenzel, EPL **57**, 571 (2002).
18. M. H. M. Wegewis, H. Schoeller, and W. Wenzel **90**, 076805 (2003).
19. J. Park, A. Pasupathy, J. I. Goldsmith, C. Chang, Y. Yaish, J. R. Petta, M. Rinkoski, J. Sethna, H. D. Abruna, P. L. McEuen, and D. C. Ralph, Nature **417** (2002).
20. W. Liang, M. P. Shores, M. Bockrath, and J. R. L. . H. Park, Nature **417**, 725 (2002).
21. J. Heurich, J.-C. Cuevas, W. Wenzel, and G. Schoen **88**, 256803 (2002).
22. J. Heurich, F. Pauli, J. Cuevas, W. Wenzel, and G. Schoen ((submitted)).
23. R. Smit, Y. Noat, C. Untiedt, N. L. andM.C. van Hermert, and J. van Ruitenbeck, Nature **419**, 906 (2002).
24. J. Cuevas, A. L. Yeyati, and A. Martín-Rodero **80**, 1066 (1998).
25. D. Papaconstantopoulos, *Handbook of the band structure of elemental solids* (Plenum Press, New York, 1986).
26. R. Landauer, IBM J. Res. Develop. **1**, 223 (1957).
27. P. Hay and W. Wadt, J. Chem. Phys. **82**, 299 (1985), lAN2DZ, ECP.
28. A. Becke, J. Chem. Phys. **98**, 5648 (1993), b3LYP.
29. J. Kua and W. A. G. III, J. Phys. Chem. B **102**, 9492 (1998).
30. B. O. Roos, K. Andersson, and M. P. Fülscher, Chem.Phys.Letters **192**, 5 (1992).
31. T. Hashimoto, S. Nakano, and Hirao **104**, 6244 (1996).
32. P. S. H. Keiter and W. Wenzel, J. Comput. Chem. **20**, 1559 (1999).
33. P. S. W. Wenzel, *Improved Implmentation and Application of the Individually Selecting Configuration Interaction Method* (Springer, 2002), p. 215.

Computer Science

Prof. Dr. Christoph Zenger

Institut für Informatik
Technische Universität München
Boltzmannstraße 3
D-85748 Garching

Computer Science is one of the basic disciplines involved in all High Performance Computing projects. In this section only three articles are collected in which Computer Science plays a major role. Two of them are concerned with long term projects having been presented in previous issues of this series.

In "Efficient and Object-Oriented Libraries for Particle Simulations" recent advances in the numerical simulation of astronomical phenomena are reported by the researchers of the very successful "Sonderforschungsbereich 382". The close cooperation of physicists and computer scientists leads to impressive results.

The "SKaMPI"-project is an established benchmark for MPI-implementations. Recent extensions are presented in the paper "SKaMPI – including more complex communication patterns".

The third contribution "Performance analysis using the PAR-Bench benchmark system" did not appear in previous volumes and is devoted to the investigation and optimisation of the performance of High Performance Computing Systems.

As in the last years computer science oriented projects in the evolving field of bioinformatics or large and complex data base applications are still missing. Contributions from this fields are highly encouraged for future research at the HLRS computer systems.

Towards a Holistic Understanding of the Human Genome by Determination and Integration of Its Sequential and Three-Dimensional Organization

Tobias A. Knoch

Deutsches Krebsforschungszentrum (DKFZ),
Im Neuenheimer Feld 280, D-69120 Heidelberg, Germany
E-mail: TA.Knoch@DKFZ-Heidelberg.de or TA.Knoch@taknoch.org

In Cooperation With: F. Bestvater[1,2], C. Cremer[6], T. Cremer[4], K. Fejes-Toth[1,6], A. Friedel[4], M. Göcker[9], A. Kellerer[4], R. Lohner[5], K. Monier[8], J. Langowski[1], P. Lichter[1], P. Quicken[4,7], J. Rauch[6], K. Rippe[1,6], E. Spiess[1], K. Sullivan[8], W. Waldeck[1], M. Wachsmuth[3], T. Weidemann[1]
[1]Deutsches Krebsforschungszentrum, Heidelberg, Germany; [2]Febit GmbH, Mannheim, Germany;
[3]Leica Microsystems GmbH, Mannheim, Germany; [4]Ludwig-Maximillians Universität, Munich, Germany;
[5]Supercomputing Center Karlsruhe, Universität Karlsruhe, Karlsruhe, Germany;
[6]Ruprecht-Karls Universität, Heidelberg, Germany; [7]Symantec Education Services, Ismaning, Germany;
[8]The Scripps Institute, La Jolla, California, USA; [9]Universität Tübingen, Tübingen, Germany

Summary. Genomes are one of the major foundations of life due to their role in information storage, process regulation and evolution. However, the sequential and three-dimensional structure of the human genome in the cell nucleus as well as its interplay with and embedding into the cell and organism only arise scarcely from the unknown, despite recent successes e. g. in the linear sequencing efforts and growing evidence for seven genomic organization levels. To achieve a deeper understanding of the human genome the structural, scaling and dynamic properties in the simulation of interphase chromosomes and cell nuclei are determined and combined with the analysis of long-range correlations in completely sequenced genomes as well as the analysis of the chromatin distribution *in vivo*: This integrative approach reveals that the chromatin fiber is most likely folded according to the Multi-Loop-Subcompartment (MLS) model in which the chromatin fiber bents into 63–126 kbp big loops aggregated to rosettes connected by again 63–126 kbp linkers. The MLS model exhibits fine-structured multi-scaling and predicts correctly the transport of molecules by moderately obstructed/anomalous diffusion. On the basic sequence level, genomes show fine-structured positive long-range correlations, allowing classification and tree construction. This, DNA fragment distributions after carbon ion irradiation and on the highest structural level, the nuclear morphology visualized by histone autofluorescent protein fusions *in vivo*, agrees again best with the MLS model. Thus, the local, global and dynamic characteristics of cell nuclei are not only tightly inter-connected, but also are integrated holisticly to fulfill the overall function of the genome.

1 Introduction

In human cells the genetic information controlling most processes from the cellular level, over embryogeneses to cognitive ability, manifests in a diploid set of 23 DNA molecules, the chromosomes. They consist of $\approx 7\times10^9$ base pairs (bp) storing around 1.4×10^{10} Bit or 1.75 GByte, and add to a molecular length totalling ≈ 2 m, which are kept in comparably small cell nuclei with typical diameters of $\approx 10\,\mu$m or volumes of $500\,\mu m^3$. This corresponds to a compaction factor of $\approx 2\times10^5$. Consequently, beyond pure compaction, the structuring of the genetic information in several organizational levels seems obviously to allow on the one hand sufficient performance during information transcription in interphase and on the other hand replication of the information and segregation of the chromosomes into the daughter cells during mitosis. Additionally, the abundant mutations need to be continuously found, controlled and repaired to avoid the inevitable course of entropy.

Considering the huge length and time scales, which bridge 10^{-9}–10^{-5} m and 10^{-10}–10^4 s, the genetic information of the human genome involves seven organizational levels according to current believe (Fig. 1): the DNA double helix (i), winds around a protein complex forming the nucleosome (ii), which condenses irregularly to the 30 nm chromatin fiber (iii), which is folded into chromatin loops (iv), which aggregate to chromosomal subdomains (v), which constitute a chromosome (vi), being non-randomly arranged in the cell nucleus (vii).

The DNA double helix and the nucleosome structure are known to atomic precision, but already the detailed nucleosome conformation in the 30 nm chromatin fiber is still debated. The latter holds even more for the higher-order structures having born many a hypothesis: Whereas light microscopic studies by Rabl (1885) and Boveri (1909) proposed territorial chromosomes with a hierarchical, self-similar organization of chromatin fibers in the late 19 th century, electron microscopy suggested thereafter a random interphase chromatin organization in the models of Comings (1968, 1978) and of Vogel & Schroeder (1974). To explain the high condensation degree of metaphase chromosomes and their stainable ideogram bands, chromatin loops attached to a nuclear matrix scaffold were suggested in the Radial-Loop-Scaffold model by Paulson & Laemmli (1980). According to Pienta & Coffey (1984) these loops persisted in interphase forming stacked rosettes in metaphase. Microirradiation and fluorescence *in situ* hybridization (FISH) finally proved a territorial organization of chromosomes, of their arms, and of subchromosomal domains and led to the Inter-Chromosomal Domain (ICD) model hypothesizing an interchromosomal channel network (Zirbel *et al.*, 1993, Cremer & Cremer, 2001). For the intraterritorial chromatin folding, the chromonema fiber (CF) model by Belmont & Bruce (1994) postulated a helix hierarchy, whereas in the Random-Walk/Giant-Loop (RW/GL) model 1–5 Mbp loops are attached to a non-protein backbone (Sachs *et al.*, 1995) and in the Multi-Loop-

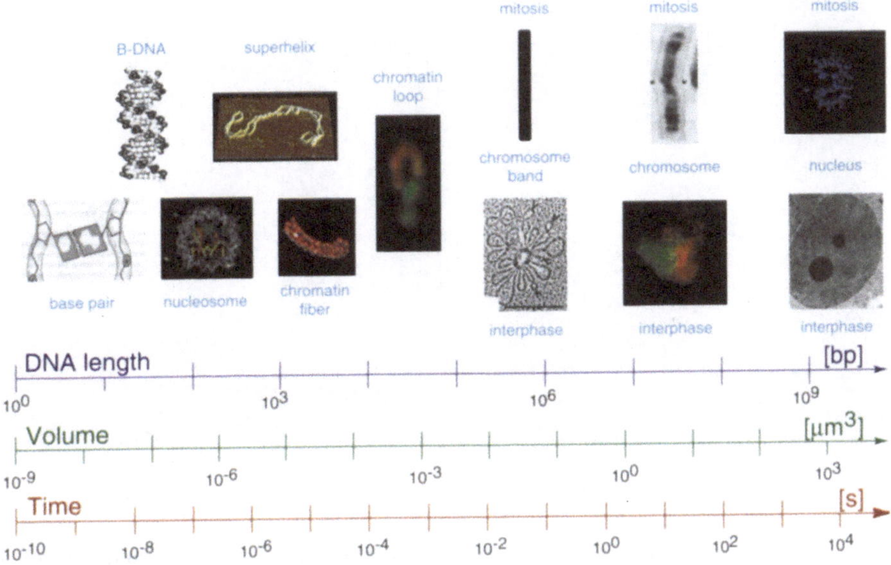

Fig. 1. *Overview on the Size and Time Scaling of the Human Genome.* The scaling and the levels of organization range over 9 decades for base pairs, 12 decades for the volume ($\approx$4 length decades) and 14 decades for the time: At the initial stage base pairs are formed composing the DNA double helix (Voet & Voet, 1995) which winds around the histone core complex building the nucleosome (callotte model according to Luger *et al.*, 1997), which condense into the 30 nm chromatin fiber (simulation image, Wedemann & Langowski, 2002). The DNA double helix forms also superhelices (scanning force microscopic image of plasmid DNA, with courtesy of K. Rippe, Division Molecular Biophysics, Kirchhoff Institute for Physics, Ruprecht-Karls-Universität, Heidelberg, Germany). The next compaction step consists of chromatin loops (fluorescence *in situ* hybridization (FISH) image, with courtesy of P. Fransz, Swammerdam Institute for Life Sciences, BioCentrum Amsterdam, Amsterdam, The Netherlands) possibly forming rosettes (electron microscopic image, Reznik *et al.*, 1990), which make up interphase chromosome arms and territories (FISH image, with courtesy of S. Dietzel) and the metaphase ideogram bands (giemsa chromosome staining from Alberts *et al.*, 1994). 46 chromosomes compose the human nucleus and are decondensed in interphase (electron microscopic image, with courtesy of K. Richter, Division Molecular Genetics, Deutsches Krebsforschungszentrum, Heidelberg, Germany) and condensed as shown for separated metaphase plates (Fig. 10 B).

Subcompartment (MLS) model 60–120 kbp loops form rosettes connected by a similar linker (Münkel & Langowski, 1998; Münkel *et al.*, 1999).

However, despite all these successes, the sequential and three-dimensional organization as well as their connection are still known only rudimentarily. Therefore, aspects from all nuclear scales as well as simulations and experiments were combined to determine and integrate the sequential and three-dimensional organization of the human genome (Knoch, 2002).

Fig. 2. *Volume Rendered Images of Simulated Chromosome Models for Chromosome XV.*
From the typical startconfiguration for simulations with the form and size of a metaphase chromosome (above), interphase chromosomes in thermodynamical equilibrium are decondensed by Monte Carlo (MC) steps, followed by relaxing Brownian Dynamics (BD) steps. In a Random-Walk/Giant-Loop (RW/GL) model, with 5 Mbp loops and 378 kbp, the large loops intermingle freely not forming distinct features like in the Multi-Loop-Subcompartment (MLS) model (8×10^4 MC, 10^3 BD, left). In an MLS model with both 126 kbp loops and linkers the rosettes form subcompartments as separated organizational and dynamic entities (5×10^4 MC, 10^3 BD, middle). In a RW/GL model with 126 kbp loops and 63 kbp linkers, the small loops neither intermingle freely nor form distinct subcompartments (9×10^4 MC, 10^3 BD, right). Obviously, the MLS model can easily be transformed in both RW/GL models by minor topological changes, indicating possible biological mechanism or the origin of possible experimental preparation artefacts. Consecutive loops of the RW/GL or MLS rosettes are painted in red and green. The fiber diameter is 30 nm in all images.

2 Simulation of Single Chromosomes and Cell Nuclei

To investigate the folding of the 30 nm chromatin fiber into chromosome territories, their morphology and experimental distinguishability, single chromosomes based on the Multi-Loop-Subcompartment (MLS) model and the Random-Walk/Giant-Loop (RW/GL) topology, were simulated for various loop and linker sizes. The 30 nm chromatin fiber was modelled as a polymer chain with stretching, bending and excluded volume interactions. A spherical boundary potential simulated the confinement of other chromosomes and the nucleus. Monte Carlo and Brownian Dynamics methods were applied to generate chain configurations at thermodynamic equilibrium (Fig. 2). These simulations of single chromosomes were extended to nuclei of diploid human cells containing all 46 chromosomes, to determine the chromosome arrangement and the related microscopic morphology, besides the validation of the results

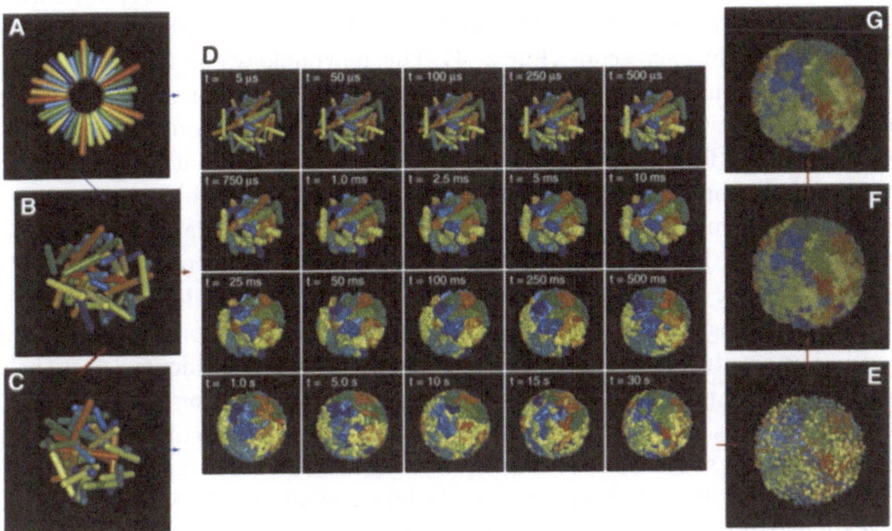

Fig. 3. *Startconfigurations and Decondensation into Interphase of Simulated Cell Nuclei.*
Metaphase chromosomes were placed in a metaphase plate (A) or as cylinders randomly (C) into a spherically constrained potential before optional relaxing by simulated annealing avoiding unphysical configurations (B). The cylinders were split into spheres and decondensed into interphase by $\leq 6\times10^6$ Brownian Dynamic (BD) steps of 5 µs, i.e. 30 s (D). Then the 30 nm chromatin fiber was placed into the spheres with scaled down chain segments avoiding concatenation (E) and softly relaxed with 10^3 BD steps linearly increasing from 0.01–0.5 µs (F). Segment sizes were 300 nm or smaller that chromatin loops consisted of ≥ 4 segments. Finally the resolution was increased to 50 nm segments, and relaxed further by 5×10^3 BD steps of 0.5 µs to ensure proper morphologic and quantitative analysis (G). Shown for an MLS model with 126 kbp loops and linkers and 5 µm nuclear radius.

of the simulation of single chromosomes. The chromatin fiber was simulated as in the case of single chromosomes. Since the computer power increased by a factor of 46, in this case simulated annealing and Brownian Dynamics methods as well as a four step decondensation procedure from metaphase (Fig. 3) were applied to generate interphase configurations again at thermodynamic equilibrium (Fig. 4). The simulation program VirtNuc as well as various helper and analysis programs were written in the object oriented C++ language and parallelized using the message passing interface (MPI) standard. By application of a very efficient linked cell algorithm for the calculation of the physical interactions and by implementation of dynamic load balancing, the parallelization efficiency has been improved to 95 % using 16 processors. The simulation and analysis of single chromosomes, totalled $\approx$ 96,000 CPUh ($\approx$ 11 years) on a single R6000 processor with 60 MHz and those of whole nuclei totalled $\approx$ 260.000 h ($\approx$ 30 years) on a single R6000 processor with 120 MHz.

Both the MLS and the RW/GL model form chromosome territories with different morphologies: The MLS rosettes result in distinct subcompartments visible with light microscopy (Fig. 2–4). This morphology and the size of these subcompartments agree with the morphology found by expression of histone autofluorescent protein fusions (see below) and FISH experiments. In contrast, the big RW/GL loops lead to a homogeneous chromatin distribution. Even small changes of the model parameters induced significant rearrangements of the chromatin morphology as can be seen in simulated electron microscopic (EM, Fig. 7 A&C) and confocal laser scanning microscopic (CLSM) images (Fig. 4). Thus, pathological diagnoses of e. g. cancer based on the nuclear morphology, might be related to structural changes on the chromatin level. The position of chromosome territories in interphase depends on their metaphase location (Fig. 3), and suggests a possible origin of current experimental findings. Only the MLS model leads to a low overlap of chromosomes, their arms and subcompartments, again in agreement with experiments. The chromatin density distribution in CLSM image stacks of the MLS model but not the RW/GL model reveals a bimodal behaviour in agreement with recent experiments (Weidemann, 2002). Review and comparison of experimental to simulated spatial distance measurements between genomic markers as function of their genomic separation also favour an MLS model with loop and linker sizes of 63–126 kbp (Fig. 5).

3 Scaling of the Simulated Chromatin Topology and Nuclear Morphology

In order to characterize the huge time and length scale bridging several levels of packaging of the genome, the scaling behaviour of the 30 nm chromatin

Fig. 4. *Morphology of Cell Nuclei by Rendering, EM and CLSM Images.* All differences of chromosome models 63-63-MLS^3 (AI), 63-252-MLS^3 (AII), 126-252-MLS^3 (AIII), 126-126-MLS^4 (BI), 84-126-MLS^4 (BII), 126-126-MLS^5 (C), 1320-RW/GL^6 (D), are visible in the three-dimensional rendering of the actual 30 nm chromatin fiber (α), the electron microscopic (EM) image (β), the EM image chromosome map (γ; homologous chromosome painting legend below), the confocal laser scanning microscopic (CLSM) image with a high resolution 100× 1.4 oil immersion PL APO objective (δ), the CLSM image with a lower resolution 60× 1.2 water immersion PL APO objective (ε; colour intensity coding below), and the CLSM image chromosome map (ϕ): The rosettes of the MLS model form subcompartments being visible as separated organizational and dynamic entities (A, B, C). In contrast, the large loops of the RW/GL model intermingle freely, neither forming distinct features like in the MLS model (D), nor forming clearly separated chromosome territories due to high overlap. The EM and CLSM images were normalized to highest intensity, representing absolute no intensities. Nomenclature: L_S-L_{IS}-model^r; L_S: loop size; L_{IS}: linker size; r: nuclear radius.

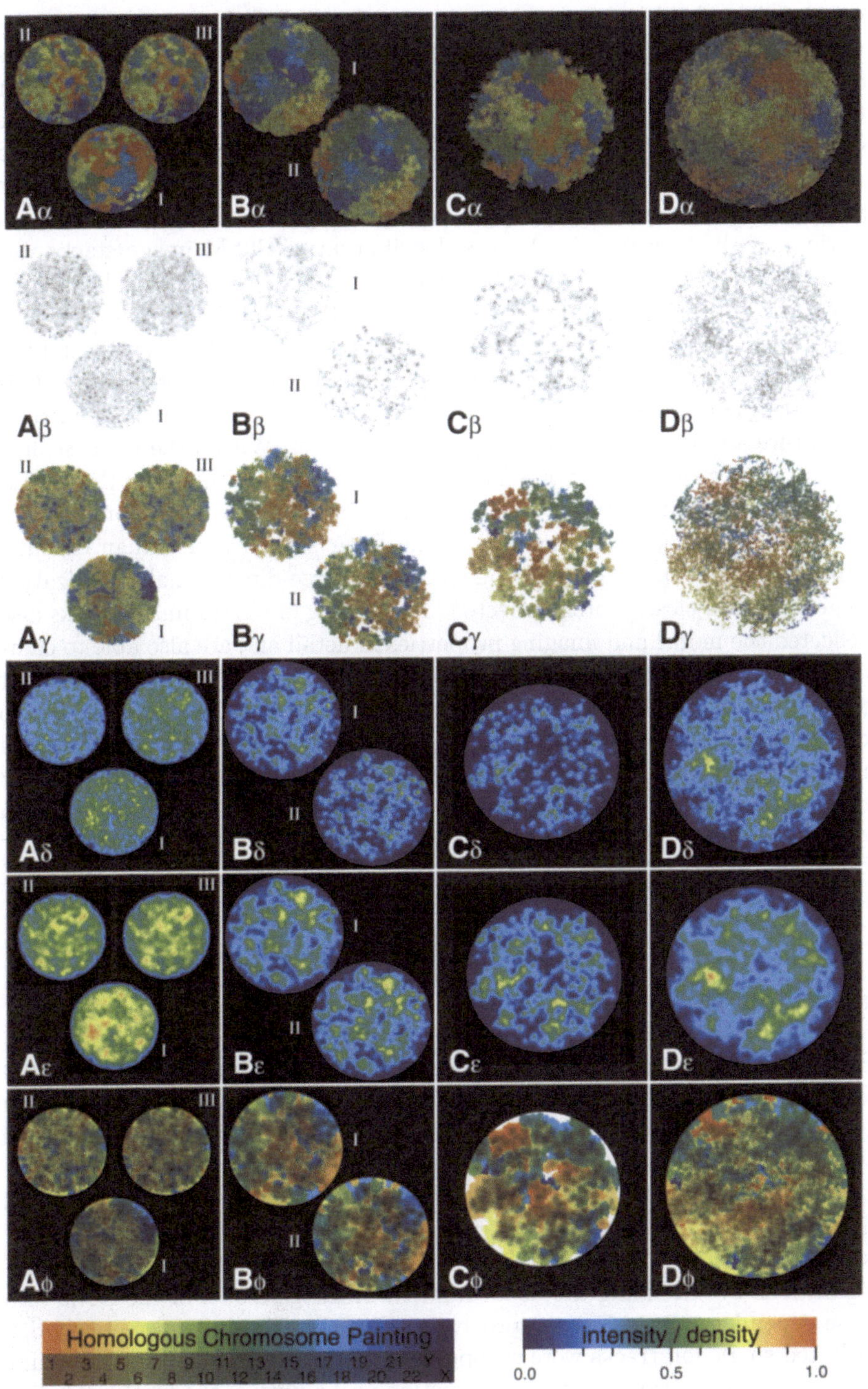

Aα
Bα
Cα
Dα
Aβ
Bβ
Cβ
Dβ
Aγ
Bγ
Cγ
Dγ
Aδ
Bδ
Cδ
Dδ
Aε
Bε
Cε
Dε
Aφ
Bφ
Cφ
Dφ
Homologous Chromosome Painting
1 3 5 7 9 11 13 15 17 19 21 Y
2 4 6 8 10 12 14 16 18 20 22 X
Intensity / density
0.0
0.5
1.0

fiber topology (Fig. 7 A-C) and the scaling behaviour of the morphology of simulated confocal laser scanning microscopic (CLSM) image stacks was determined (Fig. 4). Both were obtained from simulations of single chromosomes and whole nuclei using the RW/GL and MLS chromatin fiber topologies. For the analysis, the following scaling/fractal dimensions were calculated: The scaling of the one-dimensional axis of the 30 nm chromatin fiber was analysed with the exact spatial-distance and yard-stick dimensions and its voluminous scaling was investigated by the pure box-counting dimension. The scaling behaviour of the (inverse-, iso-) mass distribution of CLSM image stacks was investigated with the weighted box-counting, lacunarity and local mass dimensions. To support the latter, additionally, the local diffuseness, skewness and kurtosis distributions were calculated. The scaling of the chromatin fiber revealed different power-law behaviours on different scales (Fig. 6 A&B). This multi-scaling is created by the random walk behaviour of the fiber, the globular nature and the arrangement of loops or rosettes. Within the multi-scaling regime a fine structure was present for the MLS model arising from the rosette loops. A similar fine-structured multi-scaling behaviour was also found in the correlation behaviour on the level of the DNA sequence of human chromosomes (see below). Thus, the sequential and three-dimensional organization of genomes are closely interconnected. The scaling of CLSM image stacks also reflected the model and imaging properties in detail and are also able to characterize the morphology found in experiments using histone autofluorescent protein fusions *in vivo* (Fig. 6 C-E). Thus, the chromatin fiber topology is also closely connected to nuclear morphology. Therefore, scaling analyses of the nuclear morphology are a suitable approach to differentiate between different cell states, e. g. during the cell cycle, due to malignancy, in apoptosis or in response to drugs. Consequently, the scaling behaviour shows that all nuclear organization levels are connected.

4 Simulation of the DNA Fragment Distribution after Ion Irradiation

Since the chromatin fiber is inhomogeneously distributed and exhibits fine-structured multi-scaling behaviour in cell nuclei also the length distribution of DNA fragments after irradiation with an inhomogeneous nuclear radiation dose distribution and thus the sites of double strand breakage depend on the spatial arrangement of the chromatin fiber (Friedl, 1994; Knoch, 2000). Quantitatively, the fragment distribution of carbon ions was calculated by simulating irradiation with carbon ions of single chromosome interphase configurations by extending the usual random breakage model of the DNA to a dose dependent mixed Poisson processes (Fig. 8 B): The inhomogenously distributed single ion tracks were mapped against the chromatin fiber assuming an energy distribution and thus destruction probability of the track according to Kraft/Scholz and Chatterjee. The resulting fragment distributions revealed

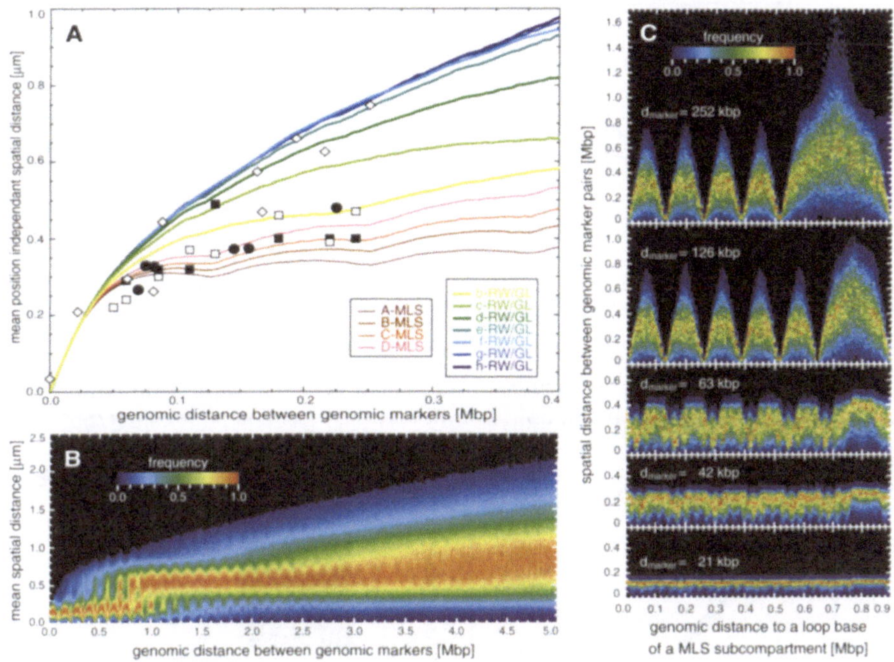

Fig. 5. *Simulated Spatial Distance Dependencies and Comparison to Experiments.*
Simulated spatial distances between position independent placed genomic markers as function of their genomic distance reflect the chromatin fiber topology in detail and yields the better agreement with the MLS model and experiments the better their latter preparation and imaging methods are (A, single chromosome simulations). Thin lines are A-D MLS models with 126 kbp loops and 63, 126, 189 and 252 kbp linkers. Thick lines: a-h RW/GL models with 0.252, 0.504, 1.0, 2.0, 3.0, 4.0 and 5.0 Mbp. Data are: full circles: Knoch (1998, 1999, 2000) and Rauch (1999); data from Monier (1997): full squares: fibroblasts 11q13, open squares: lymphocytes 11q13; open rhombi: Trask *et al.* (1989). The independent spatial distance distributions (B, for a 126-126-MLS[5] nucleus) reveal loop introduced oscillations up to ≈3 Mbp and distance spikes for loop size multiples due to the higher probability of markers located in the linker. The visible shifts into new rosettes (e. g. 0.75 Mbp) and the oscillation decrease suggests higher rosette than loop flexibility concerning phase space distributions. The distributions of position dependent spatial distances of a marker set with different genomic separations as function of their shift (5.2 kbp) relative to a loop basepoint of a rosette show characteristic oscillations and correlations as expected for fixed loop and rosette structures (C, for a B-MLS model).

better agreement for the Chatterjee track assumption with experiments and represent the chromatin fiber topology not only in its general behaviour but by the appearance of a fine structure also in its detail, e. g. loop size or loop arrangement in rosettes (Fig. 8 A). Thus, the fine-structured multi-scaling of the chromatin folding is present also in the fragment distribution. A compar-

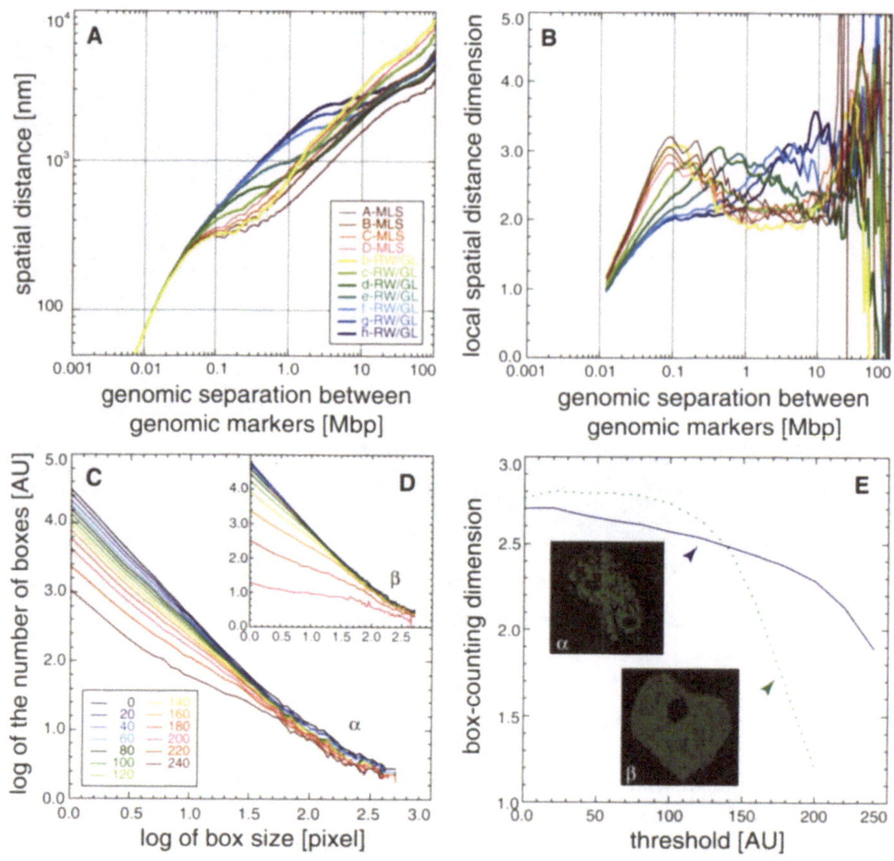

Fig. 6. *Scaling of the Chromatin Fiber of Simulated Chromosomes and of the Nuclear Morphology in vivo.*

In simulated chromosomes the spatial-distance function (A, double log plot of Fig. 5 A) and its local slope the spatial-distance dimension (B) shows a fine-structured multi-scaling behaviour of the chromatin fiber topology below the cut-off greater ≈ 80 Mbp or ≈ 8 μm due to the finite chromosome size. The general multi-scaling behaviour is characterized by an increase from 1.0 on small scales characterizing the stiff chromatin segments, over values ≈ 2.0 as for the random walk of the segments to a maximum of 3.0 stating the ring-shaped loops of both the MLS and the RW/GL model and globular state of the rosettes of the MLS models. In the MLS model thereafter again local dimensions ≈ 2.0 are reached describing the random organization of the rosettes relative to each other. The fine structure is attributable to the loops aggregated in rosettes for MLS models and reflects the models in detail. The weighted box-volume function of the nuclear chromatin morphology in H2A-YFP expressing cells (Fig. 10) shows distinct power-law behaviour for two nuclei for different intensity thresholds (C, D). Its slope the box-counting dimension as function of these thresholds differs clearly for both nuclei. Thus not only the morphology can be quantified but in comparison with the simulated scaling behaviour also an MLS model is more likely supported.

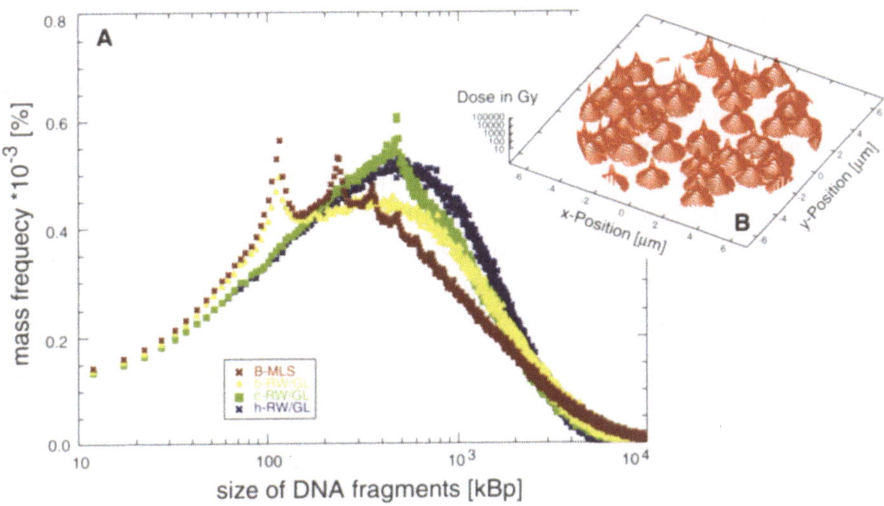

Fig. 7. *DNA Fragmentation after Irradiation with Carbon Ions.* Simulation of the length distribution of DNA fragments after irradiation with carbon ions with an inhomogeneous nuclear radiation dose distribution and thus the sites of double strand breakage (B) depend on the spatial arrangement of the chromatin fiber and reflect the used interphase topology of simulated chromosomes like loop size or loop arrangement in rosettes in detail (A). Best agreement of the simulations with experiments is reached for the MLS model.

ison to experimental fragment distribution results again in best agreement with an MLS topology with loops and linkers of 63–126 kbp.

5 Dynamics in Simulated Interphase Cell Nuclei

To determine the impact of the three-dimensional genome organization on molecular mobility, the accessability of nuclear loci and the hypothesis of the Inter-Chromosomal Domain (ICD) model, the diffusion of spheres was simulated by Brownian Dynamics in computer generated nuclei with an MLS chromatin fiber topology. The tracers interacted with the static fiber by an excluded volume potential. Visual inspection of the morphology of simulated chromosomes or nuclei revealed big spaces allowing high accessibility to nearly every spatial location. A channel like network for molecular transport between chromosome territories, as postulated by the ICD model, was not apparent in the simulations (Fig. 7 A-C). The big spaces are supported by estimating the nuclear volume occupied by chromatin of $\leq 30\,\%$, leaving $\geq 70\,\%$ of space for diffusion with an average mesh spacing of 29–82 nm for nuclei of 6–12 µm diameter. This agrees with the simulated mean displacement for 10 nm sized particles of ≈ 1–2 µm within 10 ms. Therefore, the diffusion of biological relevant tracers is only moderately obstructed. The anomaly parameter characterizing the degree of obstruction ranged from 2.0 (obstacle free diffusion) to 4.0 (Fig. 7 C), in agreement with experiments in which the obstruction of fluorophores was measured as function of the chromatin density labelled with histone autofluorescent protein fusions *in vivo* (Fig. 7 C; Wachsmuth *et al.*, 1998, Wachsmuth, 2001; Wachsmuth *et al.*, 2003). The degree of obstruction was proportional to the nuclear density, the fiber diameter, the interaction hardness and the tracer size. Different fiber topologies had no effect on the average particle displacement. Consequently, molecules and proteins might reach every nuclear location by energy independent diffusion without a special channel like network without necessity of an ICD model, which can therefore be refuted.

6 Long-Range Correlations in DNA Sequences

The sequential organization, i. e. the relations within DNA sequences, and its connection to the three-dimensional organization of genomes was investigated by correlation analyses of 113 completely sequenced chromosomes of 0.5×10^6 to 3.0×10^7 bp from Archaea, Bacteria, *Arabidopsis thaliana, Saccharomyces cerevisae*, *Schizosaccharomyces pombe*, *Drosophila melanogaster* and *Homo sapiens*. The analyses are based on the concentration profile of single bases along the DNA sequence (Knoch *et al.*, patent): The square root of the mean-square deviation between the concentration of bases c_l in a window of length l and the concentration c_L of bases in the entire DNA sequence with length L was calculated

$$C(l) = \sqrt{\langle (c_l - c_L)^2 \rangle_{s=L-l-1}} \,.$$

For a fractal self-similar sequence like a random walk the concentration fluctuation function $C(l)$ shows power-law behaviour whose exponent characterizes the scale dependent degree of correlation. To avoid numeric instabilities infinitely exact calculation tools provided by the GNU multiple precision package GMP were implemented. Calculations were performed on PCs and an IBM SP2, using in total ≈ 5000 h CPU time. On the latter the analyses were split into jobs of a few minutes computing single or few windows, thus being an extremely efficient filler for the unavoidable gaps in batch mode of big parallel machines. The analyses would also be predestined for grid computing, e. g. a screen saver application.

All sequences revealed long-range power-law correlations almost on the entire observable scale (Fig. 9). The local correlation coefficient shows close to random correlations on the scale of a few base pairs, a first maximum from 40–3400 bp (for *Arabidopsis thaliana* and *Drosophila melanogaster* divided in two submaxima), and often a region of one or more second maxima from 10^5–3×10^5 bp. This multi-scaling behaviour was species specific. Computer generated random sequences assuming a block organization of genomes reproduced such multi-scaling. Within this multi-scaling behaviour an additional

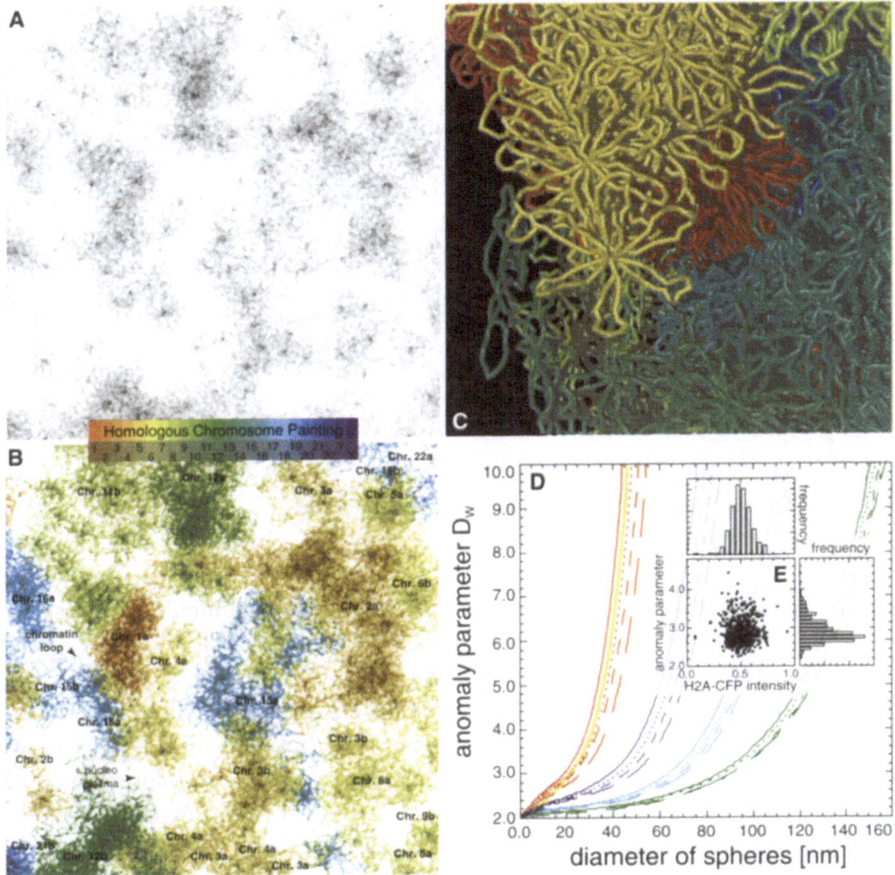

Fig. 8. *Dynamic Properties of Simulated Nuclei: Diffusion of Particles.* The simulated electron microscopic image (A) and chromosome map (C) for a 126-126-MLS[5] nucleus (deliberately unrelaxed, thus the voids between chromosome territories are an artefact) and the detailed rendered view of a 126-126-MLS[6] nucleus demonstrate the low overlap of chromosome territories and the MLS rosettes. The mean spacing between chromatin fibers ranges from 50–100 nm, thus small chemical substances as nucleotides or ATP molecules reach every location in the nucleus and most relevant proteins or protein subunits should only be obstructed moderately, regarding the scale. This refutes the Inter Chromosomal Domain (ICD) hypothesis. The anomaly parameter characterizing the degree of obstruction of diffusing spheres in simulated nuclei (D), agrees with this semi-quantitative view and is proportional to the nuclear radius and therefore density (nuclear radii of 3, 4, 5, and 6 µm are red, blue, light blue and green). Changes of the chromatin fiber diameter leads to a smaller proportionality (diameters of 25, 30, 35 and 40 nm are solid, dotted, dashed, and long dashed). This agrees with the small obstruction and its minor dependence of the chromatin morphology, determined *in vivo* for Alexa–568 fluorescence molecules by two colour fluorescence correlation spectroscopy (FCS) as function of the chromatin density labelled by H2A-CFP fusionproteins (E, Wachsmuth, 2001; the anomaly parameter is bigger than expected due to binding effects of Alexa–568).

fine structure is present and attributable to the codon usage in all except the human sequences. Here it is connected to nucleosomal binding. Computer generated random sequences assuming the codon usage and nucleosomal binding agree with these results. Mutation by sequence reshuffling destroyed all correlations, thus their stability seems evolutionary tightly controlled and connected to the spatial genome organization on large scales. This is supported by the scaling behaviour of the chromatin topology (see above). The correlation behaviour was used to construct trees, which were similar to the corresponding phylogenetic trees for β-Tubulin genes of *Oomycetes* and Eukarya genomes. For Archaea and Bacteria tree construction led to a new classification system with four major tree branches/classes. In summary, these findings suggest a complex sequential organization of genomes closely connected to their three-dimensional organization.

7 "Chromatin Alive": *In vivo* Labelling of the Chromatin Morphology

The *in vivo* morphology and dynamics of chromatin is difficult to assess by electron microscopy, fluorescence *in situ* hybridization (FISH) and *in vivo* stains since these methods require fixation or produce artefacts. To overcome these limitations a novel *in vivo* technique for chromatin labelling was established (Knoch *et al.*, patent): DNA vectors encoding the fusion proteins of all histones H1.0, H2A, mH2A1.2, H2B, H3, H4 and the autofluorescent proteins CFP, GFP, YFP, DsRed1 DsRed2 were developed and expressed stably in HeLa, LCLC103H, Cos7 and ID13 cells. 2.6 to $\approx 20\,\%$ of the nucleosomes carry a label. No apparent influence of the cell cycle status, the proliferation rate or the AFP fluorescent excitation/emission spectra, but recently a somewhat increased nucleosomal repeat length was detected (Weidemann, 2002). With this approach the structure and dynamics of histones, nucleosomes, chromatin, chromosomes and whole nuclei during cell cycle, differentiation, and apoptosis could be investigated artefact-free *in vivo* (Fig. 10). The interphase morphology showed globular structures as predicted by the Multi-Loop-Subcompartment model. All stages of mitosis as well as apoptosis were clearly distinguishable (Gil-Parado *et al.*, 2002, 2003). Deacetylase inhibitors led to a smoothing of the interphase morphology. With this *in vivo* chromatin label the interphase morphology and changes thereof could be investigated by quantitative scaling and statistical analyses (Fig. 6 C-E). The technique could also be applied for cell culture control and counterstaining, or *in organo* and *in organismo* by creation of transgenic animals.

This now widely used *in vivo* method led also to the discovery of construct conversions in simultaneous co-transfections (Bestvater *et al.*, patent, 2002), a convenient and widely used approach in multicolour labelling experiments *in vivo*, using green fluorescent proteins with distinct spectral characteristics. These co-transfections can cause false positive results due to con-

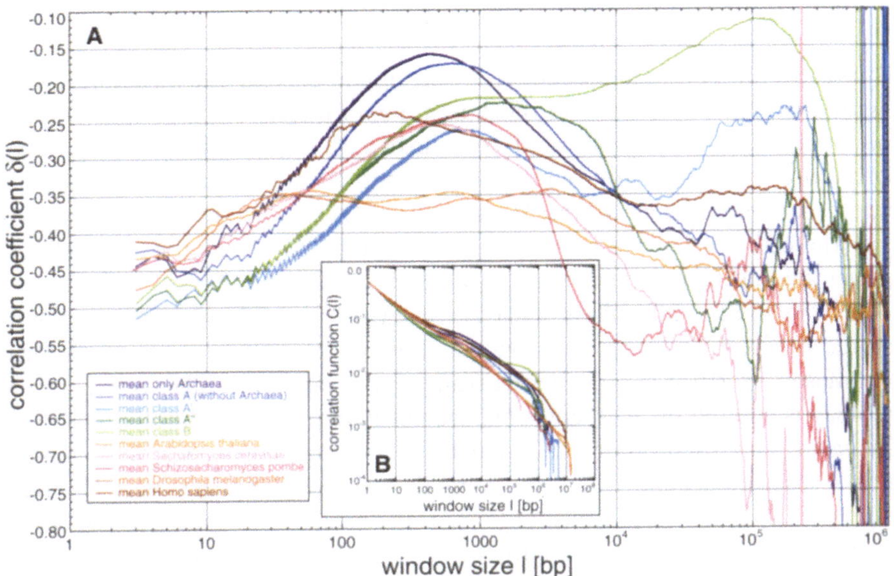

Fig. 9. *Fine-Structured Multi-Scaling Long-Range Correlations in Completely Sequenced Genomes.*

The average over the single analysed chromosomes (A) for each of the Eukarya genomes, the Archaea and the Bacteria classes A, A', A" and B, reveals that only *Homo sapiens* does not show the zig-zag pattern due to the codon usage, although it shows a nucleosome dependent fine structure not present in the other sequences. All genomes show a maximum between window sizes of 100–1000 bp of which only the maxima of *Homo sapiens* is connected to the nucleosome. The classes A' and B show a second maximum after a decrease of $\delta(l)$ with very high correlations for window lengths of $\approx 10^5$ bp. Beyond, only *Homo sapiens* shows a distinct second peaked maximum which is washed out in the average but existing in the of the single analysed human chromosomes and indicating a looped and clustered chromatin topology. For comparison the means of the concentration fluctuation function for the same averages are shown (B).

version of their spectral properties. Standard transfection result in $\approx 8\,\%$, depending on the treatment of the DNA up to $26\,\%$, of the cells expressing altered fusion proteins. This could lead to severe misinterpretation of the results. The conversion is independent of the transfection method and the cell type. The results show that conversion is based on homologous recombination/repair/replication (RRR) events occurring between the nucleotide sequences of the fluorescent proteins. Conversion can be avoided by consecutive transfection or by fluorescent constructs with low sequence similarities. The appearance of conversion allows to easily exchange spectral properties in fusion proteins, to create libraries or to assemble DNA fusion constructs *in vivo*. The detailed quantification of the conversion rate could be used to investigate RRR processes in general.

8 Conclusions

To achieve a deeper understanding of the human genome the structural, scaling and dynamic properties in the simulation of interphase chromosomes and cell nuclei were determined and combined with the analysis of long-range correlations in completely sequenced genomes as well as the analysis of the chromatin distribution *in vivo*: Integration of these different aspects suggests that the cell nucleus can be viewed as an optimized bioreactor in which the sequential and three-dimensional organization coevolved:

Advancing the interphase nucleus from the cellular level reveals a globular morphology. Staining of the single chromosomes reveals that these form chromosome territories, which are arranged non-randomly. The globular morphology is created by aggregates of chromatin loops within the chromosomes. Increasing the resolution reveals that the underlying chromatin fiber consists of nucleosomes around which the DNA is wound. Analysing the DNA base pair sequence shows a complex organization which can be linked to the codon usage, the nucleosome and the chromatin fiber topology on larger scales. Thus, every structural level of nuclear organization is connected and represented in all the other levels. Features present on one scale are reflected on other scales, and changes on one scale might either reflect or induce changes on other scales. These structural links are best described by scaling analyses.

Beyond the structural also the dynamics of the three-dimensional organization itself, i. e. chromosomes or chromatin loops, or the mobility of particles inbetween is scale dependent. Chromosomes or large protein complexes move slowly, in contrast to small and highly mobile molecules. Due to the low volume occupancy of the three-dimensional topology the mobility of medium sized molecules is only moderately obstructed. Thus, most molecules and proteins can reach nearly every location in the nucleus by simple diffusion very quickly and can commit to their function. Therefore, the dynamics is also closely connected to the underlying or surrounding structure, i. e. structural changes shape also the accessibility by molecules.

Consequently, the local, global and dynamic characteristics of cell nuclei, are tightly inter-connected, obviously due to their coevolution. Beyond, however, this view of the nucleus as an entity stresses, that its function can only be fulfilled by the integrated whole and that the information for processes from the cellular level, over embryogenesis to cognitive ability is present in this integrated whole holisticly.

Acknowledgements

The Supercomputing Center Karlsruhe (SCC; grant ChromDyn), and the Computing Facility of the Deutsches Krebsforschungszentrum (DKFZ) are thanked for access to their IBM SP2s. This work was supported by the Bundesministerium für Bildung und Forschung (BMBF) under grant 01

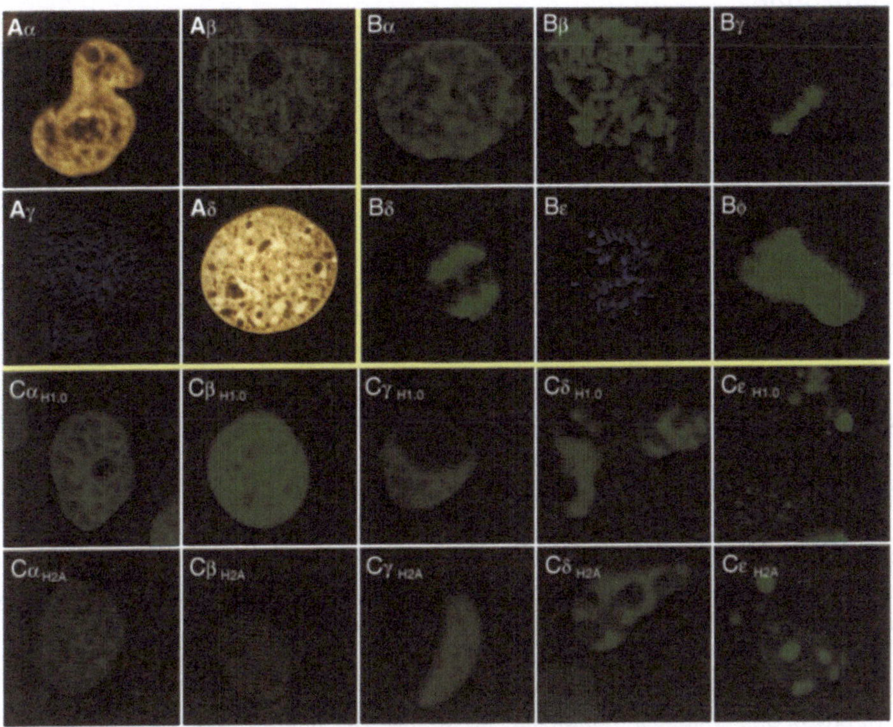

Fig. 10. *In vivo Labelling of the Chromatin Morphology in Interphase, Mitotic and Apoptotic Nuclei.*
Expression of H2A-YFP fusionproteins reveals different interphase morphologies and thus different genome organizations in human HeLa-H2A-YFP (Aα), the primate Cos7-H1-GFP (Aβ), the human LCLC103H-H2A-CFP (Aγ) and the mouse ID13-H2A-YFP (Aδ) cell lines. During mitosis all stages of the cell division are visible: At the beginning of prophase chromosomes start condensing (Bα) into cylindrical chromosomes (Bβ), pair in prometaphase and arrange in the metaphase plate in metaphase (Bγ). Then the chromosomes are draged appart into the daughter cells in anaphase (Bδ; Bε, H2A-CFP), before decondensation again into interphase during telophase (Bϕ). The induction of apoptosis by the deacetylation inhibitor Sodiumbutyrat in Hela cells expressing H1.0-YFP or H2A-YFP indicates interphase changes and apoptosis stages until total cellular destruction: In contrast to control nuclei (Cα), treated nuclei show rapidly a lower granularity (Cβ), before apoptosis sets in with the transformation to "apoptotic half moon" shaped nuclei and first chromatin agglomerations (Cγ). Then the chromatin condenses totally (Cδ) until apoptotic fragmentation of nuclei (Cε). Image sidelength: 20 μm: Aα Aβ; 25 μm: Aδ, Bα, Bβ, Bϕ; 30 μm: Aγ; 35 μm: Bγ, Bδ, Bε, C.

KW 9602/2 (Heidelberg 3D Human Genome Study Group, German Human Genome Project). T. A. Knoch was kindly provided with a dissertation grant of the Deutsches Krebsforschungszentrum (DKFZ).

References

Alberts, B., Bray, D., Lewis, J., Raff, M., Roberts, K. & Watson, J. D. Molecular Biology of the Cell. 3rd ed., *Garland Publications Inc., New York & London*, ISBN0-8240-7283-9, 1994.

Bestvater, F., Knoch, T. A. & Spiess, E. Nachweisverfahren für homologe Rekombinationsereignisse. *German Patent Application 10116826.8* and *International Patent Application PCT/DE02/01207.*

Bestvater, F., Knoch, T. A., Langowski, J. & Spiess, E. GFP-Walking: Artificial construct conversions caused by simultaneous cotransfection. *BioTechniques 32(4)*, 844-854, 2002.

Boveri, T. Die Blastomerenkerne von *Ascaris meglocephala* und die Theorie der Chromosomenindiviualität. *Archiv für Zellforschung 3*, 181-268, 1909.

Comings, D. E. The rationale for an ordered arrangement of chromatin in the interphase nucleus. *Am. J. Hum. Genet. 20*, 440-460, 1968.

Comings, D. E. Mechanisms of chromosome banding and implications for chromosome structure. *Annu. Rev. Genet. 20*, 440-460, 1978.

Cremer, T. & Cremer, C. Chromosome territories, nuclear architecture, and gene regulation in mammalian cells. *Nat. Rev. Gen. 2(4)*,292, 2001.

Friedl, A. A. Development of simulation methods and their application in the qualitative and quantitative analysis of DNA double strand breaks in lower and higher Eukarya. *Dissertation*, Ludwig-Maximilians Universität, Munich, Germany, 1994.

Gil-Parado, S., Fernàndez-Montalvàn, A., Assfalg-Machleidt, I., Popp, O., Bestvater, F., Holloschi, A., Knoch, T. A., Auerswald, E. A., Welsh, K., Reed, J. C., Fritz, H., Fuentes-Prior, P., Spiess, E., Salvesen, G. & Machleidt, W. Ionomycin-activated calpain triggers apoptosis: A probable role for Bcl-2 family members. *J. Biol. Chem. 277(30)*, 27217-27226, 2002.

Gil-Parado, S., Popp, O., Knoch, T. A., Zahler, S., Bestvater, F., Felgenträger, M., Holoshi, A., Fernàndez-Montalvàn, A., Auerswald, E. A., Fritz, H., Fuentes-Prior, P., Machleidt, W. & Spiess, E. Subcellular localization and subunit interactions of over-expressed human µ-calpain. *J. Biol. Chem.*, 2003. (in press)

Knoch, T. A., Waldeck, W., Müller, G., Alonso, A. & Langowski, J. DNA-Sequenz und Verfahren zur *in vivo* Markierung und Analyse von DNA/Chromatin in Zellen. *German Patent Application 10013204.9-44* and *International Patent Application PCT/DE01/01044.*

Knoch, T. A., Münkel, C. & Langowski, J. Three-dimensional organization of chromosome territories and the human cell nucleus - about the structure of a self replicating nano fabrication site. *Foresight Institute - Article Archive*, Foresight Institute, Palo Alto, CA, USA, http://www. foresight.org/, 1-6, 1998.

Knoch, T. A. Dreidimensionale Organisation von Chromosomen-Domänen in Simulation und Experiment. (Three-dimensional organization of chromosome domains in simulation and experiment.) *Diploma Thesis*, Faculty for Physics and Astronomy, Ruperecht-Karls Universität Heidelberg, Heidelberg, Germany, 1998 and TAK Press, Tobias A. Knoch, Mannheim, Germany, ISBN 3-00-010685-5, 2003.

Knoch, T. A., Münkel, C. & Langowski, J. Three-Dimensional Organization of Chromosome Territories and the Human Interphase Nucleus. *High Performance Scientific Supercomputing*, Wilfried Juling (ed.), Scientific Supercomputing Center (SSC) Karlsruhe, University of Karlsruhe (TH), 27-29, 1999.

Knoch, T. A., Münkel, C. & Langowski, J. Three-dimensional organization of chromosome territories in the human interphase nucleus. *High Performance Computing in Science and Engineering 1999*, Egon Krause and Willi Jäger (eds.), Springer Berlin-Heidelberg-New York, ISBN 3-540-66504-8, 229-238, 2000.

Knoch, T. A. Approaching the Three-Dimensional Organization of the Human Genome. *Dissertation*, Ruprecht-Karls Universität Heidelberg, Heidelberg, Germany, and TAK Press, Tobias A. Knoch, Mannheim, Germany, ISBN 3-00-009959-X (soft cover, 3rd ed.), ISBN 3-00-009960-3 (hard cover, 3rd ed.), 2002.

Knoch, T. A., Göcker, M. & Lohner, R. Methods for the analysis, classification and/or tree construction of sequences using correlation analysis. *US Patent Application 60/436.056.*

Luger, C., Mäder, A. W., Richmond, R. K., Sargent, D. F. & Richmond, T. J. Crystal structure of the nucleosome core particle at 2.8 A resolution. *Nature 389*, 251–260, 1997.

Monier, K. Cartographie linéaire et tridimensionnelle du génome humain par hybridation *in situ* fluorescente et imagerie microscopique digitale. *Dissertation*, Institut Albert Bonniot, Université Joseph Fourier Grenoble I, Grenoble, France, 1997.

Münkel, C. & Langowski,J. Chromosome structure described by a polymer model. *Phys. Rev. E 57 (5-B)*, 5888-5896, 1998.

Münkel, C., Eils, R., Zink, D., Dietzel, S., Cremer, T. & Langowski, J. Compartmentalization of interphase chromosomes observed in simulation and experiment. *J. Mol. Biol. 285 (3)*, 1053-1065, 1999.

Paulson, J. R. & Laemmli, U. K. The structure of histone-depleted metaphase chromosomes. *Cell 12*, 817-828, 1980.

Pienta, K. J. & Coffey, D. S. A structural analysis of the role of the nuclear matrix and DNA loops in the organization of the nucleus and chromosome. *J. Cell. Sci. Suppl. 1*, 123-135, 1984.

Rabl, C. Über Zellteilung. *Morphologisches Jahrbuch 10*, 214-330, 1885.

Rauch, J. Spektrale Präzisionsdistanzmikroskopie zur Untersuchung der 3D-Topologie ausgewählter DNA-Punktmarker. *Dissertation*, Faculty for Physics and Astronomy, Ruprecht-Karls Universität Heidelberg, Heidelberg, Germany, 1999.

Reznik N. A., G. P. Yampol, Kiseleva, E. V., Khristolyubova, N. B. & Gruzdev, A. D. Possible functional Structures in the chromomere, *Nuclear Structure and Function.* by Harris, J. R. & Zbarsky, I. B. (eds.), Plenum Press, New York - London, 27-29, 1990.

Sachs, R. K., van den Engh, G., Trask, B., Yokota, H. & Hearst, J.E. A random-Walk/Giant-Loop model for interphase chromosomes. *Proc. Nat. Acad. Sci. USA 92*, 2710-2714, 1995.

Trask, B.J., Allen, S., Massa, H., Fertitta, A., Sachs, R., van den Engh, G. & Wu, M. Studies of metaphase and interphase chromosomes using fluorescence *in situ* hybridization. *Cold Spring Harb. Symp. Quant. Biol. 58*, 767-775, 1993.

Voet, D & Voet, J. G. Biochemistry. 2nd ed., *John Wiley & Sons Inc.*, 1995.

Vogel, F. & Schroeder, T. M. The internal order of the interphase nucleus. *Humangenetik 25 (4)*, 265-97, 1974.

Wachsmuth, M, Waldeck, W. & Langowski, J. Anomalous diffusion of fluorescent probes inside living cell nuclei investigated by spatially-resolved fluorescence correlation spectroscopy. *J. Mol. Biol. 298 (4)*, 677-686, 2000.

Wachsmuth, M. Fluorescence fluctuation microscopy: Design of a prototype, theory and measurements of the mobility of biomolecules in the cell nucleus. *Dissertation*, Faculty for Physics and Astronomy, Ruprecht Karls Universität Heidelberg, Heidelberg, Germany, 2001.

Wachsmuth, M., Weidemann, T., Müller, G., Urs W. Hoffmann-Rohrer, Knoch, T. A., Waldeck, W. & Langowski, J. Molecules between binding and diffusion: A quantitative approach continuous photobleaching. *Biophys. J. 84 (5)*, 3353–3363, 2003.

Wedemann, G. & Langowski, J. Computer simulation of the 30-nanometer chromatin fiber. *Biophys. J. 82(6)*, 2847-2859, 2002.

Weidemann, T. Quantitative investigation of distribution, mobility, and binding of fluorescently labeled histones *in vitro* and *in vivo* with fluorescence fluctuation microscopy. *Dissertation*, Faculty for Physics and Astronomy, Ruprecht-Karls Universität Heidelberg, Heidelberg, Germany, 2002.

Zirbel, R. M., Mathieu, U. R., Kurz, A., Cremer, T. & Lichter, P. Evidence for a nuclear compartment of transcription and splicing located at chromosome domain boundaries. *Chrom. Res. 1 (2)*, 93-106, 1993.

Efficient and Object-Oriented Libraries for Particle Simulations

S. Ganzenmüller[1], M. Hipp[1], S. Kunze[2], S. Pinkenburg[1], M. Ritt[1], W. Rosenstiel[1], H. Ruder[2], and C. Schäfer[2]

[1] Wilhelm-Schickard-Institut für Informatik, Universität Tübingen
[2] Institut für Astronomie und Astrophysik, Universität Tübingen

Summary. We present two libraries for the parallel computation of particle simulations. One is the object-oriented library sph2000 written in C++, the other is ParaSPH, a library written in C, that supports hybrid architectures (clustered SMPs). They are portable and performant on a variety of parallel architectures with shared and distributed memory. We give details of the object-oriented design of sph2000, the parallelization of ParaSPH for hybrid architectures using MPI and OpenMP and discuss the speedups of the codes. Further, we give three examples of applications based on these libraries, which simulate protoplanetary discs, colliding rubber rings and the injection of diesel into a combustion chamber.

1 Introduction

In the Sonderforschungsbereich 382 there are several particle codes for the simulation of astrophysical problems. In the last years there was a strong effort to develop fast parallel particle libraries to support these applications. However, there is still a need for larger simulations needing more memory and computing power. To meet this requirements we are continuously investigating to reduce the parallel overhead and the serial parts of our codes, that limit the overall speedup. For the increasing number of hybrid architectures, a parallelization combining threads and message passing is a promising way to reduce the parallel overhead.

Our main interest is to develop efficient libraries for particle simulations, which are portable to all important parallel platforms. Therefore access to several different parallel architectures is crucial for us to verify our concepts. Here we can profit from the large number of parallel architectures the HLRS provides. For the development of the libraries, the parallel machines are not

* This project is funded by the DFG within SFB 382: *Verfahren und Algorithmen zur Simulation physikalischer Prozesse auf Höchstleistungsrechnern* (Methods and algorithms to simulate physical processes on supercomputers).

needed for large production runs, but the applications on top of the libraries have a great demand for parallel computing power.

We are mainly working with three parallel machines. The Hitachi SR8000 and the Cray T3E, both installed at the HLRS in Stuttgart and the Kepler Cluster installed in Tübingen. The Hitachi is a 8-way SMP machine coupled with a fast communication network. The Cray is a conventional parallel machine with one 450 MHz Alpha processor per node and a fast 3D–torus communication network. Kepler is built of commodity hardware and has dual SMP Pentium III nodes with 650 MHz processor speed and a switched full–bisection-bandwidth Myrinet network.

1.1 The SPH Method

One main focus in the Sonderforschungsbereich 382 are particle simulations. An important particle simulation method is Smoothed Particle Hydrodynamics (SPH). SPH is a grid-free numerical Lagrangian method for solving the system of hydrodynamic equations for compressible and viscous fluids. It was introduced in 1977 by [3] and [6] and has become a widely used numerical method for astrophysical problems. Due to the mesh-less nature of the method SPH is especially suited for free-boundary problems. Another advantage of SPH is the ability to handle large density gradients well. Rather than being solved on a grid, the equations are solved at the positions of the so-called particles, each of which represents a mass element with a certain volume, density, temperature, etc. while moving with the flow according to the equations of motion.

2 The Libraries

2.1 A Hybrid Parallel Library for Particle Simulations

Why Hybrid?

The dominant part of the TOP500 super computers are so called hybrid parallel architectures, a combination of shared memory nodes with message passing communication between the nodes. Two large hybrid machines at the HLRS in Stuttgart are the Hitachi SR8000 and the NEC SX5.

Hybrid parallelization is the combination of two programming models optimized for hybrid parallel architectures. It uses a thread based programming model for parallelization on the nodes together with message passing parallelization between the nodes.

Obviously, the share of common data structures on one SMP node can reduce the amount of memory used. But more important is the reduction of the communication. Every parallel implementation has a maximum speedup limited by its serial part. In our non trivial parallel codes the dominant serial part is the communication. A hybrid implementation reduces the amount of communication, because some data is shared on the SMP node and thus the communication is implicit. But also the communication itself is faster. A small test shows this for the MPI *Allgather* call with a 200 MB data array on the Hitachi SR8000. K is the number of nodes. The numbers are the time for 50 MPI *Allgather* calls in seconds.

	K=1	K=2	K=4	K=8
MPI intra/inter	15.8	49.2	90.2	116.8
MPI inter	9.3	14.6	19.5	19.7

One can see, that the MPI *Allgather* with pure inter-node communication (comparable to the hybrid programming model) is much faster than the MPI call where MPI is also used for intra-node communication.

OpenMP

To keep the implementation portable and support the majority of hybrid architectures, we chose OpenMP, a wide-spread standard for shared memory parallelization. Compilers for OpenMP are available on all important hybrid plattforms including Hitachi SR8000, NEC SX5 and Linux (Intel C++ or Portland Group compilers). OpenMP has other advantages over explicit thread programming, for example POSIX threads. It annotates sequential code with compiler directives (pragmas). This makes an incremental parallelization possible and keeps the code portable, since non-OpenMP compilers ignore the directives. POSIX threads require to implement parallel sections in separate functions and to write wrappers (sometimes called jackets) for functions with more than one argument, since the POSIX threads API supports only one argument. OpenMP, on the other hand, allows to join and fork threads at arbitrary positions in the source code.

Hybrid Development

The hybrid implementation is based on the ParaSPH library for particle simulations. ParaSPH is written in C and parallelized with MPI. The library separates the parallelization from the physics and numerics code. The interface between the library and the application is optimized for particle simulations. The library provides an iterator concept to step through all particles and their neighbors and later communicates the results of this time step. In parallel mode, the library transparently distributes the work amongst all processors. Every local iterator processes only a subset of all particles. The code performs well on machines with a fast message passing network. We tested the code on Cray T3E, Hitachi SR8000, IBM SP and the Kepler cluster.

In the hybrid implementation, OpenMP is used for the inner node parallelization. MPI is still used for inter-node communication. Tests showed, that it is necessary to optimize the load balancing. Therefore, we introduced a two stage balancing. The "old" load balancer for distributed memory is only used for a coarse load balancing between the nodes. For the fine balancing on the node we provide two new load balancers. The user can choose between a fixed load balancing and a dynamic master-worker load balancing. The master-worker algorithm promises the best load-balancing, especially for inhomogeneous problems but has disadvantages in cache utilization. Therefor the static load balancing is usually faster.

We used three different OpenMP features (omp parallel, omp barrier, omp threadprivate) together with some functions of the OpenMP library. All over, we used 9 *parallel* pragmas, one *barrier* pragma and one *threadprivate* pragma to parallelize about 95 percent of the code. We also needed two additional locks to protect internal structures.

Experiences on Hitachi SR8000 and Kepler

The first OpenMP version of ParaSPH on Hitachi SR8000 frequently called sections with a *critical* region and showed a bad performance. Explicit locks instead of critical regions improved the performance only a little. Enhancing the fixed load balancer with a lock free implementation fixed the problem.

On the Linux platform, we used Intels C++ compiler. To port the code we had to replace the *threadprivate* directives by explicit memory allocation for every thread, because the *threadprivate* directive triggers a bug in the compiler.

Performance results

On both architectures, we measured the speedup (figure 1) of a SPH simulation with 300 000 particles and about 80 interaction partners per particle. This medium sized simulation requires about 900 megabytes of memory on one node. One time step needs about 23 seconds on one Hitachi node (8 CPUs) Typical production runs need 1 000 or even more time steps resulting in several hours of computation time on eight CPUs and about two days on one CPU.

We compared speedups of the pure MPI and the hybrid parallelization for different numbers of processors. On Kepler with only two CPUs per SMP node, we observe almost no difference between the pure MPI and the hybrid parallelization. The Hitachi with 8 CPUs per node shows a big difference between the pure MPI and the Hybrid version especially for large node numbers.

There are three main reasons for the performance impact of the hybrid version. First, the code is not fully annotated with OpenMP instructions. There is a small serial part (about 5%), which is not present in the pure MPI version. This shows a performance decrease especially on the Hitachi SR8000 for large numbers of OpenMP threads. The second reason may be a lesser efficient cache utilization and the third reason is the OpenMP overhead itself (thread creation, locking of critical sections).

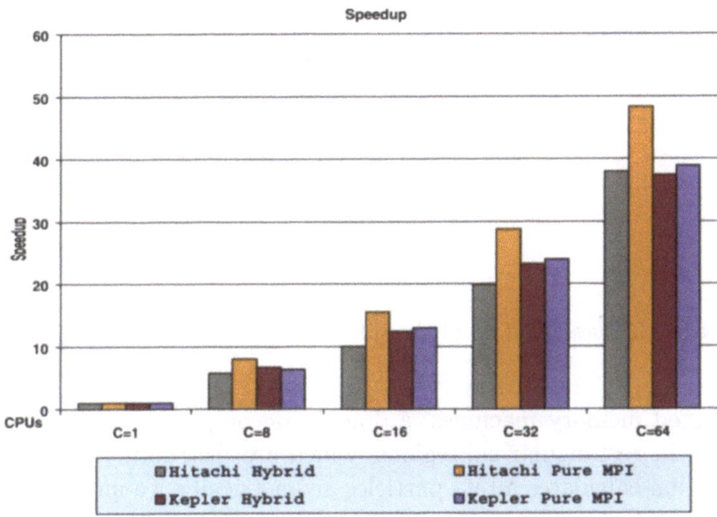

Fig. 1. Speedups using different parallelization strategies on Hitachi SR8000 and Kepler. The hybrid parallelization is significantly slower on the SR8000 compared to the pure MPI version especially for large node numbers.

2.2 Object-Oriented SPH

An Object-Oriented Library for Particle Simulations

The SPH simulation program sph2000 is implemented on top of an object-oriented particle simulation library written in C++. We wanted to prove the feasibility of the object-oriented approach in the performance-critical domain of particle simulations. First results are shown in this section. Our principle subject is the development of an object-oriented library for particle simulations. The object-orientation brings a well structured design of a library with classes modeling the elements of the problem domain. Design patterns [2] help to organize the classes to get a clear and efficient design. One main concern in the design was the strict decoupling of the parallelization and physics. These aspects increase the extendibility, maintainability and reusability of the code.

Another goal was to simplify the configuration of simulations, which mainly means to configure a simulation run after the compilation at runtime by reading a parameter file. To avoid conditional compilation with preprocessor directives, as it is often seen in C libraries, the strategy pattern, which is based on the object-oriented concept of polymorphism is used. With this pattern the program instantiates objects at runtime due to the configuration parameters. One such strategy is the communication shown in figure 2.

An abstract base class defines the interface for the communication classes, concrete subclasses define the communication code for different machine architectures. sph2000 runs on architectures with shared and distributed memory.

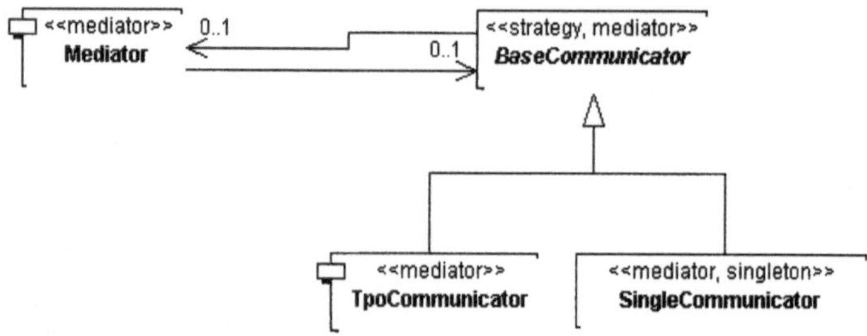

Fig. 2. Class diagram of the sph2000 communication classes.

On distributed memory machines, a domain decomposition splits the simulation space in rectangular subregions, which are distributed one per node. Each subregion calculates all its particles geometrically. To interact with the particles in the subregion of the neighbors, proxy particles are copied and sent between the neighbor subregions. Both communication codes, the SingleCommunicator and the TpoCommunicator classes are compiled but due to a configuration parameter only one will be loaded and instantiated. The same strategy pattern is used for the kernel function, the integrator, the physical quantities and the handling of the subregions of the simulation domain.

The mediator pattern is used twice to decouple parallelization, the numerics and the physics. First, the communicator class mediates between the nodes. It handles the whole communication. If neighboring nodes need proxies, the communicator requests them from the subregion mediator.

Inside a node, a subregion mediator class controls the execution of the calculation classes as shown in figure 3. It is responsible for a correct order of calculation and communication and decouples the calculation classes by keeping the objects from referring to each other explicitly. This concept separates the object interactions and class dependencies from the physical classes. The

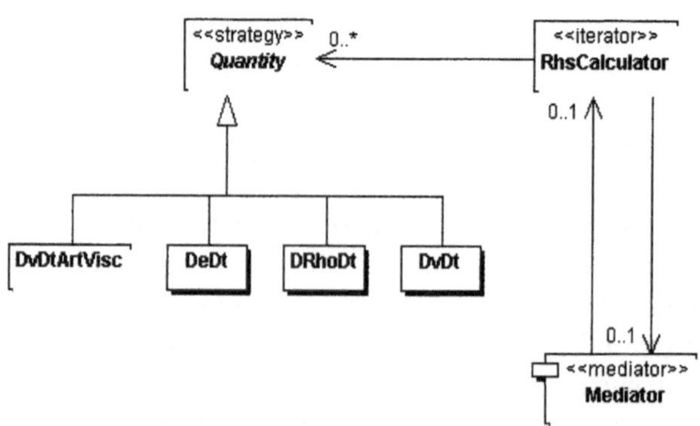

Fig. 3. Simplified class diagram of the sph2000 calculation classes.

result of object-oriented techniques with design patterns is a library, in which classes have clear and strictly separated responsibilities. Different methods can be interchanged without influencing other code. Extensions, for example a new physical effect, are delimited to the corresponding strategy.

Performance Results

We ran simulations with the object-oriented SPH code sph2000. The speedups can be found in figure 4.

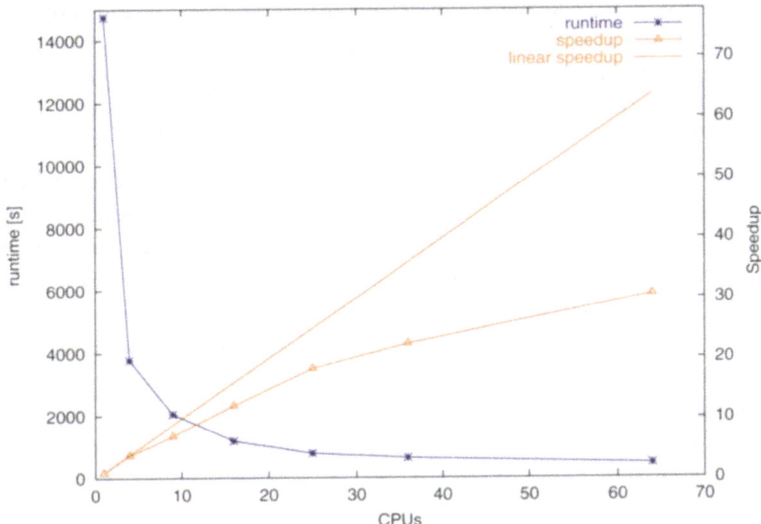

Fig. 4. Time and speedup on Kepler for 100 000 particles with sph2000.

The measuring took place on the Kepler cluster, simulating 100 000 particles on 1, 4, 9, 16, 25, 36 and 64 nodes. We plan to port the C++ program to the Cray T3E to compare the performance of the object-based communication with the Kepler results.

3 Applications

Although its main application still lies in the astrophysical area, the SPH approach is used nowadays also in fields of the material sciences, for modeling multiphase flows (Diesel injection problem, see [9] for details), and the simulation of brittle solids (e.g. [1]). To give an short overview and to show the advantages of the method, we present two examples of calculations, which were done using ParaSPH and one example done using sph2000. The first example is an astrophysical application, the simulation of a protoplanet embedded in a protoplanetary accretion disc, the second is a collision of two rubber rings and the last example is the injection of diesel into an air filles chamber.

Simulations of protoplanetary discs

Since the discovery of the first extrasolar planet in 1995 [7], the interest in planet formation theory has grown intensely. Most of the over one hundred new found planets feature attributes that differ completely from those of the planets of the solar system. These planets have masses in the range of 0.4 to 11 Jupitermasses, and large eccentricities up to 0.67, while the most massive planet in our solar system is Jupiter and the highest eccentricity is 0.25. Nowadays planet formation theories need not only to explain the formation of the solar system, but the formation of planetary systems with these different features.

It is generally believed that planets form in circumstellar discs, which have formed out of a gravitational collapse of a dense molecular cloud. The typical lifetime of such a disc is between 10^6-10^7 years. During this time material migrates constantly radially inwards in the disc and finally accretes onto central protostar. The dust particles in the disc grow through a hit-and-stick mechanism and finally combine into larger bodies, which are eventually able to build planetesimals. Planetary sized objects form out of these planetesimals by collision processes. These objects grow further by accretion of gas and dust, and build the giant planets. A more detailed description of planet formation can be found e.g. in [10].

The high masses of the new found planets give rise to more questions. The tidal interaction between a massive protoplanet and the protoplanetary disc can lead to a so-called gap formation at the orbital region of the planet. This gap formation has severe impact on the accretion rate on the planet, and can probably prevent accretion through this gap completely depending on the mass of the protoplanet. In numerical grid-based simulations it was shown by [5] that the gap formation process does not inhibit accretion in the case of a jovian protoplanet and material can still reach the planet.

By the use of supercomputers, we were now able to study the protoplanet-disc system with SPH. The evolution of the surface density during the first tens orbits of the protoplanet is shown in figure 5 for a Jupiter-sized planet at an orbital radius of 5.2 AU and a central star with one solar mass. Already after ten orbits, the decrease of density at the orbital region of the planet is visible. After forty orbits the gap is completely formed, and the accretion onto protoplanet diminishes. In order to achieve a high enough spatial resolution of the accretion on the protoplanet, more than 360 000 particles have been used for this simulation.

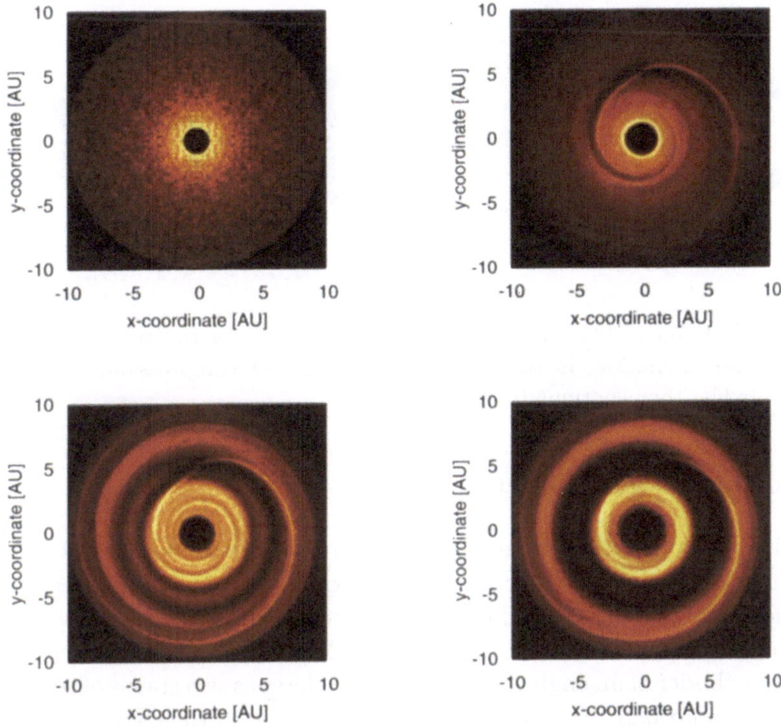

Fig. 5. Color-coded plots of the surface density in the case of a jovian protoplanet at 5.2 AU away from the protostar. The protostar is located in the center of the disc and has one solar mass. The surface density is shown for four different times: at the beginning of the simulation, after one, ten and forty-two orbits of the protoplanet. The circle represents the approximation for the Roche lobe of the planet.

Simulation of colliding rubber rings

Although the SPH method has its roots in the astrophysical area, recent enhancements lead to the application of SPH in solid body physics. By introducing an additional elastic stress tensor, the SPH method can be successfully extended to model elastic solid bodies (see [4] for details).

As an example we present the simulation of the collision of two rubber rings, which is a typical test problem for elastic solid body codes: Two rubber rings with the initial density ϱ_0 collide with 6% of the sound speed c_0. The shear modulus μ, which characterizes the elastic behavior of the material, is set to 0.22 $\varrho_0 c_0^2$ in this problem. The results of the simulation are presented in figure 6. The rings bounce back off each other as expected and continue to oscillate.

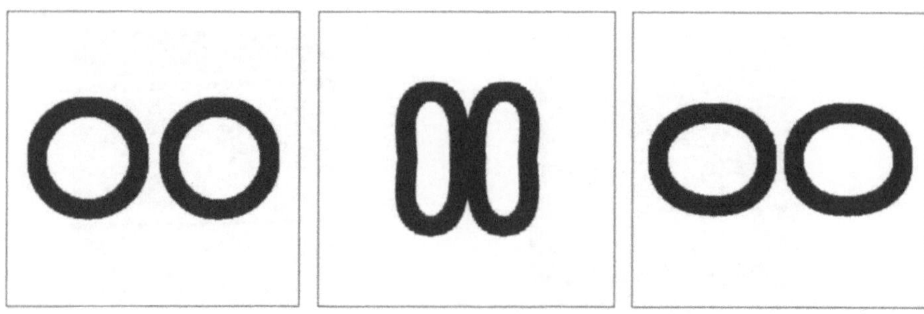

Fig. 6. Collision of two rubber rings. The pictures show the rubber rings at the start of the simulation, in the moment of the highest compression, and after the collision when they continue to oscillate.

Simulation of diesel injection

Diesel engine manufactures are interested in an optimal injection of the diesel into the combustion chamber. A perfect mixing of diesel and air means means an efficient use of the fuel and therefore reduces emissions. To achieve this, the breakup of the diesel jet must be examined and understood. When injected into the cylinder of an engine, the diesel jet undergoes two stages of breakup. In the *primary breakup* large drops and filaments split off the compact jet. These turn into a spray of droplets during the *secondary breakup.* This secondary process is well known and can be modeled as a spray, but the understanding of the primary breakup is only in the initial stages. The physical effects that might influence the primary breakup are the pressure forces in the interface region of diesel and air, instabilities of the jet induced by cavitation inside the injection nozzle, surface tension and turbulences.

4 Future work

4.1 Parallel I/O

The computation of our massive parallel application showed a lack in performance due to sequential I/O. For example, writing the data of 10^6 particles in sph2000 in each integration step limits the maximum speedup of the parallelization to 25. Moreover, visualizations of physical simulations typically require hundreds of gigabyte. As a result, the I/O bottleneck limits the scalability of the parallelization significantly. The constantly growing gap between processor and hard disk performance during the last decades makes this problem even worse and requires the use of parallel I/O systems. Distributed hard disks in special I/O nodes, which use the underlying communication network for data exchange are the most common way in massive parallel systems to

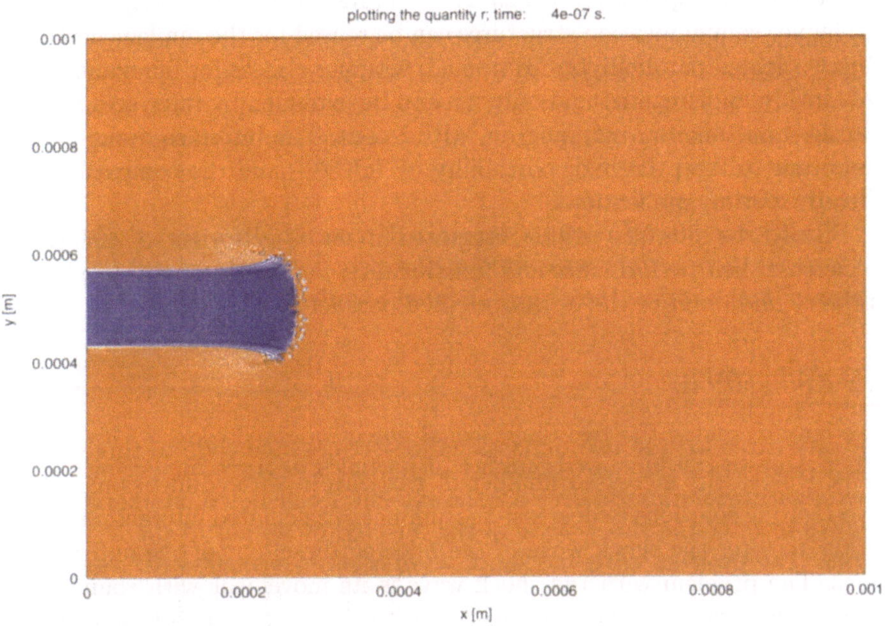

Fig. 7. Injection of diesel in an air filled chamber, simulation with 200.000 particles. The picture shows the injected diesel after $7.2e-07$ s. First drops are already splitted off the jet.

overcome the problem. Thus, the application can profit from the aggregated performance of all hard disks available in the system.

However, special parallel I/O interfaces have to be implemented and used by the application to benefit from the whole hard disk performance. The most often used interface for parallel I/O is MPI-IO [8]. One implementation of this part of the widely accepted and used standard MPI 2 is ROMIO [11], which is setup on top of the file system using an abstract device interface for I/O (ADIO) to achieve high portability (e.g. NFS, PIOFS, HFS, XFS).

In implementing parallel I/O two major problems rise: First, the application has to be ported on the parallel I/O interface of MPI 2 to enable the use of higher hard disk throughput. Second, besides a parallel implementation sequential I/O also has to be implemented for keeping the portability of the interface. Both leads to a more complex and larger implementation. Moreover, MPI 2 only supports procedural interfaces for parallel I/O, which makes it hard to be used in object-oriented applications. Therefore, concepts for object-oriented parallel I/O have to be developed, which is a very difficult task due to irregular distributed object structures in main memory and dynamic memory sizes of the objects.

To close the gap between object-oriented software development and procedural communication we developed an communication library for parallelizing object-oriented applications called TPO++. Its concepts for mapping objects on simpler communication structures can be reused for the implementation of object-oriented parallel I/O. Additional, wrapper classes for file management are already implemented and only have to be extended by interfaces for data transfer and function mapping on MPI 2 calls. The interface design is very important to keep the high portability of TPO++ and the compatibility to already existing applications.

Finally, the interface will be integrated in our object-oriented applications to increase their performance and functionality. Verifications will then help to optimize and improve the object-oriented parallel I/O interface.

4.2 Applications

We have extended the two dimensional object-oriented code to three dimensions. Due to the increased number of particles in three dimensions, we are investigating solutions to decrease the amount of calculated interactions. First, we try to filter the shock wave of the injected diesel out of the outer air regions. The problem with the shock wave is its movement with sound speed towards the system boundaries together with the interference of its reflections with the diesel jet. So, the cylinder size must be chosen big enough to prevent this artificial interference. The second step will be the online generation of air particles according to the motion of the jet. The idea is to calculate only air regions, which are affected by the diesel jet. Another field of work is the development of new physical models of the surface tension and the cavitation. To implement these quantities efficiently, we need new concepts for the parallelization, for example to adjust the neighborhood of particles.

5 Summary

The application of object-oriented development methods has improved the quality of our simulation codes. While having a performance, which is comparable to procedural approaches, an object-oriented implementation is easier to maintain and extend, f.ex. to add new physics.

For massive parallel applications, scalability is always an issue. Besides improving the parallel algorithms, we currently follow two other approaches to reduce the serial parts: A hybrid parallelization promises to reduce the communication cost, because intra-node data can be shared, and parallel I/O can reduce the time spent for persistency. However, the study of ParaSPH on Hitachi SR8000 shows that improving the performance with an hybrid parallelization is not trivial. The main obstacle for the use of parallel I/O is the lack of an object-oriented interface in current standards.

References

1. W. Benz and E. Asphaug. Catastrophic Disruptions Revisited. *Icarus*, 142:5–20, November 1999.
2. Erich Gamma, Richard Helm, Ralph Johnson, and John Vlissides. *Design Patterns: elements of reusable object-oriented software.* Addison-Wesley, 1995. 22nd Printing July 2001.
3. R. A. Gingold and J. J. Monaghan. Smoothed particle hydrodynamics: Theory and application to non-spherical stars. *Monthly Notices of the Royal Astronomical Society*, 181:375–389, 1977.
4. J. P. Gray, J. J. Monaghan, and R. P. Swift. SPH elastic dynamics. *Computer Methods in Applied Mechanics and Engineering*, 190(49-50):6641–6662, 2001.
5. W. Kley. Mass flow and accretion through gaps in accretion discs. *Monthly Notices of the Royal Astronomical Society*, 303(4):696–710, March 1999.
6. L. B. Lucy. A numerical approach to the testing of the fission hypothesis. *The Astronomical Journal*, 82(12):1013–1024, 1977.
7. M. Mayor and D. Queloz. A Jupiter-mass companion to a solar-type star. *Nature*, 378:355–359, 1995.
8. MPI-IO Committee. *A Parallel File I/O Interface for MPI.* Online. URL: http://lovelace.nas.nasa.gov/MPI-IO, 1996.
9. F. Ott and E. Schnetter. A modified SPH approach for fluids with large density differences. *ArXiv Physics e-prints*, pages 3112–+, March 2003.
10. M. A. C. Perryman. Extra-solar planets. *Reports on Progress in Physics*, 63(8):1209–1272, 2000.
11. R. Thakur, E. Lusk, and W. Gropp. Users Guide for ROMIO: A High-Performance, Portable MPI-IO Implementation. In *Technical Memorandum ANL/MCS-TM-234*, Mathematics and Computer Science Division, Argonne National Laboratory, 1998.

SKaMPI – Including More Complex Communication Patterns

Michael Haller and Thomas Worsch

LIIN, Universität Karlsruhe, Germany, ⟨*lastname*⟩@ira.uka.de

Summary. SKaMPI is now an established benchmark for MPI implementations. In autumn 2002 the development of the "new SKaMPI" has started in three major directions: (i) extension of the benchmark to cover more functions of MPI and a redesign of the benchmark allowing it to be extended more easily (thus matching requests from SKaMPI users); (ii) construction of a collection of important algorithm kernels which are not supported by core MPI collective operations; (iii) development of a simulator which (at least partially) supports the simulation of one MPI implementation while using another MPI implementation (possibly running on a different kind of machine).

In the present paper we give an overview of the new SKaMPI and describe fast implementations of `MPI_Alltoall` and of an algorithm for transposing multi-dimensional matrices as first instances of SKaMPI-Alg.

1 Introduction

"The MPI standard defines a set of powerful collective operations useful for coordination and communication between many processes. Knowing the quality of the implementations of collective operations is of great interest for application programmers. In particular, one has to decide, whether to use predefined collective operations, which usually lead to more readable programs, or to implement collective operations by using point-to-point primitives. Similarly, it is often unclear, whether to use complex collective operations, like `MPI_Reduce_scatter`, or to use more primitive collective operations (like in this case `MPI_Reduce` and `MPI_Scatterv`)."

The above text [7] describes one of the major motivations for the development of SKaMPI, the Special Karlsruher MPI-Benchmark [5].

SKaMPI (`http://liinwww.ira.uka.de/~skampi/`) measures the performance of an MPI implementation on a specific underlying hardware. By providing not simply one number, but detailed data about the performance of each MPI operation, a software developer can judge the consequences of design decisions regarding the performance of the system to be built.

The text also indicates, and benchmark results from a wide variety of machines indeed show, that more often than it should be there is for example a collective operation which is implemented in a suboptimal way, asking for self-made replacements. The development of new implementations for collective operations covered by MPI and for more complex collective operations not covered by MPI is thus a natural next step.

This becomes more difficult as the target hardware becomes less homogeneous. In many modern parallel machines one already has at least two layers of communication: a network of nodes consisting of shared memory coupled processors. Grid computing adds at least a second (significantly slower) network connecting parallel machines. Testing algorithms for the grid is still very tedious work because real "grid hardware" (let alone software) is rare and not easy to get access to. Thus a simulator which allows to run tests without the real hardware is desirable.

These are the motivations for the three major directions of work for the new SKaMPI.

The rest of this paper is organized as follows. In Section 2 we give a short overview over the components for the new SKaMPI.

The main contribution of this paper (see [2]) is a fast algorithm for transposing 3-dimensional matrices using a fast implementation of `MPI_Alltoall`. These are first examples of fast algorithms which will become part of SKaMPI-Alg. In Section 3 we describe an implementation of a fast algorithm for transposing matrices. It can make use of an improved `MPI_Alltoall` exploiting shared memory for intra-node communication in parallel machines with SMP nodes as a clearly distinguished layer of the communication hierarchy (like e.g. IBM pSeries 690 or NEC SX series); this is the topic of Section 4. In Section 5 we show performance numbers for these implementations.

We conclude this paper in Section 6.

2 Overview of the New SKaMPI

2.1 SKaMPI-Bench

SKaMPI is now an established benchmark for MPI implementations. The goals for its new incarnation are threefold.

First, work has started to include more MPI functions in it. In particular, until now virtual topologies have been neglected completely. This is due to the fact that "benchmarking" an implementation of virtual topologies does not fit very well into the standard framework of looping over a set of parameter values.

Second, for all non-trivial MPI functions the current implementation of SKaMPI does one loop varying one parameter (e.g. message size *or* number of processes) while keeping all others fixed. There are cases where it would be more convenient to have an easy way to get measurement data for e.g. all pairs of parameter values (e.g. message size *and* number of processes). This requires changes throughout SKaMPI, starting from the format for the configuration file and not ending at the heuristics used to detect anomalous "jumps" in the run of measurement curves and refining its measurements at those critical points.

Third, SKaMPI should be extendible. This is already a must if one wants to include the measurement of self-written routines (e.g. intended to replace the native implementation of a collective operation). But there are even users who want to use much of SKaMPI for benchmarking OpenMP implementations.

Of course all of these improvements should not sacrifice any of the already existing features of SKaMPI, nor should its stability and portability suffer from them.

2.2 SKaMPI-Alg

Sometimes there is a need for efficient implementations of e.g. collective operations. It may be that SKaMPI-Bench revealed that the native implementation of the corresponding MPI function on a machine is suboptimal. Or it may happen that a more complex collective operations not covered by MPI is needed.

SKaMPI-Alg will contain algorithms for both types of problems. First examples are described below. The fast implementation of `MPI_Alltoall` described in detail in Sections 4 exploits shared memory on SMP nodes coupled by a communication network.

The fast algorithm for transposing a three-dimensional matrix (see Section 3) uses MPI data types in a clever way to avoid the time consuming copying of memory areas.

Other useful tools which e.g. arose as byproduct of the development of SKaMPI-Bench will also be made available separately. For example the new method for benchmarking collective operations in MPI [7] introduced in the current version of SKaMPI-Bench uses a method to approximate "Global MPI Time". This tool will not only be useful for SKaMPI-Sim but possibly for other users, too.

2.3 SKaMPI-Sim

The simulation of one MPI implementation for one hardware by using another MPI implementation on another hardware in order to obtain a reasonable prediction of the timing behavior of the former is impossible to do in full generality.

But, first experiments with a prototype currently under development at our research group indicate that for example for the reasonably small but still useful subset of collective MPI operations one can hope for a tool that does its job sufficiently well. If successful, SKaMPI-Sim would not only allow to develop collective algorithms for hardware which is (almost) not accessible (like several parallel machines in a grid). One could for example also try to estimate the benefits of replacing the existing communication network of a machine by a faster one.

3 A Fast Algorithm for Transposing Matrices

Many scientific and technical computations handle data organized in form of matrices. When performing climate and weather simulations for example, data are arranged in large distributed, three-dimensional matrices, according to the three dimensions of space. Some of these applications strongly depend on the ability to efficiently transpose such matrices.

From a theoretical point of view, there is nothing spectacular in computing the transpose of a distributed matrix. In practice, the situation is more difficult because this operation turns out to be communication intensive.

3.1 Basics

Throughout this paper, we refer to the three-dimensional matrix partitioning shown in Figure 1. In direction of the z axis, the matrix remains unpartitioned. In the xz plain and the yz plain we have striped partitioning whereas in the xy plain the matrix is checkerboard partitioned.

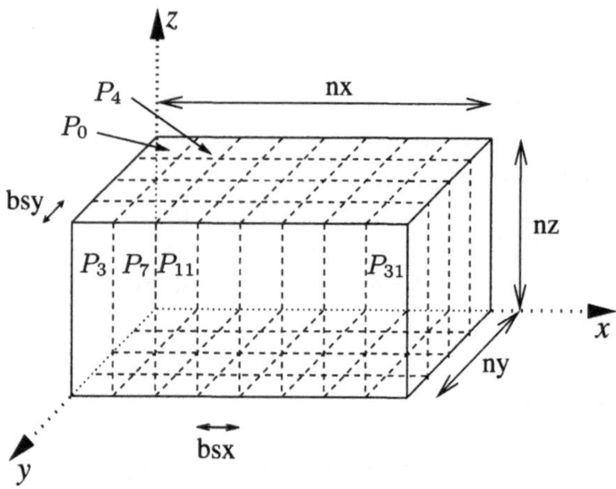

Fig. 1. 3D matrix partitioning

The following definition is an extension of the two-dimensional case. As an example, we define the transpose in the xz plain. Analogously the transposes in the xy plain and the yz plain can be defined.

Definition 1. *Let* $A = (a_{i,j,k}) \in \mathbb{R}^{nx \times ny \times nz}$, $nx, ny, nz \in \mathbb{N}$ *be a matrix. Then we define* $A^T_{xz} = (a_{i,j,k})^T_{xz} := (a_{k,j,i}) \in \mathbb{R}^{nz \times ny \times nx}$ *to be the* transpose *of A in the xz plain.*

3.2 The Naive Transposing Algorithm

Figure 2 shows a straightforward algorithm to compute the transpose of a (two-dimensional) striped partitioned matrix. We have embedded a three-dimensional coordinate system to make clear that this is also our situation with the three-dimensional partitioning (Figure 1) when projected to the xz plane.

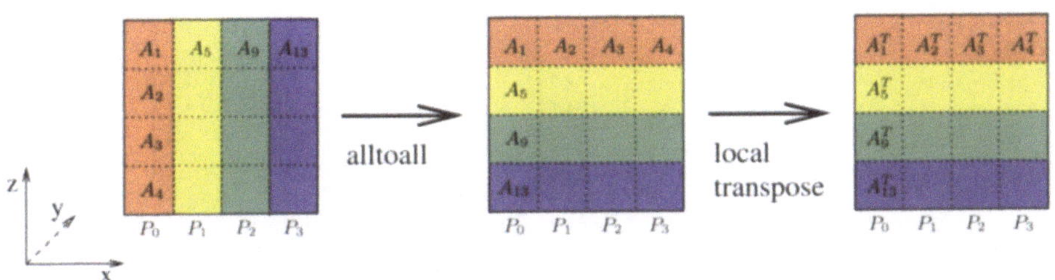

Fig. 2. 2D distributed matrix transposition

For a single processor its sub-matrix has the shape `(1:nz, 1:bsy, 1:bsx)` in FORTRAN 90 notation. The transpose in the xz plain will have the shape `(1:nx, 1:bsy, 1:bsz)`.

Thus an algorithm for the computation of A^T_{xz} can be outlined as follows:

1. **Do in parallel** `ny/bsy` times along the y axis
2. **(Preparation)** Build coherent blocks of `bsx*bsy*bsz` elements for the `Alltoall` operation (i. e. locally reorder elements).
3. **(Global transposition)** Call `MPI_Alltoall` using the primitive data type of the matrix entries (i. e. exchange the blocks among the processes in the xz plane).
4. **(Local transposition)** Locally transpose (rearrange) the blocks.
5. **End do**

3.3 The Data Type Oriented Variant

Steps 2 and 4 in the algorithm lead to temporary copies of the sub-matrices. We can avoid these if we model the data distribution in the sub-matrices by means of customized MPI data types.

In detail, two data types are constructed corresponding to the two data type place-holders in the prototype of `MPI_Alltoall`: a send type (to make step 2 superfluous) and a receive type (as a replacement for step 4).

In the global transpose, we have to exchange blocks of `bsz*bsy*bsx` elements each. We now describe the construction of the send data type.

Figure 3 illustrates the construction. In a first step, we create a single contiguous "column" of elements by means of the `MPI_Type_contiguous` constructor (left part of Figure 3). Then, we build a two dimensional array of such columns using the `MPI_Type_create_hvector` API function (middle part). Finally, we assemble these "planes" to form a three-dimensional data structure. This is achieved by a call to `MPI_Type_create_hvector`. We assume the matrix elements have the data type `double`.

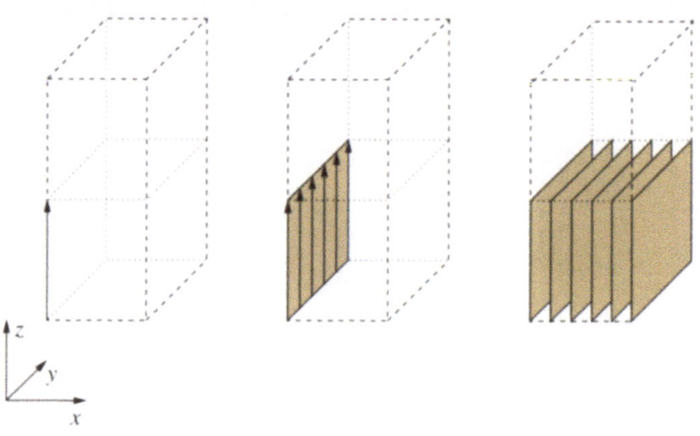

Fig. 3. The three steps of constructing the send data type

```
MPI_Type_contiguous( bsz, MPI_DOUBLE, &column_type );
MPI_Type_hvector( bsy, 1, nz * sizeof(double),
                  column_type, &plane_type );
MPI_Type_hvector( bsx, 1, nz*bsy * sizeof(double),
                  plane_type, &send_type );
```

The receive type can be constructed in an analogous way. The only thing to keep in mind is the different destination arrangement of the elements (notice the changes in the third parameter of the `MPI_Type_hvector()` calls designating the data offsets):

```
MPI_Type_hvector( bsz, 1, nz*bsy * sizeof(double),
                  MPI_DOUBLE, &type1 );
MPI_Type_hvector( bsy, 1, nx * sizeof(double),
                  type1, &type2 );
MPI_Type_hvector( bsx, 1, sizeof(double),
                  type2, &recv_type );
```

In section 5, we will compare the running times of the algorithm variants in a concrete example.

4 A fast `MPI_Alltoall`

There exist several improvements to certain MPI implementations which try to be aware of hierarchies (e. g. SMP clusters or distributed computing over WAN) in communication performance. If based on vendor specific MPI versions, then detailed information on such software often is not disclosed to the public. However, some other efforts are built on non-profit MPI variants such as MPICH [1], and information on them has been published.

For example, in [3] a modification to MPICH for a two-level communication hierarchy is presented and results for matrix multiplication and the `MPI_Bcast` primitive are shown.

Huse [4] thoroughly discusses SMP aware optimizations for point-to-point and collective communication routines on clusters with the SCI interconnect, whereas Träff in [6] focuses on algorithms for all-to-all communication on Giganet based SMP clusters.

4.1 TurboMPI's Solution

TurboMPI is an optimized implementation of MPI collective communication operations. It is available only within IBM for testing purposes. For this reason, there are no publications on TurboMPI that could be cited. But since our improved implementation for `MPI_Alltoall` was inspired by the design of TurboMPI we present a schematic architectural view in Figure 4.

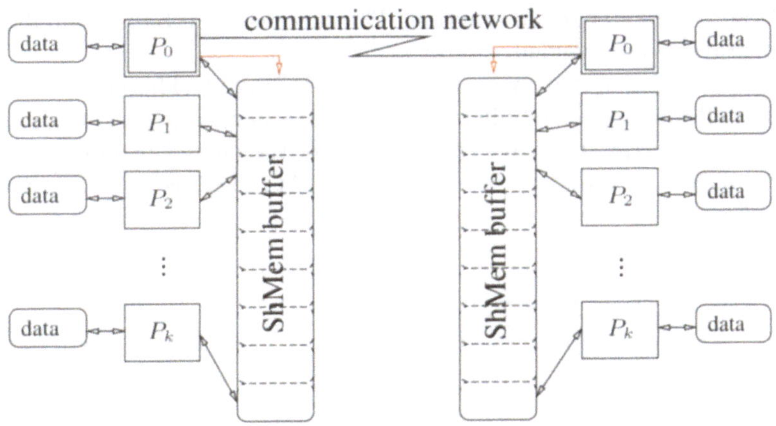

Fig. 4. TurboMPI's architecture

In TurboMPI the processes located on one physical node are combined in one or more groups which we call *virtual nodes*. One process in each virtual node is called the *master process* (here: P_0). This designated process handles the communication between the virtual nodes in order to avoid contention at the network devices.

Instead of sending all the messages from each source process to each destination, the communication between the virtual nodes is processed in a bundled fashion. To this end, for each virtual node an additional shared memory buffer is allocated. All processes of the same virtual node have their own data slot in this buffer. They attach it to their address space and copy their private data to and from this slot. Between the virtual nodes, the master processes exchange the contents of the whole shared memory buffers at a time, i. e. only few large messages are exchanged between the virtual nodes. This communication is split up into `MPI_Isend` and `MPI_Irecv` calls that are issued in a loop (a *flat factor algorithm* as defined in [6]). All in all, TurboMPI optimizes in favour of bandwidth.

The size of the virtual nodes can be adjusted. Thus it is possible to have several master processes on one physical node in order to use the network devices to full capacity.

4.2 Using Local Shared Memory and Data Types

Figure 5 depicts the architecture of our improved `MPI_Alltoall` routine (which we will henceforth call `FastAlltoall`). The main changes affect the processes' data buffers. In `FastAlltoall`, there exists no additional per-node shared memory buffer. Instead, the per-process message buffers already have the shared memory property. The master processes still exchange bundled messages between the virtual nodes, but these are now constructed on the fly using MPI data type constructors, in particular the `MPI_Type_create_hindexed` function. The effect is, as in Subsection 3.3, that copy operations are avoided.

There is one major drawback with `FastAlltoall`: The user must make sure that all MPI message buffers are shared memory segments. Additionally, the master processes must be able to identify the shared memory buffers of their "children". Therefore, the usage of `FastAlltoall` is not transparent to the application (as it was in TurboMPI).

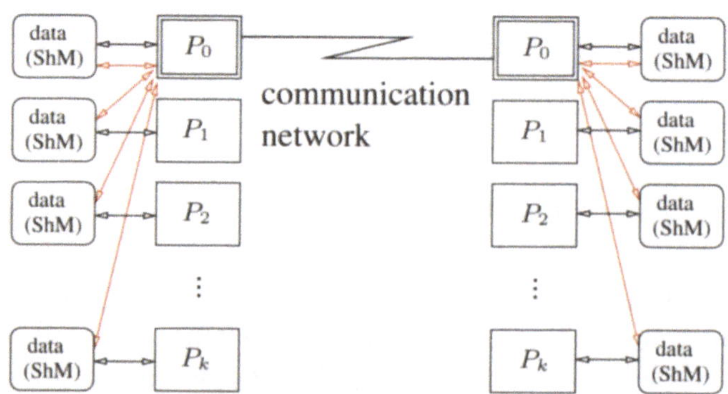

Fig. 5. An improved usage of shared memory for `MPI_alltoall`

5 Results

This section presents some performance numbers we were able to measure on the 32-way SMP system IBM pSeries 690 (1.3 GHz clock frequency). The system was configured as four 8-way logical partitions and behaved like four 8-way SMP nodes connected by two IBM SP2 switch adapters.

5.1 How Fast Is FastAlltoall?

Figure 6 shows the relative throughput of **FastAlltoall** compared to the standard implementation. The communication throughput of the standard call was scaled to 1 in order to compensate the enormous range (several orders of magnitude) of the communication times.

The relative throughput is computed by the formula $T(x) = \frac{t_{\mathtt{std}}(x)}{t_{\mathtt{fast}}(x)}$, where $t_{\mathtt{std}}(x)$ and $t_{\mathtt{fast}}(x)$ designate the execution times of the standard **Alltoall** routine resp. **FastAlltoall** depending on the message size x.

We observe that if each virtual node comprises all 8 CPUs on one physical SMP node (fast all2all 1 in the diagram), for most message sizes **FastAlltoall** is not fast at all. When the message size is small, the administrative overhead is dominant. For large messages the fact that only one process per physical node handles communication is the bottleneck.

When using 2 or 4 virtual nodes per physical node (fast all2all 2 resp. 4 in the diagram), the administrative overhead gets even larger (and the throughput worse). But after a break-even point between 2 and 4 kB, **FastAlltoall**

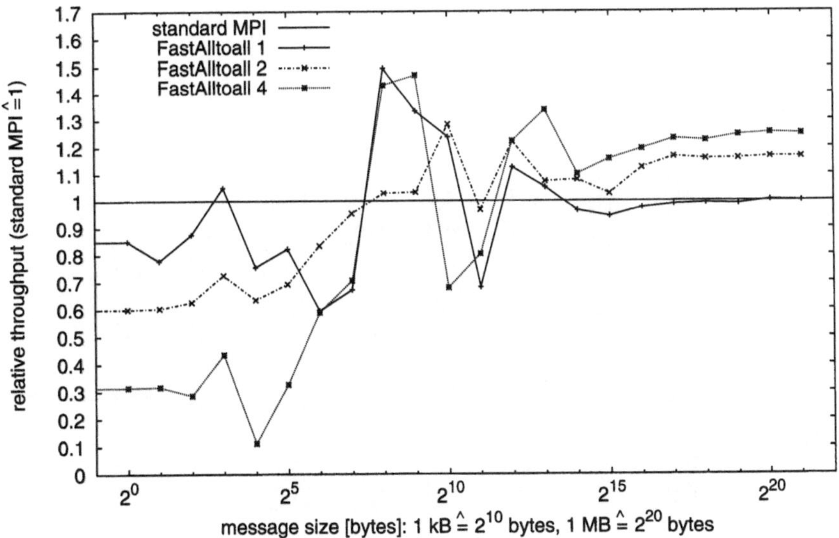

Fig. 6. Speed comparison to standard MPI

performs better than the standard implementation; it outperforms standard `Alltoall` by 20 percent or more.

5.2 Overall Performance

Figure 7 shows absolute running times of different configurations of the matrix transpose for the example

$$nx = 768, \quad ny = 768, \quad nz = 256, \quad bsx = 96, \quad bsy = 192.$$

The matrix entries had the type `REAL*8/double` and were chosen at random.

The "naive" label refers to the naive transpose algorithm that uses the system (standard) implementation of MPI. "DT" marks the timing for the improvement due to the usage of data types (still using the standard implementation).

The "F" variants additionally use the `FastAlltoall` implementation. Finally, the numbers (1,2,4) designate the number of master processes per physical SMP node.

The Figure makes clear that just the usage of customized data types (DT) improves the running time by about 20%. `FastAlltoall` alone (Fnaive x) accelerates the running time of the basic algorithm by nearly 30%. If we combine `FastAlltoall` and the datatype-optimized algorithm using 2 master processes per node (FDT 2), we achieve an improvement of around 38% compared to the basic algorithm using standard MPI.

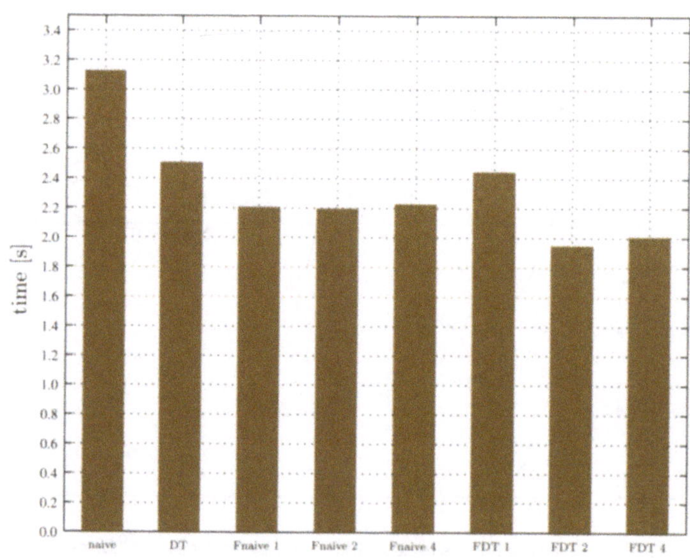

Fig. 7. Timing results

It should be noted that the improvement obtained by using MPI data types depends on the MPI implementation and the hardware used (in the case above IBM's native MPI implementation on a p690). For example on a NEC SX-5 the "naive" algorithm for transposing matrices is faster than the one using data types by a factor of approximately 2.

6 Conclusion

6.1 Fast `MPI_Alltoall` and Matrix Transposing

The preceding pages have shown that customized MPI data types represent a powerful instrument for the programmer to avoid copy operations. Together with the shared memory concept they also allow to optimize for better bandwidth utilization.

With these two concepts, a performance improvement for a particular application of up to 38 percent can be accomplished although `FastAlltoall` is built on top of the "high level" primitives `MPI_(I)Send` and `MPI_(I)Recv` and does not exploit machine specific hardware details in a non-portable way.

6.2 Use of Parallel Machines at the HLRS

The use of the machines at the HLR Stuttgart is an inevitable and invaluable help for the development of portable software. The widespread use of SKaMPI-Bench all over the world can be attributed to the fact that it compiles and runs everywhere where MPI is available.

One of the ideas used was and is that SKaMPI-Bench is distributed as one C source file which only has to be compiled. This is not possible for the new parts of SKaMPI. Therefore it will be more difficult to implement them in such a way that they can be used easily on a wide variety of systems. Early experiences with SKaMPI-Sim show that this makes it even more important to be able to do test on many different parallel platforms. Furthermore it will help to explore the limitations of SKaMPI-Sim.

6.3 Outlook

The extensions of SKaMPI-Bench for two-dimensional measurements and virtual topologies are under development. The redesign for extendibility is in its first phase.

SKaMPI-Alg has been started. First algorithms as described above are there. More will follow.

A first prototype for SKaMPI-Sim is currently under development and will enter an extensive testing phase in the second quarter of 2003.

All parts will be made available under the GPL and we are accepting contributions wherever we see them fit.

References

1. William Gropp, Ewing Lusk, Nathan Doss, and Anthony Skjellum. High-performance, portable implementation of the MPI Message Passing Interface Standard. *Parallel Computing*, 22(6):789–828, 1996.
2. Michael Haller. Parallele Transposition dreidimensionaler Felder im High Performance Computing. Diploma thesis, Fakultät für Informatik, Universität Karlsruhe, 2003.
3. Parry J. Husbands and James C. Hoe. MPI-StarT: Delivering network performance to numerical applications. In ACM, editor, *SC'98: High Performance Networking and Computing: Proceedings of the 1998 ACM/IEEE SC98 Conference*. ACM Press and IEEE Computer Society Press, 1998.
4. Lars Paul Huse. MPI optimization for SMP based clusters interconnected with SCI. In Jack Dongarra, Peter Kacsuk, and N. Podhorszki, editors, *Proceedings EuroPVM/MPI*, volume 1908 of *Lecture Notes in Computer Science*, pages 56–63. Springer-Verlag, 2000.
5. Ralf H. Reussner, Peter Sanders, and Jesper Larsson Träff. SKaMPI: a comprehensive benchmark for public benchmarking of MPI. *Scientific Programming*, 10(1):55–65, 2002.
6. Jesper Larsson Träff. Improved MPI all-to-all communication on a Giganet SMP cluster. In Dieter Kranzlmüller, Peter Kacsuk, Jack Dongarra, and Jens Volkert, editors, *Proceedings EuroPVM/MPI*, volume 2474 of *Lecture Notes in Computer Science*, pages 392–400. Spinger-Verlag, 2002.
7. Thomas Worsch, Ralf H. Reussner, and Werner Augustin. Benchmarking collective operations with SKaMPI. In Egon Krause and Willi Jäger, editors, *High Performance Computing in Science and Engineering '02*, pages 491–502. Springer-Verlag, 2002.

Performance Analysis Using the PARbench Benchmark System

Andreas Kowarz

Technische Universität Dresden – Zentrum für Hochleistungsrechnen
Mommsenstr. 13 , 01062 Dresden , kowarz@zhr.tu-dresden.de

1 Introduction

This project is not meant to find a solution to any known problem in the field of science or technique. It's rather a question of the evaluation of some high performance computer systems used for this purpose. These systems located at the High Performance Computing Center Stuttgart were analyzed using the PARbench benchmark system. Special attention was dedicated to scheduling system within the scope of this analysis. Conclusions on the optimization of the examined computer systems become possible that way.

2 PARbench

To make correct statements about the system behavior of a certain examined computer possible, it is necessary to get PARbench work in a dedicated environment. That means it has to place the only user processes besides the necessary operation system tasks. PARbench is able to simulate the working condition typically occurring in computing centers or similar facilities on its own. It executes one or several work loads under strict logging of all data concerning the time flow. These certain work loads are generated by PARbench with special parameters in mind whereby the parameters can be set nearly free by the user. Each work load is meant to represent a real user job. Conclusions on the behavior of special system components become possible by analysis of the differences between the captured data of different runs.

In contrast to many other benchmarks PARbench uses a system of synthetic kernels. In general, a part of a program is understood as a kernel if its function has the highest meaning for the overall design. For instance, the kernel of an operation system is, without meeting any demands regarding to user comfort, those routines all other programs are built of, thus ensuring the usability of a computer system. In the context of benchmark systems, loops of a program consuming the main part of the user time can be treated as such kernels in particular. Typically those kernels are burden system components under observation in a special way. It doesn't matter which components are burdened by a special kernel. The overall constructed set decides the general usefulness.

The synthetic benchmark kernels PARbench uses to model its work loads are featured by two main properties. On the one hand, they are short enough in time to give PARbench the possibility to combine them in a given time frame usually made of less seconds. On the other hand is the set of all kernels able to represent the overall parameter space given by the examined system. The term synthetic is derived from an other property of the kernel set. In most cases they don't lay claim to the solution of any known problem of science or technique.

As of the current version the benchmark system, PARbench owns seventeen main kernels. One of these kernels is specialized in execution of input/output operations exclusively. The remaining sixteen burden the memory system and the floating point units primarily. By changing a certain parameter, the amount of input/output operations of all kernels can be varied. By modifying the size of the underlying matrices, one can change the problem size in addition. Consequently, 340 different kernel versions are available in all.

Due to the design of PARbench for main usage on high performance computer systems the sixteen kernels not exclusively using input/output operations are available as parallelized versions too. Parallelization was made by use of OpenMP, which is the easiest way of using several processors at present due to its wide acceptance[1] and good usability, at least when we are talking about parallelizing a program for a few CPU's in a short time frame.

3 The NEC Azusa

The first examined computer system within the scope of this work is the Intel Itanium Server Azusa constructed by the company NEC. This server is intended for operation at the back bone of the Internet and for enterprise computing. It is built on the Itanium processor architecture developed by Intel. The system at the High Performance Computing Center Stuttgart owns 16 CPU's and 32 MB main memory which are evenly distributed among four nodes. A RedHat Linux derivative is used as an operating system, mainly adjusted regarding to the scheduling algorithm and the memory management.

3.1 Management of serial work loads

The possibility to use several processors doesn't lead to the use of parallel programs inevitably. It's a great convenience to process different requests out

[1] OpenMP an MPI (Message Passing Interface) are the two most important ways of parallelizing a given program at present

of the net on different CPU's just for the usability of servers. There's no doubt about the crosstalk of such processes. It's only a question about the kind of influence and the order of magnitude.

Memory conflicts

A special job was generated as basis for this experiment leading to a high memory access rate under extensive bypassing the cache structure. The overall work load was executed after the special job was reproduced to comply with the number of processors used for the actual run. Table 1 shows the summary of all run time data.

Table 1. Average run time depending on the number of used CPU's at the NEC Azusa

Number of concurrent jobs	average CPU time in seconds
1	54.94
2	54.95
4	54.95
8	56.26
12	67.42
16	103.89

As one can clearly see, the greater the number of concurrently used CPU's, the higher the necessary computing time initiating at the number of eight CPU's is. The main reason is the high rate of memory accesses not covered by the caching system. These accesses are addressing the same memory hardware, in many cases leading to time consuming conflicts. Thus, the memory rate decreases and the user time rises.

The fact of the nearly constant CPU times until four concurrent jobs take place may be surprising upon first view. This behavior is a result of the Node Affine Scheduler[2] as well as the hardware structure of the NEC Azusa. The scheduler was developed to meet the requirement of a nearly optimal start distribution of all tasks, among other things. That leads to one processed job per node for this special case. As a result of the sufficient main memory per node, there is no necessity to access the remote memory of any other node, and memory conflicts will consequently be avoided.

The average run time of a job is increasing with the number of additional jobs on the same node. What complexity this increase has depends on the memory access behavior and the cache miss rate in particular. Thus, the run

[2] For further details on the Node Affine Scheduler by Erich Focht see: http://home.arcor.de/efocht/sched/

time deterioration of about 2 measured for this experiment doesn't have a universal meaning.

Scheduling with overload conditions

While the scheduling is of lower meaning in situations where the system is less or at least not fully loaded, the rating changes with arising overloads. If there are more jobs than processors, the scheduling system decides if and when a job gets the desired resources. That way it is responsible for the assigned waiting times, in other words, for the degree of fairness each single program is applied to.

Figure 1 shows the result after the execution of sixteen serial jobs on sixteen processors. The overall waiting time is with a value of 7.43 seconds extremely small. The main reason is collisions between operating system on user jobs. The fact of increasing user times for some jobs speaks for it, too, due to the change of cache contents by the operating system.

In addition, Figure 2 shows a situation with fifty percent more jobs than processors available. The system behavior in such overload situations depends on the scheduling algorithm, primarily. The Node Affine Scheduler developed for the NEC Azusa in Stuttgart tries really hard to hold and finish a task on the processor where it was created. This feature makes sure that the difference between local and distant memory access, which corresponds approximately to the factor 1.6, will only be accepted if the gain that emerges from another node justifies it. Figure 2 shows this pattern, which is interesting from a fairness angle. The short waiting time and the considerably unchanging CPU time justifies, however, this manipulation.

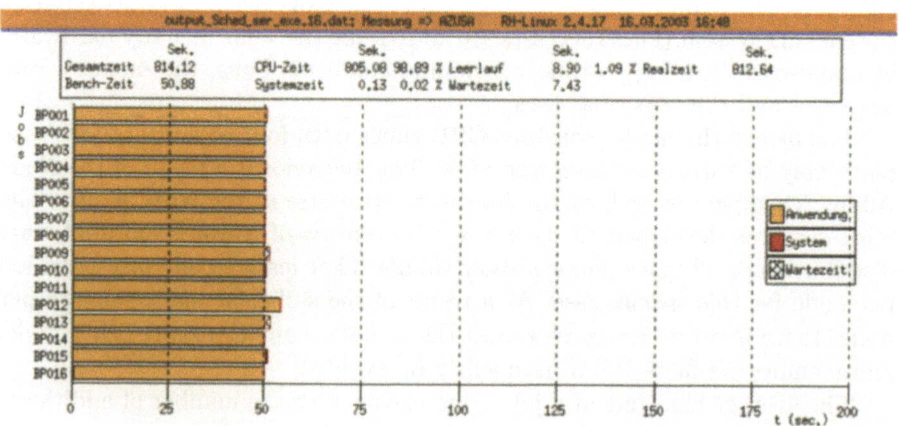

Fig. 1. Azusa under 100 percent serial load

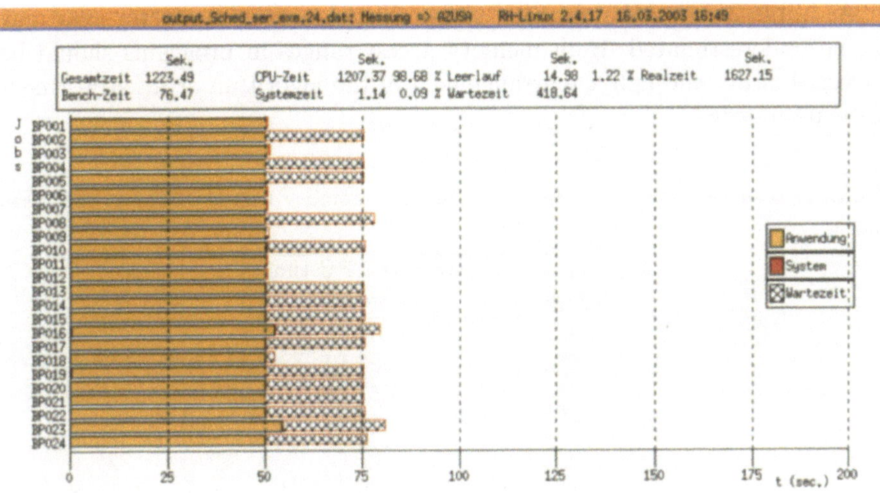

Fig. 2. Azusa under 150 percent serial load

Unfavorable Memory Access Patterns

Not only does the PARbench offer an investigation of the handling of several programs, but also the possibility to investigate single workloads. The optimisation of such a program is oftentimes the preliminary step before a parallelisation. In the following, the influence of the memory access patterns on the required CPU time will be made clearer through consistent trials.

Every computing system has its own, very specific weak points. The most meaningful of the Von-Neumann-Architecture was the bottleneck between the main memory and the processor. In order to relieve this problem, caches were installed. These fast temporary memories, at present available in controlled hierarchies, use the Locality Principle in today's applications. This means that there is a likelihood that a data element that is used once is used repeatedly in frequent intervals, respective to its neighbouring elements that are soon necessary. Due to this reason, such data elements, summarized as so-called cache lines, are loaded out of the main memory in the cache and wait there to be used.

The position of a line in the cache is found in the typical set-associative cache of today as part of the main memory address. The set offers placement for colliding cache lines in the different cache parts. The number of cache parts that are in the set varies according to the producer but can generally be found in the one-digit area.

The problem with cache displacement stems from the fact that after loading a ready position in the cache, the old value is no longer available. If one accesses a program in an increment of a power of two to the data, then the probability is very high that the cache behaviour will unfavourably drop out.

As Table 2 shows the memory access range is a runtime factor that may not be underestimated. High memory access ranges in programs should be corrected either through a rearrangement of the algorithm or an adjustment of the data order.

Table 2. Run time data of jobs with different memory access range at the NEC Azusa

memory access range	average CPU time in seconds
1	50.07
2	53.10
4	57.52
8	62.26
16	67.79
32	75.73
64	90.98
128	123.89
256	187.35

3.2 Mixed Workloads

Dependent on the type of application of the Azusa server, parallel jobs are being executed too. Especially in situations in which more processors are demanded than are available, the scheduling plays an essential role. Due to the fact that no computing system can afford more than 100 percent, it is doubtful that the parallel processing under overload conditions would produce noticeable differences.

In order to better view such a circumstance, the trial of Fig. 1 under variable conditions was once again carried out. This time eight of the serial jobs were carried out parallel with four threads (see Fig. 3).

Although the trials remained constant, there were some changes made in the execution. The most important difference is the increase in waiting time to 373.33 seconds, from the original value of 7.43 seconds. In order to help explain these findings, it is important to view two considerable problems.

For one, the scheduling algorithm requires CPU time for itself, which burdens the user programs with an additional waiting time. Another aspect to consider is that the current scheduling algorithm of the Azusa does not treat the threads in common which belong together by the parallel program. This results in situations which necessitate a synchronisation of threads that are not all inevitably active. This synchronisation requires at least two time slices instead of the one that is actually necessary.

As the figures make clear, it is more favourable in such a situation and as a practical strategy to avoid a parallel execution. The expected decrease of the dwell time of a few of the jobs in the system cannot be achieved.

4 NEC SX-4 and SX-5

The Vector computer systems SX-4 and SX-5 of the NEC company belong to the same development line, due to their affiliation. As one can guess from the name, the SX-5 is the successor of the SX-4. The test system that was used has 32 (SX-4) and 16 (SX-5) processors, respectively.

4.1 Investigations with Serial Workloads

A few important statements about the characteristics and behaviour of the class of processor that are being investigated are viewed most easily through an example of absolute serial workloads. The trials that were carried out on the NEC Azusa were fit in their parameters to the SX-5 and were carried out on both test systems afterwards. Due to the fact that both processors must bring the same work capacity, a direct comparison will be possible.

Memory Architecture

Just as in the case of the NEC Azusa, this experiment offers the disclosure of runtime delays that arise from the conflicts of memory access. The measured run times for both computer systems are depicted in Table 3.

The results of the SX-5 are clear-cut; although the time lag of concurrent jobs does not reach the degree of the Azusa due to processor conflicts, it is with a factor of around 1.16 not irrelevant. The reason for the relatively low change in runtime can be found in the good memory architecture, which also happens to affect the price of such a system.

The results of the SX-4 appear next to be confusing. This shows that even with such a close relationship between two architectural systems, the

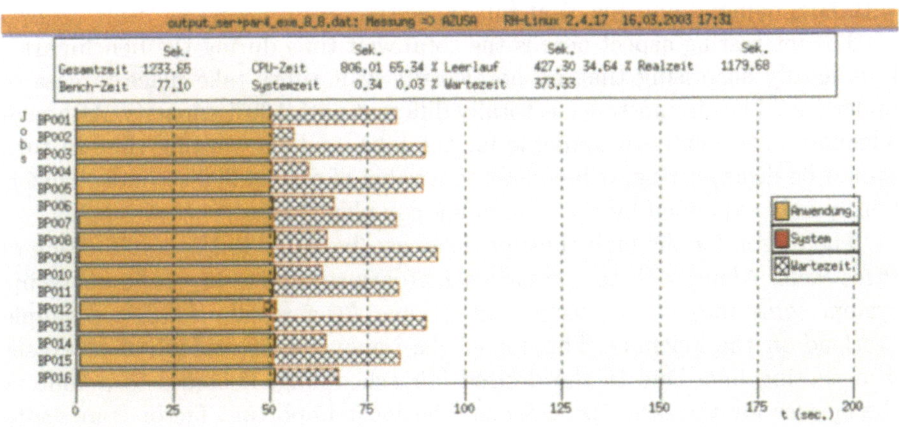

Fig. 3. Eight serial and eight parallel jobs with four processors each at the NEC Azusa

Table 3. Run time data of concurrent jobs with a high memory access rate at NEC SX-4 and SX-5

Number of concurrent jobs	average CPU time in seconds	
	SX-4	SX-5
1	154.24	59.92
2	154.26	61.65
4	154.20	63.30
8	154.44	66.10
16	154.53	69.86
32	154.91	

behaviour of the system can be vastly altered. The carrying out of the benchmark that generates the highest memory rate capacity for the SX-5 does not inevitably force the same effect for the SX-4. The value lies principally only in two thirds of the measured maximums.

The influence of the runtime of concurrent jobs on both vector calculators becomes less important and only has to be considered closely in regard to runtime optimisation, if all or close to all jobs work with extremely high memory access rates.

Scheduling with Overload

The commercial use of many high performance computer systems forces their possible 100% utilization. Efficient scheduling is very important in such situations. Only if the scheduler is in the position to divide the tasks efficiently will the actual capability be optimized.

Because the behaviour of both machines practically matches, the facts here only become clear with an example of the SX-4. Figure 4 shows the NEC SX-4 with 50% demand and Fig. 5 at full capacity.

The interesting aspect here is the total wait time during the benchmarks. It is hardly surprising that no noteworthy wait times take place in case of underload. The circumstance is totally different with 100% capacity. Although adequate processors are available for the jobs, and the waiting time in turn should be disappearing, it increases to a value of almost 900 seconds. That is hardly only explained by the scheduling complexity.

The reason for the high runtime delay can be found in the combination of three facts. Along with the actual user programs, processes of the operating system come into use in certain situations. As a result, there is a simple overload on the machine. This forces the removal from storage of processes in a waiting line. Due to the desired fair scheduling, it comes to a context changeover for the other processors. The most important factor comes into play. Every context changeover forces the saving of the entire vector register set. This affects around 140 KByte data per processor. Finally, this process creates the high wait times.

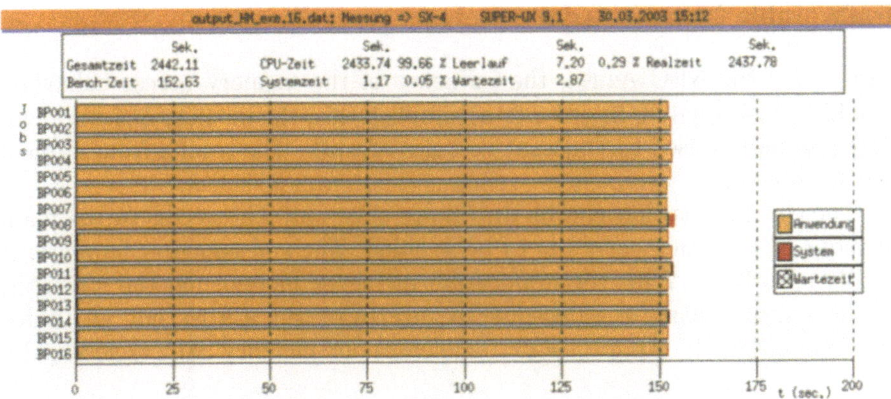

Fig. 4. SX-4 with 50% serial load

Due to this wait time, it is not generally sensible to use both vector calculators in load situations higher or the same as 100%. The batch system that is used in many areas for daily work should avoid these situations through targeted runtime delays.

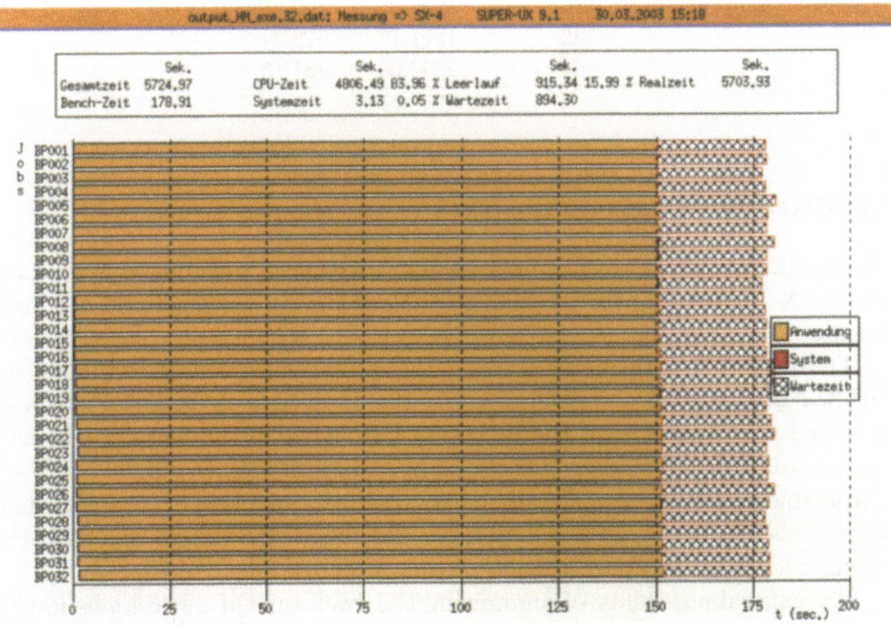

Fig. 5. SX-4 with 100% serial load

Unfavorable Memory Access Rates

Just as for the NEC Azusa, the influence of the memory access behaviors should be investigated. In order to accomplish this, the memory access ranges were gradually raised by constant volumes of work. The results from Table 4 become obvious.

The conclusions that can be drawn are clear-cut. The access to the main memory in large increments is to be avoided. The consequence, however, is hardly surprising. With the application of bigger increments, the strengths of the vector calculator, consequently, are bypassed. Either not all vector elements are used, or a few, in extreme cases one memory bank is accessed.

Table 4. Run time data of jobs with different memory access range at SX-4 and SX-5

memory access range	average CPU time in seconds	
	SX-4	SX-5
1	154.24	59.92
2	155.41	60.37
4	161.37	108.75
8	305.84	213.46
16	605.75	409.58
32	1211.17	522.34
64	2392.04	887.15

4.2 Mixed Workloads on the SX-4

The results pertaining to the application of PARbench with mixed workloads for the SX-4 make clear the possibility of also achieving speed gains in overload situations (see Fig. 6).

The dwell time of the parallel jobs in the system is made noticeably smaller. This was only possible, however, because the dwell time of the serial jobs was increased. The lessening of the necessary CPU time of the parallel jobs has two causes. First, the memory access is bypassed through the higher number of available vector registers for the job. Second, the adoption of several caches of the scalar units can cut down on time. The high total waiting time is again a consequence of the frequent context changeover of the CPU's.

The general possibility of minimizing the dwell time of certain jobs in the system in the case of overload situations should be implemented carefully. The attained advantages of these jobs become possible due to the disadvantages of the others.

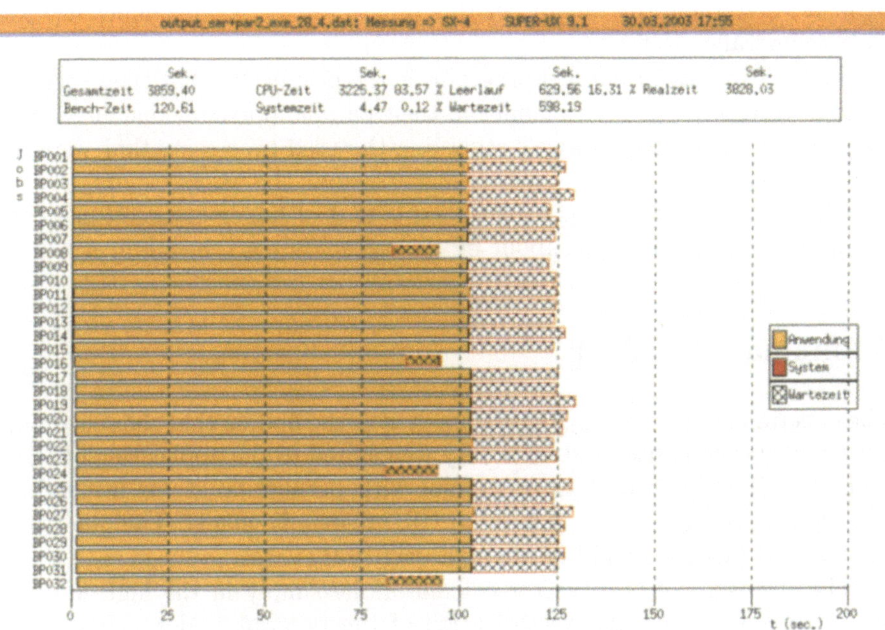

Fig. 6. SX-4 with 28 serial and 4 parallel jobs with two threads each

4.3 Mixed Workloads on the SX-5

During the trials with serial and parallel jobs on the SX-5, the application of the PARbenchmark system encountered a few problems. The used GANG-scheduling of the computer system controls a parameter for every task, and this parameter marks the number of the scheduling objects that belong together. This parameter can currently only be changed by the administrator and the batch system and is handed down from the father to child processes.

The scheduling system uses this parameter in such a way that the appropriate number of CPU's is reserved for the current time slice. The actual existence of these tasks are, however, not verified. For example, if during the carrying out of 16 serial jobs out of a shell with the parameter 16 all of the jobs are executed on a processor, then the rest will remain idle.

The problem with parallel programs is found in a similar form. If the parameter is set too low, then not all of the threads will be viewed as belonging together. If the parameter is omitted, then free processors will be blocked. Furthermore, the problem is worsened by other characteristics of parallel programs. These include in particular the typical user programs of the SX-5.

An exact analysis of Amdahl's Law agrees with the practical considerations that every parallel program possesses a serial part. Due to the fact that this part is expressed with unchanging scheduling parameters, free processors are at least blocked. A mitigating factor in this context seems only to be the fact or hope, that the serial portion of this program is omitted significantly less than the parallel part.

Another quality of parallel programs is the peculiarity that they very seldom proceed in a balanced manner in their parallel portion. That inevitably leads to the same disadvantages regarding the used processors that were described before.

It can be speculated that a considerable part of the empty runtime of the computer system is brought about by this aspect of the scheduling system.

5 Summary/Outlook

The experiments on the NEC Azusa resulted in the finding that the scheduling system of this computer system is very well suited to the application of serial programs. It is important to be careful with parallel programs in overload situations in regards to the number of CPU demands offered. The number of the inserted threads should be thought over in this case.

The strengths of both vector calculators lie simply in the high memory access ranges. At optimal vectorization of the program codes, a good flow rate can be achieved from almost completely separate problems from the ones being solved. Problems arise as soon as the system is conducted at less than overload. The remapping of tasks carried out by the scheduler with regards to the processors leads to high wait times for the completed user program. The increase in the time slice length can lessen the problem.

All of the investigated computer systems have in common that they have a higher need for CPU time for jobs that are being carried out with the application of unfavorable memory access patterns. In general, the way to avoid this problem lies in the hands of the user. It should be made sure that the user is well-acquainted with this difficulty.

The depicted results show a work in progress that this project has investigated. They are shaped not only in complexity, but also in documentation to the requirements of this report. Further reaching and more detailed interpretations as well as additional test series will be available with the soon-to-be released thesis covering this topic.